The Time Domain in Surface and Structural Dynamics

NATO ASI Series

Advanced Science Institutes Series

*A Series presenting the results of activities sponsored by the NATO Science Committee,
which aims at the dissemination of advanced scientific and technological knowledge,
with a view to strengthening links between scientific communities.*

The Series is published by an international board of publishers in conjunction with
the NATO Scientific Affairs Division

A Life Sciences
B Physics

Plenum Publishing Corporation
London and New York

C Mathematical
and Physical Sciences
D Behavioural and Social Sciences
E Applied Sciences

Kluwer Academic Publishers
Dordrecht, Boston and London

F Computer and Systems Sciences
G Ecological Sciences
H Cell Biology

Springer-Verlag
Berlin, Heidelberg, New York, London,
Paris and Tokyo

The Time Domain in Surface and Structural Dynamics

edited by

Gary J. Long

Department of Chemistry,
University of Missouri-Rolla, Rolla, U.S.A.

and

Fernande Grandjean

Department of Chemistry,
University of Liège, Liège, Belgium

Kluwer Academic Publishers

Dordrecht / Boston / London

Published in cooperation with NATO Scientific Affairs Division

Proceedings of the NATO Advanced Study Institute on
The Time Domain in Surface and Structural Dynamics
Il Ciocco, Italy
June 15-26, 1987

Library of Congress Cataloging in Publication Data

NATO Advanced Study Institute on the Time Domain in Surface and
    Structural Dynamics (1987 Il Ciocco, Italy)
    The time domain in surface and structural dynamics / edited by
Gary J. Long and Fernande Grandjean.
        p.   cm. -- (NATO ASI series. Series C, Mathematical and
physical sciences ; vol. 228)
    "Proceedings of the NATO Advanced Study Institute on the Time
Domain in Surface and Structural Dynamics, Il Ciocco, Italy, June
15-26, 1987"--T.p. verso.
    "Published in cooperation with NATO Scientific Affairs Division."
    Includes indexes.
    ISBN 978-94-010-7819-1
    1. Solids--Surfaces--Congresses. 2. Structural dynamics-
-Congresses. 3. Time-domain analysis--Congresses. 4. Nuclear
magnetic resonance spectroscopy--Congresses.   I. Long, Gary J.,
1941-   . II. Grandjean, Fernande, 1947-   . III. North Atlantic
Treaty Organization. Scientific Affairs Division. IV. Title.
V. Series NATO ASI series. Series C, Mathematical and physical
sciences ; no. 228.
QC176.8.S8N39 1987
530.4'1--dc19
                                                        88-3064
                                                        CIP

ISBN 978-94-010-7819-1   ISBN 978-94-009-2929-6 (eBook)
DOI 10.1007/ 978-94-009-2929-6

Published by Kluwer Academic Publishers,
P.O. Box 17, 3300 AA Dordrecht, The Netherlands.

Kluwer Academic Publishers incorporates the publishing programmes of
D. Reidel, Martinus Nijhoff, Dr W. Junk, and MTP Press.

Sold and distributed in the U.S.A. and Canada
by Kluwer Academic Publishers,
101 Philip Drive, Norwell, MA 02061, U.S.A.

In all other countries, sold and distributed
by Kluwer Academic Publishers Group,
P.O. Box 322, 3300 AH Dordrecht, The Netherlands.

Table of Contents

Preface

About two years ago, while studying the dynamic properties of $Fe_3(CO)_{12}$, we realized that there was virtually no single source of information on the structural dynamics of materials. The time domain of different dynamic structural processes covers many orders of magnitude and may be investigated by numerous, vastly different, experimental techniques. Indeed, the subject seemed appropriate for a NATO Advanced Study Institute at which we could bring together chemists, physicists, metallurgists, and bioscientists using the various techniques for the study of sundry time sensitive materials. The actual Advanced Study Institute, which met in Il Ciocco, Italy, from 14 to 26 June 1987, was, in fact, a dynamic experience for those of us involved. Now we have come to the final phase, the communication of the results of this Advanced Study Institute to the general scientific community. In so doing, we hope to provide in one place a convenient source of information on dynamics at the surface and within a solid state material.

The beautiful mountainous setting of Tuscany and especially the idyllic surroundings of Il Ciocco provided an ideal venue for the Advanced Study Institute. Our field trip to Pisa linked our topic to the history of time measurement through a visit to the Pisa cathedral where, presumably, Galileo first conceived the isochronism of the simple pendulum and its use in time measurement. History tells us that Pisa was also the birthplace, in 1170, of Leonardo Fibonacci, much less well known then Galileo, but perhaps just as important to modern science. Fibonacci, through his book, 'Liber Abaci' introduced and popularized the use of Arabic numerals throughout Europe.

The Advanced Study Institute was made possible through the generous support of the Scientific Affairs Division of NATO. This support is acknowledged with thanks by the editors and the participants. The additional financial support of the United States National Science Foundation, the Centro Nazionale di Ricerca, Italy, Carlo Erba Instrumentation, and the Tourism Office of the Province of Lucca was also beneficial to the overall success of the ASI. During the organization of the ASI, the editors greatly appreciated the extensive support and helpful ideas of Peter Day of Oxford University and Umberto Russo of Padova University, both of whom served on the international organizing committee. The further organizational help of Christopher Dobson of Oxford University, Raymond Andrew of the University of Florida, and Renee Diehl of Liverpool University insured the success of the ASI. Further, we express our thanks to Bruno Giannasi and the gracious staff of Il Ciocco, who made our stay so pleasant. In the preparation of this manuscript for publication, we wish to thank Ms. S. Maquet and P. Schmitz and Mr. J. Cloes for their technical assistance.

November 1987

Gary J. Long

Fernande Grandjean

List of Contributors and Participants

Silvio Aime (L,C)

Istituto di Chimica Generale e Inorganica,
Universita degli Studi di Torino,
Corso Massimo d'Azeglio 48, I-10125 Torino, Italy

E. Raymond Andrew (L,C)

Department of Physics, University of Florida,
215 Williamson Hall, Gainesville, FL 32611, U.S.A.

Akkus Baki (P)

Istanbul Universitesi, Fen Fakultesi Fizik Bolumu,
TK-34459 Istanbul, Turkey

T. L. Beck, P. A. Braier, R. S. Berry (C)

Department of Chemistry, University of Chicago,
5735 South Ellis Avenue, Chicago, IL 60637, U.S.A.

William S. Bennett (L,C)

Abt. Wittmann, Max-Planck Institut fur Molekulare Genetik,
Ihnestrasse, 73, D-1000 Berlin 33,
Federal Republic of Germany

K. Beshah, R. Ebelhauser, T. S. Huang, E. T. Olejniczak

D. M. Rice, D. J. Siminovitch, R. J. Wittebort (C)

Francis Bitter National Magnet Laboratory
Massachusetts Institute of Technology
Cambridge, MA 02139, U.S.A.

Carel Boekema (L,C)

Department of Physics, San Jose State University,
San Jose, CA 95192, U.S.A.

Claude Bostoen (P)

U.I.A., Universiteitsplein 1, B-2610 Wilrijk, Belgium

Tilman Butz (L,C)

Physik Department E15, Technische Universitat Munchen,
James Franck Strasse, D-8046 Garching,
Federal Republic of Germany

Enrico Campari (P)

Dipartimento di Fisica, Universita di Bologna,
Via Irnerio 46, I-40126 Bologna, Italy

Edward W. Castner (P)

Department of Chemistry, University of Chicago,
5735 South Ellis Avenue, Chicago, IL 60637, U.S.A.

Nigel J. Clayden (L,C)

ICI, Chemistry and Polymers Group, P.O. Box 90,
Wilton Middlebourgh, Cleveland TS6 8JE, England

Heidi L. Davis (P,C)

Department of Chemistry, University of Chicago,
5735 South Ellis Avenue, Chicago, IL 60637, U.S.A.

Peter Day (O,L,C)

Inorganic Chemistry Laboratory, Oxford University,
South Parks Road, Oxford OX1 3QR, England

Jose Dianoux (L,C)

Institut Laue Langevin, Avenue des Martyrs, B.P. 156 X,
F-38042, Grenoble Cedex, France

Renee D. Diehl (L,C)

Department of Physics, University of Liverpool,
Liverpool L69 3BX, England

Judith M. Fiddy (P)

Department of Physics, University of Liverpool,
Liverpool L69 3BX, England

Dino Fiorani (L,C)

Istituto di Struttura Elettronica, CNR,
Area della Ricerca di Roma, Via Salaria Km 29.5,
Casella Postale 10, I-00016 Monterotondo Stazione (Roma)
Italy

Paul Stephen Fowles (P)

Department of Physics, University of Liverpool,
P.O. Box 147, Liverpool L69 3BX, England

R. Gerardin, E. Millon, J. F. Brice, O. Evrard (C)

CNRS, Laboratoire de Metallurgie, F-54042 Nancy Cedex,
France

Then Gerhard (P)

Karlsruhe Universitat, Institut fur Kern Physik
Postfach 3640, D-7500 Karlsruhe,
Federal Republic of Germany

Wolfgang Girnus (P)

Institut Physikalische Chemie, Strehlowweg 36,
D-2900 Hambrug 52, Federal Republic of Germany

Andre Gourdon (P)

Laboratoire de Chimie des Metaux,
Universite Pierre et Marie Curie,
Place Jussieu 4, F-75252 Paris Cedex, France

Fernande Grandjean (O,L,C)

Institut de Physique, Universite de Liege, B5,
B-4000 Sart-Tilman, Belgium

Robert G. Griffin (L,C)

National Magnet Laboratory, NW 14-5113,
Massachusetts Institute of Technology,
77 Massachusetts Avenue, Cambridge, MA 02139, U.S.A.

Jacques Guillin (P)

Institut fur Physik, Medizinische Universitat Lubeck,
Ratzeburger Allee 160, D-2400 Lubeck,
Federal Republic of Germany

Fleming Y. Hansen (L,C)

Fysisk-Kemisk Institut, Technical University of Denmark,
DK-2800 Lyngby, Denmark

Samar S. Hasnain (L)

Science and Engineering Research Council
Daresbury Laboratory, Warrington, Cheshire WA4 4AD
England

Mohamedally Kurmod (P)

Inorganic Chemistry Laboratory, Oxford University
South Parks Road, Oxford OX1 3QR, England

Jean-Pierre Launay (L,C)

Laboratoire de Chimie des Metaux de Transition,
Universite Pierre et Marie Curie, Place Jussieu 4,
F-75230 Parix Cedex 05, France

Gerard Le Caer (P,C)

CNRS, Laboratoire de Metallurgie,
F-54042 Nancy-Cedex, France

Gary J. Long (O)

Department of Chemistry, University of Missouri-Rolla
Rolla, MO 65401, U.S.A.

B. Malaman, B. Ech-Chahed, C. Gleitzer (C)

Laboratoire de Chimie du Solide Mineral, UA 158,
Universite de Nancy I, B.P. 239
F-54506 Vandoeuvre-les-Nancy Cedex, France

Paolo Matteazzi (P)

Ist. Chimica Industriale, Via Marzolo 9,
I-35100 Padova, Italy

Pierre Mihailovic (P)

Laboratoire de Chimie et Electrochimie
des Materiaux Moleculaires, ESPCI, F-75005 Paris, France

Steen Morup (L,C)

Laboratory of Applied Physics II,
Technical University of Denmark, Building 307,
DK-2800 Lyngby, Denmark

Basaran Reha (P)

Istanbul Universitesi, Fen Fakultesi Fizik Bolumu,
TK-34459 Istanbul, Turkey

Umberto Russo (O,L,C)

Dipartimento di Chimica Inorganica,
Universita degli Studi di Padova,
Via Loredan, 4, I-35131 Padova, Italy

Cindy Schauer (P)

Department of Chemistry, Northwestern University,
Evanston, IL 60201, U.S.A.

John Stevens (P)

University of Liverpool, P.O. Box 147
Liverpool L69 3BX, England

Haskell Taub (L,C)

Department of Physics, University of Missouri-Columbia,
Columbia, MO 65211, U.S.A.

Wolfgang Troeger (P)

Physik-Department E15, Technische Universitat Munchen,
James Franck Strasse, D-8046 Garching,
Federal Republic of Germany

Tamara A. Ulibarri (P)

Department of Chemistry, University of California,
Irvine, CA 92717, U.S.A.

Simon D. Waddington (P)

Department of Physcis, University of Liverpool,
P.O. Box 147, Liverpool L69 3BX, England

Hans W. Weber (P)

Karlsruhe Universitat, Institut fur Kern Physik
Postfach 3640, D-7500 Karlsruhe,
Federal Republic of Germany

Wolter Wegener (P)

SCK-CEN, Boeretang 200, B-2400 Mol, Belgium

Robert D. Young (L,C)

Department of Physics, Loomis Laboratory of Physics,
University of Illinois at Urbana-Champaign,
1110 West Green Street, Urbana, IL 61801, U.S.A.

Claude M. E. Zeyen (P,C)

Institut Laue Langevin, Avenue des Martyrs, B.P. 156 X,
F-38042 Grenoble Cedex, France

(O) Organizer, (L) Lecturer,
(P) Participant, (C) Contributor

CHAPTER 1

TIMESCALES OF PHYSICAL MEASUREMENTS:
SOME ELEMENTARY CONSIDERATIONS

Peter Day
Oxford University
Inorganic Chemistry Laboratory
South Parks Road
Oxford OX1 3QR
England

ABSTRACT: Some elementary considerations are given of the timescales
that influence the applicability of physical methods for studying
dynamical processes in molecules and solids. Four different measures
of the timescale relevant to each technique are described, and examples
are used to show how these limit the usefulness of the methods in
practical situations.

1. INTRODUCTION

The electronic and chemical processes taking place in atoms, molecules
and condensed phases of interest to chemists and physicists encompass
an immense range of timescales. From the excitation of inner shell
electrons in atoms to the dilatation of groups of sidechains
accompanying the binding of a chemical substrate to an enzyme or
protein is a timescale difference of at least ten orders of magnitude
in seconds. (Not to mention that some chemical substitution reactions,
e.g. in low spin d^6 metal complexes, have half lifetimes of weeks to
months). Consequently we need not feel surprised at the massive array
of disparate methods, based on many different physical phenomena, that
has been deployed in an effort, first to monitor, and then to
understand them. The methods in question may involve the elastic or
inelastic scattering of particles like electrons, neutrons or photons.
They may require the capture or release of a photon, with subsequent
excitation or de-excitation of the system. The excited state
frequencies concerned range over more than ten orders of magnitude in
Hz. Given this diversity, the factors influencing the timescales of
the processes that can be probed are equally diverse. In this brief
introductory chapter we therefore wish to indicate some general
considerations, applicable to all the methods reviewed at this ASI,
that can be used to estimate the effective timescale of phenomena that
can be investigated by each.

G J Long and F Grandjean (eds), The Time Domain in Surface and Structural Dynamics, 1–6
© 1988 by Kluwer Academic Publishers

2. TIMESCALES: SOME FACTORS

In every experimental technique there are numerous different timescales that need to be taken into account in assessing its applicability. We shall summarize four of these.

2.1 'Contact' time of radiation

Clearly one can only use a particle or a radiation to probe the physical state of a system during a period of time in which the probe and the sample are in proximity with one another. Thus the shortest timescale that can be associated with any physical technique is set by the time taken for the photon to traverse the atom or molecule in which it is going to be absorbed or, in the case of an electron or X-ray in a diffraction experiment, the timescale of the elementary scattering process. The electromagnetic radiation will be travelling at about 3.10^{10} cm.s^{-1} and our atom or molecule has dimensions from 1Å upwards. Consequently the 'contact' time during which information can be transferred from the molecule to the radiation is between 10^{-16} and 10^{-19} s. As we shall se below, this is a much shorter time than the molecule needs to undergo, say a vibration or rotation, so one might think that a diffraction experiment would give a frozen snapshot of the molecule. However, while that might be true of an individual scattering event, such an event will not give a signal capable of being detected. In practice, therefore, the information has to be gathered by averaging over many scattering events, on a total timescale determined by the photon flux and sensitivity of the detection system. This point is amplified in 2.4 below.

2.2 Excited state lifetime

Any spectroscopic technique that involves the creation of an excited state by the absorption of electromagnetic radiation has associated with it a relaxation time that may be interpreted as the time taken for the system to undergo the transition between ground and excited states or, in the case of an emission process, the lifetime of the excited state. Many and various are the kinds of excited states that form the object of spectroscopy. For example, nuclear states figure in nuclear magnetic resonance (NMR) nuclear quadrupole resonance (NQR) and Mossbauer spectroscopy. Transitions between electronic states provide the basis of electron paramagnetic resonance (EPR), electronic absorption and emission spectroscopy and photoelectron spectroscopy (PES), while quantized vibrational and rotational states are accessible in absorption.

Within these broad subdivisions into classes of excited states, however, there are some obvious differences to mention. Among the techniques based on nuclear energy levels, for instance, NMR arises from transition between the Zeeman-split sublevels of the nuclear ground state, and hence is sensitive only to the properties of that state. In contrast, the process involved in the creation of a Mossbauer transition is the emission of a γ-ray as a nuclear excited state decays to the ground state. Consequently the properties of both states come into question. Similarly, EPR only involves transitions between the Zeeman levels of a single electronic state (nearly always the ground state, though in photo-excited experiments it can be an excited state). Electronic and PE spectroscopy concern two electronic states, in the latter case, one being the continuum.

TABLE 1. Timescales of spectroscopic techniques

	Frequency of excited state $(s^{-1})E$	Relaxation time(s) Δt	Linewidth (s^{-1}) ΔE
NMR (solution)	10^8	10	10^{-1}
ESR (solution)	10^{10}	10^{-5}	10^5
Microwave spectroscopy (gas phase molecular rotation)	10^{11}	10^{-4}	10^4
Infrared spectroscopy (gas phase molecular vibration)	10^{14}	10^{-8}	10^8
Ultraviolet spectroscopy (liquid phase electronic excitation)	10^{16}	10^{-14}	10^{14}
Mossbauer (γ-ray) spectroscopy	10^{19}	10^{-8}	10^8

In general, the relaxation time Δt associated with a technique is long if the energy separation between the states involved in the spectroscopic transition is small. Table 1 lists the orders of magnitude of excited state energies and typical relaxation times for a number of spectroscopic techniques. From this Table one can see that Mossbauer spectroscopy is an exception to the above generalization. This is because the effective relaxation time is given by the lifetime of the nuclear excited state, which has to be exceptionally long if a useful spectrum is to be recorded. The minimum spectral linewidth ΔE obtainable in any transition is determined by the uncertainty principle $\Delta E . \Delta t \sim$ h because this determines the uncertainty in the energy of the excited state. Given that h is approximately 10^{-34} Js, the typical relaxation times listed in Table 1 result in the linewidths shown, which are given in frequency units of Hz or s^{-1} for convenience of comparison. Clearly NMR linewidths in solution can be exceptionally narrow, yielding resolutions $\Delta E/E$ of 10^{-9} while electronic spectroscopy in solution is only capable of resolution of about 10^{-2}. The latter is an illustration of the fact that Δt, and hence ΔE is rarely at the limit determined by the uncertainty principle, but depends on interaction of the atom or molecule with its environment. A simple example would be collisional deactivation in the gas or liquid phase.

2.3 Lifetime of the chemical species

To use techniques like those in Table 1 to study dynamical processes the next important timescale to consider is the lifetime of the species being observed, in relation to the relaxation time Δt. The species under observation may evolve with time either because it is engaged in a chemical reaction (i.e. an intermolecular exchange of groups) or because an intramolecular scrambling is taking place. If the rate of the transformation is comparable to Δt the spectral lines are broadened and an averaging occurs. The simplest situation occurs when a molecule contains two chemically distinct groups, each of which contributes its own distinct transition to the overall spectrum, but which interchange on some timescale. Two limiting regimes can be distinguished. If the frequency separation (Hz or s^{-1}) between the peaks is much larger than the frequency at which the two species are exchanging, (i.e. exchange is slow) then two separate peaks are observed. As the exchange frequency increases, these broaden and coalesce until, when exchange is fast, a single peak occurs at the mean frequency of the other two. For example in an NMR experiment one might easily find two proton resonance frequencies separated by 100 Hz. For exchange times much greater or much less than 0.01s, respectively one and two peaks appear in the spectrum. A convenient way of passing from one regime to the other for an activated exchange process is to vary the temperature. By this means, ^{1}H NMR of solutions has been used very extensively to monitor intramolecular exchange in organic and organometallic chemistry. In the latter case the name 'fluxionality' has been coined. A typical example is the cis-trans isomerization of $(C_5H_5)(CO)Fe(CO)_2$-$Fe(C_5H_5)(CO)$ shown in Figure 1.

On the other hand spectroscopic techniques involving higher frequency excitations, like infrared or electronic absorption spectra, can only be used in this way to follow much more rapid exchange reactions. For instance a 1 cm^{-1} interval, which would represent quite a reasonable resolution in an infrared spectrum ($\Delta E/E$ between 10^{-3} and 10^{-4} say) corresponds to 10^{10} Hz. Thus only exchange processes occurring on a timescale much shorter than 10^{-10} s, would influence the appearance of two vibrational transitions separated by this interval. Averaging of vibrational frequencies has, however, been used to set a lower limit to the rates of intra- and intermolecular electron transfer reactions.

Because of its relatively long relaxation time (Table 1), Mossbauer spectroscopy is the other main technique, after NMR, for studying exchange reactions though since it is confined to the rigid solid state, the exchange of electrons rather than functional groups forms the subject matter. An early example of a temperature dependent Mossbauer spectrum spanning the range of exchange times relative to the excited state relaxation time is shown in Figure 2. Formally, the mixed valency compound Eu_3S_4 contains Eu(II) and Eu(III), and at low temperatures two separate resonances corresponding to these oxidation states are observed. With increasing temperature the peaks broaden and coalesce as the interionic electron transfer rate increases, until an average signal is obtained. The exchange rates are indicated at the side of the Figure.

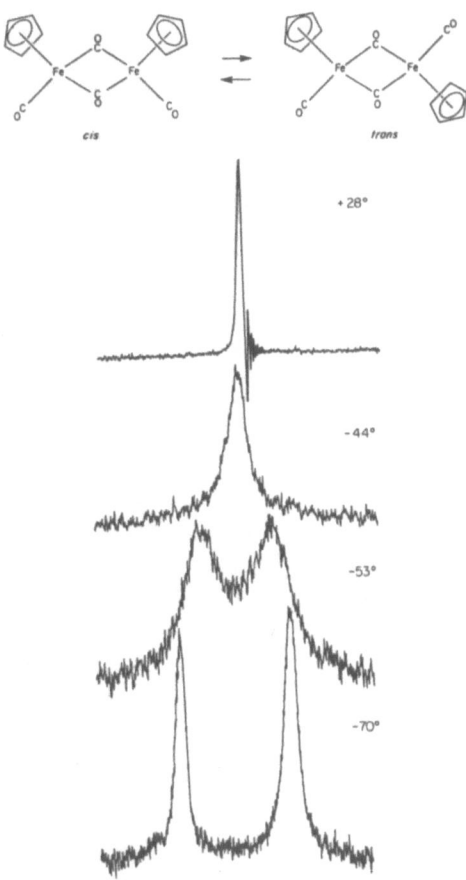

Figure 1. 100Mz proton NMR spectra of cis- and trans-bis(cyclopenta-dienyl)tetracarbonyl-di-iron at different temperatures. (After Bullitt, Cotton & Marks, J. Amer. Chem. Soc. 92 2155 (1970)).

2.4 Data collection time

Finally, although the timescale of a single scattering or spectroscopic excitation event is extremely short, as indicated in 2.1, the total length of time needed to accumulate enough signal intensity to observe may be very great. The data collection time for an X-ray or neutron diffraction experiment, for example, can be measured in days, so complete vibrational averaging will take place. Chemical reaction or decomposition over the timescale of data collection will also lead to averaging in a slowly scanned spectrum. On the other hand repetitive pulse techniques, as in NMR or time-resolved luminescence, give views of the system on timescales that depend on the single pulse length rather than the total accumulation time required to superimpose the signals from many thousands of pulses.

6

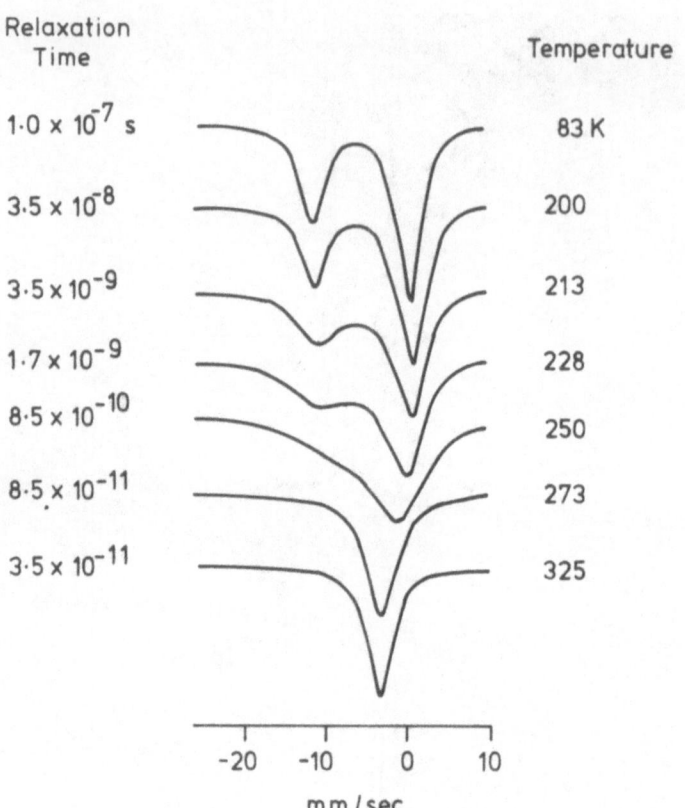

Relaxation Time / **Temperature**

1·0 × 10^{-7} s — 83 K

3·5 × 10^{-8} — 200

3·5 × 10^{-9} — 213

1·7 × 10^{-9} — 228

8·5 × 10^{-10} — 250

8·5 × 10^{-11} — 273

3·5 × 10^{-11} — 325

-20   -10   0   10

mm / sec

Figure 2. Temperature dependence of the ^{151}Eu Mossbauer spectrum of Eu$_3$S$_4$. (After Berkooz, Malamud & Shtrikman, Chem.Phys.Lett. 6 185 (1968)).

3. REFERENCES

Because of the general nature of the points made in this chapter, no detailed references are given. Books covering the range of physical methods used by inorganic chemists are:

Hill, H.A.O. and Day, P. (eds) Physical Methods in Advanced Inorganic Chemistry, London, Interscience, 1968.
Drago, R.S., Physical Methods in Inorganic Chemistry
Ebsworth, E.A.V., Rankin D.W.H. and Cradock S., Structural Methods in Inorganic Chemistry, Oxford, Blackwell 1987.
Cheetham, A.K. and Day, P., Solid State Chemistry: Techniques, Oxford University Press 1987.

CHAPTER 2

TIME : ITS HISTORY AND MEASUREMENT
BILLIONS OF YEARS WITHIN A NANOSECOND

Fernande GRANDJEAN
Institut de Physique, B5,
Université de Liège,
B-4000 Sart Tilman, Belgique

ABSTRACT. This chapter, after discussing the concept of time, presents the historical definitions of the unit of time, the second. The next section gives a the 1983 definition of the meter, a definition which is based upon the second and a defined value for the speed of light. A review of the measurements of the speed of light from 1676 to the present follows. The next section presents a summary of the development of various clocks and their accuracy for time interval measurement in clock making. Finally, the spread of time domains in the universe over 40 orders of magnitude is discussed.

1. THE CONCEPT OF TIME

The word "time" is easier to use than to define and at the beginning of this chapter, I would like to invite the reader to open his dictionary or his encyclopedia to the word "time" and to think about the different meanings found there. I opened the Oxford English Dictionary and Le Larousse.

In the Oxford English Dictionary, I found several meanings among which the most appropriate for our subject are (1) duration, indefinitely continued existence and, (2) amount of time as reckoned by conventional standards, point of time, especially stated in hours and minutes of the day.

In Le Larousse, I found "La durée mesurée par la succession plus ou moins longue des jours et des nuits" and "portion déterminée de la durée". Of course, the French language does not simplify the problem as the same word "temps" means "time" and "weather".

In spite of the difficulty in defining the time, it is clear from everyday experience that time is a notion which plays a role in many different fields of human knowledge. Furthermore, from the definitions mentioned above, it appears that there are clearly two aspects of time which are of interest to the scientist. The first is the chronological aspect of time, or the instant of time which is typically defined with a calendar and a clock. I will call this the "absolute time". The second aspect is the interval of time, which I will call the "relative time". The following examples will make this clear. If you must catch a plane or a train, you are interested in the instant of time; for instance 9:15 am on Sunday, 13 June 1987. If you are a boxing referee, you are interested in the interval of time a boxer spends on the floor, for instance, 5 seconds. Both aspects require the definition of a unit of time, the second, s, in order to make measurements, but only the measurement of absolute time requires the definition of a time scale with minute, hour, day, month, year relative to an initial point in time.

7

G. J. Long and F. Grandjean (eds.), The Time Domain in Surface and Structural Dynamics, 7–17.
© *1988 by Kluwer Academic Publishers.*

In this chapter, I do not plan to discuss in detail semi-philosophical aspects of time, which are closely related to physics, but I would like to mention them briefly here. A physicist observes various cyclic motions, and arbitrarily selects one as a reference and associates with the period of the motion, a unit of time. Hence, he has a clock which he can use to study other phenomena. In Newtonian mechanics, time exists independent of motion, a concept which was strongly questioned in the 18th century. Also, in Newtonian mechanics, all observers can communicate information with infinite speed signals. Newtonian time is of course abandonned in special relativity where the speed of light is a limit and where two observers in relative motion measure different intervals of time for the same event. This is known as "the time dilation". In general relativity, the rate of an atomic clock will be affected by the gravitational field. Thus, a photon emitted by an atom at the surface of the sun, with a given frequency, will be observed with a slightly smaller frequency on the surface of the earth. A similar effect may be observed on earth, by the Mössbauer effect, which permits the measurement of a very small frequency shift between upgoing and downgoing photons in the gravitational field. The theoretical relative shift is given by $2gd/c^2$ where g is the gravity acceleration, d the distance travelled by the photons in the gravitational field and c the speed of light. For a distance of 22.5 m, an experimental shift of 0.94 ± 0.10 times the theoretical one was observed.[1]

A last general remark comes from our everyday experience. The irreversible nature of time is not included in the laws of mechanics, but appears in the second law of thermodynamics.

2. UNIT OF TIME, THE SECOND

The unit of time is based upon natural processes. Until 1967 the unit of time in SI units, the second, was defined from first, the earth's rotation and later from the earth's revolution. Three different astronomical time scales are in use.

True solar time: The highest altitude of the sun at a certain location defines the instant 12:00, noon, in the true solar time of the location (true local time). The length of the true solar day is the time elapsed between two consecutive passages of the sun at the highest altitude.

Mean solar time: The highest altitude of a ficticious "mean sun" defines the instant 12:00, noon, in the mean solar time of a given location. The mean solar time is proportional to the earth's angle of rotation (related to the position of the "mean sun"). It is a uniform time, and when the angle of rotation of the earth has increased by 2π, a mean solar day has elapsed. The mean solar time of the zero Greenwich meridian is referred to as Universal Time (TU). Until 1956, the second was defined as 1/86,400 of a mean solar day.

Due to the obliquity of the ecliptic and the ellipticity of the earth's orbit, the true solar time deviates from the mean solar time by up to + 16.4 and - 14.3 minutes in the course of one year. The difference is referred to as the equation of time.

Ephemeris time: By 1956, it was apparent that, because of irregularities in the earth's rotation, the mean solar day varies with the seasons and through the centuries. Its length increases by 45.10^{-9} s or 45 ns each day. Thus, in 1956, the definition of the second was changed and the second of the ephemeris time became 1/31,556,925.9747 times the duration of the tropical year at the beginning of 1900. The tropical year is the time elapsed between two consecutive passages of the sun at the vernal point, the intersection of the celestial equator and the ecliptic plane, or the spring equinox (Fig. 1).

The ephemeris time is in principle uniform and based upon the study of the motion of the planets.

Although the ephemeris time is uniform, the accuracy of its determination is very poor and not satisfactory for the needs of modern physics.

The 1967 definition of the second

The 1967 definition of the second states that: "the second is the duration of 9,192,631,770 periods of the radiation corresponding to the transition between the two hyperfine levels of the ground state of the cesium-133 atom". This specific number of periods was chosen so that the new SI second corresponded as closely as possible to the ephemeris second. Current cesium standard clocks have uncertainties of less than 10^{-13}. This atomic second serves as a basis for International Atomic Time (TAI).

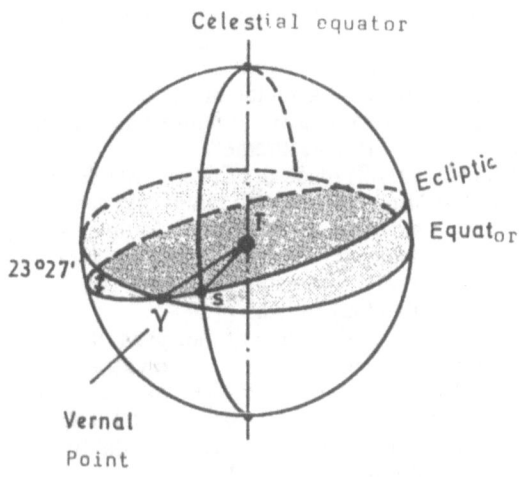

Fig.1. Vernal point. The celestial equator is obtained by extending the equatorial plane of the Earth (T). The ecliptic is the plane described by the sun (S) in its apparent motion, about the Earth.

3. THE SPEED OF LIGHT AND THE REDEFINITION OF THE METER

The high accuracy of frequency measurements, in the infrared region achieved with the first cesium standard clocks was improved and then a similar accuracy obtained for visible light. The rapid development of lasers, within the past twenty years, lead in 1983 to an interesting redefinition of the meter based on a fixed speed of light and the definition of the second [2].

Le Comité International des Poids et Mesures now recommends that the meter be realized by one the following methods:

(a) by means of the length, l, of the path travelled in vacuum by an electromagnetic plane wave in a time, t. This length is obtained from the measured time t, using the relation ship, $l = ct$, and the value of the speed of light in vacuum, $c = 299,792,458$ m/s.

(b) by means of the wavelength in vacuum, λ, of an electromagnetic plane wave of frequency f; this wavelength is obtained from the measured frequency f, using the relationship, $\lambda = c/f$, and the value of the speed of light in vacuum, $c = 299,792,458$ m/s.

(c) by means of one of the radiations from the list below, whose stated wavelength in vacuum, or whose stated frequency, can be used with the uncertainty shown, provided that the given specifications and accepted good practice are followed;

and that in all cases any necessary corrections be aplied to take into account the actual conditions such as diffraction, gravitation, or imperfection in the vacuum.

The committee recommendation is accompanied by a list of radiations produced either by stabilized lasers or by conventional spectral lamps. It is not reproduced here because it is not directly related to our purpose, but is summarized elsewhere [2].

With this definition, the meter becomes a derived unit based upon the second and the value of the speed of light in vacuum, fixed by agreement and exactly equal to 299,792,458 m/s. Therefore, the meter is defined with the accuracy of the frequency standard which is an improvement by a factor of 10^3 over the previous definition. Because the speed of light has become a basic number in the unit system, it seems worthwhile to review the history of its measurement.

Most likely, Galileo (1564-1642) made the first attempt, helped by his assistant to measure the speed of light. Both, equipped with lamps and covers, stood on opposite sides of a valley. Galileo's assistant was to uncover his lamp as soon as he saw Galileo uncover his. Galileo hoped to measure the delay due to the light's journey across the valley and back. As they were trying to measure an interval of time of a few tenths of a millisecond, it is not surprising that they failed.

Figure 2 shows the evolution of the value of the speed of light from 1676 to 1980 [3]. The first two successful measurements were based on astronomical observations. In 1676, Roemer (Figure 3) observed anomalies in the observed times of the eclipse of a moon of Jupiter (J) with the relative position of the earth (E1, E2) in its orbit around the sun (S) and correctly ascribed this to the time needed by the light to travel across the earth's orbit. He found 214,000 km/s. In 1726, Bradley observed an aberration in the apparent position of stars depending upon the direction of the earth's motion on its orbit. He correctly ascribed this aberration to the finite speed of light and deduced a value of 301,000 km/s.

The first successful measurement by a refined "Galilean" method, i.e. with a chopped light beam was made by Fizeau in 1849 when he obtained 313,000 km/s. Fizeau used a rapidly rotating toothed wheel to modulate a light source and to determine the arrival time of the light back at the wheel after reflection from a distant miror. The same basic idea was used by Foucault in 1862, who replaced the toothwheel by a rotating miror and found 298,000 km/s. Later these two methods were refined by several scientists including Michelson, in the 19th century and gave values from 298,500 to 300,400 km/s. The other measurements during the 19th century were either obtained from electromagnetic waves along wires yielding values from 297,600 to 300,300 km/s or from the ratio of electrostatic to electromagnetic units, giving values from 280,900 to 310,800 km/s. At the end of the 19th century, it was clear that the speed of light was slightly less than 300,000 km/s.

During the first half of the 20th century, various optical, electrical, light wave modulation, radio wave modulation, and interferometric methods were applied. Only a few selected measurements will be discussed herein. One of the most famous optical small oscillations (10 to 15 degrees) of a simple pendulum is independent of the amplitude of the oscillations. However, it was only in 1650, that Christian Huygens studied the complex pendulum and, observing its isochronism, had the idea of using this property in building clocks. Different types of escapement were built to regularly drive the pendulum and the pendulum prevailed for three centuries in clocks construction.

Transportation and travel, first by sea and later by rail, strongly contributed to the improvement of the accuracy of time measurement. Indeed, in 1760, on a sailing to the Caribbean, a positional accuracy of 3 to 4 km was reached after six weeks. This corresponds to a stability of the chronometer of six seconds over a period of one month. In Great Britain, railroad travel brought about another problem at the beginning of the 19th century. Not only were accurate chronometers necessary, but the whole country had to have a common time to make sense of the schedules and to avoid crashes on the one-way sections of the track. The problem was solved when legal time replaced local time and the Greenwich meridian was chosen as a zero hour reference point for Universal Time (TU), the time still in use.

The best free-penduluum clock was the Shortt free-pendulum clock[5] installed in 1925 in Greenwich. The first clocks to surpass the performance of this clock were based on the quartz crystal oscillator[6]. Quartz is very useful in time keeping because its mechanical properties are not strongly affected by temperature and it shows the piezoelectric effect. Mechanical vibrations of the crystal produce oscillating voltages between metal electrodes applied to its surface. If these electrodes are connected to a suitable link between the input and output of an amplifier, the amplifier supplies the energy required to maintain the vibrations at one of the natural resonant frequencies of the crystal. With such a system, an accuracy of 4×10^{-7} s can easily be achieved and maintained. Quartz clocks were used by the Greenwich time service from 1939 to 1964.

It was realized by the early 1960's that the earth was not a very good time keeper for the definition of the second. Fortunately, many periodic atomic phenomena of great stability had been discovered and lead in 1967 to a redefinition of the second. Atomic clocks were then born. They reach an accuracy of one part in 10^{13} and Figure 4 shows the progress achieved in accuracy from the pendulum clock to the atomic clock[7]. The first cesium-133 clock was put in operation in 1955 by Essen and Parry and a schematic diagram[5] is shown in Figure 5. Cesium atoms are evaporated in a vacuum oven and pass through a series of slits to form a beam of isolated atoms with equal numbers of atoms in the two relevant energy states. A strong localized non uniform magnetic field separates the two states into two divergent beams. These beams then pass through a resonant cavity which concentrates and confines the micro-wave energy fed to the cavity from an amplifier. The micro-wave frequency of 9192 MHz is supplied by a quartz oscillator. The cavity is placed in a weak uniform magnetic field whose axis is perpendicular to the beam. After passing through the cavity, the atoms enter a second region of non-uniform field and are again deflected according to their magnetic state. A hot wire is placed to intercept only those which have changed state during their passage through the cavity. The hot wire ionizes them so that they may be detected electronically. The frequency of the quartz oscillator is controlled by the output signal from this hot wire. Time markers are produced by the quartz oscillator as in a quartz clock.

Atomic clocks are delicate, expensive and complex to operate, so that they cannot be conveniently used in all laboratories. In order to make time signals accessible to all laboratories, the atomic clocks of a few laboratories are used to control the carrier frequency of radio broadcasters. With a rather simple electronic device, anyone can have at home a frequency having the stability of international atomic frequency standards. Table 1 lists a few time signal transmitters in operation in Europe[8].

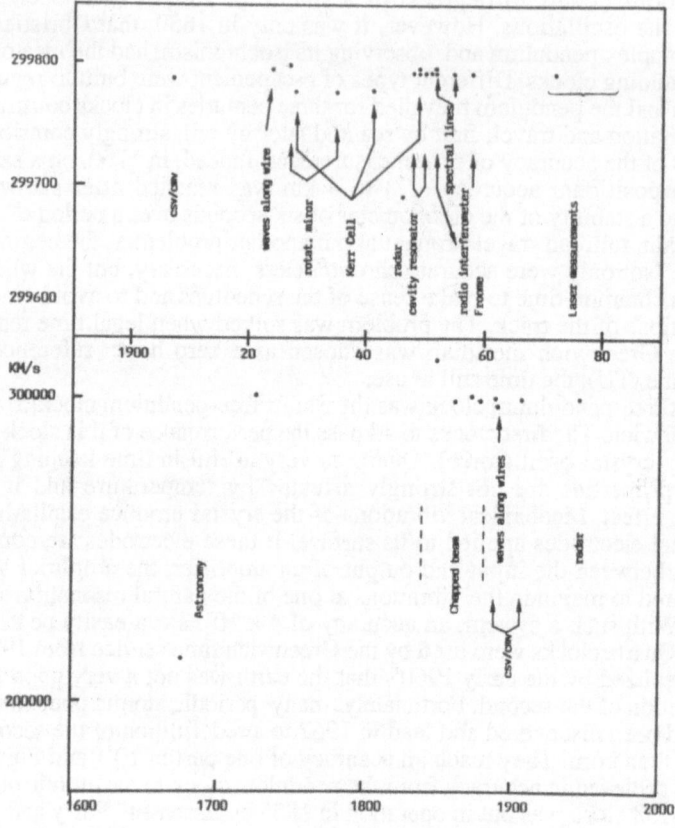

Fig.2. Measurements of the speed of light from 1676 to 1980.

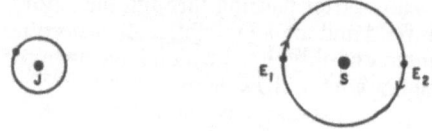

Fig.3. Roemer deduced the speed of light by observing a moon of Jupiter.

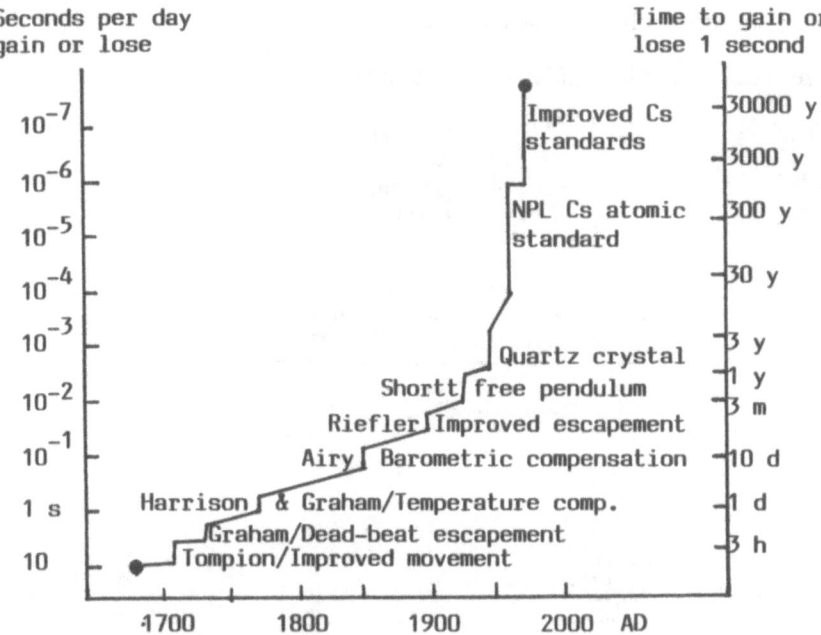

Fig. 4. The increasing accuracy of precision clocks (order of magnitude).

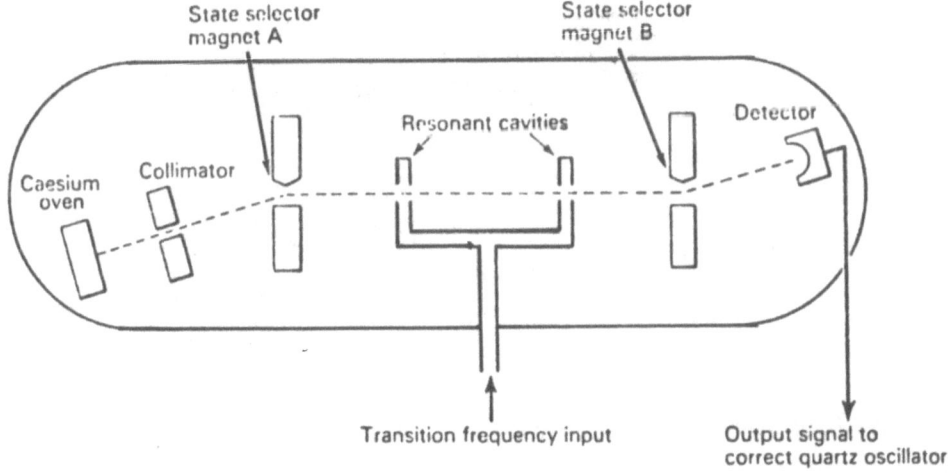

Fig. 5. Cesium tube.

14

Table 1. Time signal transmitters in Europe.

Transmitter	Frequency (kHz)	Country
DCF77	77.5	Germany
HBG	75	Swiss
MSF	60	England
OMA	50	CSSR

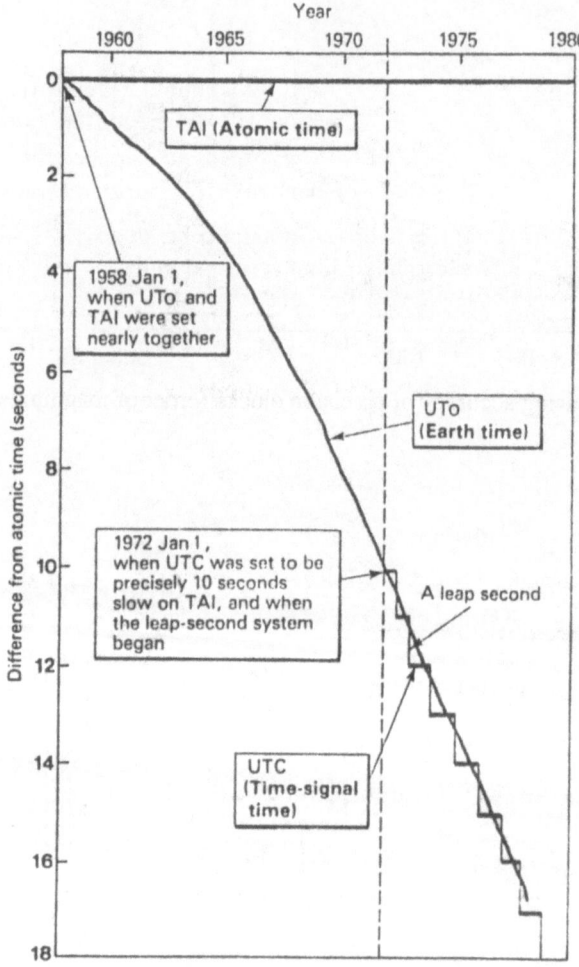

Fig.6. Time signals and the leap second.

What time do our clocks read? This is really a problem of relating the highly accurate cesium-133 clock to chronological time.

measurement is the one planned by Michelson and finished in 1935 by Pease and Pearson who found 299,774 ± 11 km/s. Among the electrical measurements, a value of 299,795 ± 30 km/s was obtained in 1923 by Mercier who used modern high frequency techniques to measure the speed of waves along a wire. During World War II, with the development of radar and its use for geodesic surveys, values of c ranging from 299,792 to 299,794 were obtained in agreement with a measurement of 299,792.5 ± 2 km/s by Esser in 1950.

With the development of microwave technology, it became possible to measure the speed of light by measuring the wavelength and the frequency of the radiation. Froome made the most notable of these measurements in 1958, with 30GHz, 1cm, radio waves, and obtained a value of 299,792.5 ± 0.1 km/s; a value which remained the most accurate until 1972. In 1972, Luther and Bay completed the last determination of c with a modulation technique, but , for the first time using a laser beam. Their value was five times more accurate than Froome's measurement and was 299,792.462 ± 0.018 km/s.

With the advent of the laser and the improvement in frequency measurements, measurements of the speed of light via the relation $c = \lambda f$, where c is obtained by measurement of both the wavelength, λ, and the frequency, f, of stabilized lasers became possible. Figure 2 shows several recent determinations of the speed of light [4]. These lead to the 1973 recommendation of the 'Comité consultatif pour la définition du mètre', that the speed of light be conventionally fixed at 299,792,458 m/s with an uncertainty of about 4×10^{-9}. Ten years later in 1983, the 17th Conférence Générale des Poids et Mesures abolished the former definition of the meter and promulgated the new one, defining the meter in terms of the second and fixing the value of the speed of light.

4. CLOCKS

The problem of time measurement goes back to the beginning of humanity and still has the double aspect of measuring an absolute chronological time, and a relative interval of time.

The first solution was obvious and used astronomical phenomena such as, day and night, phases of the moon and seasons of the year. If a high accuracy is not required, these phenomena are good time-keepers. As far as we know, it seems that the custom of dividing day and night, each into six equal parts is very ancient and may have resulted from the facility of dividing a circle into six equal parts. Even with a primitive compass, a further division by two is equally elementary. This explains the twelve hour faces on our clocks. It was not until the 14th century that astronomers introduced twenty-four hour clocks to distinguish between daytime and nighttime hours. Hence, until nearly the middle ages, the hours did not have a fixed duration because they resulted from the division of the daylight period which varies with latitude and, at a given latitude, with the season of the year.

The sundial from prehistory to middle ages was the most common instrument for time measurement. Its major inconvenience was the absence of time indication in absence of a bright sun. For measuring an interval of time, instruments similar to eggtimers were used. Such instruments were good enough to measure short time intervals but could not be used as clocks or as time keepers, even for a few days.

In the early middle ages, mechanical clocks were used as time keepers. In these clocks, the rotation of toothed wheels was driven by a falling weight. Various systems were developed to produce a regulated rotation of the wheels.

In 1583, Galileo made a major discovery when he realized that the period of the

Our clocks read Coordinated Universal Time, TUC. This time differs from International Atomic Time (TAI) by an integral number of seconds, TUC=TAI-n (1s) where n is an integer. What is the difference between Universal Time (TU) and Coordinated Universal Time (TUC)? Universal time, derived directly from astronomical observations is noted TU_o and becomes TU_1 when corrected for various irregularities. TUC cannot differ from TU_1 by more than 0.9s. When the deviation exceeds 0.9s, TUC is adapted to TU_1 by adding or substracting 1s preferably at midyear or at the end of the TUC year. Figure 6 shows how Coordinated Universal Time differs from International Atomic Time[7]'

5. TIME DOMAINS IN THE UNIVERSE

Table 2 shows different time intervals in the universe and the span from the shortest to the longest time interval is greater than 40 orders of magnitude. In the course of this school, we will be dealing with time intervals which are mainly situated in the lower part of the table, below 1 second and we will see how each technique is able to detect motions operating in different time domains.

Table 2. The Range of Time Intervals in the Universe

seconds

10^{18}	Age of Universe
10^{15}	Age of Earth
10^{12}	Earliest men
	Age of Pyramids
10^9	Lifetime of a man
10^6	1 year = 3.156×10^7 s
	1 day = 8.64×10^4 s
10^3	Light travels from Sun to Earth
1	Interval between heartbeats
10^{-3}	Period of a sound wave
10^{-6}	Period of a radio wave
10^{-9}	Light travels 1 ft
10^{-12}	Period of a molecular vibration
10^{-15}	Period of an atomic vibration
10^{-18}	Light travels an atomic diameter
10^{-21}	Period of a nuclear vibration
10^{-24}	Light travels a nuclear diameter

REFERENCES

1. R.V. Pound and G.A. Rebka Jr., Phys. Rev. Lett. 1960, **4**, 337.
2. P. Giacomo, Am. J. Phys.1984, **52**, 607.
3. K.M. Baird, 'Speed of light, Historical Review to 1972', in *Quantum Metrology and Fundamental Physical Constants*, Eds. P.H. Cutler, A.A. Lucas, NATO-ASI series Plenum Press, p. 143-164, 1983.
4. K.M. Evenson, 'Frequency measurements from the microwave to the visible, the speed of light and the redefinition of the meter', in *Quantum Metrology and Fundamental Physical Constants*, Eds. P.H. Cutler, A. A. Lucas, NATO-ASI series

Plenum Press, p. 181-208, 1983.

5. Roger Stevenson, 'Mechanical and electrical clocks at Greenwich', in *Greenwich Time and the Discovery of the Longitude*, Appendix III,p. 205-219, Oxford University Press, 1980.

6. John Pilkington, 'Modern precision clocks', in *Greenwich Time and the Discovery of the Longitude*, Appendix IV, p. 220-226, Oxford University Press, 1980.

7. D. Howse,*Greenwich Time and the Discovery of the Longitude*, Chapter 7, p. 173-187, Oxford University Press, 1980.

8. Gerhard Becker, 'Time scales-Production and Distribution', in *Quantum Metrology and Fundamental Physical Constants,* Ed. P.H. Cutler, A.A. Lucas, NATO-ASI series Plenum Press, p. 109-142, 1983.

CHAPTER 3

GALILEO, HIS LIFE IN PISA AND HIS SCIENTIFIC INSTRUMENTS

Umberto Russo
Dipartimento di Chimica, Metallorganica ed Analitica
Via Loredan, 4
I-35131 Padova
Italy

ABSTRACT. Our trust in scientific instruments, which today forms the basis of most scientific research, originated with Galileo Galilei. A discussion of the cultural, scientific, and social atmosphere of the sixteenth century yields an understanding of the central role of Galileo in the development of science. His family surroundings and ancestry are briefly described as is his life up to 1592, when he left Pisa to join the University of Padua as a Professor of Mathematics. His development and construction of two instruments, the military and geometric compass, the predecessor of the modern slide rule, and the telescope, are discussed. The compass was developed to help common people perform complex calculations. It may be considered the first realization of a modern computing machine. The telescope was mainly used by Galileo as a scientific instrument and made possible his famous astronomical discoveries. Unfortunately, these discoveries led him into trouble with the Catholic Church. The practical use of these instruments is described in detail in order to stress the ingenuity with which Galileo solved many technological problems.

This NATO Advanced Study Institute is dedicated to the information we can obtain on time dependent phenomena by means of instrumental techniques. We take for granted that instruments show just a part of the universe, enlarged, or reduced, so that we may investigate it in detail far beyond that directly available to our senses. The results of these investigations are accepted as reality. This trust in instruments is now part of our experience, a part of our mind set which we no longer even question. But this has not always been the case, and actually is a rather recent acquisition, occuring within the last 400 years. This trust was developed at the beginning of the seventeenth century and resulted mainly from the work of one great scientist.

The end of the sixteenth century was quite unusual because it was preceeded by a large number of scientific discoveries and technical innovations, and was further characterized by great cultural and social changes. In those years, interest in natural phenomena increased after centuries of neglect. A new curiosity arouse about man's surroundings which was no longer encompassed in the old, abstract theories of the middle ages. The world was now considered an open laboratory within which all human experience should be investigated. Because of this renewed interest in nature, learned people began again to read the classic books written by the great Latin and Greek scientists and scholars. These books were republished and often translated into modern languages. Hence the works of Plato, Euclid, Pythagoras, and Plinius

19

G. J. Long and F. Grandjean (eds.), The Time Domain in Surface and Structural Dynamics, 19–30.

experienced a revival, which was further enhanced by the discovery and rapid diffusion of printing. In 1447 in Mainz, Germany, the first printing house was established by J. Gutenberg. This revolutionary technology spread very quickly throughout Europe, and in Italy, Venice and Florence became the most important publishing centers. But the introduction of printing was not the only technological innovation of this period. Other technologies resulted in enormous progress in mining, metallurgy, transportation, architecture, and other related fields. This development created new wealth and, as a consequence, new social classes with particular interests. Again Venice and Florence became famous for their bankers, traders, merchants, and financiers. All these "nouveaux riches" competed with the old nobility to subsidize all kinds of art, such as painting, sculpture, architecture, music, and literature. Michelangelo (1475-1564), Leonardo da Vinci (1452-1519), Bramante (1444-1554), Ariosto (1474-1533), and Monteverdi (1567-1643) are some of the Italian artists who benefited from this wealth. In conclusion, these years were characterized by rapid cultural growth, many scientific discoveries, deep transformations in human thought, and social advancement.

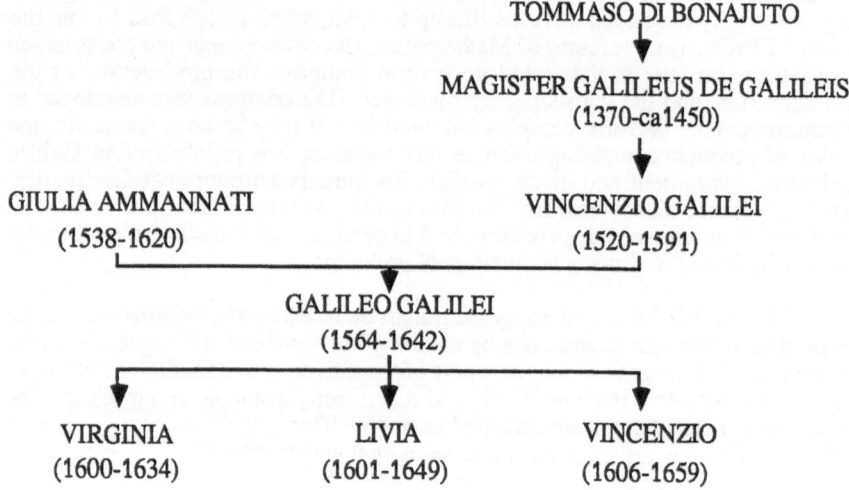

Figure 1. Some members of the family of Galileo Galilei

On 15 February 1564 Galileo Galilei was born in Pisa, the last descendant of an ancient and noble Florentine family. Among his ancestors, some of which are reported in Figure 1, we find Tommaso di Bonajuto, who was a member of the government that restored democracy to Florence in 1343, and Magister Galileus de Galileis, a famous physician and "Gonfaloniere di Giustizia", a Minister of Justice in the Florentine political system. As a recognition of his importance, he was buried in the ancient family sepulchral monument in Santa Croce, a church in Florence where many great Italian men have been buried. Unfortunately, such an honor was denied Galileo until 1734. Galileo's mother, Giulia Ammannati, was an introverted woman and an overbearing mother who did not leave good memories with her son. His father, Vincenzio, was a genial man, a skilful musician and a great lute player. He applied himself principally to theoretical musical problems, and also to literature, classical languages, and even to mathematics. In summary, the familiar surroundings of Galileo were surely culturally advanced, but, without doubt, lacked extensive parental affection. This lack of parental affection added to his quarrelsome and suspicious nature, and also affected some of his decisions such as to place, at an early age, his daughters, Virginia and Livia, in a nunnery near Florence.

Galileo was born in Pisa, as reported in Figure 2, because his family had moved there a few years earlier because of financial difficulties which had forced his father to go into commercial trade. Galileo's family was quite numerous. He had two brothers and four sisters. Among them were Virginia, Michelangelo, a famous musician, and Livia, all of whom played important, but negative roles in his unhappy family life. He was first tutored in Pisa by Jacopo Borghini. In 1574, his family returned to Florence, probably because by that time the family financial difficulties had been partially solved, and a respectable life in Florence was once again possible.

GALILEO GALILEI

1564	Birth in Pisa
1574	Return to Florence
1581	Enrollment as a medical student in the Pisa University
1583	Mathematical instruction by Ostilio Ricci
1583	Discovery of the isochronal pendulum oscillation
1585	Return to Florence
1586	"La Bilancetta" ("The Small Balance")
1588	Application at Bologna University
1589	Professor of Mathematics at Pisa University
1590	"De Motu" ("About Movement")
1590	"Contro il Portar la Toga" ("Against Wearing the Gown")
1591	Death of Vincenzio Galilei
1592	Professor of Mathematics at Padua Univesity

Figure 2. A short outline of the life of Galileo Galilei until his move to the University of Padua.

In the sixteenth century, Florence was the capital of a wealthy, active, and independent state under the rule of the Medici family. It was famous throughout Italy, and probably most of Europe, for the presence of the greatest artists, painters, sculptors, architects, authors, craftsmen, engineers, scientists, bankers and traders. Florentine society promoted craftsman's shops in which advanced instruments and manufacturing technologies were developed.

Galileo attended the school of the famous monastery of Santa Maria at Vallombrosa, and in 1578 unsuccessfully applied for a scholarschip at the University of Pisa. We do not have much information on Galileo's activity in Florence, but no doubt he visited, and probably worked in some of those craftsman's shops. His interest in technology and its practical applications to scientific discoveries, that was to become so important in his later career, may have been born in these shops.

In 1581 he enrolled at the University of Pisa, as a student of medicine in the Artists School. He was forced into this decision by his father, who, in such a way, hoped to solve forever the family economic problems. After four long and apparently wasted years as a medical student, Galileo left Pisa and the University without a degree. But, in spite of this failure, these years were crucial to his future as a scientist. The definitive turning point in his life occurred in 1583, when, at home for the summer holidays and without his father's knowledge, he received mathematical instruction from Ostilio Ricci, a skilful mathematician and a student of Niccolo Tartaglia. Ricci taught Galileo a mathematics dedicated to technical applications, rather than a general mathematics based on abstract hypotheses. Ricci was also an ardent student of the famous classic mathematicians and an admirer of Archimides. Up to 1583, Galileo's interest has wandered with superficiality over science, as if looking for something to which to devote himself. From 1583 on, mathematics occupied a prominent place in his life. Further mathematics itself will become a powerful method for this study of the deepest secrets of nature.

1583 was also important because it marked Galileo's debut in the scientific world. He immediately became a major scientific figure when he discovered the law of isochronal pendulum oscillation and moreover realized that the isochronal oscillation was not perfect. He was only 19. He was unaware that, in the tenth century, Ibn Junin, an Arab scientist, had obtained the same result. Huygens (1629-1695), some fifty years later, proved that isochronism is typical not of the arc of a circle, but of a cycloide. Two aspects of this fundamental discovery should be mentioned. First the circumstances of the discovery. The observation of the motion of a lantern in the Pisa cathedral would suggest such a general law only to a person with an excellent mathematics background and an excellent observational ability. Second, he immediately stressed the connection of this discovery with its practical applications to time measurement, especially when short intervals are measured, such as the human pulse rate.

As already mentioned, in 1585 Galileo returned to Florence without a degree or a job, but with clear ideas about his future within the scientific world. In the following year, he improved the hydrostatic balance, described in his first scientific publication, "La bilancetta" (The Small Balance). This is clearly an immature work, in which Ricci's influence is evident from the presence of an obvious Archimedean inspiration. In the following two years, he discovered several theorems on the center of gravity of solid bodies. These discoveries were known only among his friends, the circle of which was widening to include important people, such as the marquis Guidobaldo del Monte. In 1587, during a visit to Rome, Galileo was introduced to the Jesuit mathematician, Christoph Clavius, and, in 1588, was invited by the Florentine Academy to lecture on the mathematical aspects of Dante's Inferno. He began to look for a job, obviously in the academic world, and in 1588 applied for the chair of mathematics at the University of Bologna. The Bologna faculty preferred Antonio Magini, probably for a pretended superiority in astronomy. One year later, his friendships proved useful and Cardinal Francesco del Monte, brother of Guidobaldo, helped him to obtain the chair of mathematics at the University of Pisa, with a contract for three years and a salary of sixty scudi per year. In comparison with the salary of Mercuriale, Professor of Medicine, it was very low, sixty scudi in comparison with two thousand, but this was his first sure income. Moreover the professorship meant his reentry as a professor in that same university from which he was unable to graduate. On 11 December 1589, he gave his inaugural lecture. Galileo remained in Pisa for the three years of the contract, until 1592.

During his professorship in Pisa, he wrote many manuscripts collected in a treatise, known as "De Motu" (About Movement), that represents the beginning of his struggle with Aristotelian physics. This treatise deals with the theory of falling bodies derived from the buoyancy principle of Archimedes and the law governing the equilibrium of weights on inclined planes. The discrepancy between the calculated and the experimental velocity of descent led him to many fundamental discoveries about acceleration and infinitesimal calculus. This treatise is also the beginning of the scientific methodology revolution that is the greatest glory of Galileo.

In Pisa his difficult character became appearent. He was completely unable to tolerate the narrow university mentality bound to the old Aristotelian schemes. Moreover, the tradition bound university faculty was unable to accept any change or new ideas. Intolerant towards these limits, Galileo published "Contro il Portar la Toga" (Against Wearing the Gown), a jocular poem that shows his displeasure with the university faculty and resulted in much hostility from his colleagues. In 1591, Vincenzio, his father, died, and his death loaded Galileo with all the economic responsibilities of the family. Now he needed a higher salary, but his difficult relations with the university made this unlikely. Furthermore, because of his criticism of a scheme for the dredging of the harbor of Leghorn, his relations with the Medici family, the patrons of the university, were also strained. Again his friend Guidobaldo del Monte came to his rescue. The chair of mathematics at the University of Padua was vacant after the death of Giuseppe Moletti in 1588. Galileo obtained a salary of 180 francs per year for four years. Without regrets, he left Pisa and on 7 December 1592 he gave his inaugural lecture in Padua.

Galileo left his original and eternal mark on many fields of science and culture. His concept of science led him to some revolutionary ideas. For instance, an important aspect of his work is the attention he devoted to the practical applications of scientific discoveries and to their diffusion beyond a narrow scientific circle. To pursue these objectives, many of his most important scientific works were written in Italian, in contrast to the normal use of Latin in scientific work throughout Europe. Latin was used as a "lingua franca" that would guarantee comprehension everywhere, as does English today. Unfortunately, by the sixteenth century, the common people had forgotten Latin, and preferred the national languages. The concept of science as a community effort, carried out by wider and wider groups of scientists, together with Galileo's interest in the education of the common man led to his use of Italian. Perhaps Galileo did not realize the revolutionary meaning of these ideas, but the political and religious rulers surely did. This was one of the reasons for the Church pursuing a ruthless and endless war against Galileo. The Galilean writing style is important for its added accuracy and precision, in marked contrast with that of the seventeenth century. In a way we can say that modern scientific literature began with Galileo's work, which had a very precise qualitative and quantitative meaning. The ambiguity that was common in other scientific writings was absent in those of Galileo. His prose was clear and concise; far from the complex style of his contemporary scientists, for instance Kepler, whose writings were verbose, pompous, and strictly bonded to the literary rules of the Baroque period. The use of the vernacular is often considered a minor contribution of Galileo to the development of science. In contrast, his use of instruments in scientific research is often considered a major contribution.

Clearly there is no need to emphasize the importance of instruments in modern scientific research, but it is important to remember that, four centuries ago, few instruments were available to scientists. Also the psychological attitude of scientists towards these instruments was quite different from that of today. In the time of Galileo, there were instruments such as simple balances, ovens, and compasses, and astrolabes, solar quadrants, and devices for astronomic observations. The introduction of printing greatly contributed to the diffusion of knowledge about these research instruments and by the end of the sixteenth century technicians, engineers, and craftsmen, able to make such complex instruments, were numerous in England, France, Italy, Holland, and Germany. New social and political conditions contributed to the development of these instruments and to the birth of a new class of craftsmen, highly specialized in the construction of these technologically sophisticated devices. Strong motivaton for this development came from the new military necessities, mainly connected with the evolution of artillery, and naval exploration. In Italy, Venice and Florence became the most important centers in which art, science, and technology grew and together contributed to the progress of each other. Galileo assiduously visited first the craftsman's workshops in Florence and then the shipyards in Venice.

Among the many instruments that Galileo developed and brought to near perfection is the "geometrical and military compass". It is an instrument for calculations, which Galileo developed in 1597 by improving upon the sector built by his friend Guidobaldo del Monte. It is a simple and inexpensive aid to the problems of drafting and design. In its basic form it consists of two legs connected at one end by a movable joint as illustrated in Figure 3. The two legs have equal lengths and numerical scales. Galileo became interested in this instrument for practical reasons. With the help of a simple compass, many complex geometric problems could be easily solved by uneducated people. The instrument was developed with the aim of helping soldiers to find the appropriate charge for an artillery weapon with a given bore and with different shells. Shortly this instrument obtained new scales, and hence new uses of a more and more scientific and theoretical nature. So the first two scales of Figures 3 were devoted to the problems of the artillery officer to "make the caliber". The first scale gives the volume for equal weight of different materials and the second provides cube roots. The third scale gives square roots, the fourth divides a line into any number of equal parts, the fifth draws a given regular polygon on a line of given length, the sixth gives, for any polygon, the side of any other polygon with the same area, the seventh inscribes any regular polygon in a circle. Eventually, another scale was added to give the area of any circular segment. In the end, this instrument solved many of the mathematical current problems.

Galileo wrote many handbooks for the compass, all of them in Italian. He always stressed that, by means of a computing compass, even the most complex problems became easy for common people, who no longer had to ask the learned people for help. From the time of this instrument, 1598, up to the present, the idea of a "computing device" remained practically unchanged. As with the military compass, modern electronic computers help scientists to carry out calculations that otherwise are too timeconsuming and tedious. Further they give all people a computing capability otherwise impossible.

The above applications, practical and scientific, are found also in the development of the telescope, that is shown in Figure 4, and of the microscope, even if the approach of Galileo to these instruments was completely different. His mathematical and geometric training, necessary for the development of the military compass, was extremely good. In contrast, his knowledge of optics, indispensable in building a good telescope, was very poor. In spite of this lack of background in optics, the improvements he brought to the telescope are simply astonishing. The evolution of lens manufacturing from simple glasses used to enlarge or reduce objects, to spectacles, and then to telescopes and microscopes was long and tedious. The use of lenses was known from ancient times, but for centuries their low quality prevented their scientific use. Only the progress in their chemical composition and the highly improved construction techniques developed in the fourteenth and fifteenth centuries, especially in Venice, allowed their use as something more than a toy. In 1316 spectacles were sold for six soldi in Bologna. Clearly lens quality was then good enough for daily use. In England, Leonard Digges (1510-1558) and John Dee (1527-1608) performed the first experiments with optical instruments and were able to make distant objects appear nearby. A telescope with a concave and a convex lens was exhibited at the 1608 Frankfurt exhibition. On 2 October 1609, J. Lippershey (1570-1619), a Dutch lens-grinder, applied for a patent on a device to make distant

25

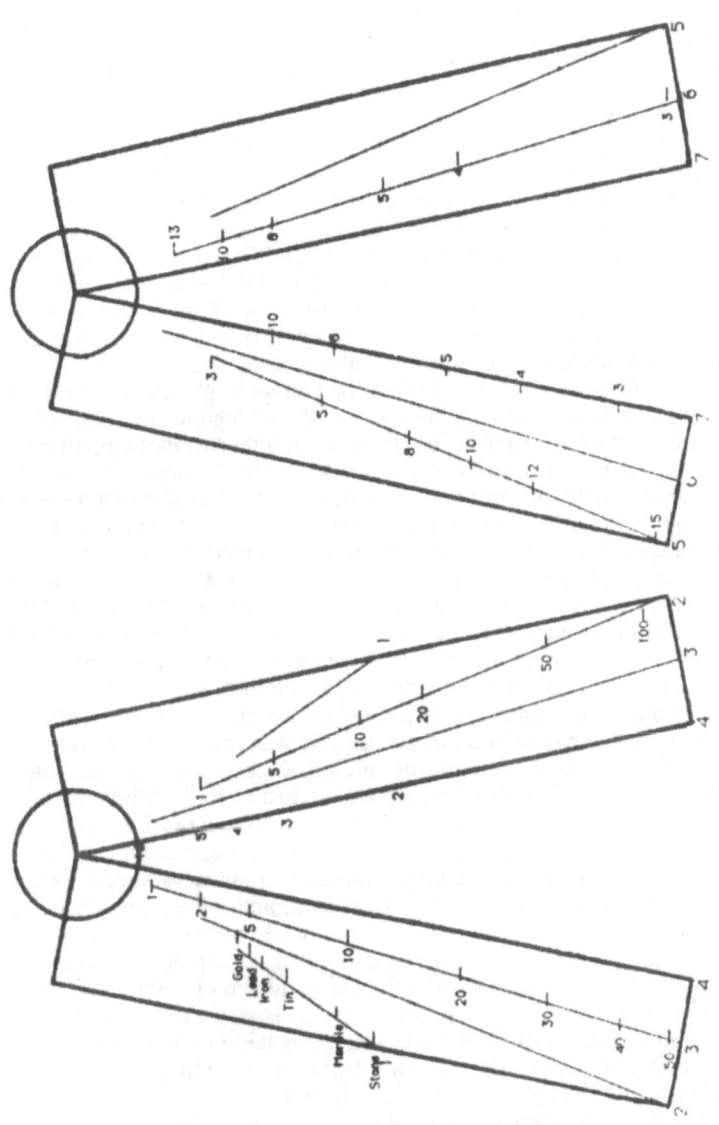

Figure 3. A schematic drawing of Galileo's "military and geometric compass" showing the computing scales.

objects appear closer. Two of these telescopes were given as a gift to the king of France and in April 1609 similar instruments were sold in Paris. In the summer of 1609 T. Harriot (1560-1621) pointed his telescope at the moon and began its mapping.

Sarpi, a Venetian priest and friend of Galileo, knew of the patent and informed Galileo, who immediately began his improvement of the instrument. His first telescope, a lead tube 2.9 meters long with a diameter of 42 mm, was equipped with a plano-convex objective and a plano-concave eyepiece. It had a magnification power of three diameters. On 25 August 1609, Galileo, together with the Doge and several members of the Venetian senate, ascended the San Marco belltower in Venice and immediately demonstrated the practical, economic, and military applications of a telescope similar to that shown in Figure 4. In the same year, he pointed his telescope at the sky and, within a few years, made many fundamental discoveries, some of which were described in "Sidereus Nuncius" (The Sidereal Messenger) which was published in Venice in March 1610. His discovery of the mountains and valleys of the moon, the satellites of Jupiter (Figure 5), the shape of Saturn, the phases of Venus, the structure of the Milky Way, and finally the spots on the sun soon became topics of discussion among scientists, philosophers, and theologians.

Galileo's discoveries were more than enough to undermine the foundations of the ethical and scientific dogma upon which the philosophical, religious, and hence political power of the time were based. Even though the work of Copernicus had been published some fifty years earlier in 1543, early in the seventeenth century the earth was still considered to be the center of the universe and the heavens were still considered to be motionless and perfect spheres. For this reason the mountains on the moon and the spots on the surface of the "perfect" sun were a dramatic revolution. In a few years, with observations of the sky through his simple telescope, Galileo destroyed the physics of Aristotle which the Church considered the basis of its philosophy. As a consequence, the use of any scientific instrument was criticized. How is it possible to be sure that what a telescope shows but which cannot be touched is true? Can we be sure that it actually exists? It is interesting to remember the opinion of a cardinal on the solar spots: "I have spent all the night looking for these spots, but I couldn't find any!" He could do this because he was reading the works of Aristotle.

Galileo's trust in the telescope clearly had an enormous impact on his astronomical research. He trusted what he saw, just as today we believe in what our instruments show us in the infinitely small or large worlds. Galileo, with an instrument that today seems as simple as a toy, proved that scientific theories, which were centuries or millennia old and deep-rooted in humanity, were not eternal. When even a simple telescope shows something that contradicts the normally accepted ideas, a scientist cannot close his eyes, as did so many when they saw the 1618 comets. In contrast, the theories must be changed or at least adapted to include the new observations. Only continuous verification, as provided by natural events, give value and credibility to scientific ideas and theories. It is also useful to remember that, in 1612 in Florence, thanks to the accuracy and precision with which Galileo used his simple telescope, he could make the first sighting of Neptune, which was officially discovered only in 1795 by Johann Galle in Berlin.

The measurements made with the telescope of Galileo were affected by an error often less than 10 arc seconds. Only his ingeniousness permitted such a precision. In fact a micrometer scale, that is indispensable in making quantitative measurements, could not be used because Galileo's telescope generated a virtual image and not a real image. But in spite of this, the notes on the position of Neptune as compared to that of Jupiter, made by Galileo in December 1612 and January 1613,

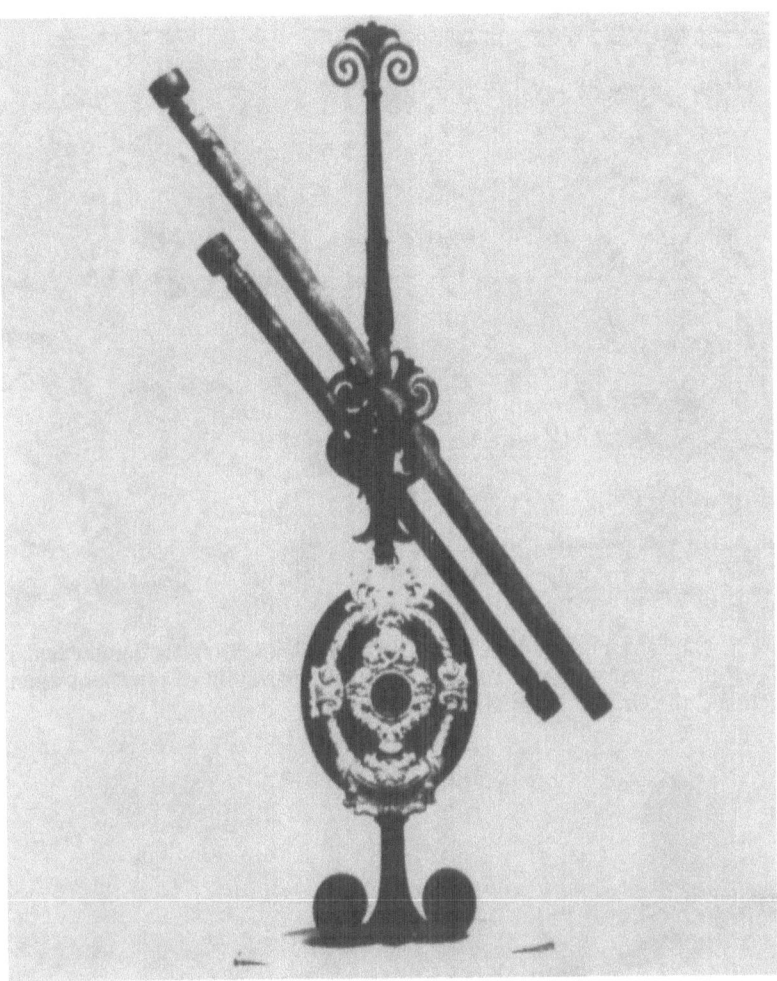

Figure 4. Galileo's telescope.

question the accuracy of the currently accepted orbit of Neptune. On the basis of notes written by Giovanni Alfonso Borelli, one of Galileo students, his measurement apparatus can now be described and is illustrated in Figure 6. The Galilean micrometer was an accurately ruled grid mounted next to the end of the telescope. Galileo looked through the telescope with one eye and through the grid with the other, superimposing Jupiter and its satellites on the grid. Then he rotated the grid until one ruling was aligned with the plane of the satellites and moved the grid toward or away from his eye until the planet fits exactly within four squares of the ruling, as shown in Figure 7. In such a way an accurate quantitative system of measurement was obtained. In December 1612 and January 1613 he made many recorded observations of Jupiter and its satellites and charted also the position of a fixed star shown in the lower left of Figure 7. We know today that this star is Neptune. The discrepancy between Galileo's reported positions of Neptune and the currently calculated ones is almost sixty arc seconds, a value considered too large by the standards of celestial mechanics. This discrepancy could be due to either the presence of an undiscovered planet beyond Neptune or to an error in the orbital period of Neptune.

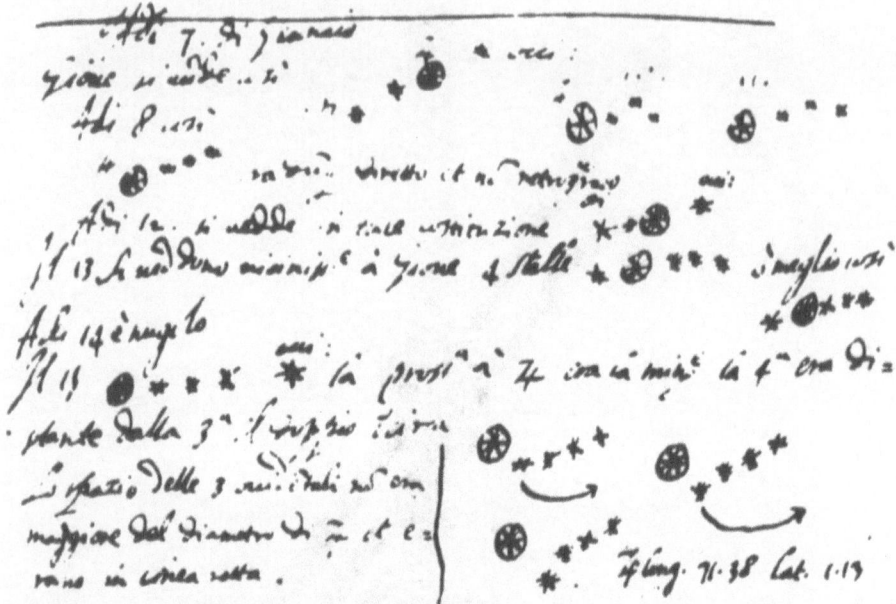

Figure 5. Galileo's notes on the first sighting of the moons of Jupiter on 7 January 1610. (Reproduced with permission from the department of rare books and special collections, the University of Michigan library)

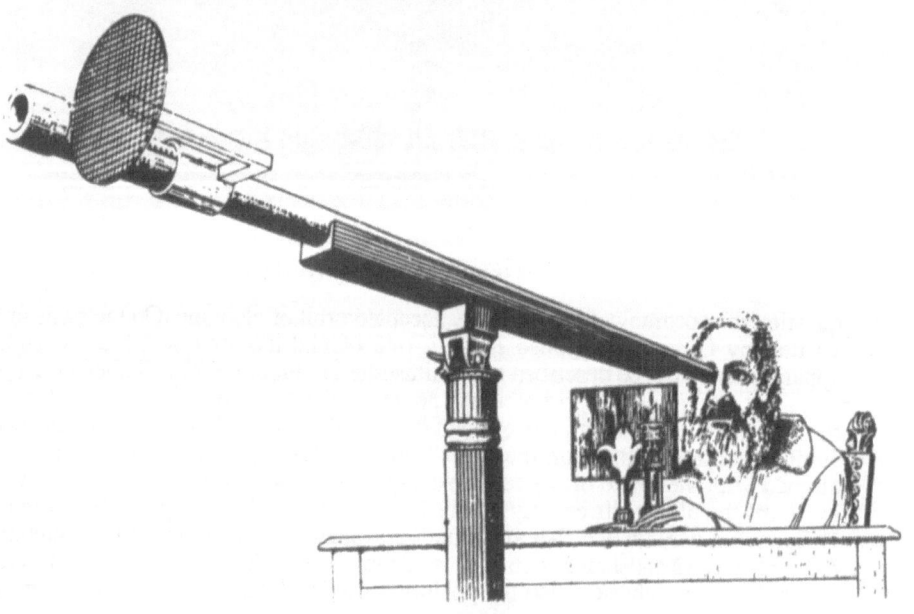

Figure 6. Galileo working with his telescope. (Reproduced from ref.5 Copyright (1980) by Scientific American, Inc. all rights reserved).

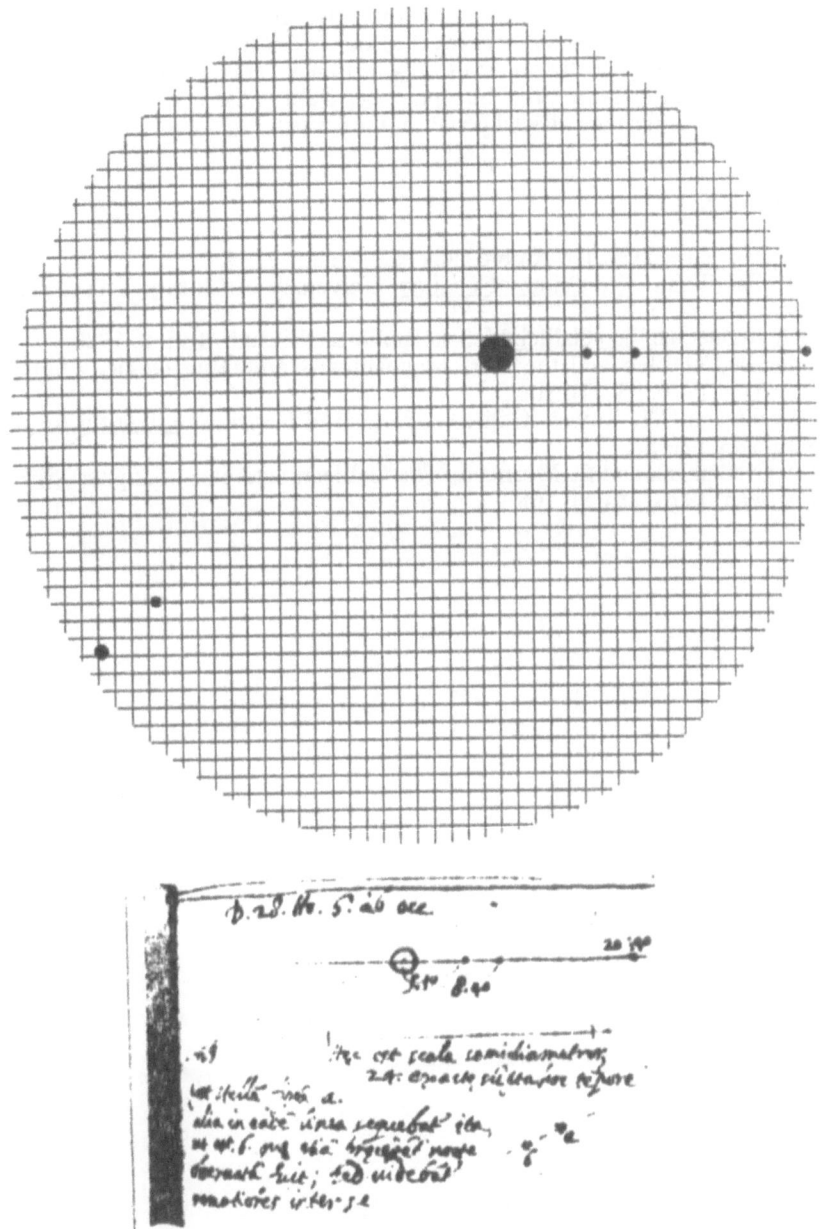

Figure 7. A representation of Galileo's micrometer scale and his notes on the first sighting of Neptune.(Reproduced from ref.5 Copyright (1980) by Scientific American, Inc. all rights reserved).

A long path connects Galileo's simple instruments to our modern complex instruments. The progress made in less than five centuries is truly astonishing. Nevertheless the high technological level of our instruments must not let us forget who first realized that scientists must use any instrument at their disposal to investigate the secrets of nature. To put this into practice, Galileo did not hesitate to fight an uneven struggle against the official power of the Church.

REFERENCES

1. A. Banfi, Galileo Galilei, Ambrosiana ed., Milano, 1949.
2. G. DeSantillana, The Crime of Galileo, University of Chicago Press, Chicago, 1955.
3. L. Geymonat, Galileo Galilei, Piccola Biblioteca Einaudi, Torino, 1980.
4. S. Drake, 'Galileo and the First Mechanical Computing Device', Scientific American, 1976, 234(4), 104.
5. S. Drake and C.T. Kowal, 'Galileo's Sighting of Neptune', Scientific American, 1980, 243(12), 52.
6. D.P. Cruikshank and D. Morrison, 'The Galilean Satellites of Jupiter', Scientific American, 1976, 234(5), 108.

CHAPTER 4

TIME DOMAIN NMR AND MAGIC ANGLE SPINNING

E. Raymond Andrew
Departments of Physics, Radiology and
Nuclear Engineering Sciences
University of Florida
Gainesville
Florida 32611
U.S.A.

ABSTRACT. Rapid specimen rotation about an axis inclined at the
'magic angle' of 54°44' to the Zeeman field direction can remove many
sources of broadening from the NMR spectrum of a solid and enable
finer features to be revealed. In this chapter the basic principles
of NMR are first introduced with especial reference to the time
domain. Next, the principles of magic angle spinning are given and
the effects on the chemical shift interaction, the magnetic dipolar
and pseudo-dipolar interactions, both homonuclear and heteronuclear,
and electric quadrupolar interactions are examined. The anisotropic
parts of these interactions are removed from the central spectrum and
appear as spinning sidebands. The interactions that remain are the
isotropic shifts and J couplings as in isotropic fluids. When applied
to metals the anisotropy of Knight shift is removed, isotropic Knight
shifts may be measured with precision and the Ruderman-Kittel interaction may be determined in the presence of a much larger dipolar
interaction. Magic angle spinning may be used on its own, and may also
be successfully combined with multiple-pulse and cross-polarization
NMR methods to obtain high-resolution NMR spectra of solids.

1.   INTRODUCTION TO NMR

Nuclear Magnetic Resonance (NMR) is a branch of spectroscopy in
the radiofrequency range of the electromagnetic spectrum, mainly prac-
tised between 10^6 and 10^9 Hz with periods in the time domain between
10^{-9} and 10^{-6} s. Since the first discoveries of NMR were reported in
1946 by Bloch, Hansen and Packard and by Purcell, Torrey and Pound,
the phenomenon has become an indispensable technique for the determination of molecular structure and dynamics in liquids and solids and
more recently has become the basis of a new method of imaging which is
gaining widespread use in medicine.

If an atomic nucleus of magnetic moment μ and spin I is placed in
a magnetic field $\vec{H}$ it has (2I+1) eigenvalues of energy $-m\mu H/I$, running

G. J. Long and F. Grandjean (eds.), The Time Domain in Surface and Structural Dynamics, 31–48.
© 1988 by Kluwer Academic Publishers.

with equal spacing from +μH to -μH, where m is the nuclear magnetic
quantum number. If an electromagnetic field is applied of such a frequency that the quanta hν exactly match the spacing between the adjacent energy eigenvalues a resonant exchange of energy takes place
between the magnetic nucleus and the electromagnetic field. This
requires

$$h\nu = \mu H/I \tag{1}$$

and the process is called nuclear magnetic resonance (NMR).

From a classical standpoint a spinning magnetic nucleus in a
magnetic field experiences a torque which causes it to precess. The
angular frequency ω of precession is given by Larmor's Theorem

$$\omega = YH \tag{2}$$

where Y is the nuclear gyromagnetic ratio. Remembering that Y is
μ/IħI, we see that the classical Larmor precessional frequency given by
equation (2) is identical with the NMR frequency derived from quantum
theory in equation (1). This happy agreement enables us to describe
many features of NMR in classical terms rather than in the more
rigorous but more difficult quantum mechanical theory.

Nuclear magnetic resonance was first detected in individual molecules in molecular beam experiments and later in bulk matter. If say
1 ml of solid or liquid are placed in a magnetic field, the nuclei are
distributed over the magnetic eigenstates in proportions determined by
the Boltzmann factor at the temperature of the the sample. We are, in
fact, dealing with a weak nuclear paramagnetic assembly whose magnetization is only large at temperatures well below 1K. Nevertheless, the
preponderance of nuclei in the low eigenstates enables an NMR signal
to be observed when the irradiating frequency satisfies equations (1)
and (2).

NMR has been observed for over a hundred nuclear species. All
elements of the periodic table have at least one magnetic isotope,
though sometimes the natural abundance is very weak. NMR has been
observed in matter in all its forms: solids, liquids, gases, liquid
crystals, adsorbed and occluded phases, metals, alloys, semiconductors, dielectrics, glasses, ceramics, polymers, biopolymers, cellular
systems, plants, vegetables, fruit, animals and whole human beings.

Most regular NMR applications to solids and liquids use magnetic
fields in the range from 1T to 14T, although outside this range work
has been done in lower fields down to the earth's magnetic field ($5x
10^{-5}$ T) and in the much higher hyperfine fields of transition and rare
earth elements. The favorite NMR nucleus ^{1}H, the proton, has an NMR
frequency of 42.6 MHz in 1T and 600 MHz in 14T; there are projects
afoot pursuing a goal of 1000 MHz for proton NMR. With the exception
of the triton ^{3}H, all other nuclei have a lower NMR frequency in a
given field, often an order of magnitude lower. Most NMR work is,
therefore, carried out as stated earlier at radiofrequencies in the
range 10^6 to 10^9 Hz. In the time domain the NMR period (Larmor
period) is thus in the range 10^{-9} to 10^{-6} s. Work in the earth's

magnetic field extends the domain of operations to Larmor periods of 10^{-3} s. However, as we shall see, there are many other times which enter into a description of NMR phenomena which extend the interest of the NMR practitioner well beyond this range at both ends.

Application of a pulse of resonant NMR electromagnetic radiation to a nuclear spin system nutates the nuclear magnetization $\vec{M}$ away from the direction of $\vec{H}$, and if the magnetization is turned through 90° it is said to be a 90° pulse. The precessing nuclear magnetization induces an e.m.f. in the receiver coil surrounding the specimen; this e.m.f. may be observed and is the NMR signal. Because the signal decays to zero after the exciting pulse has been shut off it is called the Free Induction Decay (FID).

After the 90° pulse the magnetization is not in equilibrium with its surroundings; it decays to zero and regrows longitudinally along $\vec{H}$, a process called nuclear relaxation. To a first approximation, these relaxation processes are often exponential. The regrowth of the magnetization along the field direction is characterized by a time constant T_1, the longitudinal relaxation time (sometimes called the spin-lattice relaxation time from its significance in solids). The decay of the transverse magnetization which generates the FID is characterized by the time constant T_2, the transverse relaxation time (sometimes called the spin-spin relaxation time). These two relaxation times T_1 and T_2 and their relationship to the dynamical properties of the materials in which they are observed, are of immense significance to NMR scientists.

Observed values of the relaxation times T_1 and T_2 vary enormously. Observed values of T_1 may range from 10^{-6} s to 10^6 sec (hours); T_2 from 10^{-6} s to 10 s. In pure liquids T_1 and T_2 are often equal; for example, for ^1H in pure water both are about 4 s. However, in ice (solid water) at low temperatures T_1 may be hours and T_2 may be as short as 10^{-5} s. These differences in behavior are rather well understood and are related to the correlation time τ_c of the nuclear motions in the liquid or solid which generate the relaxation mechanisms. Molecular correlation times in liquids range from 10^{-12} s for a mobile liquid to 10^{-4} s for a viscous liquid, becoming much longer in a glass. Measurement of T_1 for solids over a wide temperature range gives great insight to the rate of molecular motion in solids and enables molecular motions typically in the range $10^4 - 10^{12}$ Hz to be measured, corresponding in the time domain to a range from 10^{-4} to 10^{-12} s.

It is often helpful to view NMR phenomena in a frame of reference rotating relative to the laboratory at the Larmor frequency. This frame is called the rotating reference frame. Suppose we apply a 90° pulse and turn the magnetization $\vec{M}$ into the plane at right angles to $\vec{H}$ and then apply a resonant radiofrequency magnetic field $\vec{H}_1$, typically 10^{-3} T, parallel to $\vec{M}$. This a situation which is particularly appropriate for viewing in the rotating reference frame because in this frame both $\vec{M}$ and $\vec{H}_1$ are at rest and the dominant influence on $\vec{M}$ is $\vec{H}_1$. In this situation we may study nuclear relaxation in the rotating frame and obtain a rotating frame relaxation time $T_{1\rho}$ which is especially sensitive to slow molecular correlation times in the time domain up to 1 s with an overall coverage of τ_c by T_1 and $T_{1\rho}$ together from 10^{-12} s to 1 s.

Measurements in the rotating reference frame require the continuous application of a fairly strong resonant radiofrequency field which can present experimental difficulties. An alternative approach is to disturb the equilibrium in the dipolar system and watch it recover using a suitable pulse sequence. The return of the dipolar system to equilibrium is characterized by yet another time, the dipolar relaxation time T_{1D}, which, like $T_{1\rho}$, is sensitive to slow molecular motions in the range 1 to 10^5 Hz. Both T_{1D} and $T_{1\rho}$ provide information on slower molecular motions in solids extending the time domain for relaxation studies from 10^{-12} s to 1 s.

To measure the FID, excitation of the specimen by a resonant 90° pulse is all that is required. To measure T_1 it is customary to invert the nuclear magnetization with a 180° pulse and inspect its recovery with a subsequent 90° pulse. To measure T_2 one may start with a 90° pulse followed by a 180° pulse which generates a spin echo. The decay of these echoes gives T_2, though to avoid complications due to diffusion, a sequence of 180° pulses is usually applied. To measure $T_{1\rho}$ and T_{1D} more complex pulse sequences are required. More sophisticated NMR measurements call for still more complicated pulse sequences. It will be seen, therefore, that the whole operation of NMR is carried out in the time domain. In general, a series of radiofrequency pulses of appropriate length, phase and spacing is applied in the time domain, and their response is an NMR signal in the time domain, which is recorded.

The FID following a single 90° pulse describes the simplest NMR response of the specimen in the time domain. Application of a Fourier transform describes this response in the frequency domain and gives an NMR spectrum. For water the transverse relaxation time T_2 describes the exponential decay of the proton NMR signal and its Fourier transform gives an NMR spectrum consisting of a single Lorentzian line of width about 0.1 Hz. On the other hand, from an organic liquid the ^{1}H or ^{13}C NMR spectrum generally consists of many well-resolved spectral lines. These separately resolved resonances have two origins.

First, each nucleus is surrounded by molecular electrons which slightly screen the nucleus from the applied magnetic field and slightly shift its resonance frequency. This is called the chemical shift because it differs from one molecular group to another in the molecule, giving rise to distinctly different resonant frequencies. Second, there is an interaction between magnetic nuclei in adjacent groups via the molecular electrons. This is called spin-spin coupling and generates spin multiplets in the spectrum. The spectrum thus consists of lines which reflect the groups present in the molecule and finer multiplets which reflect the way the groups are assembled in the molecule; in this way the spectrum provides a characteristic signature of the molecule. Such a spectrum is therefore extremely valuable in chemical analysis and in molecular structure determination and is currently being applied to molecules of increasing complexity such as proteins and nucleic acids.

High-resolution NMR spectra of liquids are usually analyzed in the frequency domain after Fourier transformation from the FID in the time domain. The spectra are inhomogeneous, different parts arising from different molecular frequency sources which can be separately recognized and assigned. Although the same information is present in the FID in the time domain, NMR spectroscopists would probably find it very hard to analyze and understand the NMR response simply by examining the time domain signal. Even so, some NMR spectroscopists consider that the Fourier transform may not be the most efficient means of unravelling the NMR information from the FID, and alternative procedures including direct comparisons of observed and postulated FID's in the time domain are under evaluation.

The overall length of the FID determines the width of the individual spectral lines and thus the resolution in the spectrum; this is determined by the inherent T_2 of the resonance lines and by instrumental factors such as the uniformity of the magnetic field to which can be assigned an effective transverse relaxation time T_2^*. The effects of residual inhomogeneity can be considerably reduced by spinning the specimen and so averaging the field inhomogeneities. To be effective the period of rotation must be less than T_2^*, typically of order 10^{-1} sec. The ultimate resolution is of order 0.1 Hz; the ultimate T_2 is of order of seconds. Since chemical shift frequency differences are proportional to the applied field H, instrument manufacturers are continually striving to increase the operating field and frequency in order to open up the spectrum of a liquid and improve its resolution.

It would be equally valuable to obtain such high-resolution NMR spectra from solid specimens also, but usually the NMR spectrum from a solid is very brief with short T_2. An excellent example is provided by water whose proton NMR linewidth at room temperature is about 0.1 Hz, while for ice at low temperatures it is of order 10^5 Hz, six orders of magnitude broader. The well understood origin of this substantial difference in behavior lies in the static anisotropic interactions to which the nuclei in the solid are subject. In mobile fluids the rapid isotropic motions of the nuclei average the anisotropic interactions and effectively remove them, yielding a rich, high-resolution NMR spectrum.

In solids too there is often sufficient motion of the nuclei to narrow the NMR spectra, and sometimes the spectrum is sufficiently narrowed to resolve a high-resolution NMR spectrum. An example is provided by polycrystalline P_4S_3 at 420 K, which is 26 K below the melting point, which exhibits a doublet from the basal phosphorous nuclei and a quartet from the apical nucleus (Andrew, Hinshaw, et al., 1978). However, in general sufficient motion is not present in solids and one procedure is to emulate Nature by imposing a suitable fast motion on the assembly of nuclear spins. It is difficult to impose an isotropic motion on a solid specimen, but in fact this is not necessary. It is sufficient to rotate the specimen about a single well-chosen axis, and we turn to this subject in the next section.

Meantime we close this section concerned with an introduction to NMR by mentioning some books for further reading: Abragam (1961), Andrew (1969), Fyfe (1983), Harris (1983), Haeberlen (1976), Mehring (1976), and Slichter (1978).

2. MAGIC ANGLE SPINNING

By spinning a solid we impose a time dependence on all anisotropic nuclear interactions within the solid and if the rotation is fast enough we achieve an averaging of these interactions. It is found theoretically and has been verified experimentally, that the averaging is most efficient when the angle β between the axis of rotation and the magnetic field direction has the value $\cos^{-1}(1/\sqrt{3})$ or 54°44'. Because of its exceptional properties this angle is called the magic angle.

2.1 Magic Angle Spinning and Dipolar Interactions

We begin by examining the effect of rapid specimen rotation on nuclear dipolar interactions in a solid, since they are always present in some degree and were historically the first interactions to be considered and removed; moreover, this gives good insight to the operation of the method.

The truncated dipolar interaction Hamiltonian, for all nuclear pairs i,j in the solid is

$$\mathcal{H}_d = \sum_{i<j} \frac{1}{2} \gamma_i \gamma_j \hbar^2 r_{ij}^{-3} (\vec{I}_i \cdot \vec{I}_j - 3 I_{iz} I_{jz})(3\cos^2\theta_{ij} - 1) , \qquad (3)$$

where γ_i, γ_j are the nuclear gyromagnetic ratios, $\vec{r}_{ij}$ is the internuclear displacement, and θ_{ij} is the angle between $\vec{r}_{ij}$ and the Zeeman field $\vec{H}_o$, which is directed along the z axis in the laboratory frame. The NMR spectrum is in principle calculated by treating $\mathcal{H}_d$ as a perturbation on the Zeeman term $\mathcal{H}_z$, where

$$\mathcal{H}_z = -\sum_i \gamma_i \hbar \vec{I}_i \cdot \vec{H}_o \qquad (4)$$

Since the isotropic average $\overline{\cos^2\theta_{ij}} = \frac{1}{3}$, it follows from (3) that the isotropic average of the dipolar Hamiltonian, $\overline{\mathcal{H}_d} = 0$, and rapid isotropic motion in fluids therefore eliminates the dipolar interaction from the NMR spectrum. To be effective the characteristic time of the motion τ_c must be much less than T_2, a condition which is strongly satisfied in mobile liquids.

If a solid specimen, whether monocrystalline, polycrystalline or amorphous, is rotated uniformly with angular velocity $\vec{\omega}_r$ about an axis inclined to $\vec{H}_o$ at angle β, every internuclear vector $\vec{r}_{ij}$ in the solid describes a motion illustrated in Fig. 1. The angle θ_{ij} becomes time-dependent, and the factor $(3\cos^2\theta_{ij}-1)$ in (3) runs through a range of values that may be both positive and negative. The average value may be made zero by judicious choice of β.

Expressing $\cos\theta_{ij}$ in terms of the other angles of the system,

$$\cos\theta_{ij} = \cos\beta \cos\beta'_{ij} + \sin\beta \sin\beta'_{ij} \cos(\omega_r t + \phi_{oij}) , \qquad (5)$$

we find after rearrangement that

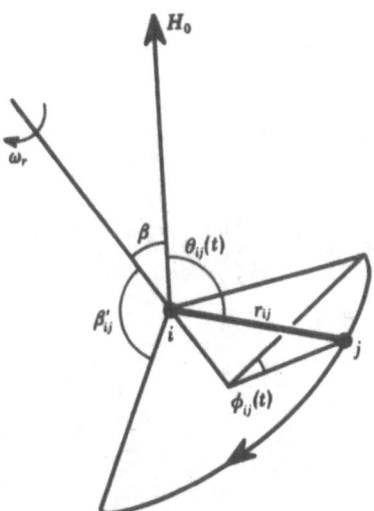

Figure 1. Diagram illustrating the motion of a typical internuclear
vector r_{ij} when a solid is rotated with angular velocity ω_r about an
axis inclined at angle β to H_0.

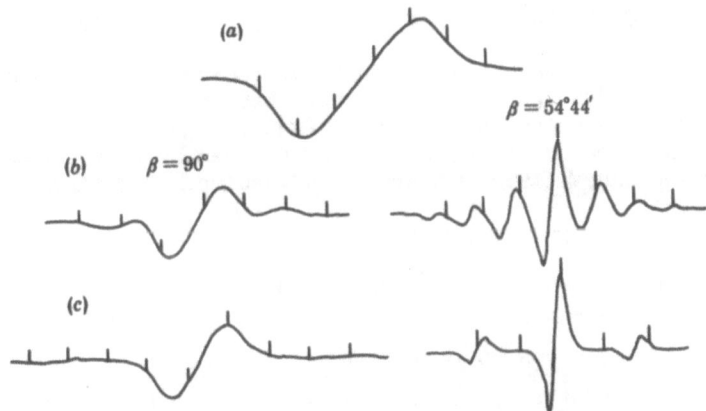

Figure 2. The effect of rapid rotation of a solid on its dipolar-
broadened spectrum. ^{23}Na NMR derivative spectra of monocrystalline
sodium chloride. Markers are at 800 Hz intervals. (a) Static crystal
(b), (c) recordings with the crystal spinning at 800 and 1600 Hz
respectively. The left hand pair of recordings were obtained with the
rotation axis perpendicular to the magnetic field. The right hand pair
were obtained with the rotation axis at the magic angle of 54°44' to
H_0, (Andrew, Bradbury and Eades 1958 a,b, 1959).

$$\mathscr{H}_d(t) = \sum_{i<j} \frac{1}{2} \gamma_i \gamma_j \hbar^2 r_{ij}^{-3} (\vec{I}_i \cdot \vec{I}_j - 3I_{iz}I_{jz}) \{ \frac{1}{2}(3\cos^2\beta - 1)(3\cos^2\beta'_{ij} - 1)$$
$$+ \frac{3}{2} \sin 2\beta \, \sin 2\beta'_{ij} \, \cos(\omega_r t + \phi_{oij})$$
$$+ \frac{3}{2} \sin^2\beta \, \sin^2\beta'_{ij} \, \cos 2(\omega_r t + \phi_{oij}) \} . \tag{6}$$

Notice that the first term in the curly bracket is constant, while the second and third terms are periodic with zero mean value. It is convenient to divide $\mathscr{H}_d(t)$ into two parts, its mean value $\bar{\mathscr{H}}_d$ and the remainder, which is periodic, with zero mean:

$$\mathscr{H}_d(t) = \bar{\mathscr{H}}_d + (\mathscr{H}_d - \bar{\mathscr{H}}_d) . \tag{7}$$

The first term on the right hand side is constant and gives a reduced dipolar interaction and the narrowed spectrum, while the second term, which is periodic in ω_r and $2\omega_r$, gives rise to rotational sidebands at multiples of ω_r. We note that in the special case of $\beta = \pi/2$, that since $\sin 2\beta = 0$, only even order sidebands contribute. We see, therefore, that

$$\bar{\mathscr{H}}_d = \frac{1}{2}(3\cos^2\beta - 1) \sum_{i<j} \frac{1}{2} \gamma_i \gamma_j \hbar^2 r_{ij}^{-3} (\vec{I}_i \cdot \vec{I}_j - 3I_iI_j)(3\cos^2\beta'_{ij} - 1) . \tag{8}$$

If we now compare (3) and (8) we see that for polycrystalline and amorphous material the spectrum retains an identical shape, but is reduced in width by the scale factor

$$F(\beta) = | \frac{1}{2}(3\cos^2\beta - 1) | . \tag{9}$$

The second moment (mean square width) of the spectrum is reduced by $F^2(\beta)$. On the other hand one can show that the second moment of the whole spectrum, including sidebands, is unchanged, expressing the expected invariance of the second moment of the whole spectrum (Andrew, Bradbury and Eades, 1958a,b; Andrew and Newing, 1958; Andrew and Jenks, 1962).

We note some particular values of $F(\beta)$:

for $\beta = 0$	$F(\beta) = 1$
$\beta = \pi/2$	$F(\beta) = 1/2$
$\beta = \cos^{-1}(1/\sqrt{3}) = 54°44'$	$F(\beta) = 0$.

Thus, rotation of the specimen about $\vec{H}_o$ has no effect, rotation about an axis normal to $\vec{H}_o$ halves the spectral width, while rotation about an axis making the magic angle of $54°44'$ should reduce the average dipolar broadening to zero. It was at the AMPERE Congress here in Pisa in 1960 where we presented some of our early results that the late Professor C.J Gorter of Leiden asked a question about the apparently magic properties of this special angle which led us to refer to it as the 'magic angle' thereafter.

These predictions were found to be borne out experimentally. The first experiments were made with ^{23}Na NMR in sodium chloride. A single crystal was used to prevent any significant quadrupolar broadening due to strains or other defects in this cubic crystal. The second moment from the static crystal agreed within experimental error with the value of 0.55 G^2 calculated from the theory of Van Vleck (1968). When the crystal was first rotated about an axis perpendicular to $\vec{H}_o$, the spectrum was indeed halved in width with sidebands appearing at $2\omega_r$ (Andrew, et al., 1958a,b), as illustrated in Fig. 2. Then, by rotating the crystal at the magic angle, the central line did indeed narrow very sharply with spinning sidebands appearing at multiples of ω_r (Andrew, et al., 1959), also shown in Fig. 2. These spectra were taken before the days of Fourier transform NMR spectrometers and appear as derivative spectra as was customary from wide-line NMR spectrometers at that time. The variation of the width of the central spectrum followed the reduction factor $F(\beta)$ (Eq. (7)) quite well, as shown in Fig. 3. The linewidth at the magic angle was 200 Hz, largely determined by field inhomogeneity of the wideline spectrometer; magic angle linewidths as narrow as a few hertz have since been obtained by using high-resolution magnets. Early verification of these ideas was independently made by Lowe (1959), who used a crystal of calcium fluoride and a sample of polytetrafluoroethylene.

A good example of the removal of dipolar broadening by magic angle spinning is provided by the ^{27}Al NMR spectrum of polycrystalline aluminium metal (Andrew, Hinshaw and Tiffen, 1973, 1974a). As illustrated in Fig. 4, the spectrum of the static material is about 10 kHz broad. Magic angle spinning at 7.7 kHz reduced this width to about 400 Hz.

It will be seen for the examples in Figs. 3 and 4 that effective line narrowing by magic angle spinning requires the rotation rate to be of the same order as the spectral linewidth. The central line then narrows and the rotational sidebands are detached. In the time domain this requirement becomes $T_r < T_2$, where $T_r = 2\pi/\omega_r$ is the period of rotation, essentially equivalent to the requirement $\tau_c < T_2$ for motional narrowing by molecular reorientation, discussed in Section 1.

2.2 Magic Angle Spinning and Anisotropic Shift Interactions

The chemical shift introduced in Section 1 is an anisotropic quantity reflecting the anisotropy of the electron distribution around the nucleus. In a single crystal the chemical shift is therefore observed to depend on the orientation of the crystal in the magnetic field $\vec{H}_o$. Consequently, the spectrum from polycrystalline or amorphous specimens is broadened by the distribution of shift values exhibited. On the other hand, in a mobile liquid the isotropic reorientations of the molecules average the shifts to their isotropic mean values, giving sharp lines in the spectrum from each nuclear environment.

In metals the shifts are usually larger and of opposite sign, resulting from the interaction of the conduction electrons. This is called the Knight shift. In non-cubic metals the shift is anisotropic, leading to spectral broadening in polycrystalline metals.

We shall now see that magic angle spinning has the effect of removing the anisotropy of both these interactions also.

The chemical shift interaction for a particular nucleus in a non-metal may be written:

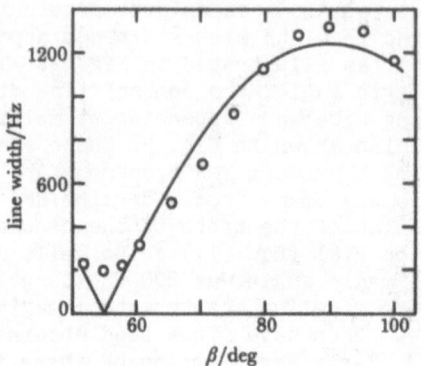

Figure 3. Variation of linewidth of ^{23}Na NMR spectrum of a rotating crystal of sodium chloride with the angle β. The full line is the theoretical curve given by equation (9); the circles are the experimental observations (Andrew, Bradbury and Eades 1959).

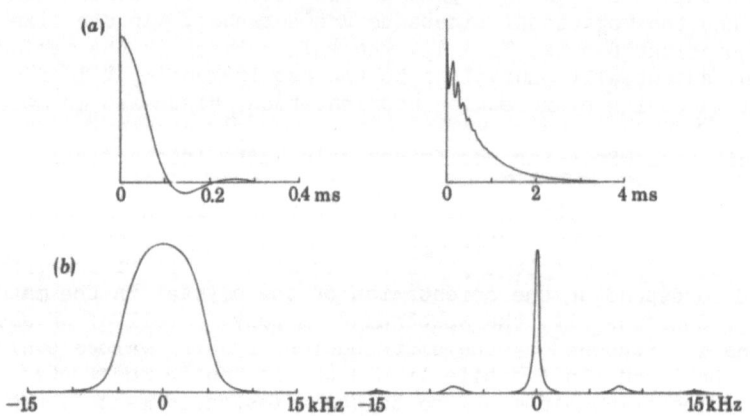

Figure 4. ^{27}Al free induction decays (a) and Fourier-transformed NMR spectra of polycrystalline aluminum. The left-hand diagrams refer to the static specimen; the right-hand diagrams refer to the specimen spinning about the magic axis at 7.7 kHz. Two rotation sidebands are seen on either side of the narrowed central line (Andrew, Hinshaw and Tiffen 1973, 1974a).

$$\mathcal{H}_s = \gamma\hbar(\vec{I} \cdot \underline{\sigma} \cdot \vec{H}_o) , \tag{10}$$

where $\underline{\sigma}$ is the chemical shift tensor of the nucleus. In metals the shift interaction is conventionally defined with opposite sign:

$$\mathcal{H}_s = -\gamma\hbar(\vec{I} \cdot \underline{K} \cdot \vec{H}_o) , \tag{11}$$

where $\underline{K}$ is the Knight shift tensor of the nucleus. Apart from this difference in sign the behavior of $\underline{\sigma}$ and $\underline{K}$ is the same in what follows.

We note that $\underline{\sigma}$ and $\underline{K}$ are second rank tensors, but unlike the dipolar interaction tensor, which is a traceless, axially symmetric tensor, $\underline{\sigma}$ and $\underline{K}$ are not traceless and are not necessarily axially symmetric.

Since the components of the shift tensor are small compared with unity we need only retain σ_{zz} and re-write (10) as

$$\mathcal{H}_s = \gamma\hbar I_z \sigma_{zz} H_o . \tag{12}$$

If the principal values of $\underline{\sigma}$ are σ_p $(p = 1,2,3)$ and the direction cosines of its principal axis with respect to $\vec{H}_o$ are λ_p, then

$$\sigma_{zz} = \sum_p \lambda_p^2 \sigma_p \tag{13}$$

Since the isotropic average of each λ_p^2 is 1/3, the average value of σ_{zz} in a normal fluid is

$$\overline{\sigma_{zz}} = \frac{1}{3} \, \text{tr} \, \underline{\sigma} = \sigma , \tag{14}$$

where σ is the scalar chemical shift encountered in high-resolution NMR spectra of fluids.

When a rigid array of nuclei in a solid is rotated with angular velocity ω_r about an axis inclined at angle β to $\vec{H}_o$ and at angles χ_p to the principal axes of $\underline{\sigma}$ we have (cf. Eq. (5))

$$\lambda_p = \cos\beta \cos\chi_p + \sin\beta \sin\chi_p \cos(\omega_r t + \psi_p) . \tag{15}$$

From (12), (13) and (15) we see that a time dependence is imposed on $\mathcal{H}_s$, and as with the dipolar interaction we may decompose $\mathcal{H}_s$ into its mean value $\overline{\mathcal{H}_s}$ and terms periodic in ω_r which generate spinning sidebands. Substituting (15) into (13) and taking the time average we find for each nucleus

$$\overline{\sigma_{zz}} = \frac{3}{2} \sigma\sin^2\beta + \frac{1}{2} (3\cos^2\beta - 1)\sum_p \sigma_p \cos^2\chi_p . \tag{16}$$

Consequently, when β is the magic angle, $\cos^2\beta = 1/3$ and $\sin^2\beta = 2/3$, and $\overline{\sigma_{zz}}$ reduces to the scalar isotropic value σ and the shift anisotropy is removed from the NMR spectrum for every nucleus in the specimen whatever the degree of anisotropy or asymmetry of its shift tensor, and whatever the orientation of its principal axes (Andrew and

Wynn, 1966; Andrew, 1971).

The time-averaged shift in (16) may be re-expressed in terms of the anisotropy parameter $\delta = \sigma_3 - \sigma$ and the asymmetry parameter $\eta = (\sigma_2 - \sigma_1)/\delta$ of each nucleus (Lippmaa, et al., 1976). Substituting these parameters in (16) and rearranging we get:

$$\bar{\sigma}_{zz} = \sigma + \frac{1}{2}(3\cos^2\beta - 1)\delta\{\frac{1}{2}(3\cos^2\theta' - 1) + \frac{1}{2}\eta\sin^2\theta'\cos2\phi'\} , \quad (17)$$

where θ' and ϕ' are spherical polar angles relating the specimen rotation axis to the principal axes of the shift tensor. Expressed in this form we see in a direct manner how the ubiquitous Legendre factor $\frac{1}{2}(3\cos^2\beta - 1)$ controls the shift anisotropy in the time-averaged Hamiltonian $\bar{\mathcal{H}}_s$ for every nucleus in the specimen.

If β is adjusted to be close to the magic angle, but not exactly equal to it, we see that the asymmetry is reduced by the scale factor $F(\beta)$ of Eq. (9). In this way information concerning the shift tensor is retained in the spectrum while still enabling the individual resonances to be resolved from each other.

An experimental example of the removal of shift anisotropy by magic angle spinning is provided by cadmium, a metal with a hexagonal crystal structure. The ^{111}Cd NMR spectra of a polycrystalline specimen of cadmium metal are shown in Fig. 5 (Andrew, et al., 1974a,b). The spectrum of the static material shows the asymmetric profile characteristic of an axially symmetric shift tensor. Dipolar broadening is small in this material on account of the low isotopic abundance (12%) and the low magnetic moment ($-0.6~\mu_N$); the Knight shift anisotropy is the dominant source of NMR broadening in the powder. Figure 6 also shows the spectra for three spinning rates, and provided the first experimental demonstration of the removal of shift anisotropy by magic angle spinning.

A new feature is to be noticed in Fig. 5. Unlike the dipolar broadened spectra, where the rotation rate must be comparable with the linewidth to get substantial narrowing, here we see that the line breaks up almost immediately into an array of satellites each of narrow breadth. The essential difference here is that the line is inhomogeneously broadened. Each element of the spectrum of the static specimen arises from crystallites of a particular orientation, each with small intrinsic width. The situation is closely analogous to the NMR line from a liquid broadened by an inhomogeneous magnetic field; each volume element in the sample contributes its own portion to the spectrum. The intrinsic width of the spectrum may then be observed (Maricq and Waugh, 1977) and the decay envelope of such echoes gives the intrinsic width of each shifted spin packet.

2.3 Magic Angle Spinning and Anisotropic Interactions

In sections 1.1 and 1.2 we have shown from first principles that fast spectrum rotation about an axis inclined at an angle β to $\vec{H}_0$ introduces a factor $(3\cos^2\beta - 1)/2$ into the time-averaged Hamiltonian representing the dipolar interaction and also into that representing the anisotropic shift interaction. In fact, this is a quite general

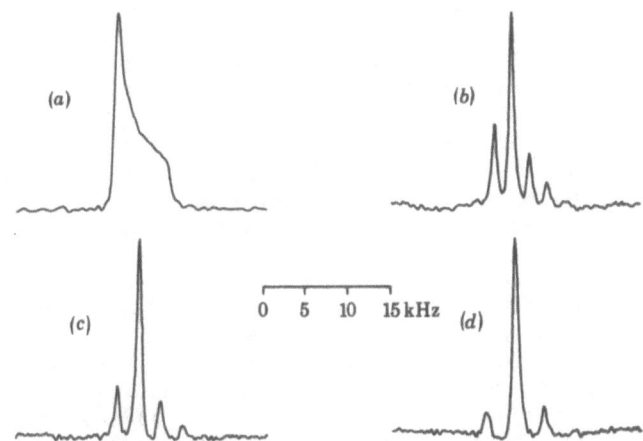

Figure 5. The removal of shift anisotropy by magic angle spinning.
^{111}Cd NMR spectra of polycrystalline cadmium metal (a) static, and
rotating about the magic axis at (b) 2.1, (c) 2.6 and (d) 3.6 kHz
(Andrew, Hinshaw and Tiffen 1974 a,b).

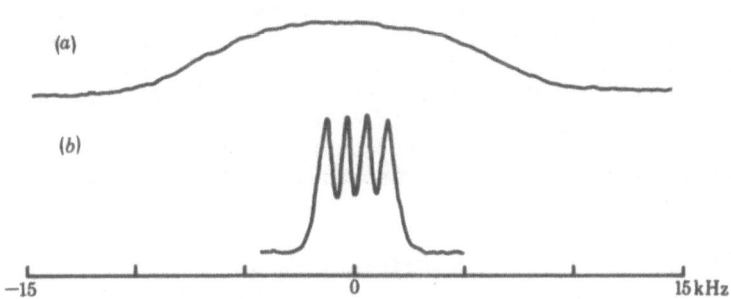

Figure 6. A spin multiplet resolved in the solid state by magic angle
spinning. ^{19}F NMR spectra of polycrystalline K As F$_6$. (a) Static
specimen; (b) specimen spinning at 5.5 kHz, displaying quartet structure
due to J coupling between ^{19}F and ^{75}As nuclei (Andrew, Farnell and
Gledhill 1967).

result. Similar calculations from first principles show that the same factor appears in the time-averaged Hamiltonian representing the anisotropic electron-coupled, spin-spin or pseudo-dipolar interactions (Andrew and Farnell, 1968) and in that representing the first order electric quadrupolar interaction (Cunningham and Day, 1966), and that magic angle spinning should therefore remove all these anisotropic interactions. Although all these cases were originally examined individually they may actually be treated in a common formalism which covers the effect of magic angle spinning on all symmetric second rank tensor interactions (Haeberlen, 1976; see also Andrew, 1981).

The removal of anisotropic broadening interactions by magic angle spinning frequently reveals fine structure previously obscured. An example is shown in Fig. 6, which shows the ^{19}F NMR spectrum of polycrystalline hexafluoroarsenate, $K As F_6$ (Andrew, Farnell and Gledhill, 1967). The spectrum of the static material is about 15 kHz broad. Magic angle spinning at 5.5 kHz largely removed the anisotropic sources of broadening (dipolar, pseudo-dipolar and chemical shift) and resolved a spin quartet with an isotropic coupling constant of the ^{19}F nuclei to the ^{75}As nucleus in the octahedral AsF_6 lines in the solid. This was in fact the first spin multiplet to be resolved in the NMR spectrum of a solid.

The removal of spectral broadening due to nuclear quadrupole interactions is not so easy to demonstrate experimentally since quadrupole interactions tend to be large. However, an unequivocal test was carried out (Tzalmona and Andrew, 1974) using two polycrystalline non-cubic caesium salts, Cs_2SO_4 and Cs_2CO_3. The caesium nucleus ^{133}Cs, which has 100% abundance and spin 7/2 has an unusually small electric quadrupole moment, -3×10^{-27} cm^2. Consequently, the quadrupole broadening in these salts was of order 10^3 Hz. Moreover, since the nearest-neighbor separation of the caesium atoms is about 0.4 nm, and there are no other abundant magnetic nuclei in these solids, the dipolar contribution to the linewidth is an order of magnitude smaller than the quadrupolar contribution. A rotation rate of 3 kHz about the magic axis reduced the linewidth by a factor of 18 in the sulphate and a factor of 7 in the carbonate, showing that magic angle spinning does substantially remove quadrupolar broadening in both solids. A later, more striking demonstration was provided by ^2D NMR (Ackerman, Eckman and Pines, 1979).

3. MAGIC ANGLE SPINNING IN METALS

All the sources of NMR spectral broadening found in non-metals are found in metals also, and we have seen in Figs. 4 and 5 that magic angle spinning successfully removes dipolar broadening and anisotropy of the Knight shift. This enables measurements of the isotropic Knight shift of metals to be obtained with much improved precision. In the past the accuracy had been severely restricted by the difficulty of precisely locating the centre of the broad resonance line.

For metallic copper, the isotropic Knight shifts for ^{63}Cu and ^{65}Cu relative to CuBr were found to be 2392.9 ± 1.2 ppm and 2395.1 ± 1.6 ppm, respectively, an order of magnitude improvement in precision

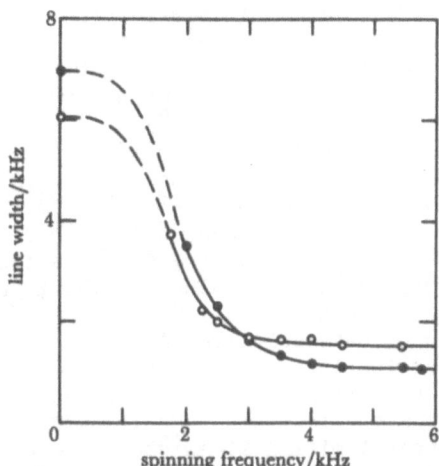

Figure 7. NMR linewidths for copper metal as a function of spinning frequency. [63]Cu filled circles, [65]Cu open circles (Andrew, Carolan and Randall 1971b).

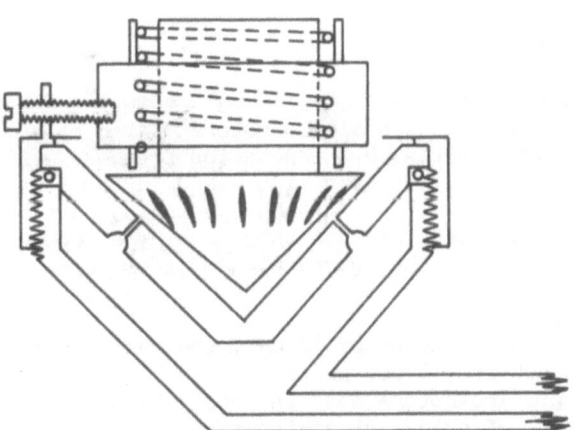

Figure 8. Turbine assembly for magic angle spinning (Andrew et al. 1969).

on previous measurements (Andrew, Carolan and Randall, 1971a). The error limits refer to a 99% confidence level. The small difference in the Knight shifts is therefore not statistically significant. For s electrons any fractional difference in Knight shift for the two isotopes should be close to their hyperfine structure anomaly which is small for the copper isotopes (1.5×10^{-4}).

For aluminum, magic angle spinning at 7.7 kHz reduced the linewidth by a factor 25 and enabled an improvement in precision of the isotropic Knight shift to be made by a corresponding factor to 1640 ± 1 relative to $AlCl_3$ solution (Andrew, Hinshaw and Tiffen, 1973).

For cadmium removal of broadening due to anisotropy of the shift and dipolar interactions enabled values of the isotropic Knight shift of 4321 ± 7 ppm for ^{111}Cd and 4324 ± 7 ppm for ^{113}Cd to be obtained relative to cadmium nitrate solution (Andrew, et al., 1974a,b).

Another interaction is revealed by fast magic angle spinning in metallic copper. It is seen in Fig. 7 that while magic angle spinning substantially reduces the ^{63}Cu and ^{65}Cu linewidths, for rotation rates above 3 kHz a constant limiting linewidth is reached that is evidently due to a rotationally invariant interaction (Andrew, Carolan and Randall, 1971b). By comparing the second moments of the residual lines and finding them to be in the inverse ratio of the abundances of the two copper isotopes, it was shown that these residual breadths were due to the Ruderman-Kittel electron-coupled interactions. A value of J = 230 ± 10 Hz was extracted for the nearest neighbor coupling constants, in reasonable agreement with theory, and consistent with the value for silver (Andrew and Hinshaw, 1973). Spin coupling constants could then be calculated for remoter neighbors.

4. SPINNERS

In the pioneering days of magic angle spinning rotation rates up to 13 kHz were achieved with simple rotors driven by compressed gas. An example is shown in Fig. 8. A rotation frequency of 13 kHz corresponds to 780,000 revolutions per minute and in the time domain the period of revolution is 77 μs. At this spinning rate the peripheral velocity of a 19 mm rotor exceeds the velocity of sound in air and it was necessary to propel the rotors with compressed helium gas; the velocity of sound in helium is 2.7 times higher than in air (Andrew, Farnell, et al., 1969).

At a rotation frequency of 10 kHz, the peripheral velocity of a 19 mm rotor is 600 ms^{-1} (2000 km/hr^{-1}), and the peripheral acceleration is 4×10^6 g. The rotors must therefore be made of strong materials. Rotors made of nylon, glass-fibre and carbon-fibre reinforced materials and sapphire have been used. In recent years frequencies in excess of 20 kHz have been achieved by using rotors of much smaller diameter. Commercial instruments usually have slower spinning rates up to about 5 kHz which generally suffices for nuclei other than ^{1}H and ^{19}F.

5.   FINAL REMARKS

Magic angle spinning is an excellent means of providing high-resolution NMR spectroscopy of solids for nuclei other than ^{1}H and ^{19}F. However, for these two nuclei it is hard to satisfy the condition $T_r <$ T_2 for effective narrowing since T_2 may be as short as 10 µs. In these circumstances multiple pulse irradiation methods can be more effective since pulse cycle times of this order can be achieved. However, multipulse methods do not remove broadening due to anisotropy of the chemical shift nor do they remove heteronuclear dipolar and pseudo-dipolar interactions.   It is therefore necessary to combine the technique with magic angle spinning to obtain sharp, highly-resolved spectra from polycrystalline and amorphous materials, which cover most solid materials of practical interest.   Such a combination is called Combined Rotation and Multiple Pulse Spectroscopy (CRAMPS).

Another important procedure is the use of cross-polarization methods to provide high-resolution NMR spectra of rare spins such as ^{13}C, ^{15}N or ^{29}Si in solids. Under strong irradiation the magnetization of abundant spins in the solid, for example protons, is transferred to the rare spins and their heteronuclear anisotropic interactions are suppressed. Because the rare spins are so dilute the homonuclear dipolar interactions are weak.   Nevertheless, in polycrystalline and amorphous materials substantial broadening due to anisotropic shift interactions remain.   Here again high-resolution NMR spectra from solid specimens are only obtained by combination with magic angle spinning of the specimen which removes the anisotropic shift broadening.

Multiple quantum spectra also have a residual breadth arising from chemical shift anisotropy.   These residual sources of broadening may again be removed by magic angle spinning providing yet another avenue for high-resolution NMR spectroscopy from solid specimens.

REFERENCES

Abragam, A. (1961).   The Principles of Nuclear Magnetization.   Oxford:
        Clarendon Press.
Ackerman, J.L., Eckman, R. and Pines, A. (1979).   Chem. Phys. 42, 423.
Andrew, E.R. (1969).   Nuclear Magnetic Resonance.   Cambridge:
        University Press.
Andrew, E.R. (1971).   Prog. NMR Spectr. 8, 1.
Andrew, E.R. (1981).   Proc. Roy. Soc. A 299, 505.
Andrew, E.R., Bradbury, A. and Eades, R.G. (1958a).   Nature 182, 1659.
Andrew, E.R., Bradbury, A. and Eades, R.G. (1958b).   Arch. Sci.
        (Geneva) 11, fasc. spec., 223.
Andrew, E.R., Bradbury, A. and Eades, R.G. (1959).   Nature 183, 1802.
Andrew, E.R., Carolan, J.L. and Randall, P.J. (1971a).   Phys. Lett. A
        35, 435.
Andrew, E.R., Carolan, J.L. and Randall, P.J. (1971b).   Phys. Lett. A
        37, 125.
Andrew, E.R. and Farnell, L.F. (1968).   Molec. Phys. 15, 157.
Andrew, E.R., Farnell, L.F., Firth, M., Gledhill, T.D. and Roberts, I.
        (1969).   J. Magn. Reson. 1, 27.

Andrew, E.R., Farnell, L.F. and Gledhill, T.D. (1967). Phys. Rev. Lett. **19**, 6.

Andrew, E.R. and Hinshaw, W.S. (1973). Phys. Lett. A **43**, 113.

Andrew, E.R., Hinshaw, W.S., Hutchins, M.G. and Jasinski, A. (1978). Proc. Roy. Soc. A **364**, 553.

Andrew, E.R., Hinshaw, W.S. and Tiffen, R.S. (1973). Phys. Lett. A **46**, 57.

Andrew, E.R., Hinshaw, W.S. and Tiffen, R.S. (1974a). J. Magn. Reson. **15**, 191.

Andrew, E.R., Hinshaw, W.S. and Tiffen, R.S. (1974b). Proc. 18th Ampere Congress, Nottingham, 325. Amsterdam: North-Holland.

Andrew, E.R. and Jenks, G.J. (1962). Proc. Phys. Soc. **80**, 663.

Andrew, E.R. and Newing, R.A. (1958). Proc. Phys. Soc. **72**, 959.

Andrew, E.R. and Wynn, V.T. (1966). Proc. Roy. Soc. A **291**, 257.

Bloch, F., Hansen, W.W. and Packard, M.E. (1946). Phys. Rev. **69**, 127.

Cunningham, A.C. and Day, S.M. (1966). Phys. Rev. **152**, 287.

Fyfe, C.A. (1983). Solid State NMR for Chemists. Guelph: CFC Press.

Haeberlen, U. (1976). High-resolution NMR in Solids (Adv. Magn. Reson. Suppl. 1). New York: Academic Press.

Harris, R.K. (1983). Nuclear Magnetic Resonance Spectroscopy. London: Pitman.

Lippmaa, E., Alla, M. and Tuherm, T. (1976). Proc. 19th Congress Ampere, Heidelberg, 113.

Lowe, I.J. (1959). Phys. Rev. Lett. **2**, 285.

Maricq, M. and Waugh, J.S. (1977). Chem. Phys. Lett. **47**, 327.

Mehring, M. (1976). High-resolution NMR Spectroscopy in Solids. Berlin: Springer-Verlag.

Purcell, E.M., Torrey, H.C. and Pound, R.V. (1946). Phys. Rev. **69**, 37.

Slichter, C.P. (1978). Principles of Magnetic Resonance. Berlin: Springer-Verlag.

Tzalmona, A. and Andrew, E.R. (1974). Proc. 18th Ampere Congress, Nottingham, 241. Amsterdam: North-Holland.

Van Vleck, J.H. (1948). Phys. Rev. **74**, 1168.

ACKNOWLEDGEMENT

The author is happy to acknowledge support from the National Institutes of Health, Grant Numbers 1 P41 RR02278, 5R01 CA 42283 and 110 245512.

CHAPTER 5

CHEMICAL, MOLECULAR AND SPIN DYNAMICS

N.J. Clayden
ICI plc, P.O. Box No. 90, Wilton Centre, Middlesbrough,
Cleveland, TS8 6JE, England.

ABSTRACT. NMR is a versatile technique for studying dynamic processes in the solid state. In the following account the application of NMR methods to the study of chemical, molecular and spin dynamics over the wide range of correlation times 10^{-3}s to 10^{3}s will be illustrated.

1. INTRODUCTION

A major strength of NMR is the ability to investigate dynamic processes occurring in the solid state. Although dynamic processes may be studied in their own right we can also expect that a study of dynamics will help to understand aspects of structural chemistry. Broadly speaking dynamic processes can be classified into one of three areas namely chemical, molecular and spin dynamics. Chemical dynamics results in changes in the molecular framework whilst molecular dynamics leave the molecular structure intact but alter the average conformation or position of the molecules. Spin dynamics are reflected in the nuclear spin magnetisation either internally by relaxation or externally by applied perturbations. The presence of one type of dynamic process does not rule out another type indeed it is clear that spin dynamics leading to nuclear spin relaxation are always present. It is worth emphasising that the utility of NMR lies in the wide range of processes which can be studied. Examples of the many types which can be found are listed below.

1. Molecular or atomic reorientation at a lattice site with and without an attendant change in structure (1,2).
2. Molecular or atomic reorientation throughout the lattice, diffusion (3).
3. Intramolecular rearrangement (4,5).
4. Intermolecular rearrangement.
5. Chemical reaction (6).
6. Nuclear spin dynamics, spin diffusion and magnetisation transfer (7,8).
7. Bulk mechanical spinning (9).
8. Radiofrequency pulse, rotation of the nuclear spin in spin space (10).

G J Long and F. Grandjean (eds.), The Time Domain in Surface and Structural Dynamics, 49–63.
© 1988 by Kluwer Academic Publishers.

The wide applicability of NMR can be attributed to a number of factors.

Two of the most important reasons are; 1) the presence of an NMR active nucleus for most elements and 2) the range of NMR parameters that can be measured and which give information on different timescales for the motion as illustrated in Figure 1. A corollary of this is that in order to study a process of a given correlation time the appropriate parameter and hence NMR experiment must be chosen.

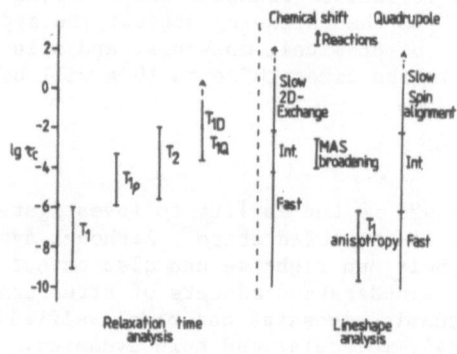

Figure 1. Relationship between the correlation time for a molecular motion and NMR parameters.

It is evident from Figure 1 that the NMR experiment is sensitive to dynamic processes covering a wide range in rates. The fast motion limit is determined by the Larmor frequency and may be increased by working at higher static magnetic field strengths thus the T_1 minimum at 600 MHz is at 1.67×10^{-9}s. At the other extreme the slowest processes which can be studied and give a correlation between initial and final states are limited by the nuclear spin T_1 and the acquisition time of the experiment. This sets a maximum correlation time of ca. 100s. Slower processes, in particular chemical reactions, may be studied out to 10^3s by simply collecting NMR spectra as a function of time, however, no direct correlation between the initial and final states is then possible.

Although the emphasis of this account is on high resolution studies, in the sense of the observation of resolved resonances corresponding to different atoms in the structure, low resolution methods have historically played the major part in NMR studies (11). High resolution can be achieved in a number of ways, use of single crystals, magic angle spinning, multiple pulse sequences and selective isotropic enrichment.

2. COHERENT AVERAGING

When magic angle spinning and multiple pulse sequences are used it is necessary to bear in mind that these impose a spin dynamics on the system as the aim of bulk mechanical spinning and multiple pulse sequences is to bring about a coherent averaging of the nuclear spin interaction in order to average the internal Hamiltonians to zero. Consequently if the correlation time of the motion is similar to the time constant in the coherent averaging then the effectiveness of the averaging is lessened and broadening of the resonances will take place owing to the reintroduction of the internal Hamiltonian. Hence, broadening can be expected if the motion is in the frequency ranges shown below;

1.	MAS	Spinning speed	1-5kHz (12)
2.	High power decoupling	Decoupler power	55 kHz (13)
3.	Multiple pulse	Pulse cycle time	20 kHz (14)

3. CHEMICAL DYNAMICS

The principle of using NMR to follow chemical dynamics is straightforward. The isotropic resonances or indeed anisotropic components in the MAS spectrum are used as fingerprints for the different species and their intensities are monitored as a function of time. As an example consider the hydration of calcium silicates where the reaction involves the polymerisation of an orthosilicate species (15-17).

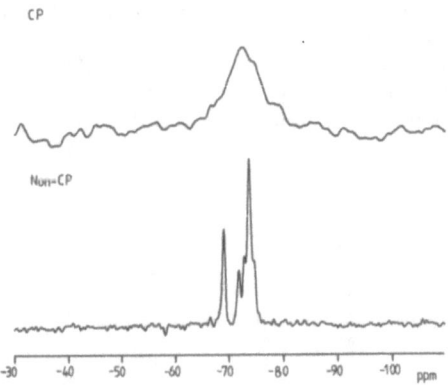

Figure 2. Comparison of the cross-polarisation and single pulse excitation ^{29}Si MAS NMR spectra of a hydrated calcium silicate after stopping the reaction during the induction period.

Three different timescales can be followed; i) the first 5 hours, known as the induction period where cross polarisation has to be used to enhance the signal intensity from the small amount of product, ii) 6-24 hours, iii) months and years.   Following the reaction during the induction period is an interesting example of using cross-polarisation not only to enhance the rare nucleus sensitivity but also to achieve selectivity in the species seen.   Thus in the hydration the reactant is a pure calcium silicate which does not respond to H to Si cross polarisation but the product is a hydrated silicate of the type $SiO_2(OH)_2^{2-}$ which does appear.   This allows a very small concentration of product to be monitored and identifies the initially formed silicate during the induction period as an orthosilicate, see Figure 2.   Later in the reaction after the induction period and up to 24 hours only a dimeric species is observed (16).   Further chain polymerisation is seen with increasing time but even a year after the initial hydration no evidence is seen for a branched polymer.

4.   MOLECULAR DYNAMICS

There are a number of ways in which dynamic processes are apparent in an MAS NMR spectrum.

Two examples are as follows:-

1. The presence of molecular motion is reflected in the NMR spectrum of K penicillin G by the broadening of the resonances corresponding to the C2,C3,C5 and C6 in comparison with the spectrum of Na penicillin G, Figure 3 (18).   The broadening of the resonances in the case of K penicillin G arises because of an aromatic flip about the C1-C4 axis.   This could be caused by either a classical exchange broadening or through the ring flip interfering with the coherent averaging.   Both effects have been seen in pencillin systems.

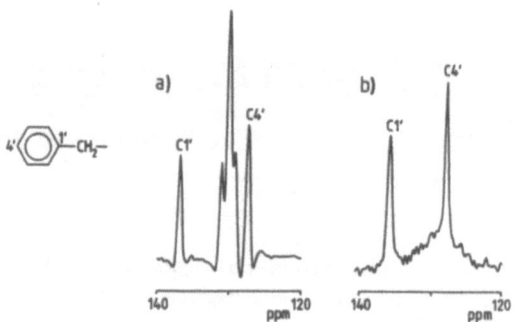

Figure 3. Aromatic regions of the 50.32 MHz ^{13}C CPMAS NMR spectra of a) sodium penicillin G and b) potassium penicillin G.

2. For ferrocene intercalated in alpha, beta and gamma cyclodextrin the motion is apparent from the sharp ferrocene resonance and in the long decay time in the dipolar dephasing or Non Quaternary Suppression experiment, Figure 4. In general chemical shift heterogeneity is seen in solid state NMR spectra leading to broad resonances hence a sharp resonance suggest that molecular motion must be present to average this chemical shift heterogeneity. The long decay times in the dipolar dephasing experiment show that motion of the ferrocene within the cyclodextrin cavity is reducing the dipolar coupling. Moreover, the differing decay profiles demonstrated that the ferrocene dynamics can not be the same in the alpha, beta and gamma cavities.

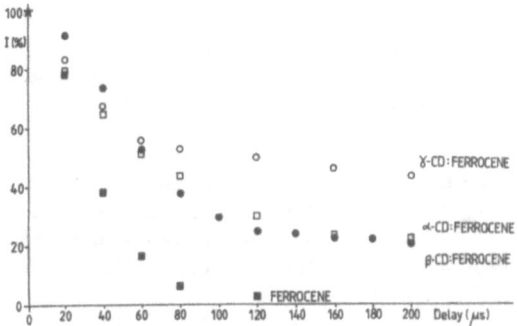

Figure 4. Signal intensity in a dipolar dephasing experiment as a function of the dephasing period.

Both of these examples illustrate qualitative aspects of molecular motion in typical CP MAS experiments. Dynamics may also be inferred from a number of other observations such as a change in the spectrum with time, the presence of scalar coupling to F e.g. in PF_6^- groups, averaged tensor components and for the static samples a reduction in the second moment. For the most part the information is only qualitative and for a quantitative study the experiments must be directed towards this end.

5. QUANTITATIVE STUDY OF DYNAMIC PROCESSES

Dynamic processes can be studied in two basic ways by NMR.

1. Lineshape analysis (19)
2. Nuclear spin relaxation times (20,21)

I shall concentrate on lineshape analysis and will not consider the use of nuclear spin relaxation times. The most important nuclear spin interactions in lineshape analysis are the chemical shift anisotropy and the quadrupole interaction with dipole-dipole coupling of less importance. The reason for this is that the dipole-dipole coupling is normally multi-centred leading to a relatively featureless lineshape. As a consequence lineshape analysis is restricted to the measurement of the second moment and following its variation as a function of temperature. Despite this limitation molecular motion can be studied in some detail (22,23). Note that when the dipolar coupling is restricted to a single spin $\frac{1}{2}$ pair the lineshape is analogous to the ^2H case and thus amenable to a more sophisticated analysis.

In lineshape analysis it is necessary to distinguish three different timescales for the motion in relation to the magnitude of the nuclear spin interaction corresponding to fast intermediate and slow exchange. In the slow limit, $\Delta.\tau_c < 1.0$, where τ_c is the correlation time for the motion and Δ is the magnitude of the nuclear spin interaction, the spectrum is described simply by the static tensor components of the nuclear spin interaction. For the fast limit, $\Delta.\tau_c > 1.0$, it is necessary to calculate only the average tensor components generated by the motion. On the otherhand with intermediate exhange, $\Delta.\tau_c \sim 1.0$, a complex lineshape is observed requiring a rate formalism to describe it.

In the case of intermediate or fast exchange molecular dynamics are reflected in the partial averaging of the anistropic nuclear spin interactions. The effect of the motion is to cause a reorientation of the co-ordinate system of the nuclear spin interaction with respect to the applied magnetic field.

Two types of reorientation are normally considered.
1. Discrete motion hops or jumps.
2. Continuous rotations
It is important to note that pure translational motion does not affect the lineshape in the absence of chemical shift or quadrupolar heterogeneity.

5.1 FAST LIMIT

Essentially any change in orientation can be treated as a rotation and the effect on the nuclear spin interaction can be readily expressed by the unitary transformation:

$$V' = R.V.R^{-1}$$

Where V is the nuclear spin interaction in the PAS and thus diagonal and R is the standard rotation matrix describing the transformation to the new orientation. Evaluation of this expression is straightforward giving, in general, a non diagonal tensor. Now for a continuous motion is it necessary to integrate each tensor component over the angular range of the motion. In the case of a hopping motion each of the tensor components is calculated for the different sites using the appropriate Euler angles and the average component calculated.

The problem now is that the tensor is usually non-diagonal and thus requires diagonalisation to obtain the three tensor components which describe the averaged fast motion line shape, happily standard algorithms exist for matrix diagonalisation.

5.2 CHEMICAL SHIFT ANISOTROPY

In the case of chemical shift anisotropy a molecular motion causes a change in the orientation of the chemical shift tensor with respect to the applied field. A commonly studied problem is that of aromatic ring motion about a pseudo twofold axis where three cases are usually discussed namely a static ring, a ring undergoing 180° flips about a single axis and continuous motion about the same axis. It can be seen from the sets of three chemical shift tensor components obtained for the static, 180 flip and the continuous motion that these motions can be distinguished.

	σ_{11}	σ_{22}	σ_{33}	$\tilde{\sigma}_{11}$	$\tilde{\sigma}_{22}$	$\tilde{\sigma}_{33}$
Static	225.0	160.0	30.0	225.0	160.0	30.0
180 Flip	225.0	160.0	30.0	209.0	176.0	30.0
Continuous rotation	225.0	95.0	95.0	176.0	119.4	119.4

Note that CSA of the atom lying along the twofold axis is not affected by the 180° ring flip. A practical example is gamma zirconium phosphate exchanged with phenyl phosphate leading to a mixed organic – inorganic layered structure. A ^{13}C CPMAS NMR spectrum of the exchanged phase clearly shows the incorporation of phenyl phosphate into the lattice, Figure 5, and a comparison of the observed spinning sideband pattern with the simulated spectra for the different motions clearly demonstrates that the ring is static (24).

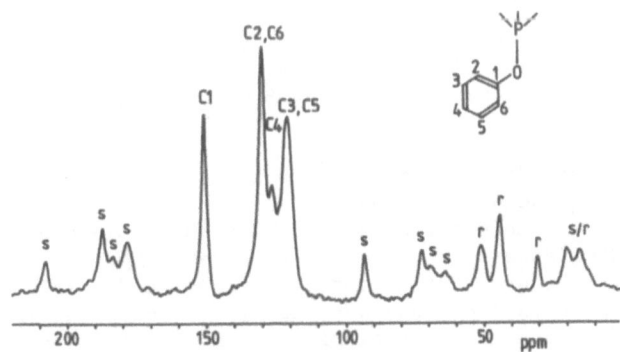

Figure 5. 50.32 MHz ^{13}C CMPAS NMR spectrum of phenyl phosphate exchanged γ-Zr(HPO4)2.2H2O. Spinning speed 2875 Hz. Sidebands are labelled s and rotor resonances r.

Having shown that the ring is not undergoing a large scale motion the question arises to what extent can low amplitude torsional oscillations be ruled out. Two simple types of potential for the torsional motion can be used i) a square potential and ii) a simple harmonic oscillator. However, independently of the potential chosen the effect of a torsional oscillation on the chemical shift anisotropy only really becomes apparent when the angular fluctuations are greater than +/- 20°. Simulations of the deuterium lineshape show that it too is not very sensitive to the small torsional motions despite the larger magnitude of the interaction. In essence the lineshape is only sensitive to large amplitude motions, small amplitude motions lead to only small changes in the lineshape. Consequently an excellent signal-to-noise ratio is necessary if lineshape analysis is to be used with small amplitude oscillations.

5.3 DIPOLE-DIPOLE COUPLING

Unlike the chemical shift and quadrupolar studies where, by the application of high power proton decoupling when necessary, the lineshape is determined solely by these interactions to first order it is often the case with dipole-dipole interactions that chemical shift anisotropy is also present. As a consequence the lineshape consists of a convolution of the two interactions. The main exception to this is the ^1H lineshape where the dipolar interactions is dominant. In low resolution studies the changes in the second moment with temperature are followed and from the reduction in the second moment the dynamics can be studied.

High resolution studies of dipolar coupling have been restricted to isolated spin $\frac{1}{2}$ - ^1H spin pairs. Two problems arise, the first is to reduce the homogeneous dipolar coupling in the system to an inhomogeneous coupling thus creating the isolated spin $\frac{1}{2}$ pair and the second is to deconvolute the dipolar and chemical shift interactions. Thus in order to use the heteronuclear dipole coupling of a ^{13}C - ^1H spin pair the trick is to make the spin pair behave as though it is isolated since strong dipolar coupling normally exists between all the protons in a solid. This is done by using multiple pulse sequences (25,27) or off resonance magic angle decoupling (26) to remove ^1H - ^1H dipolar coupling.

These sequences do not remove the heteronuclear dipole coupling but scale it by a factor dependent on the precise method used. A consequence of the scaling of the ^{13}C - ^1H dipole coupling is that slow spinning speeds are required in order to get a number of intense spinning sidebands. Furthermore, an accurate determination of the scaling factor is vital whether the dipolar coupling is being used to derive a bond length or to examine the molecular dynamics. Preferably the scaling factor should be determined before and after the experiment. In order to overcome the convolution of the chemical

shielding and dipolar spinning sidebands a two dimensional experiment
is needed whereby the dipolar spinning sidebands appear along one
axis and the chemical shift spinning sidebands along the other.

From an analysis of the dipolar spinning sideband intensities it is
possible to estimate the magnitude of the dipolar coupling and hence
any reduction occurring as a consequence of molecular action. Again
small amplitude motions will be difficult to determine since not only
are small changes going to occur in D_{33} but also an accurate scaling
factor is required. Despite these difficulties separated local field
experiments offer the opportunity for studying motion in systems for
which deuterium enrichment is not feasible.

5.4 QUADRUPOLE INTERACTIONS

Quadrupole interactions are generally speaking of significantly
greater magnitude than either dipole of chemical shift interactions.
This can lead to experimental problems concerning the proper
excitation and observation of the signal as described by Griffin (28).
For integer spins the most commonly studied nucleus is deuterium. To
the best of my knowledge no studies have been reported of dynamic
processes affecting the second-order quadrupole broadened MAS
lineshape of the central $\frac{1}{2}$↔-$\frac{1}{2}$ transition of a non-integer spin. High
resolution is achieved even in static spectra in the sense of
observing chemically inequivalent nuclei by selective isotropic
enrichment. In a similar manner to the other nuclear spin
interactions molecular dynamics causes a reorientation of the
interaction tensor, in this case the electric field gradient, leading
to a change in its magnitude and asymmetry characteristic of the type
of motion.

A straightforward example is that of d_{10}-colbaltocenium
hexafluorosphophate for which a drastic change takes place in the NMR
spectrum around 312 K. Below this temperature the structure is of low
symmetry and the ^2H powder pattern has a 65 kHz splitting consistent
with rapid motion about the cyclopentadienyl ring fivefold axis. As
the temperature is increased towards and then above 312 K so a new
sharp resonance starts to appear. This resonance continues to grow at
the expense of the broader resonance until in the high temperature
phase only this resonance is present. The sharp resonance is typical
of an isotropically reorientating molecule, with a correlation time of
nanoseconds based on the linewidth. The resulting increased symmetry
of the ion is matched by an increase in the crystal symmetry to cubic
(29).

This example illustrates that studying the dynamics of an ion can help
to understand the structure of a material.

5.4.1 TRANSLATIONAL MOTION

Bulk diffusion is not readily studied by lineshape analysis because
pure translational motion does not affect the lineshape unless there
is a significant chemical shift or electric field gradient
inhomogeneity. Then, if the diffusion causes these environments to be
experienced by the nuclear spin, an averaged value will be seen. The
main problem is that because normally there are no restraints on the
type of rotational motion occurring with the translation a simple line
narrowing will be seen once significant diffusion has set in. A more
complete model of the type of diffusion is possible only if for some
reason the rotational degrees of freedom are restricted. A good
example is that of the diffusion of xylenes in ZSM5 (30).

ZSM5 is a shape selective catalyst used for the isomerisation of ortho
and meta xylene to para xylene. The structure of ZSM5 contains a
three dimensional channel network including zig-zag channels which
allows the diffusion of the molecule in three dimensions. In the
absence of diffusion we expect to see a methyl group ^2H resonance
narrowed by fast rotation about the methyl C3 axis with a splitting
given by

$$Wq = 1/2 \ (3 \ \cos^2\theta - 1) \qquad \text{where } \theta \text{ is the angle the C-D bond makes to the } C_3 \text{ axis.}$$

Any additional rotations about this long axis will not cause further
narrowing moreover the channels in ZSM5 are not large enough for
rotation of para xylene along its long axis. Now if fast diffusion
from one channel to a perpendicular channel is possible then this will
cause an enormous change in the orientation of the new principal axis,
the methyl group threefold axis, of 90º similarly a large change in
orientation will occur if diffusion along a zig-zag channel takes
place. As a consequence we would expect to see additional narrowing.
In fact no such narrowing is seen. The molecules can not be diffusing
into the perpendicular channels or along the zig-zag channels on a
fast time scale at the temperature studied. Actually a slight
additional narrowing is seen. The reason for this is a small
fluctuation of the para xylene long axis about the axis of the
channel.

5.5 INTERMEDIATE EXCHANGE

Intermediate exchange processes cause profound changes in the NMR
lineshape which require the rate equations for the exchange to be
solved in order to obtain the expected lineshape. Thus the time
dependent fid $\underline{G}(t)$ is given by

$$\frac{d}{dt}\underline{G}(t) = \underline{G}\left(i\underline{w} + \underline{\pi}\right)$$

Where $\underline{w}$ is a diagonal matrix containing the frequencies for the

different sites undergoing exchange and $\overline{\pi}$ is a rate matrix defining
the exchange between the different sites (31,32). Access to a
reliable program has so far limited the detailed study of intermediate
exchange processes. Recently openly available programs for carrying
out such calculations have been described (33,34). As an example of
such lineshape analysis consider $Cu_2(OOCH)_4 \cdot 2$ py and a hypothetical
rotation about the fourfold axis, Figure 6. Note that despite the
changes in the lineshape the isotropic chemical shift remains
constant, consequently a dynamic process can not explain the
temperature dependence seen experimentally (35) which is due to the
introduction of paramagnetism through population of the S = 1 triplet
state.

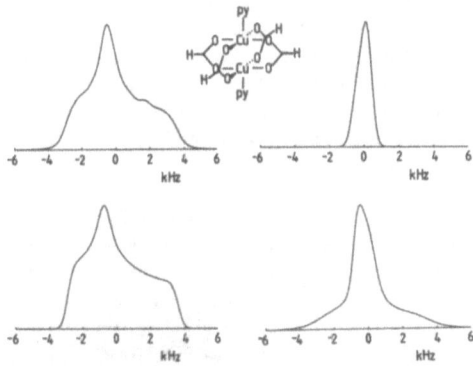

Figure 6. Simulation of the ^{13}C chemical shift powder pattern
assuming 90° jumps about the fourfold axis.

5.6 SLOW MOTIONS

An essential feature of slow motions is that they do not affect the
lineshape in a simple pulse and acquire or solid echo experiment.
Slow dynamic processes involving a change in the molecular order, for
example from an initially ordered state to a disordered one, can be
studied quite simply by collecting spectra as a function of time.
However, if a correlation between the initial and final state is
required in order to fit to a model for the process then this is
insufficient. In order to study a slower process and provide a
correlation we need a way to store the frequency labels of the initial
orientations, allow time for the reorientation to take place and
obtain the new set of frequencies. These then must be related to the
old. However, the length of time we can store a frequency label
depends on the nuclear spin relaxation times. One way to increase the
storage times is to use Zeeman order or longitudinal magnetisation
since the decay time here is given by T_1 which can be very long
indeed. The pulse sequence for carrying out such a labelling and
storage is simply the 2D-NOE/chemical exchange pulse sequence (36).
For ^2H spin alignment methods can be used (37).

Two types of the 2D-NOE experiment can be considered.

1) The exchange process involves only a change in orientation with no change in the isotropic chemical shift.
2) Exchange occurs to a new chemical shift position.

If only a change in the orientation occurs with the same isotropic chemical shift it is clear that a molecular motion can only be studied if an anisotropic component to the nuclear spin interaction is present. The experiment can be carried out using the static powder pattern (38) or with slow MAS to give a number of spinning sidebands (30). Static methods are only really feasible for a unique resonance, though they can in favourable circumstances allow the direct determination of motional parameters. For complex samples with a multiplicity of resonances the MAS method is attractive. In this experiment an exchange process leads to cross peaks in the two dimensional spectrum between the spinning sidebands of the two frequency axes. From the relative intensities of the cross peaks and diagonal peaks the exchange process can be modelled. However a number of difficulties are evident.

To avoid spurious cross peaks accurate synchronisation of the second and third pulses to the rotor period is necessary. The loss of intensity caused by the spreading out of the SSB data in two dimensions together with the reduction in signal to noise ration because of the need for intense spinning sidebands results in long accumulation times. Accumulation times may, however, be reduced by recognising that the signal in the first time domain is periodic so that the entire FID does not have to be digitised by t1 increments only one rotational echo. Then a complex spectrum is obtained requiring computer simulation to interpret and to date this has only involved spectra with one set of SSB. For long mixing times spin diffusion will give spurious cross peaks. Finally, the requirement for significant sideband intensities necessitates slow spinning which can cause overlap between sidebands and isotropic resonances for complex samples complicating the data analysis.

When the exchange takes place between environments with different isotropic chemical shifts an anisotropic component is no longer necessary, indeed it is a disadvantage because it reduces the intensity of the isotropic resonance. This has led to methods to scale the chemical shift anisotropy thereby increasing the sensitivity of the experiment (40). Again the exchange is evident in the two dimensional spectrum from the presence of cross peaks and the rate of the dynamic process can be obtained from the intensities of the diagonal and cross peaks as a function of the mixing time.

6   SPIN DYNAMICS

Spin dynamics are intrinsic to the NMR experiment and the dynamics associated with the nuclear spin relaxation times are extremely useful

for investigating molecular dynamics. Particularly interesting forms
of spin dynamics are present through the products of spin operators on
different atoms arising in dipolar coupling. These dynamics tend to
be useful for studying molecular structure rather than molecular
dynamics.

In general terms the spin dynamics can be described as the evolution
of multiple quantum coherences throughout the sample. Zero quantum
processes are present in spin diffusion for the homonuclear case and
in cross polarisation for the heteronuclear case where the transition
energies of the I and S spin can be equalised in the rotating frame.
The development of higher order coherences is clearly related to the
molecular structure through the proximity of the nuclei yielding
information on the clustering of spins (41).

Consider an example of spin dynamics which is commonly seen namely the
contact time dependence in the spin lock cross polarisation
experiment. At first the S spin magnetisation builds up with a rise
time constant of T_{IS}, it then decays with a time constant $T_{1\rho}^{H}$, the I
spin nuclear spin relaxation time in the rotating frame (31). The
interpretation of T_{IS} is complex and does not led itself to the study
of structure or dynamics despite being inversely proportional to the
second moment of the IS dipolar coupling. $T_{1\rho}^{H}$ can more clearly be
understood in terms of molecular motion and although in theory one can
obtain $T_{1\rho}^{H}$ for each S resonance the presence of rapid spin diffusion
between the I spins leads to a common decay time in most cases.

Spin diffusion is an important process as it is essential in most
pictures of nuclear spin relaxation. The physical process associated
with spin diffusion is the energy conserving, zero quantum flip-flop
transitions involving two dipolar coupled spin $\frac{1}{2}$ nuclei.

The rate of the spin diffusion depends on the magnitude of the dipolar
coupling between the nuclei and hence spin diffusion is almost
exclusively confined to neighbouring molecules. This allows the
transfer of spin diffusion to be used to study;

i) the mixing of molecules at a molecular level (42-44)
ii) the spatial arrangement of the nuclei in an extended solid

Spin diffusion can be studied in either a one-dimensional experiment
using a DANTE selective excitation pulse or by the 2D-NOE pulse
sequence with spin diffusion between nuclei occurring during the
mixing period. The features of spin diffusion or more strictly
speaking spin exchange in this case are illustrated by the example of
$(P_2W_{17}O_{61}Zn(OH_2))^{8-}$ where an isolated pair of chemically inequivalent
nuclei are held a known and fixed distance apart (45). This
introduces two problems. First, spin diffusion between chemically
inequivalent nuclei will be exceedingly slow because the flip-flop
transition is not energy conserving and some means for compensating
for the energy imbalance must be found. Secondly, MAS will suppress

the secular dipole-dipole coupling so that spin diffusion should be quenched at high enough spinning speeds. This will not be the case providing the MAS is not sufficiently fast to remove the homogeneous content to the lineshape (43). The rate constant for the spin diffusion was found to be $1.85s^{-1}$.

No spin diffusion was found for the potassium salt or when proton decoupling was employed during the mixing period demonstrating that the proton-phosphorus dipolar coupling is essential to provide the overlap of the lineshapes of the chemically inequivalent nuclei and hence allow energy conserving flip-flop transitions. For experiments based on rare nuclei such as ^{13}C an additional complication arises namely that the nearest neighbours will show a statistical distribution of the NMR active nucleus. This will tend to increase the average distance between neighbours and as a consequence spin diffusion rate constants for ^{13}C at natural abundance are much smaller around $0.01-0.02s^{-1}$ (43).

REFERENCES

1. D.E. Wemmer, D.J. Ruben and A.J. Pines, J.Am.Chem.Soc., 103, 28 (1981).
2. J. Schaefer, E.O. Stejskal, D. Perchak, J. Skolnick and R. Yaris, Macromolecules, 18, 368 (1985).
3. D. Brinkmann, M. Mali, J. Roos and R. Messer, Solid State Ionics, 5, 409 (1981).
4. H.W. Spiess, R. Grosescu and U. Haeberlen, Chem.Phys., 6, 226 (1974).
5. J.R. Lyerla, C.A. Fyfe and C.S. Yannoni, J.Am.Chem.Soc., 101, 1351 (1979).
6. J.R. Barnes, A.D.H. Clague, N.J. Clayden, C.M. Dobson C.J. Hayes, G.W. Groves and S.A. Roger, J.Mater.Sci.Lett., 4, 1293 (1985).
7. A. Abragam in "The Principles of Nuclear Magnetism", Chap V, Oxford University Press, London (1961).
8. . Pines, M.G. Gibby and J.S. Waugh, J.Chem.Phys., 59, 569 (1973).
9. E.R. Andrew, A. Bradbury and R.G. Eades, Nature, 182, 1659 (1958).
10. U. Haerberlen in "High Resolution NMR in Solids: Selective Averaging" Academic Press, New York (1976).
11. P.S. Allen in "MTP International Review of Science, Physical Chemistry: Magnetic Resonance", Eds., A.D. Buckingham and C.A. McDowell, Series 1, Volume 4, p.43, Butterworths, London (1972).
12. D. Suwelack, W.P. Rothwell and J.S. Waugh, J.Chem.Phys., 73, 2559 (1980).
13. A.N. Garroway, D.L. VanderHart and W.L. Earl, Phil.Trans. Roy.Soc.Lond., A299, 609 (1981).
14. P. Mansfield, Prog. NMR Spectroscopy, 8, 41 (1971).
15. E. Lippmaa, M. Magi, M. Tarmak, W. Wieker and A.R. Grimmer, Cem.Concr.Res., 12, 597 (1982).
16. N.J. Clayden, C.M. Dobson, C.J. Hayes and S.A. Rodger, J.C.S. Chem. Commun., 1396 (1984).

17. N.J. Clayden, C.M. Dobson, S.A. Rodger and G.W. Groves, J.Am.Ceram.Soc., submitted.
18. N.J. Clayden, C.M. Dobson, L.-Y. Lian and J.M. Twyman. J.C.S. Perkin Trans II, 1933 (1986).
19. H.W. Spiess, Colloid Polym. Sci., 261, 193 (1983).
20. H.W. Spiess in "NMR Basic Principles and Progress", Eds., P. Diehl, E. Fluck and R. Kosfeld, Volume 15, p.55, Springer-Verlag, Berlin (1978).
21. H. Pfeifer in "NMR Basic Principles and Progress", Eds., P. Diehl, E. Fluck and R. Kosfeld, Volume 7, p.53, Springer-Verlag, Berlin (1972).
22. H.S. Gutowsky and G.E. Pake, J.Chem.Phys., 16, 1164 (1948).
23. E.R. Andrew and R.G. Eades, Proc.Roy.Sc., A218, 537 (1953).
24. N.J. Clayden, J.C.S Dalton Trans., 1877 (1987).
25. S.J. Opella and J.S. Waugh, J.Chem.Phys., 66, 4919 (1977).
26. J.A. DiVerdi and S.J. Opella, J.Am.Chem.Soc., 104, 1761 (1982).
27. M.G. Munowitz and R.G. Griffin, J.Chem.Phys., 76, 2848 (1982).
28. R.G. Griffin, this volume.
29. S.J. Heyes, N.J. Clayden, C.M. Dobson and P.J. Wiseman. J. Inclusion Phenomena, 5, 65 (1987).
30. R.R. Eckman and A.J. Vega, J.Phys.Chem., 90, 4679 (1986).
31. M. Mehring in "Principles of High Resolution NMR in Solids", Springer-Verlag, Berlin (1983).
32. H.W. Spiess, Chem.Phys., 6, 217 (1974).
33. R.J. Wittebort, E.T. Olejniczak and R.G. Griffin, J.Chem.Phys., 86, 5411 (1987).
34. M.S. Greenfield, A.D. Ronemus, R.L. Vold, R.R. Vold, P.D. Ellis and T.E. Raidy, J.Magn.Reson., 72, 89 (1987).
35. T.H. Walter and E. Oldfield, J.C.S. Chem. Commun., 646 (1987).
36. J. Jeneer, B.H. Meier, P. Bachmann and R.R. Ernst, J.Chem.Phys., 71, 4546 (1979).
37. H.W. Spiess, J.Chem.Phys., 72, 6755 (1980).
38. C. Schmidt, S. Wefing, B. Blumich and H.W. Spiess, Chem.Phys.Lett., 130, 84 (1986).
39. A.E. deJong, A.P.M. Kentgens and W.S. Veeman, Chem.Phys.Lett., 109, 337 (1984).
40. G.S. Harbison, D.P. Raleigh, J. Herzfeld and R.G. Griffin, J.Magn.Reson., 64, 284 (1985).
41. J. Baum, M.G. Munowitz, A.N. Garroway and A. Pines, J.Chem.Phys., 83, 2015 (1985).
42. P. Caravetti, J.A. Deli, G. Bodenhausen and R.R. Ernst, J.Am.Chem.Soc., 104, 5507 (1982).
43. D.L. VanderHart, J.Magn.Reson., 72, 13 (1987).
44. T.A. Cross, M.H. Frey and S.J. Opella, J.Am.Chem.Soc., 105, 7471 (1983).
45. N.J. Clayden, J.Magn.Reson., 68, 360 (1986).

CHAPTER 6

EFFECTS OF MOLECULAR MOTIONS ON N.M.R. PARAMETERS OF SOLID STATE SPECTRA.

S. Aime
Faculty of Pharmacy, University of Torino
Via Pietro Giuria 9
10125 Torino
Italy

ABSTRACT. NMR is very useful for the study of dynamic properties of solids. Fluctuations in the orientation of the various interaction tensors show their effects on relaxation behaviour, lineshapes and attenuation of cross-relaxation. Several parameters are then available and the proper choice is determined by the type of solid considered ("plastic crystals", amorphous or crystalline solids, ...) and the underlying dynamic process. The availability of variable temperature apparatus on cross polaryzed-magic angle spinning spectrometers allows an impressive expansion of the range of application of solid state n.m.r. investigation.

MNR OF SOLIDS AT HIGH MAGNETIC FIELD

In an NMR experiment the major benefits associated with an increase of magnetic field strenght are the larger frequency separation of the resonances and the enhancement of the signal to noise ratio of the recorded spectrum. The technological improvment of high resolution NMR spectrometers for liquids has then continuously pursued the target to get the highest possible magnetic field[1]. However in solid state NMR this requirement may be unnecessary since the expected advantages of increased resolution and sensitivity could be balanced by the complications associated to the upsurge of the spinning side band

65

G. J. Long and F. Grandjean (eds.), The Time Domain in Surface and Structural Dynamics, 65–80.
© *1988 by Kluwer Academic Publishers.*

manyfold and by field-dependent broadening effects.

As an example of resolution improvement as the observation frequency is increased from 25.0 MHz to 67.6 MHz, I report the carbon-13 magic angle spinning spectra of crystalline $Os_3(CO)_{12}$ (isotropic region only). The spectrum recorded at higher field shows eleven resolved resonances. The magnetic inequivalence of the carbonyls (the solution spectrum gives only two equally intense absorptions) likely arises from effects due to crystal packing, i.e. the D_{3h} symmetry of the molecule is slightly distorted when it is packed in the crystal lattice[2].

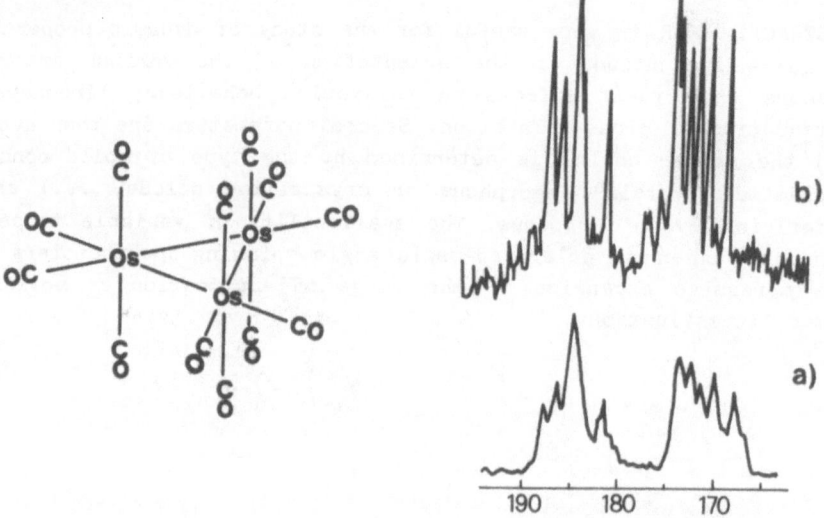

Figure 1. Magic angle spinning-^{13}C spectra of [13]CO enriched samples of $Os_3(CO)_{12}$ measured at 25.0 MHz (a) and 67.6 MHz (b) respectively.

Now the major complication induced by increasing the dc field B_o in the spectra of diluite spin systems is related to the increase of the intensity of the side bands when the shift anisotropy is considerably greater than the spinning rate.

The rotations commonly available on commercial instruments range up to 4-5 KHz and large efforts are currently made by manufactures to increase significantly this limit.

The field-dependence of chemical shift anisotropy is easily shown by considering the origin of the chemical shielding tensor σ which relates the applied magnetic field B_o to the effective field B_{eff} experienced by a given nucleus,

$$B_{eff} = (1 - \sigma)B_o$$

The magnitude of the shielding is orientation dependent and can be described by three components, σ_{ii} (i = 1, 2 and 3) in the axes system of the molecule. For instance in a single crystal a carbonyl functionality (Fig. 2) oriented perpendicular to the magnetic field will give a sharp peak characteristic of this particular orientation whereas a CO oriented parallel to the field will give a sharp signal characteristic of that particular orientation[3].

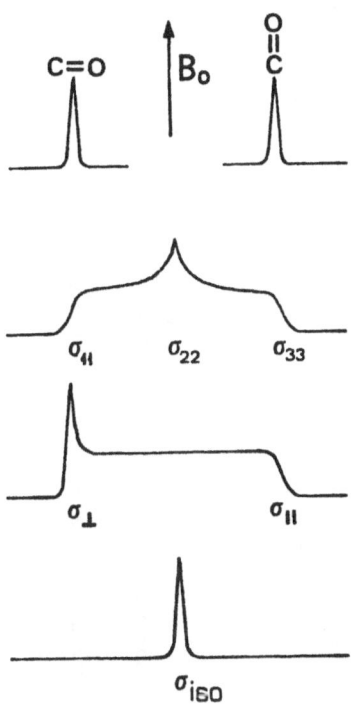

Figure 2. Schematic representation of the origin of a static spectrum for a CO functionality.

The spectrum of a polycrystalline sample corresponds to the superposition of all possible orientations. The chemical shift anisotropy is defined by ($\sigma_{33} - \sigma_{11}$). The chemical shift anisotropy (in Hz) increases with the magnetic field strength and differs, for the same nucleus, according to the electronic and structural environment of the system considered. In Tab.1 typical shielding values are reported for carbon-13 functionalities[4].

Table 1. Magnitudes of some ^{13}C chemical shift tensors.

Carbon type	Sample	Temperature(°C)	σ_{11}	σ_{22}	σ_{33}	$\Delta\sigma$
Nonprotonated aromatic	Hexamethylbenzene	-186	239	103	31	208
	Hexamethylbenzene	21	237	71	69	168
Protonated aromatic	Durene	22	211	135	35	176
Carbonyl	Acetone	-186	279	265	79	200
	Carbon monoxide	-269	-	-	-	345
Carboxyl	Oxalic acid dihydrate	ambient	249	132	109	140
	Glycine	ambient	248	130	105	143
Methylene	Polyethylene	ambient	50	36	12	38
	Malonic acid	16	62	51	19	43
Methyl	n-eicosane	-95	26	22	3	23

In liquids rapid molecular reorientation occurs and all the orientations are averaged out to give rise to a single absorption $\sigma_{iso} = 1/3(\sigma_{33} + \sigma_{11} + \sigma_{22})$. In solids, for rapidly reorienting (generally small) molecules, it is possible to reduce the broadening induced by the chemical shift anisotropy interaction provided that it becomes time dependent. In principle any broadening ($\Delta \nu$) can be narrowed if the static interaction which causes it becomes time-dependent, i.e. $\Delta \tau < 1$ where the time dependence is characterized by τ .

An example is provided by polycrystalline hexamethylbenzene (HMB). From ^{1}H second moment measurement as a function of temperature[5], it was observed that, besides a reorientation tunnelling of the individual methyl groups around their three fold axis which occurs even at very low temperatures, a drop in its profile around 170 K indicates that a molecular reorientation occurs in this temperature range . The exact nature of this molecular motion is determined by the observation of the corresponding changes induced in the chemical shift anisotropy pattern[6] of ^{13}C resonances for aromatic carbons (Fig. 3). The anisotropic pattern observed at low temperature transforms to an axially symmetric pattern at ambient temperature through the averaging of σ_{11} and σ_{22} (to give $\sigma_{\perp}$) but σ_{33} is unchanged ($\sigma_{\parallel}$).

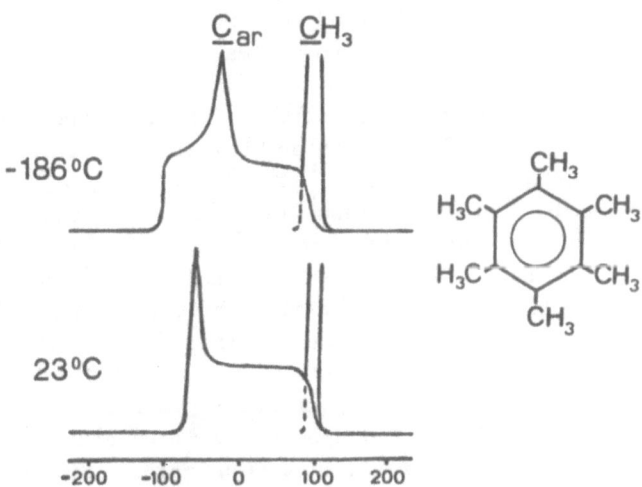

Figure 3. Schematic representation of the static carbon-13 NMR spectra of polycrystalline hexamethylbenzene at different temperatures.

This behaviour corresponds to a molecular rotation whose axis coincides exactly with one of the shielding element which remains then unchanged during the motion.

The different mobility associated with the crystalline and amorphous region of a polymer may affect the chemical shift anisotropy pattern as it has[7] been shown by Veeman and coworkers in the case of poly(oxymethylene) (Delrin).

A case showing an extensive averaging of the various broadening interactions is found in the study of plastic crystals[8] in their high temperature cubic phase where their almost complete reorientational freedom gives the very sharp absorptions normally detected on commercial high resolution spectrometers. Adamantane is the prototype example of this class of solids and the extent of narrowing[9,10], even in the absence of magic angle spinning, is so high that this compound is often employed as a reference standard for resolution tests in ^{13}C solid state work.

Sometimes in the presence of groups with large anisotropies, such as CO, the ^{13}C-spectrum recorded under conventional high resolution NMR techniques may show residual broadening whose pattern may be indicative of the occurrence of a non-isotropic molecular motion as shown for camphor[3]. Similar observations on the value and the pattern of the chemical shift anisotropy term can carry information on the nature of the interaction and the degree of mobility of species adsorbed on surfaces. These studies include the chemisorption of methanol on magnesium oxide[11], carbon dioxide on mordenite[12] and species containing phenyl groups covalently bound to silica gel[13].

In magic angle spinning experiments the line broadening associated with the static chemical shift anisotropy interaction is reduced by the macroscopic motion of the sample at the magic angle; however when the spinning speed (ω) is less than the magnitude of the static interaction, then satellites resonances (spinning sidebands) are observed at frequencies $\pm \ \omega_n$ from the central resonance.

The absorption is only partially narrowed by magic angle spinning when the spinning speed is smaller than the width of the central narrowed resonance; in polycrystalline samples however, because the static chemical shift anisotropy broadening arises from crystallites of a particular orientation, the spectral pattern breaks up into an array of satellites even for small rotation rates because the intrinsic line-width of each satellite is less than the spinning speed[3].

Suppression of spinning side bands can be obtained by using special pulse sequences developed for this purpose. The most common one is TOSS (total sideband suppression) reported by Dixon[14] and consisting of four 180° pulses prior to the acquisition.

RECOVERY OF TENSOR COMPONENTS FROM SPINNING SIDE BANDS PATTERN.

As has already been mentioned, the knowledge of chemical shift anisotropy can provide useful information on eletronic, structural, and dynamic properties not available from the σ_{iso} value alone. As has been shown above, the more direct route to its determination consists in the observation of the powder spectrum, but this procedure has the obvious limit that, in presence of several resonances falling in the same absorption region, only an average chemical shift anisotropy value becomes available.

There are at least two readily amenable methods which can be used to extract the shielding tensor components from the intensity of the side bands in a magic angle spinning spectrum:

a) Moment analysis[15].

Because the second and third moments (M_2 and M_3) of the spectrum are related to the δ and η parameters by the following equations:

$$M_2 = (\delta^2/15)(3 + \eta^2) \text{ and } M_3 = (2 \delta^3/35)(1-\eta^2)$$

where δ and η are defined as

$$\delta = \sigma_{33} - \sigma_{iso} \qquad \eta = \frac{\sigma_{22} - \sigma_{11}}{\sigma_{33} - \sigma_{iso}}$$

the evaluation of these equations leads to the determination of the shielding tensor components.

A limitation of this method is that it requires the measurement of the intensity of all side bands which may not always be done precisely.

b) Graphical Method[16]

Herzfeld and Berger derived general integral and series expansions for the intensity of the side bands. Graphical plots are provided for extracting the μ and σ parameters from the evaluation of a few sidebands.

$$\mu = (\gamma H_o/\omega_R)(\sigma_{33} - \sigma_{11})$$

$$\sigma = (\sigma_{11} + \sigma_{33} - 2\sigma_{22})/(\sigma_{33} - \sigma_{11}).$$

RELAXATION STUDIES

NMR relaxation studies are particularly valuable in the investigation
of molecular motions in liquids and solids because they are sensitive
to the time rate of variation of certain magnetic interactions.
    Several relaxation parameters may be available:

- T_1, the longitudinal relaxation time;
- T_2, the transverse relaxation time;
- $T_{1\varrho}$, the longitudinal relaxation time in the rotating frame;
- T_{1D}, the dipolar relaxation time.

    The temperature-dependent profiles of these parameters (Fig. 4)
show that the curves of log $T_{1,1\varrho,D}$ assume a typical V-shape with
different minimum points but are identical in the high temperature
limit. From these considerations it follows that, in commonly available
magnetic fields, T_1 will be sensitive to correlation times of the order
of nsec., $T_{1\varrho}$ to correlation times of the order of 0.1 msec and T_{1D} to
correlation times of the order of msec.

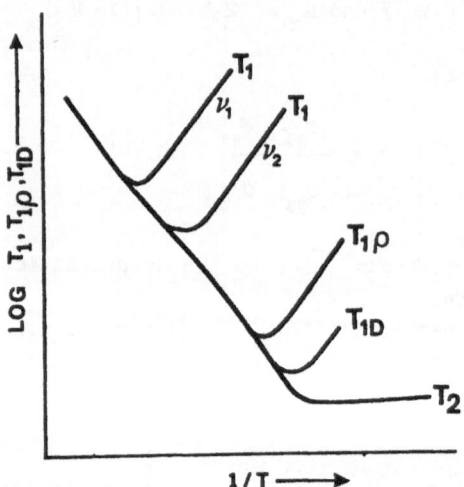

Figure 4. Profiles of relaxation times versus temperature.

    T_1 describes the energy flow between the nuclear spin system and
the lattice; as for solution n.m.r., it is defined by

$$dM_z/dt = -(M_z - M_o)/T_1$$

T_2 describes the magnetization decay in the xy plane after excitation (it is sometimes called spin-spin relaxation to enphasize the fact that direct interactions between spins can cause relaxation of M_x and M_y without energy transfer to the lattice). It is proportional to the inverse of the linewidth.

$T_{1\varrho}$ can be measured in the so called spin locking experiments in which the magnetization is maintained aligned with the radio frequency field B_1. Since the magnetization was developed in B_o, it is very large with respect to B_1 so M_y will decay with time to a value $(B_1/B_0)M_o$:

$$M_y(\tau) = M_o \exp(-\tau/T_{1\varrho}).$$

$T_{1\varrho}$ differs from T_1 because it depends on the spectral density governed by $a_1 = \gamma B_1$ (corresponding to order of tens of KHz) whereas T_1 is determined by $\omega_o = \gamma B_o$ (corresponding to order of tens of MHz)[17].

When we consider abundant nuclei (e.g. [1]H) the strong dipolar interactions result in all of the nuclei having a single common set of relaxation times. For example, consider a polymer containing methyl groups which are mobile due to internal rotation. If the motion is in the appropriate frequency range, T_1 of methyl protons will be considerably shortened. Because it is likely that dipolar couplings between methyl protons and other protons in the system are strong enough for flip-flop transitions to be efficient, the relaxation is transferred (spin-diffusion) to all the protons in the polymer. A single T_1 (^{1}H) will be observed which will be longer than the theoretical value for the methyl group by a factor corresponding roughly to the ratio of the total number of protons to the number of methyl protons.

As for motional narrowing of the chemical shift anisotropy discussed above, "plastic" crystals have been throughly investigated by spin lattice relaxation times at variable temperature.

When more magnetically active nuclei are available (for instance ^{1}H, ^{13}C, ^{14}N in succinonitrile) it is eventually possible to monitor the rotation of different vectors and get a more complete picture of the motions occurring in the plastic crystalline phase[18,19].

Several molecules have been investigated mainly by carbon-13 NMR spectroscopy using commercial Fourier transform instruments for liquid samples because the linewidths are relatively small. The observed T_1's increase as the temperature increases and appear continuous across the solid ↔ liquid phase transition. The different contributions (dipole-dipole, spin rotation, chemical shift anisotropy) to the observed relaxation rates are obtained by the experimental techniques (nuclear Overhauser enhancement, two field effects), and equations commonly used in working out T_1 calculations for liquid samples.

From the temperature dependence of the T_1's, a rotational activation energy can be calculated.

The knowledge of proton T_1 and $T_{1\varrho}$ is also useful for determining the best conditions for a cross-polarization experiment. $H-T_{1\varrho}$ determines the optimum contact time whereas $H-T_1$ determines the optimum recycle delay.

Extensive ^{13}C studies of T_1 and $T_{1\varrho}$ have been carried out for several polymers in order to obtain information on the distribution of frequencies of their motions. T_1 and $T_{1\varrho}$ for amorphous and crystalline components (which give rise to a single ^{13}C peak) of Delrin[20] are reported in Tab. 2.

Table 2. Carbon-13 spin-lattice relaxation for polyoxymethylene

	Amorphous	Crystalline
$T_{1\varrho}$	17.5 ms	0.6 ms
T_1	0.75 s	15 s

Observation Frequency = 45 MHz; Spin Locking Frequency = 25 KHz

The use of perdeuteriopolymers[21] allowed Schaefer and coworkers to study the interfacial region surrounding the phase-separated rubber domains. They measured proton $T_{1\varrho}$ (as observed by high resolution carbon-13 NMR) and intermolecular cross-polarization transfer rates.

Among the natural macromolecules, from T_{1c} and T_{2c} measurements for crystalline cellulose I it has been shown that glucose units are in two magnetically inequivalent environments[22].

Maricq and Waugh[15], in a detailed discussion of a variety of phenomena occurring in the "slow spinning" regime, have shown that very slow random molecular motions ($\tau_c \lesssim$ 0.1 sec) appear to be accessible by analyzing the rotational spin-echoes. Previously information on such motions were available only for the distortions of powder pattern line shapes in stationary samples. Furthermore, Rothwell and Waugh[23] showed that random motions of a S-spin dipolar coupled to an unlike I-spin (r.f. decoupled) can be evaluated by

$$\frac{1}{T_2} = \frac{4\gamma_I^2 \gamma_s^2 \hbar^2}{15\, r^6}\, I\,(I+1)\left(\frac{\tau_c}{1+\omega_I^2\, \tau_c^2}\right)$$

for which T_2 takes the minimum value when $\omega_I \tau_c \simeq 1$.

MAGIC ANGLE SPINNING AT VARIABLE TEMPERATURE

Experiments at variable temperature are a well established procedure in high resolution NMR techniques for liquid samples. They allow, by lineshape analysis, the direct study of a variety of dynamic processes whose E_a ranges from 3-4 up to 20-22 Kcal/mole.

By the introduction of narrowing techniques in the NMR investigations of solids, it was appropriate to extend variable temperature operation to spectrometers operating in cross polarization-magic angle spinning[24]. It may be noted that, in constrast to experiments on liquid samples a much wider temperature range may be available for solid samples; in particular very low temperature studies of low-energy intramolecular motions can be carried out.

A variety of processes may be followed by variable temperature studies of solids. These include molecular motions, conformational equilibria, chemical exchange processes, reactive chemical intermediates, molecular association/dissociation, and compounds that are liquid or gases at ambient temperature.

Sometimes changes in temperature are required to overcome the occurrence of unfavourable relaxation time ($T_{1\rho}$, T_1) at a given temperature. This may result in an inefficient cross polarization. This has been observed in the carbon-13-cross polarization-magic angle spinning spectrum of poly(ethylene) oxide[25], which shows a minimum in its proton $T_{1\rho}$ at room temperature. On decreasing the temperature $T_{1\rho}$ becomes sufficiently long to allow polarization transfer to occurr to a significant degree.

MOTIONAL EFFECTS DETECTED BY VARIABLE TEMPERATURE SPECTRA.

An example of this kind of motional effects is provided by $Ru_3(CO)_{12}$[26] (Fig. 5). The broadening observed for the resonances assigned to the axial set of CO ligands in the spectrum at +20°C may be associated with a distribution of ruthenium-carbon bond distances and angles. As the temperature is increased to +45°C, a localized oscillatory motion occurs which is responsible for the narrowing of these resonances without causing a complete equilibration inside the set. Because the detected motions occur at rate comparable to the frequency separation of the signals, changes in linewidth of a few tens Hz mean that we are actually following very low frequency motions in this class of compounds.

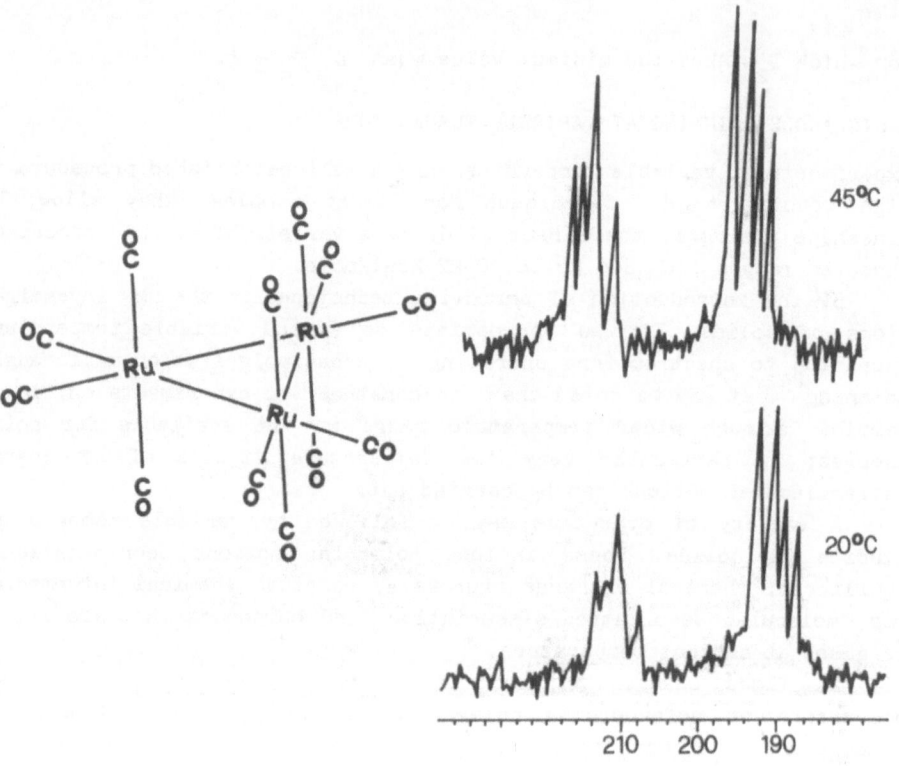

Figure 5. Carbon-13 magic angle spinning spectra of $Ru_3(CO)_{12}$ (isotropic region only) at different temperature.

CHEMICAL EXCHANGE

Exchange processes in the solid state may be analyzed by the same approach used for liquids. For instance organometallic molecules such as pentacarbonyl(cyclooctatetraene)diiron[27] and tetracarbonylbis(cyclo-octatetraene)triruthenium[28] have been shown to be fluxional in the solid state as they are in solution.

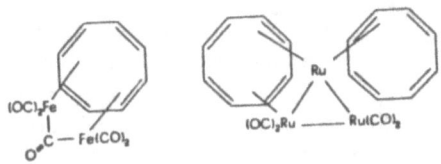

Among metal carbonyl caomplexes, where stereochemical non-rigidity is a rather general property in solution, several systems have already been considered[29,30]. In Fig.6 the variable temperature-magic angle spinning carbon-13 NMR spectra of $Co_2(CO)_8$[31] illustrate that the "frozen" structure, which was not detected in solution, is observed in the low temperature limiting spectrum. As the temperature is increased a CO scrambling process is taking place whose rate becomes high enough, in the spectra at high temperature, to lead to a full average of chemical shift anisotropy interaction as shown by the disappearance of the spinning side bands.

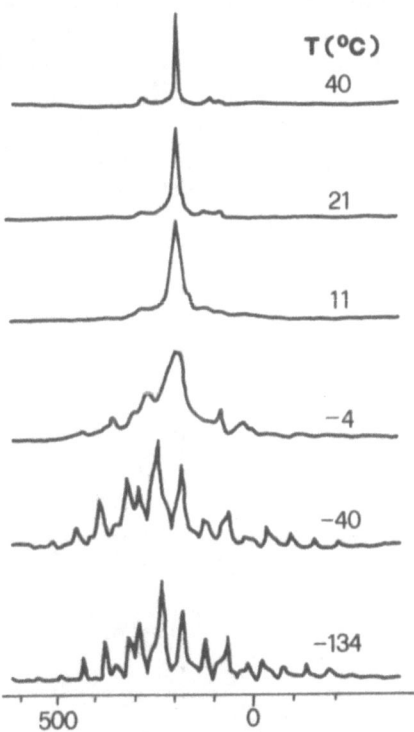

Figure 6. Magic angle spinning-variable temperature carbon-13 NMR spectra of $Co_2(CO)_8$ recorded at 50 MHz.

An interesting case of unexpected complication of a simple exchange has been found in the study of the variable temperature behaviour of naphthazarin B[32] whose cross polarized-magic angle spinning spectra show a fast intramolecular exchange process occurring at +25°C and a "frozen" structure at -160°C. At intermediate temperatures, however, the observed spectra are the superposition of these two limiting spectra in different proportions depending on the temperature. It is suggested that the sample presents a second-order phase transition between a low temperature phase (where no exchange is occurring) and a high temperature phase (where exchange is taking place).

CRYSTALLOGRAPHIC DISORDER AND NMR SPECTRAL PATTERN.

High resolution variable temperature studies can offer a clue to the understanding of the relationship between crystallographic disorder and reorientational jumps occurring in solids.

In this context suitable examples are provided by the binary metal carbonyl cluster complexes, $Fe_3(CO)_{12}$ and $Co_4(CO)_{12}$.

The carbon-13 NMR magic angle spinning spectra (MAS) of $Fe_3(CO)_{12}$ have been recorded[29] in the range of temperature from -93° up to 31°C (Fig. 7). The six-equally intense resonances observed at ambient temperature have been accounted in terms of rotation of the iron triangle within the distorted icosahedron defined by the carbonyl ligands.

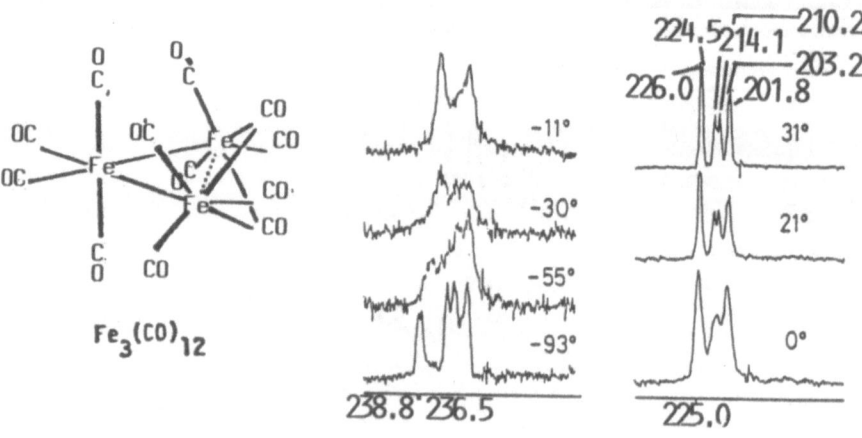

Figure 7. Variable temperature magic angle spinning carbon-13 NMR spectra of $Fe_3(CO)_{12}$ recorded at 22.6 MHz.

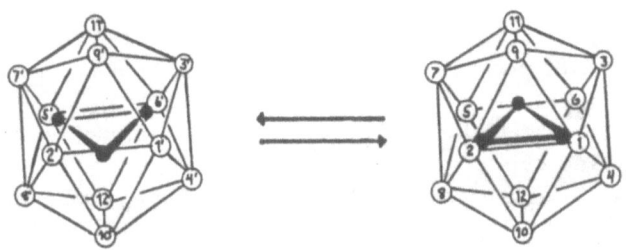

The X-ray crystal structure of $Fe_3(CO)_{12}$ is disordered and shows two molecules per unit cell related by an inversion center. The crystal structure shows the space average of these two orientations whereas the room temperature NMR spectrum is consistent with the time average of both orientations.

REFERENCES

1) R.K. Harris, Nuclear Magnetic Resonance Spectroscopy, Pitman Book Ltd., London, 1983.
2) L. Hasselbring, H. Lamb, C. Dybowski, B. Gates and A. Rheingold, Inorg. Chim. Acta, 1987, 127, L 49.
3) C.A. Fyfe, Solid State NMR for Chemists, C.F.C. Press, P.O. Box 1720, Guelph, Ontario, Canada, 1983.
4) R.E. Wasylishen and C.A. Fyfe, Ann. Rep. on NMR Spectroscopy, (G.A. Webb Ed.), 1982, 12, 1.
5) P.S. Allen and A. Cowking, J. Chem. Phys., 1967, 47, 4286.
6) A. Pines, M.G. Gibby and J.S. Waugh, Chem. Phys. Lett., 1972, 15, 373.
7) W.S. Veeman and E.M. Menger, Bull. Magn. Res., 1981, 2, 77.
8) The Plastically Crystalline State, J.N. Sherwood Ed., John Wiley and Sons, New York (1979).
9) R.E. Wasylishen and B.A. Pettitt, Can. J. Chem., 1980, 58, 655.
10) D.W. McCall and D.C. Douglass, J. Chem. Phys., 1960, 33, 777.
11) I.D. Gay, J. Phys. Chem., 1980, 84, 3230.
12) M.D. Sefcik, J. Schaefer and E.O. Stejskal in Molecular Sieves II, J.R. Katzer Ed., A.C.S. Sym. Ser., 1977, 40, 344.
13) D. Slotfeldt-Ellingsen and H.A. Resing, J. Phys. Chem., 1980, 84, 2204.
14) W.T. Dixon, J. Chem. Phys., 1982, 77, 1800.
15) M.M. Maricq and J.S. Waugh, J. Chem. Phys., 1979, 70, 3300.
16) J. Herzfeld and A.E. Berger, J. Chem. Phys., 1980, 73, 6021.
17) T.C. Farrar and E.D. Becker, Pulse and Fourier Transform NMR, Academic press, New York, 1971.
18) J.G. Powles, A. Begum and M.O. Norris, Mol. Phys., 1969, 17, 489.

19) R.E. Wasylishen and B.A. Pettitt, Mol. Phys., 1978, **36**, 1459.

20) W.S. Veeman, E.M. Menger, W. Ritchey and E. de Boer, Macromolecu-les, 1979, **12**, 924.

21) J. Schaefer, M.D. Sefcik, E.O. Stejskal and R.A. McKay, Macromole-cules, 1981, **14**, 188.

22) W.K. Earl and D.L. VanderHart, J. Am. Chem. Soc., 1980, **102**, 3251.

23) W.P. Rothwell and J.S. Waugh, J. Chem. Phys., 1981, **74**, 2721.

24) J.R. Lyerla, C.S. Yannoni and C.A. Fyfe, Acc. Chem. Res., 1982, **15**, 208.

25) D.L. VanderHart, W.L. Earl and A.N. Garroway, J. Magn. Res., 1981, **44**, 361.

26) S. Aime, to be published.

27) A.J. Campbell, C.E. Cottrell, C.A. Fyfe and K.R. Jeffrey, Inorg. Chem., 1976, **15**, 1321.

28) C.A. Fyfe, J.R. Lyerla and C.S. Yannoni, J. Am. Chem. Soc., 1979, **101**, 1351.

29) B.E. Hanson, E.C. Lisic, J.T. Petty and G.A. Iannacone, Inorg. Chem., 1986, **25**, 4062.

30) B.E. Hanson and E.C. Lisic, Inorg. Chem., 1986, **25**, 716.

31) B.E. Hanson, M.J. Sullivan and R.J. Davis, J. Am. Chem. Soc., 1984, **106**, 251.

32) W.H. Shian, E.N. Deusler, I.C. Paul, D.Y. Curtin, W.G. Blann and C.A. Fyfe, J. Am. Chem. Soc., 1980, **102**, 4546.

CHAPTER 7

DEUTERIUM NMR STUDIES OF DYNAMICS IN SOLIDS

R.G. Griffin, K. Beshah, R. Ebelhäuser, T.H. Huang[#], E.T.
Olejniczak[¶], D.M. Rice, D.J. Siminovitch[§], and R.J. Wittebort[+]
Francis Bitter National Magnet Laboratory
Massachusetts Institute of Technology
Cambridge, MA   02139
U.S.A.

ABSTRACT: Deuterium NMR has been used extensively to study the rates
and mechanisms of dynamic processes in solids. Quadrupole echo
lineshapes provide a convenient method to study processes with
correlation times $\tau_c \sim 10^{-3} - 10^{-8}$ sec, while the anisotropy of ^{2}H spin
lattice relaxation can provide information on dynamic processes in the
range $10^{-7} - 10^{-10}$ sec. Experimental methods for obtaining the spectra
are reviewed, and the utility of the technique is illustrated with
examples of systems executing two- and threefold hops in solids.

1. INTRODUCTION

The last decade has witnessed the development of a number of new nuclear
magnetic resonance (NMR) techniques for the study of solids (1,2).   In
these lectures we will discuss one of these, namely deuterium quadrupole
echo NMR spectroscopy, and examine the manner in which ^{2}H spectra can
provide information on the rates and mechanisms of dynamic processes in
solids.   In the next section we review the properties of deuterium NMR
which make it an especially attractive probe for investigating dynamic
processes.   This is followed by a discussion of static ^{2}H powder
lineshapes and the effects of motion on these spectra.   Methods for
recording these rather wide (~ 250 kHz) NMR spectra will be discussed as
well.   The current method of choice involves Fourier transformation of a
two-pulse quadrupole echo (QE).   In the static or quasi-static (fast
motion) limit, these spectra are identical to those obtained with
continuous wave (CW) techniques.   However, in the intermediate exchange
regime, the spectral lineshapes are distorted in interesting and useful
ways.   In Section 4 the application of these techniques to a number of
different systems executing discrete motion (hops) is described.
Finally, in Section 5 we discuss some recent investigations of spin
lattice relaxation which extend the timescale of ^{2}H measurements to the
$10^{-9} - 10^{-10}$ sec regime.

81

G. J. Long and F. Grandjean (eds.), The Time Domain in Surface and Structural Dynamics, 81–105.
© 1988 by Kluwer Academic Publishers.

2. BACKGROUND

2.1. Rationale

The rationale for utilizing ^{2}H-NMR in the investigation of dynamic processes in solids includes four different factors. First, the size of the ^{2}H quadrupole interaction is well matched to the time scale of many dynamic processes which occur in solids. For example, $e^2qQ/h \sim 2 \times 10^5$ Hz for ^{2}H involved in organic and inorganic molecular bonds. Thus, molecular motion which occurs on a $10^{-3} - 10^{-7}$ sec timescale will influence the deuterium spectral lineshapes. Below, we will examine several cases where motion on this timescale is present and which therefore dramatically changes the ^{2}H spectral lineshapes. Second, a wide variety of specifically deuterated molecules are now commercially available, and new methods for preparing ^{2}H-labeled systems are continually being developed. For example, almost all of the deuterated amino acids are commercially available labeled at one or more positions, and these may be biosynthetically incorporated into proteins. Nucleic acids, lipids, chemical polymers, thermotropic liquid crystals, etc., have also been ^{2}H-labeled and their spectra examined. This copious selection of deuterated samples is due to the ease of replacing ^{1}H with ^{2}H and the relative low cost of ^{2}H sources. Third, in contrast to the situation encountered with ^{13}C spectroscopy, there is little natural abundance background in ^{2}H-NMR spectra. The natural abundance of ^{13}C is 1.1%, and thus for any molecule with ~ 100 or more carbon atoms a substantial natural abundance background signal is present in its spectra. While this problem can be resolved using techniques such as difference spectroscopy, it is nevertheless a nuisance. In contrast, the natural abundance level of ^{2}H is 0.016%, and therefore the background signals are almost two orders of magnitude below those of ^{13}C and four orders below the levels obtainable with $\sim 100\%$ ^{2}H labeling. Furthermore, the primary background signal in ^{2}H spectra is very often natural abundance ^{2}H in H_2O, and this can be suppressed by utilizing ^{2}H-depleted H_2O which contains 10^{-2} of the natural abundance level. Thus, it is possible to obtain background intensities which are 10^{-6} of the ^{2}H label intensities. Fourth, over the last few years, Fourier transform quadrupole echo techniques (3) and the associated hardware have been developed which permit the observation of high quality ^{2}H-NMR spectra. These include the high power probes and amplifiers, which are mandatory for exciting spectra which are ~ 250 kHz wide. The simple two-pulse quadrupole echo sequence is used almost exclusively for these experiments and a variety of computer programs (some of which will be discussed below) have been developed for simulating various types of motional narrowing observed in the spectra.

2.2. Powder Lineshapes

Most NMR spectra are obtained from liquid samples and exhibit a single (isotropic) line for each magnetically inequivalent nucleus. Spectra of solids rotating at the magic angle are similar except that each centerband (isotropic resonance) is often flanked by rotational sidebands which carry information on the anisotropy of the chemical shift or other interaction in their intensities. In static solids, such as are considered here, the information on the anisotropy of the interaction is contained in the powder lineshape. Figure 1 shows some examples of the types of spectra which are frequently encountered. In Figures 1a and 1b are shown axially symmetric and asymmetric lineshapes, respectively, obtained from I = 1/2 nuclei (^{13}C, ^{31}P, ^{15}N, etc.) while in 1c and 1d are illustrated the corresponding lineshapes for I=1 (^2H, ^{14}N) spectra. These spectra are more accurately characterized by an asymmetry parameter η, which, for a generalized magnetic resonance interaction, R, is

$$\eta = \frac{\left| R_{11} - R_{22} \right|}{\delta} \qquad (1)$$

where

$$\delta = R_{33} - \frac{1}{3} \mathrm{Tr} \underset{\sim}{R} \qquad (2)$$

and we assume $R_{33} > R_{22} > R_{11}$ where R_{ii} are the principal values of the tensor. For the chemical shift and electric field gradient R = σ and V, respectively. When $\eta = 0$ the lineshapes shown in Figures 1a and 1c are observed, while those in 1b and 1d are for $\eta \approx 0.6$. The pattern in Figure 1e is for $\eta = 1$ and is identical for I = 1/2 and I = 1. Note that the I = 1 spectra are obtained from the I = 1/2 versions by simple reflection through the origin.

The shape of the axially symmetric patterns can be readily derived. The angular dependence of the chemical shift or quadrupole coupling is given by

$$R_{zz} = R_{11} \sin^2\theta \cos^2\phi + R_{22} \sin^2\theta \sin^2\phi + R_{33} \cos^2\theta \qquad (3)$$

For an axially symmetric tensor $R_{11} = R_{22}$ and

$$R_{zz} = (R_{||} - R_{\perp}) \cos^2\theta + R_{\perp} \qquad (4)$$

where $R_{11} = R_{22} = R_{\perp}$ and $R_{33} = R_{||}$. If we assume a uniform distribution of orientations over a sphere, then the fraction of nuclei with orientation in the solid angle Ω and $\Omega + d\Omega$, and thus with intensity I(R)dR, is

$$P(\Omega)d\Omega = I(R)dR \qquad (5)$$

Since in a powder each orientation is equally probable $P(\Omega) = 1$. Thus,

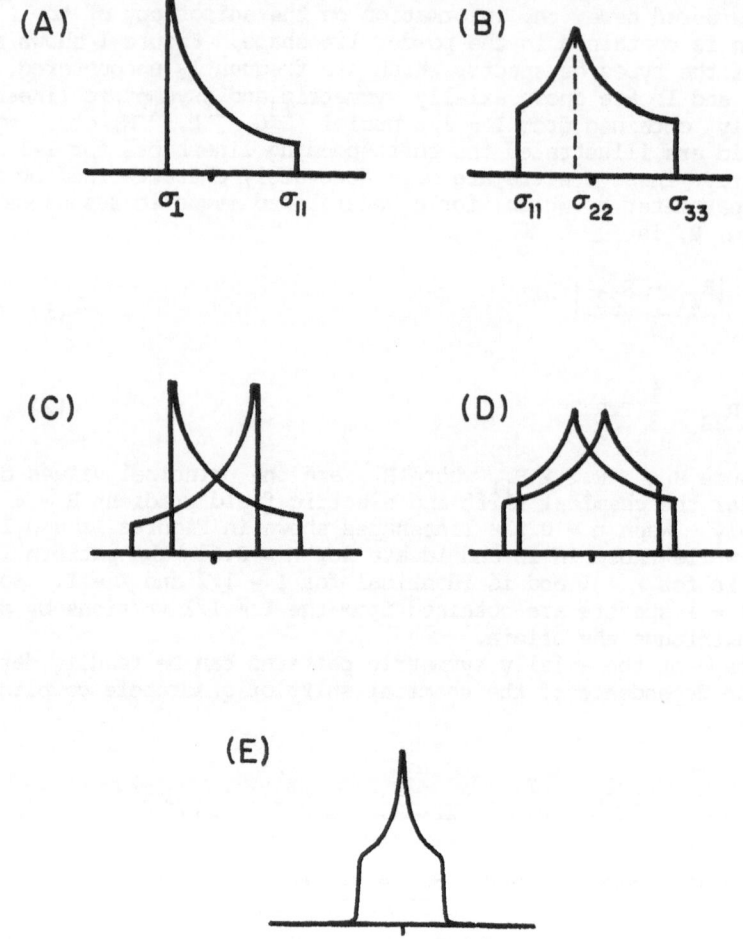

Figure 1: Powder lineshapes for an I=1/2 (chemical shift, (a) and(b))
and an I=1 (quadrupole, (c) and (d)) nucleus. (a) and (c) represent
axially symmetric (η=0) spectra, while (b) and (d) are for η ≈ 0.6. In
(e) we show the lineshape for an η=1 spectrum which is identical for
I=1/2 and I=1 nuclei. Note the I=1 spectra are obtained by simple
reflection through the origin.

$$I(R) = \left| \frac{dR}{d\Omega} \right|^{-1} \qquad (6)$$

Using (4) and (6) and $d\Omega = \sin\theta \, d\theta$ we obtain

$$I(R) = \frac{1}{2} \left[(R_{||} - R_{\perp})(R - R_{\perp}) \right]^{-1/2} \qquad (7)$$

which is the lineshape plotted in Figure 1a. The derivation of the formulae for the axially asymmetric lineshapes are more involved and we refer the reader to standard textbooks for details (1,2). A physical explanation of Eq.(7) and Figure 1a is as follows. We choose an arbitrary direction in a powdered solid, that of the laboratory field H_o, and ask what is the probability for being either parallel or perpendicular to that direction. Since the area for a given $d\Omega$ around the equator of a great sphere (perpendicular to H_o) is larger than that at the pole (parallel to H_o) the perpendicular edge of the lineshape is more intense than the parallel edge.

2.3. NMR Methodology

The currently accepted method for recording ^2H powder spectra utilizes the two-pulse quadrupole echo sequence (3) shown in Figure 2. Note that there is a 90° phase shift between the two pulses and normally the phases are cycled through all eight pairs (xy; xy; yx; yx; xy; xy; yx; yx) to suppress errors in the experiment. The signal generated by the first pulse is refocused by the second to form an echo as shown, and Fourier-transformed (FT) beginning at the echo peak. The real and imaginary parts of the decay for a palmitic acid-d_{31} sample are shown in Figure 3 together with its FT. If the imaginary channel is zero, then the spectrum will be symmetric as shown. In order to excite ^2H spectra, uniformly short pulses (1.5–2.0 μsec) are required, and the signal must be digitized at ~ 1 μsec/pt in order to accurately locate the echo peak. These and other experimental requirements are discussed in more detail elsewhere (4).

3. MOTIONAL AVERAGING OF DEUTERIUM NMR LINESHAPES

3.1. Fast Limit Lineshapes

^2H-NMR lineshapes are very sensitive to the rate and mechanism of motion occurring in solids. As an illustration of this point, we consider the ^2H-labeled aromatic ring shown in Figure 4 which can potentially undergo two types of motion about the C^β-C^γ bond. The first motion, which is often discussed in the literature, is continuous diffusion and results in a dramatic narrowing of the ^2H spectrum. In particular, the axially symmetric rigid lattice spectrum shown in Figure 5a is narrowed by an order of magnitude by this motion. In the rigid lattice the splitting between the perpendicular edges of the spectrum is D = $(3/4)(e^2qQ/h)$ = 135 kHz for an aromatic ^2H. For fast, continuous

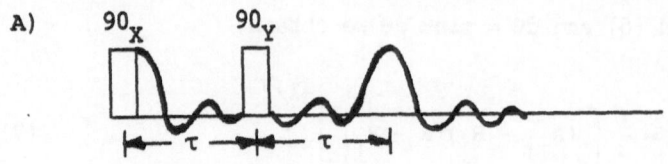

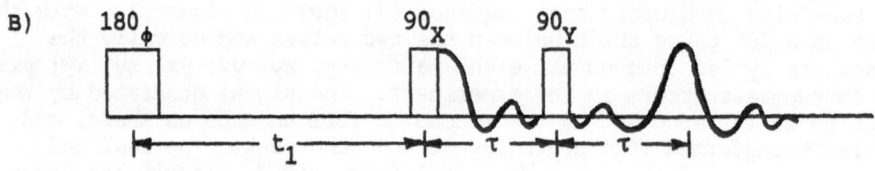

Figure 2: (A) Two-pulse sequence for exciting quadrupole echoes. (B) Inversion recovery sequence for measuring T_1 anisotropies. In both cases the spectrum is obtained as a FT of the signal beginning at the echo peak, which appears at $t=2\tau$ after the initial QE pulse. In (B) spin lattice relaxation occurs during t_1 and the lineshape which appea is "partially relaxed". The echo in (B) corresponds to a fully relaxe spectrum in the limit of $t_1 \gg T_1$

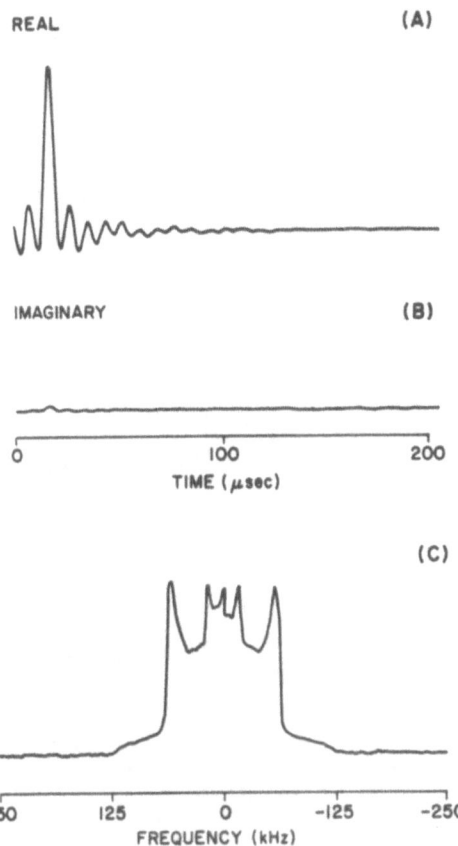

Figure 3: (A) and (B) -- Real and imaginary parts of the NMR signal generated by pulse sequence 2(A). Data collection commences prior to echo formation and the data is left-shifted before the FT. When the experiment is properly adjusted the imaginary signal is essentially zero as shown. (C) Real (absorptive) part of the complex FT of (A) and (B) showing the symmetric lineshape due to palmitic acid-d_{31}. The CD_2 groups give rise to the 124 kHz splitting, while the 41 kHz component arises from the terminal $-CD_3$.

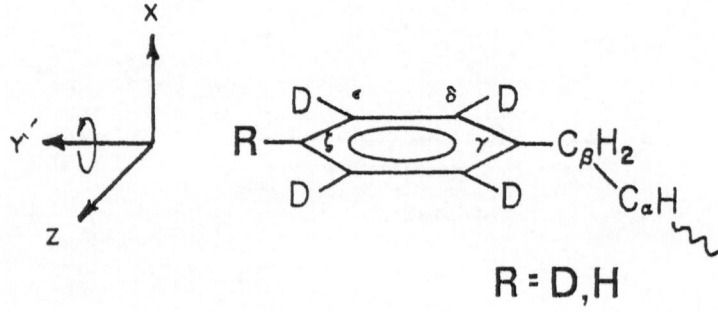

Figure 4: The aromatic sidechain of Phe-d_4(d_5)·HCl illustrating the two CD bond vectors and a Cartesian coordinate system where X is perpendicular to the plane of the ring, Y is parallel to the C^β-C^γ axis and Z lies in the plane of the ring, perpendicular to these directions. The unique component of the rigid lattice tensor is along the CD direction. For continuous diffusion about C^β-C^γ the Y direction corresponds to the unique axis of the axially symmetric motionally averaged tensor, while for discrete 180° jumps about C^β-C^γ the principal values of the axially asymmetric η=0.6 tensor appear along the XYZ directions.

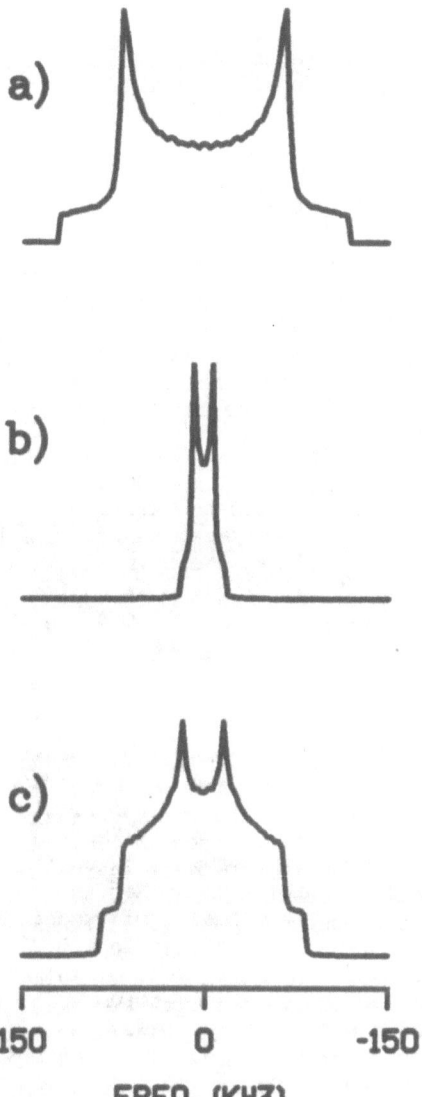

Figure 5: Theoretical ^{2}H-NMR lineshapes for Phe-d$_4$·HCl. (a) Rigid lattice spectrum exhibiting a splitting between perpendicular edges of D=135 kHz. (b) Spectrum resulting from fast continuous diffusion about the Y axis with a splitting between perpendicular edges of D/8=17 kHz. (c) Axially asymmetric spectrum arising from fast 180° jump diffusion about C$^\beta$-C$^\gamma$. The principal values of this spectrum are V$_{zz}$=5D/4, V$_{xx}$=-D, and V$_{yy}$=-D/4, yielding η=|V$_{xx}$-V$_{yy}$|/V$_{zz}$=0.6

diffusion the rotationally averaged spectrum is always axially symmetric and the unique axis of the motionally averaged tensor is coincident with the axis of motion. In this case, the separation between the parallel edges, $\Delta v^R_{Q||}$, is

$$\Delta v^R_{Q||} = \left(\frac{3eQ}{4h} \right) (3\cos^2\theta - 1)V_{||} \qquad (8)$$

where $V_{||}$ is the unique component of the axially symmetric rigid lattice tensor and θ is the angle between the diffusion axis and the unique axis. For an aromatic ring, $\theta = 60°$ and $\Delta v^R_{Q||} = -D/4 = -34$ kHz. The perpendicular edge splitting, which is easier to detect experimentally, is $D/8 = 17$ kHz. This dramatic motional narrowing is due to the fact that the CD axis, and thus the unique axis of the rigid lattice tensor, is near the magic angle with respect to the diffusion axis.

In contrast, fast, discrete diffusion by 180° twofold flips about $C^\beta-C^\gamma$ results in a dramatically different spectrum whose shape and breadth are easily understood. Since the two ring orientations are equally probable, the motionally averaged tensor will have principal axes along the XYZ coordinate system shown in Figure 4. The splitting perpendicular to the plane of the ring (X) is unaffected by the motion and remains $\Delta v_{Qxx} = -D$, whereas the splitting along the jump (Y) axis is that calculated above, $\Delta v_{Qyy} = -D/4$. Since

$$\text{Tr} \underset{\sim}{V} = 0 \qquad (9)$$

the third component must be $\Delta v_{Qzz} = 5D/4$. Thus, fast, twofold reorientation yields an $\eta = 0.6$ spectrum with a breadth about half that of the rigid lattice spectrum, and the central portion of the spectrum yields peaks which are 34 kHz rather than 17 kHz apart as in the case of continuous diffusion (5). The intermediate situation in which the ring moves in a potential well is dealt with in Ref.(6).

A few other comments concerning fast limit spectra are in order. First, a twofold aromatic ring flip results in a 120° change in orientation of the ^{2}H tensor. However, because magnetic resonance tensors are of rank two, an equivalent spectrum would result from a mechanism which produced a 60° change in orientation of the C-^{2}H bond. Thus, in the experiments described here, it is not possible to distinguish between jumps which change the tensor orientation by θ and $180-\theta$. Second, the spectral lineshapes are most sensitive to jumps larger than ~ 30°. Smaller angle excursions do not dramatically alter the parallel edge of the lineshape, and the asymmetry in the perpendicular edge is difficult to detect unless the signal-to-noise is high and derivative presentations are possible. Third, as in the case of continuous diffusion, it can be shown that fast, higher-order jump motions always yield an axially symmetric spectrum. In particular, if the motion has threefold or greater symmetry, then the fast limit spectrum will be axially symmetric, and this is observed for the threefold motion of methyl groups. Finally, we will see below that twofold hops with tetrahedral symmetry (e.g. 180° flips of an ^{2}H$_2$O molecule) lead to $\eta \simeq 1$ spectra.

3.2. Quadrupole Echo Lineshapes

In the slow ($\omega_Q\tau_c \gg 1$) or fast ($\omega_Q\tau_c \ll 1$) limit motional regimes, QE lineshapes are identical to those obtained by CW NMR techniques. However, for intermediate exchange ($\omega_Q\tau_c \sim 1$) the CW and QE lineshapes differ, and we now consider the physical basis for these differences.

In the slow or fast limit $T_2 \gg \tau$ in the echo sequence, and the magnetization is completely refocused by the second pulse. In contrast, for intermediate exchange T_2 is anisotropic and $\leq \tau$ for many orientations in the powder, so that much of the signal decays irreversibly and is not refocused. This effect is manifest in three different ways in the QE lineshapes. First, the lineshapes are distorted, as is illustrated in Figure 6, where we compare calculated spectra for the δ and ε deuterium nuclei of an aromatic ring executing twofold flips. In the right column are spectra calculated for a Bloch decay experiment, while on the left are the QE lineshapes corresponding to the same twofold hopping rates. Note in the intermediate exchange regime ($10^3 - 10^6$ sec^{-1}), that the QE spectra display intensity dips, and they appear to arise from a superposition of fast and slow hopping rings. The physical origin of this effect, as noted above, is that T_2 in the powder sample is anisotropic. Consider, for example, a CD bond in Figure 4 oriented along the direction of H_o. The twofold flip of the ring changes the molecular orientation so that the bond is now approximately perpendicular to H_o, and the resonant frequency moves approximately from the parallel to the perpendicular edge of the rigid lattice powder lineshape. Conversely, for a molecule oriented with H_o along either the X,Y or Z direction, a flip of the ring will not result in a change in frequency and for these orientations T_2 is long. For these reasons the edges at $\pm 5D/4$, $\pm D$ and $\pm D/4$ are enhanced relative to the regions between them, and this leads to the altered lineshapes shown in Figure 6.

The second effect which appears in the intermediate exchange regime is the dependence of the lineshapes on τ. This is illustrated in Figure 7 for the aromatic ring spectra. Because T_2 is comparable to τ, this dependence is strongest for $\tau < 20$ μsec. Nevertheless, it exists at longer times and can be employed to discriminate amongst various models for molecular motion. For instance, we have found it possible to simulate QE spectra of glycolipids expected to show τ-dependent lineshapes due to slow motional effects. Experimentally, this τ dependence is absent, suggesting that the rates are much faster, a fact that has recently been confirmed with T_1 measurements (12).

A unique final feature of QE spectra is that the overall spectral intensity declines dramatically when passing through the intermediate exchange region. This point is illustrated in Figure 8, again for aromatic rings executing twofold 180° flips where the theoretical line shows an intensity decay to ~ 20% of its maximum value for rates of ~ 10^5 sec^{-1}. The experimental points in the figure were obtained from two peptides containing tyrosine rings and confirm the predicted intensity losses. For higher symmetry motions — threefold or fourfold hops — the intensity drops to < 5% of its full value in the intermediate exchange region. The physical origin of this effect is again that T_2 is short in this exchange regime so much of the magnetization decays

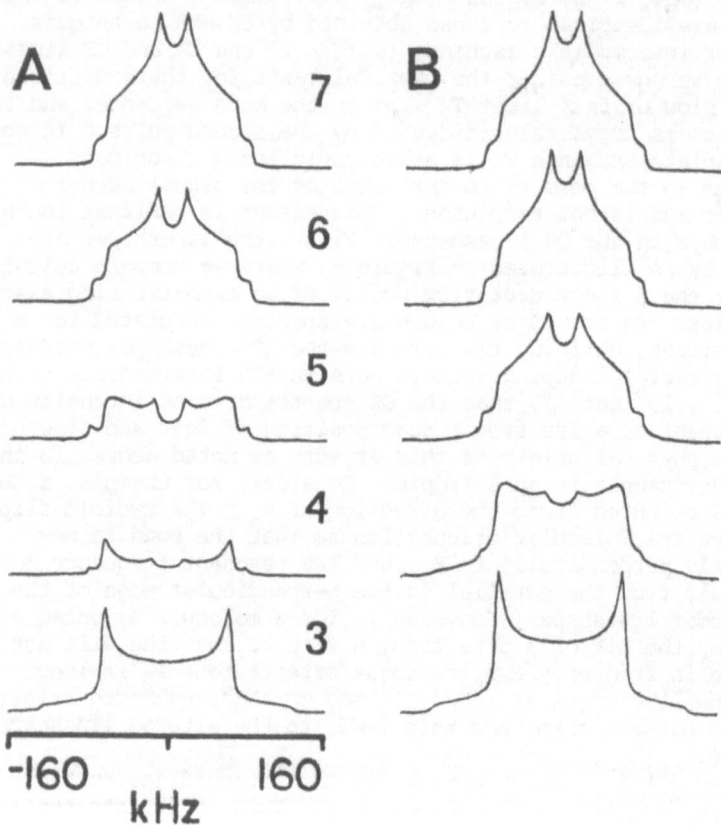

Figure 6: ^{2}H-NMR lineshapes for the δ and ε ^{2}H's of an aromatic ring as a function of the twofold flipping rate. The numbers indicate the log of the jump rate. (A) Quadrupole echo lineshapes illustrating the intensity losses and lineshape distortion. (B) Lineshapes obtained from a CW NMR or Bloch decay experiment. Notice the intensity losses which occur even when the spectra show lineshapes which appear to be in the slow or fast exchange limit -- i.e. at 10^3 sec^{-1} and 10^6 sec^{-1}.

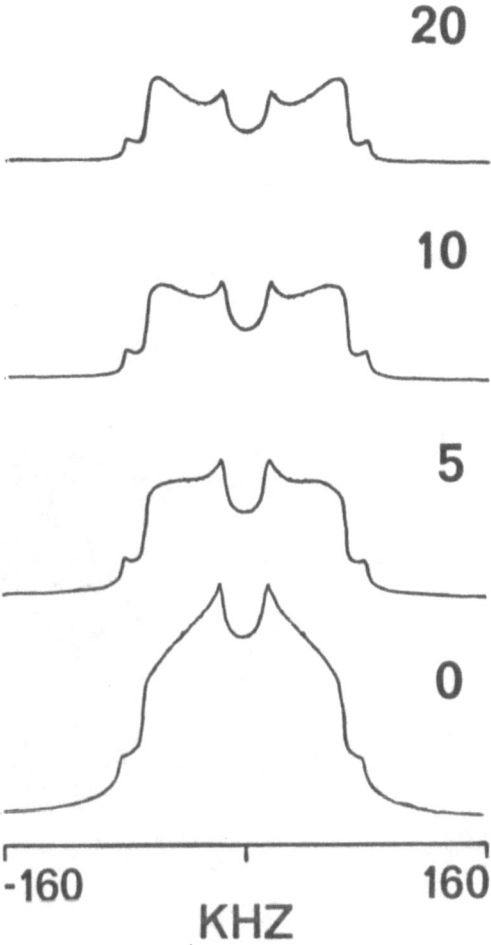

Figure 7: τ–dependence of the ^{2}H spectra of δ and ε aromatic ring deuterons undergoing twofold flips. The numbers indicate the pulse spacing (τ) in μsec. Notice the major changes in the lineshape which occur for τ < 20 μsec.

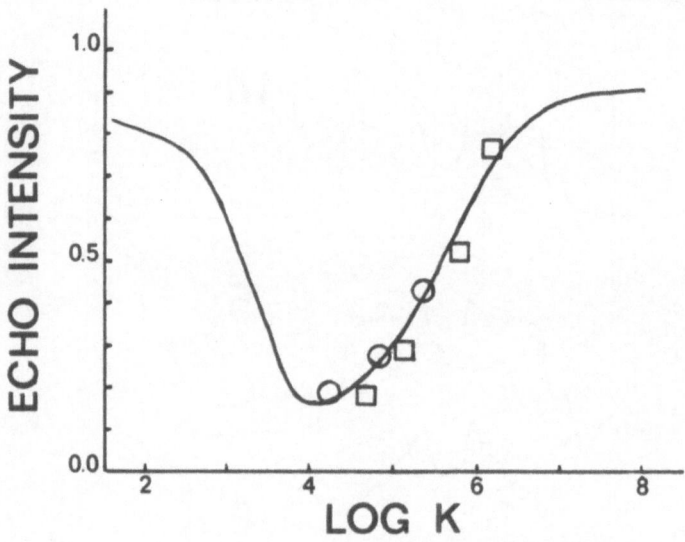

Figure 8: Calculated and experimental intensity losses in QE spectra of aromatic rings executing twofold flips. The solid line is the theoretical curve and the experimental points are for tyrosine rings in the peptides Asp–Pro–Tyr–d$_2$ and Pro–Tyr–d$_2$ (12). τ = 30 μsec.

irreversibly. The spectral effects discussed in this section have been treated in detail by a number of authors (6,9-12), and computer programs have been written to simulate lineshapes such as those shown in Figure 6. We now briefly outline these calculations.

Quite generally, the spectrum is calculated by solving the usual exchange modified Bloch equation

$$\frac{dM_i}{dt} = \sum_{j=1}^{N} (i\omega_i \delta_{ij} + R_{ij})M_j \tag{10}$$

where R_{ij} is the rate matrix describing the dynamics among the N jump sites and ω_i is the quadrupole frequency corresponding to the orientation of V_{zz} for site i specified by (θ_i'',ϕ_i''). (θ',ϕ') specifies the orientation of the crystallite with respect to the external magnetic field. Thus, we have

$$\omega = \frac{3}{4} \frac{(e^2 qQ)}{h} [P_2(\cos\phi')P_2(\cos\theta'')$$

$$- 3/4 \sin 2\theta' \sin 2\theta'' \cos(\phi'+\phi'')$$

$$+ 3/4 \sin^2\theta' \sin^2\theta'' \cos 2(\phi'+\phi'')] \tag{11}$$

The solution for the desired QE initial conditions gives the frequency domain lineshape $I(\omega,\theta',\phi')$

$$I(\omega,\theta',\phi') = \text{Re} \sum_{k=1}^{N} \frac{b_k}{\lambda_k - i\omega}, \tag{12}$$

$$b_k = a_k \sum_m a_m^* \left(\sum_n x_n^{(k)} x_n^{(m)*} \right) e^{(\lambda_m^* + \lambda_k)\tau} \tag{13}$$

and

$$a_k = \sum_i (P_i)^{1/2} x_i^{(k)}. \tag{14}$$

$x^{(k)}$ and λ_k are the N complex eigenvectors and eigenvalues of the symmetrized matrix $[i\omega_i\delta_{ij} + (R_{ij}R_{ji})^{1/2}]$. Thus, for a given crystal orientation, the spectrum is the sum of N Lorentzians. Intensity distortions of the QE spectrum ($\tau \neq 0$) are contained in the b_k terms. The real and imaginary parts of the eigenvalues, λ_k, correspond to the linewidths $(1/\pi T_{2k})$ and eigenfrequencies, respectively, of the N Lorentzians. For simulations of powder spectra, the single crystallite spectrum is numerically integrated over crystal orientations on the unit sphere.

An efficient implementation of this procedure is obtained by calculating $I(\omega,\theta',\phi')$ only for frequencies, ω, in which the intensity is substantially different from zero, i.e., a few linewidths (Re λ_k) about its eigenfrequency (Im λ_k). Numerical instabilities associated with frequency domain as opposed to time domain calculations (2) are eliminated by using the complex frequency, $\omega = \omega - i\delta$, where δ is a natural linewidth, typically about 1-5 kHz in deuterium NMR of selectively deuterated solids containing abundant protons. Furthermore, we find that a convenient quadrature for the powder average uses a set of (θ',ϕ') pointing to equal area elements on the unit sphere. (This was done simply by using constant increments in θ' and, for a given θ', the number of ϕ' increments is proportional to $\sin\theta'$). Finally, continuous dynamics are handled in this procedure by using a finite difference approximation with the requirement that the resulting rate matrix satisfies microscopic reversibility (10).

4. APPLICATIONS OF QUADRUPOLE ECHO LINESHAPES ANALYSIS

4.1. Aromatic Rings

The dynamic properties of aromatic rings have been examined in a number of different systems including amino acids, proteins, polymers, liquid crystals, and polymer model membranes. Some typical experimental results for phenylalanine-d_4·HCl are shown in Figure 9 where we plot QE spectra as a function of temperature. As can be seen from the figure, the $\eta = 0$ spectra observed at room temperature evolve to the $\eta = 0.6$ lineshapes expected from a ring executing rapid twofold flips at 170°C. From this data, one finds that the activation energy for ring flips in Phe-d_4·HCl is 17 kcal/mole (13). Similar results have been obtained for Tyr in the dipeptide Pro-Tyr (14). When the molecular size is greater — for example, in proteins such as bacteriorhodopsin (15) — the activation energy appears to decrease to ~ 10 kcal/mole. Aromatic ring motion has been observed in small proteins with ^1H spectroscopy and similar E_a's are observed. However, the accessible range of rates which can be studied in this way is much narrower than for ^2H QE spectroscopy (16,17).

4.2. D_2O Molecules

A second class of molecules exhibiting twofold flips are D_2O molecules in crystalline hydrates. CW NMR studies of these systems were first performed many years ago (18), and, more recently, we have reinvestigated one of these, $Ba(ClO_3)_2 \cdot H_2O$, because it is a convenient system for testing QE and T_1 anisotropy lineshape theories. The results of a temperature-dependent study are shown in Figure 10. Here the high temperature spectra evolve to $\eta \sim 1$ lineshapes at 18°C, where the twofold hopping rate is clearly in the fast limit (> 10^7 sec^{-1}). The shape of these spectra can be understood using arguments outlined in Section 3. From the simulations of the lineshapes we obtain $E_a = 6.1$ kcal/mole for the D_2O molecule in $Ba(ClO_3)_2 \cdot D_2O$.

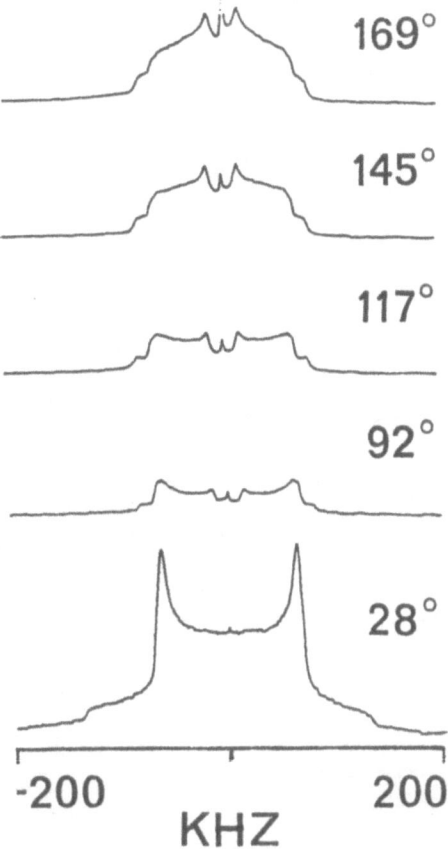

Figure 9: Experimental 45 MHz ^{2}H QE spectra of Phe-d$_4\cdot$HCl at various temperatures. The intensity losses apparent in the spectra and the lineshape distortions are predicted by theoretical calculations. The twofold hopping rates are: 28°C, slow limit; 92°C, 2.5 x 10^5 sec^{-1}; 117°C, 8.2 x 10^5 sec^{-1}; 145°C, 3.1 x 10^6 sec^{-1}; 169°C, 8.2 x 10^6 sec^{-1}.

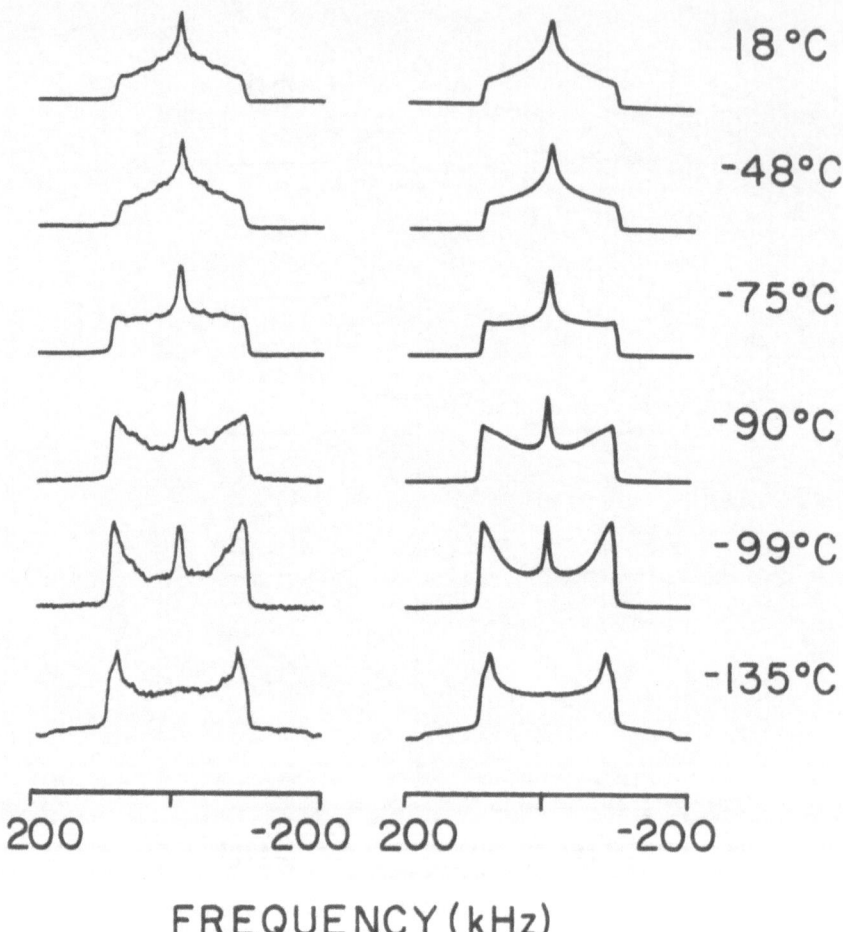

Figure 10: Experimental and calculated 61 MHz QE spectra of the D_2O molecule in $Ba(ClO_3)_2.D_2O$ as a function of temperature. The rigid lattice spectra are slightly asymmetric and evolve to $\eta \sim 1$ spectra at higher temperatures. Rates at each temperature are: $-135°C$, 7.1×10^3 s^{-1}; $-99°C$, 5.2×10^5 s^{-1}; $-90°C$, 1.5×10^6 s^{-1}; $-75°C$, 6.9×10^6 s^{-1}; $-48°C$, 3.6×10^7 s^{-1}; $18°C$, 7.5×10^8 s^{-1}

4.3. Methyl Groups

Methyl groups are ubiquitous in organic compounds and can be easily ^{2}H-labeled and therefore studied with ^{2}H-NMR methods. Some temperature-dependent ^{2}H spectra obtained from Ala-d$_3$ are shown in Figure 11, together with simulations obtained by assuming that the motion consists of threefold hops (17). In the intermediate exchange region ($-120°C < T < -75°C$) the lineshapes exhibit a triplet structure which can be understood by examining the two special orientations illustrated in Figure 12. In Figure 12a the laboratory field coincides with the C^α-C^β threefold axis and the CD bonds are oriented at 70.5° with respect to this direction. Since the threefold motion does not alter this orientation, T_2 is long and these molecules give rise to the ±40 kHz spikes which are present in the spectrum. The central portion of the triplet arises from molecules oriented as shown in Figure 12b. Here the C^α-C^β axis and one of the C^β-D vectors are at 54.7° with respect to B_o, and B_o simultaneously bisects the remaining DCD angle. Notice that an axially symmetric lineshape is observed in the fast limit since the motion has threefold symmetry. The spectral intensity losses associated with this motion have been measured and the spectral intensity was found to drop to ~ 5% of the slow or fast limit value in agreement with the theoretical predictions. In addition, the spectra exhibit the correct τ-dependence and T_1 anisotropy expected for threefold motion. The activation energy for the CD$_3$ group motion in Ala was found to be 4.8 kcal/mole, which is quite high for a methyl group and is due to the steric crowding present in this molecule; generally, E_a's for methyl groups are in the neighborhood of 3 kcal/mole.

In all of the above examples we have dealt with systems where a single motion is present which can be characterized by a single correlation time. It is reasonable to inquire if systems with multiple motions present can be investigated. For example, if two methyl groups with different hopping rates are present, can their spectra be resolved? It has recently been found that in two systems this is the case (19). In the first of these, two crystallographic forms of Ala-d$_3$ were present in the sample and exhibited very different threefold hopping rates (19). In the second, the two CD$_3$ groups in N-acetyl-dl-valine were observed to exhibit different rates and E_a's (20).

5. SPIN LATTICE RELAXATION

In the previous sections we have focused on the motional narrowing of the quadrupole echo lineshapes and have seen that this effect is most pronounced for motions in the range 10^3 - 10^8 sec^{-1}. There are, however, many systems which exhibit faster motions, and these are most easily investigated by examining ^{2}H spin lattice relaxation. There are two approaches used in these investigations, the simplest of which is the measurement of a "powder average" T_1. This can be achieved with either a saturation recovery experiment, or with an inversion recovery experiment such as that illustrated in Figure 2b. Equations can then be derived which relate T_1 for a particular motional model to a correlation time. The problem with this approach is that T_1 is a double-valued

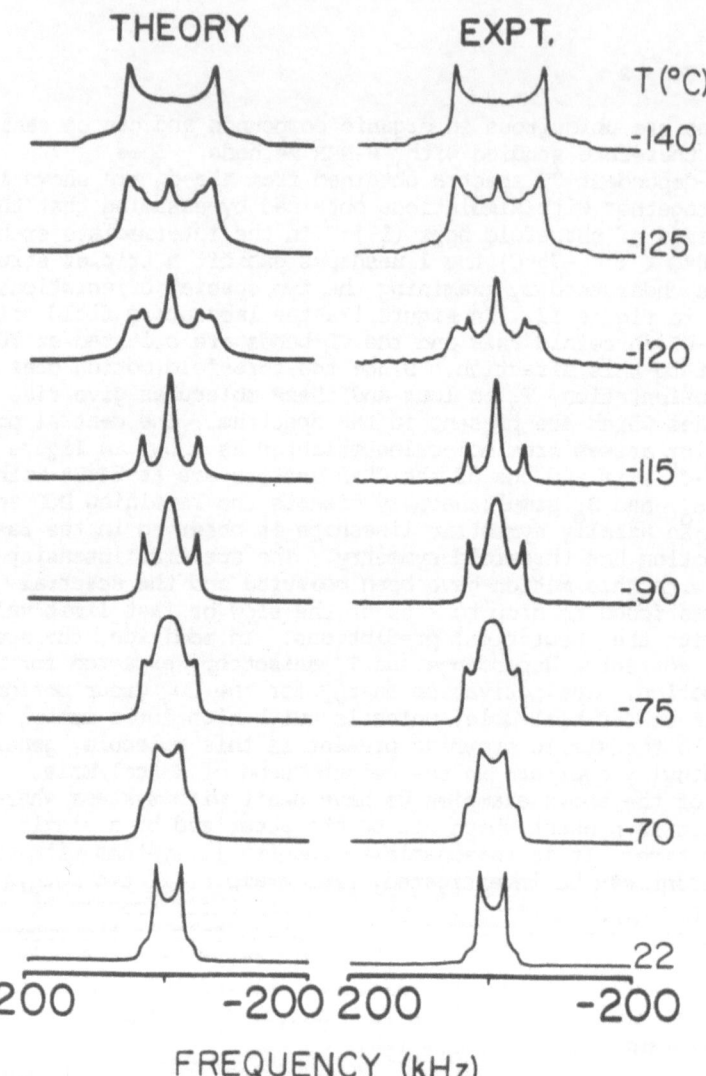

Figure 11: Temperature-dependent experimental spectra of
L-alanine-(3,3,3)d$_3$ and simulated lineshapes using a single jump rate
for each spectrum. The jump rates for the simulations and temperatures
for the corresponding experimental spectra are: −140°C, 5.5 x 10^3 s^{-1};
−125°C, 2.3 x 10^5 s^{-1}; −120°C, 3.6 x 10^4 s-1; −115°C, 6.5 x 10^4 s^{-1};
−90°C, 4.75 x 10^5 s-1; −75°C, 2 x 10^6 s^{-1}; −70°C, 3 x 10^6 s^{-1}; 22°C, 3 x
10^8 s^{-1}. The intensities of many of these spectra have been scaled up.

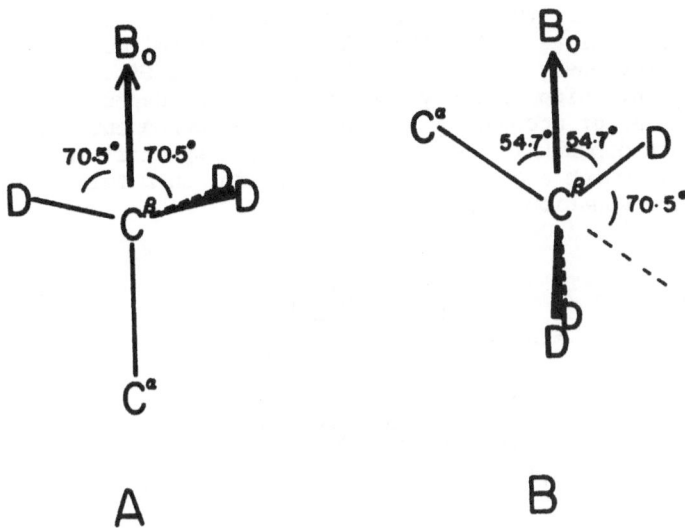

Figure 12: The two unique orientations of $C^\alpha - C^\beta - D_3$ in L-alanine that give rise to characteristic lineshapes in the intermediate exchange regime. (A) $C^\alpha - C^\beta$ is parallel to B_o and in this orientation all CD bonds make a 70.5° angle with the hopping axis and B_o. (B) $C^\alpha - C^\beta$ is at the magic angle (54.7°) with respect to B_o. All CD bonds are also at 54.7° with respect to B_o assuming a perfect tetrahedral geometry.

function, and with a single measurement of this type, it is not possible
to distinguish between the two sides of the T_1 minimum. To circumvent
this problem one can perform an inversion recovery experiment yielding
the lineshapes which are different, depending on whether the correlation
time is above, below, or near the T_1 minimum. To illustrate the utility
of studying D_2O T_1 anisotropies, we now examine results for the D_2O
molecule of $Ba(ClO_3)_2 \cdot D_2O$, which executes twofold flips in the solid.

In Figure 13 we show three sets of experimental spectra obtained at
$-90°C$, $-48°C$ and $51°C$, together with simulations. The theoretical
background necessary for these calculations has been outlined by Torchia
and Szabo (21) as well as in Ref.(6). This theoretical development
shows that the previously defined echo lineshape, $I(\omega,\theta',\phi')$, is
modified in the inversion recovery experiment by the expected factor
$1-2e^{-t/T_1(\theta',\phi')}$ after accounting for the angular-dependent T_1.
$T_1(\theta',\phi')$ in turn depends in the usual way on the spectral densities,
$J_1(\omega)$ and $J_2(2\omega)$, which are readily calculated for an N-fold jump
process using the equation:

$$J_m(\omega,\theta',\phi') = 2 \sum_{a,a'=-2}^{2} d_{ma}^{(2)}(\theta')d_{ma'}^{(2)}(\theta')$$

$$\times \sum_{n,l,j=1}^{N} \tilde{x}_1^{(0)}\tilde{x}_1^{(n)}\tilde{x}_j^{(0)}\tilde{x}_j^{(n)}$$

$$\times d_{0a}^{(2)}(\theta_l'')d_{0a'}^{(2)}(\theta_j'')$$

$$\times \cos(a\phi_l - a'\phi_j) \frac{\lambda_n}{\lambda_n^2 + \omega^2}$$

Here $\tilde{X}$ and $\tilde{\lambda}_k$ are the corresponding eigenvectors and eigenvalues of the
real symmetric matrix $(R_{ij}R_{ji})^{1/2}$ and R is the same rate matrix
specified in Section 3.2. These formulae are also corrected for power
roll-off effects due to finite widths of the inverting pulse as well of
as the quadrupole echo. The importance of these corrections can be
observed by noting that the equilibrium QE lineshape -- the top line of
each panel in Figure 13 -- is considerably different from the lineshape
obtained with inversion -- the bottom line. Notice also that the
lineshapes for each temperature are well represented by the simulations.
A complete description of this work is in preparation (22).

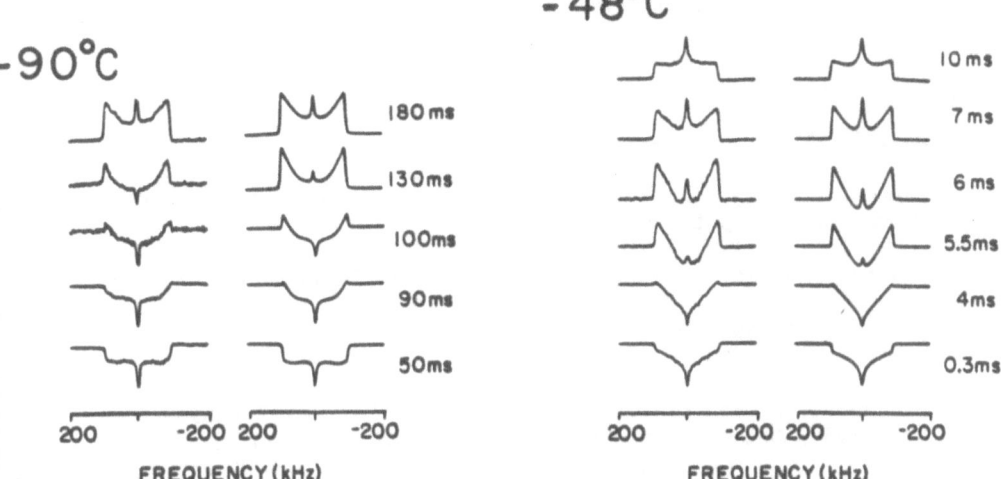

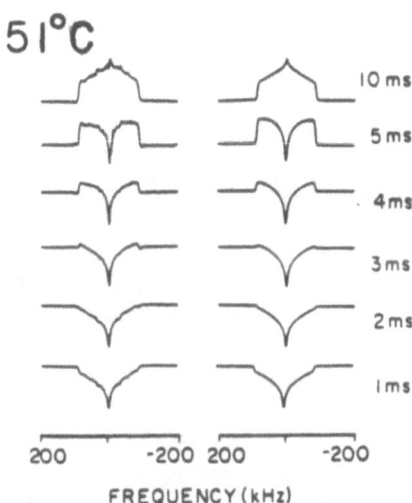

Figure 13: ^{2}H–NMR T_1 anisotropy lineshapes for Ba(ClO$_3$)$_2$.D$_2$O at –90°C, –48°C, and 51°C. These correspond to temperatures below, near, and above the ^{2}H T_1 minimum. The twofold hopping rates are 1.5×10^6 s^{-1} at –90°C, 3.6×10^7 s^{-1} at –48°C, and 1.6×10^9 s^{-1} at 51°C.

6. ACKNOWLEDGEMENTS

This research was supported by the National Institutes of Health (GM-23403, GM-23289, GM-25505, and RR-00995).

7. REFERENCES

\#   Present address: Department of Physics, Georgia Institute of Technology, Atlanta, GA 30332.

¶   Present address: Abbott Laboratories, North Chicago, IL 60064.

*   Present address: Department of Chemistry, Cornell University, Ithaca, NY 14853.

\$ Present address: Division of Chemistry, National Research Council of Canada, Ottawa, Ontario, CANADA K1A OR6.

\+   Present address: Department of Chemistry, University of Louisville, Louisville, KY 40292.

1. U. Haeberlen, 'High-Resolution NMR in Solids: Selective Averaging', Supplement 1, Advances in Magnetic Resonance, Academic Press (1976).

2. M. Mehring, 'High-Resolution NMR in Solids', Springer-Verlag, Berlin (1983).

3. J. Davis, K.R. Jeffrey, M. Bloom, M.I. Valic, and T.P. Higgs, Chem. Phys. Letters 42, 390 (1976).

4. R.G. Griffin, Methods in Enzymol. 72, 108 (1981); J. Seelig, Quart. Rev. of Biophysics 10, 353 (1977); J.H. Davis, Biochim. Biophys. Acta 737, 117 (1983).

5. D.M. Rice, R.J. Wittebort, R.G. Griffin, E. Meirovitch, E.R. Stimson, Y.C. Meinwald, J.H. Freed, and H.A. Scheraga, Jour. Am. Chem. Soc. 103, 7077 (1981).

6. R.J. Wittebort, E.T. Olejniczak, and R.G. Griffin, J. Chem. Phys. 86, 5411 (1987).

7. T-H. Huang, R. Skarjune, R.J. Wittebort, R.G. Griffin, and E. Oldfield, Jour. Am. Chem. Soc. 102, 7377 (1980).

8. R. Ebelhäuser and R.G. Griffin (to be published).

9. H.W. Spiess, and H. Sillescu, J. Magn. Reson. 42, 381 (1981).

10. P. Meier, E. Ohmes, G. Kothe, A. Blume, J. Weldner, and H. Eibl, J. Phys. Chem. 87, 4904 (1983).

11. L.J. Schwartz, A.E. Stillman, and J.H. Freed, J. Chem. Phys. 77, 5410 (1982).

12. A.J. Vega, and Z. Luz, J. Chem. Phys. 86, 1083 (1987).

13. D.M. Rice, E.T. Olejniczak, S.K. Das Gupta, J. Herzfeld, and R.G. Griffin (to be published).

14. D.M. Rice, Y.C. Meinwald, H.A. Scheraga, and R.G. Griffin, Jour. Am. Chem. Soc. 109, 1636 (1987).

15. D.M. Rice, B.A. Lewis, S.K. Das Gupta, J. Herzfeld, and R.G. Griffin (to be published).

16. I.D. Campbell, C.M. Dobson, G.R. Moore, S.J. Perkins, and R.J.P. Williams, FEBS Letters 70, 91 (1976).

17. G. Wagner, A. DeMarco, and K. Wüthrich, Biophys. Struct. Mech. 2, 139 (1976).

18. T. Chiba, J. Chem. Phys. 39, 947 (1963).

19. K. Beshah, E.T. Olejniczak, and R.G. Griffin, J. Chem. Phys. 86, 4730 (1987).

20. K. Beshah and R.G. Griffin (to be published).

21. D.A. Torchia and A. Szabo, J. Magn. Reson. 49, 107 (1982).

22. R. Ebelhäuser and R.G. Griffin, J. Chem. Phys. (submitted for publication).

CHAPTER 8

PROTEIN DYNAMICS AND THE TIME DOMAIN

Robert D. Young
Department of Physics
University of Illinois at Urbana-Champaign
1110 West Green Street, Urbana, IL 61801 USA
and
Illinois State University, Normal, IL 61761 USA

1. INTRODUCTION

1.1 Protein Reactions in the Time Domain

Proteins are dynamic systems which exhibit a wide spectrum of motions. During the last several years it has been realized that protein motions are essential to their function [1-5]. It has also become clear that motions and functional processes in proteins span an enormous range of time scales from the subnanosecond to more than a kilosecond [6,7]. Thus proteins are superb systems for exploring the dynamic relations between structure and function in the time domain.

   We have obtained a large body of information bearing on the connections between protein motions and function by investigating a simple biological process -- the binding of small ligands such as carbon monoxide (CO) and dioxygen (O_2) to many different heme proteins including myoglobin (Mb). Our experiments extend over wide ranges in time (200ps to 50ks), temperature (2 to 330K), pressure (up to about 2 kbar), pH and solvent viscosities [4,8,9]. Binding is monitored by optical and infrared spectroscopies. Our kinetic studies have led to a hierarchical model of protein states and motions with deep connections to amorphous solids and glasses [4,7,10].

1.2 Concepts of Protein Dynamics

One point is essential to the discussion which follows -- to investigate the connection between protein motions and protein function, we must look at a working protein. Several concepts which will be used in later sections are summarized below [4,10]:

   (i) States and Substates. Two equilibrium protein states are involved in the binding of CO to Mb -- deoxyMb and MbCO. Each of these states can exist in a large number of conformational substates, denoted by CS. The CS have the same overall structure, but differ in detail; they perform the same function, but possibly with different

107

G. J. Long and F. Grandjean (eds.), The Time Domain in Surface and Structural Dynamics, 107–138.
© *1988 by Kluwer Academic Publishers.*

rates. The concept of CS in proteins was first postulated on the basis of clear and compelling evidence by H. Frauenfelder and collaborators in 1974.

(ii) Equilibrium Fluctuations and Fims. A protein like Mb is much larger than any atom but is still small on a macroscopic scale. Thermodynamics implies that a protein does not have sharp values of internal energy, entropy, and volume, but that these quantities have relatively large fluctuations about their mean values [11]. At physiological temperature a resting protein fluctuates among the CS. We denote these equilibrium fluctuations by EF. Fig. 1 schematically illustrates two different protein states (MbCO and deoxyMb) with corresponding CS. The EF are transitions among the

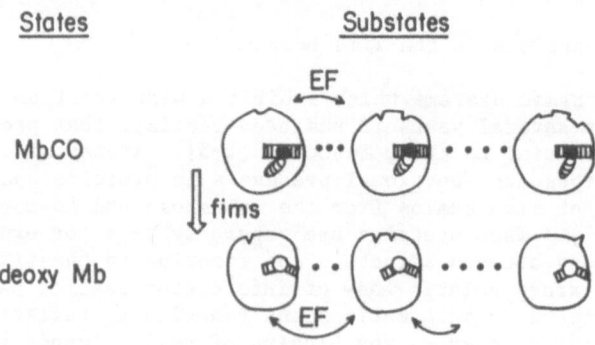

Figure 1. Schematic portrayal of states, substates, equilibrium fluctuations (EF), and functionally important motions (fims).

various CS of a given protein state. The protein action -- say the transition from MbCO to deoxy Mb -- occurs through functionally important motions, fims. In order to understand the working of a protein, we must study and understand both the EF and the fims. The EF and fims are related by fluctuation-dissipation theorems so study of the EF is also useful for understanding the functional aspects of protein reactions.

(iii) Fluctuation-Dissipation Theorem. The fluctuation of a protein at rest about its mean structure is an equilibrium phenomenon. Protein reactions, however, are dissipative leading to a state of lower energy. Fluctuation-dissipation theorems connect the rates of the EF and fims [12]. A general relation between equilibrium fluctuations and dissipative processes has been formulated by Callen and Welton [13], Kubo [14], and others [15]. The first connection between an equilibrium and nonequilibrium property was introduced by Einstein in his theory of Brownian motion when he connected the diffusion coefficient D to the friction coefficient f through the relation $D = k_B T/f$ where k_B is the Boltzmann constant and T the kelvin

temperature [16]. Experimentally EF and fims are related if the two
types of motions explore the same, or at least similar, substates.

2. BASICS OF PROTEIN STRUCTURE

2.1 Protein Structure and Flexibility

Proteins are constructed from 20 amino acids which consist of a short
covalently bonded backbone and a sidechain or residue [17]. Amino
acids form a linear primary structure which is a polypeptide chain.
The primary structure folds into a secondary structure (α-helix or
β-sheet) and ultimately into the final tertiary structure as in Fig.
2. Large EF and fims are possible because of this unique structure

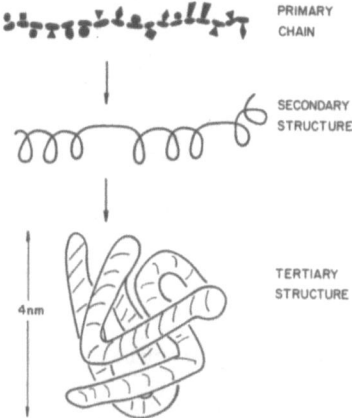

Figure 2. Schematic representation of the folding of a linear
polypeptide chain (primary sequence) into a alpha helix (secondary
structure) and ultimately into the functional tertiary structure.

since the covalently bonded polypeptide backbone is held in the folded
tertiary structure by relatively weak forces (hydrogen bonds) resulting in a great degree of flexibility. Protein flexibility can
arise because of rotations about bonds along the chain. Rotations of
several tens of degrees occur for energies of a few kJ/mol. The major
obstacle to rotations about bonds arises from the van der Waals repulsion owing to collision with the other atoms in the protein
structure. The aromatic rings of the amino acids phenylalanine,
histidine, tryptophan, and tyrosine are important in this regard.
Proline is especially important in determining the rigidity of the
polypeptide chain and is frequently found where a helix changes
direction. Hydrogen-bonding between residues and sulfur-sulfur bonds
between cysteine residues also constrain rotations.

2.2 Heme Proteins

Heme proteins are attractive to study for several reasons [18]. In
heme proteins one particular organic molecule, the heme group, which

has an iron atom (Fe) at the active site, has its behavior modified by the protein structure so that the system can perform a wide variety of tasks: Hemoglobin (Hb) transports oxygen in blood; Mb stores and transports oxygen in muscle; cytochrome c transports electrons; cytochrome P450 detoxifies substances in the liver, and peroxidases reduce hydrogen peroxide. Chlorophyll where the iron atom is replaced by magnesium is essential in the transformation of light to chemical energy.

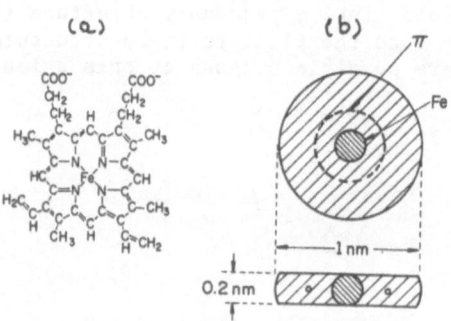

Figure 3. Two versions of the heme molecule. π indicates the pi-electron ring.

The heme group shown in Fig. 3 consists of an organic part and an Fe. The organic part protoporphyrin is made of four pyrrole groups linked by methene bridges to form a tetrapyrrole ring. Four methyl, two vinyl, and two propianate side chains are attached to the tetra-pyrrole ring. The side chains can be attached in 15 ways but only one protoporphyrin IX commonly occurs in biological systems. The Fe binds covalently to four nitrogens in the center of the protoporphyrin ring. The Fe can also form two additional bonds, one on each side of the heme plane. A schematic picture of the heme group is shown in Fig. 3b which eliminates much of the detail but retains most of the salient features. The disk is about 1 nm in diameter and 0.2 nm thick. A one-dimensional pi electron ring surrounds the Fe.

We often use Mb for our experiments and modelling [19]. Mb consists of 153 amino acids and has a molecular weight of about 17.8 kdalton. Its dimensions are about 3 nm × 4 nm × 4 nm. The x-ray structure of Mb and MbCO are known and additional x-ray data on Mb over wide ranges of temperature and pressure are becoming available [20-23]. Because Mb has a known structure, a relatively simple function, and is easily obtainable we refer to it as the "hydrogen atom of biology." However Mb still retains a significant degree of complexity which connects Mb to other complex systems including glasses and spin glasses [10,24,25].

2.3 Conformational Substates

There are several general arguments for the existence of conformational substates in proteins: (i) Proteins can have a quaside-generate ground state owing to the large fluctuations in internal energy, entropy, and volume. Large fluctuations imply many accessible states for the protein at a given temperature. (ii) The number of states Ω accessible to a system is given by the Boltzmann relation

$$\Omega = \exp(S/R) \tag{1}$$

where S is the entropy and R is the gas constant. Each amino acid in the polypeptide chain has several states of about equal energy. Suppose that $\Omega_i = 2$ for a given amino acid (molecular dynamics calculations suggest about 10) so that

$$\Omega = \prod_i \Omega_i \approx 2^{150} \approx 10^{50}. \tag{2}$$

Thus a protein the size of Mb has an enormous number of accessible states. (iii) A protein can fold into its tertiary structure in about a ms to a s. But this requires an enormous rate of sampling to find the ground state $--(10^{50}$ states/10^{-3} s$) \sim 10^{53}$ states/s. H. Frauenfelder and his collaborators realized that a solution to this problem is to postulate conformational substates so that a protein folds into any one of a very large number of substates with approximately equal energy [8]. Thus the concept of conformational substates which is crucial in interpreting the various experiments discussed in the following sections is rooted in basic theoretical arguments and structural features of proteins.

3. MODELS OF PROTEIN DYNAMICS

The binding and dissociation of a small ligand to a heme protein like Mb at first appeared to be a simple one step process [19]:

$$Mb + CO \longleftrightarrow MbCO. \tag{3}$$

Experimental and theoretical studies have, however, resulted in an increasingly complex picture of the binding processes. Here we present two simple models of protein dynamics which address different, but complementary, aspects of ligand binding -- the "single-particle" and "proteinquake" models. Before describing these two models we sketch the actual situation in Fig. 4 which shows the path of a CO molecule from the solvent S to the binding site A at the heme iron. The CO performs a Brownian motion in the solvent, moves into the protein matrix M, migrates through the protein matrix into the heme pocket B, and finally binds covalently to the Fe.

3.1 Single-Particle Model

In the "single-particle" model the protein creates an effective

112

potential in which the ligand moves. Our experimental studies of the ligand binding process show that the "single-particle" potential seen

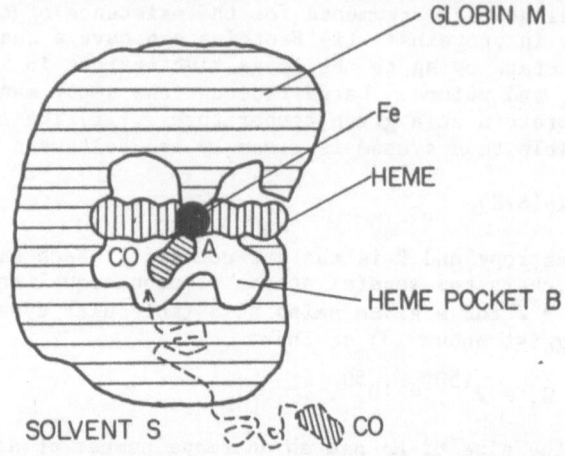

Figure 4. A schematic view of a cross section through the active site of myoglobin or other monomeric heme protein.

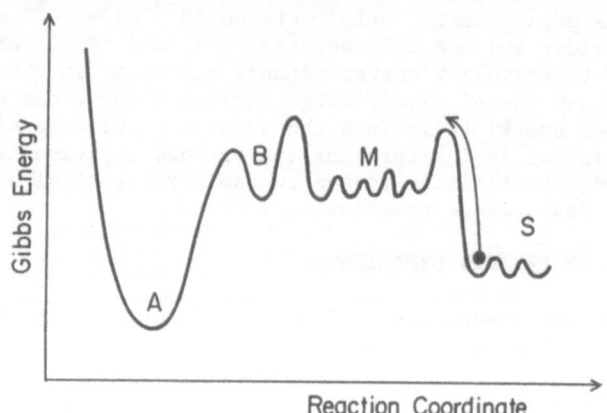

Figure 5. The "single-particle" potential for ligand binding. The bullet represents the ligand moving through a potential caused by the protein. The letters A, B, M, and S represent states defined in the text.

by the ligand appears as in Fig. 5 [4,8,26]. The general features of the potential are the same in all protein-ligand systems which we have studied, but the details can differ. The details are important, however, for biological function so their understanding on the basis of protein structure is an important goal of the work on heme proteins. The calculation of the motion of the ligand in the single-particle potential is not trivial since the ligand performs a complicated random walk and visits a given site many times before finally binding. The single-particle model is needed to understand ligand binding but does not result in a complete description since it is static and omits the motions of the protein.

3.2 Proteinquake Model

In the proteinquake model we use ligand binding as a perturbation that induces protein motions and study these motions alone [7,10]. The perturbation resulting from ligation or deligation leads to a process similar to an earthquake; A stress is relieved at the focus. The released strain energy is dissipated as waves and through propagation of deformations. There is a fortunate difference however -- in an earthquake the ratio of the released energy to the total gravitational binding energy is very small while in a proteinquake the released energy and the binding energy of the tertiary structure are roughly of the same order. The "proteinquake" model is illustrated in Fig. 6.

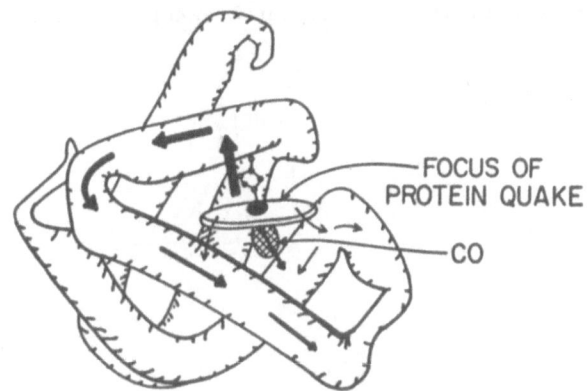

Figure 6. Proteinquake model. Binding or dissociation of a ligand at the heme iron causes a "proteinquake." See text for a complete explanation.

Upon binding of a ligand the protein is stressed; upon photodissociation, the stress is relieved and the protein finds itself in a state far from equilibrium. Return to equilibrium occurs through the proteinquake in which released strain energy is dissipated through waves and the propagation of a deformation.

   In the photodissociation of MbCO, for example, we have found evidence from a variety of experiments and also theory that the proteinquake occurs in a series of steps;

$$MbCO \xrightarrow{h\nu} Mb_4^* \xrightarrow{fim\ 4} Mb_3^* \xrightarrow{fim\ 3} Mb_2^* \xrightarrow{fim\ 2} Mb_1^* \xrightarrow{fim\ 1} deoxyMb. \qquad (4)$$

Mb_4^* to Mb_1^* are intermediate protein states involving different parts of the protein structure. The various fims depend on temperature.

114

Fim 4 occurs rapidly at 3K, fim 3 takes place near 20K, fim 2 starts above about 90K, and fim 1 begins near 200K although, as we show below, fim 1 is strongly coupled to the solvent and becomes appreciable only when the solvent is liquid. The other motions may also depend on solvent. The relevant experimental and theoretical information leading to the proteinquake model is summarized in Refs. 4 and 10. We have recently found evidence for a fifth fim 0 [7] discussed below.

4. EXPERIMENTAL STUDIES -- LIGAND BINDING AND PROCESSES IN THE HEME POCKET

4.1 Flash Photolysis

Our primary experimental technique is flash photolysis over wide ranges in time (ps to ks), temperature (2 to 330K), pressure (1 bar to about 2 kbar), pH, viscosity, and solvent state. Fig. 7 shows a schematic diagram of a typical experimental setup. In a flash

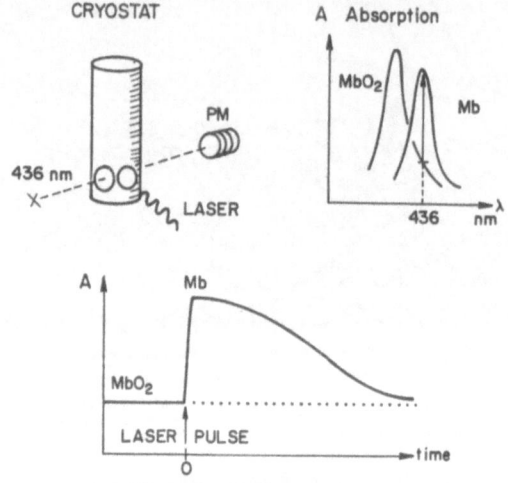

Figure 7. Typical flash photolysis setup. The protein sample (MbO_2) is placed in a cryostat and brought to the desired temperature and pressure. Abosrbance is monitored from the near UV to the mid-IR. The liganded protein sample is photodissociated with a laser pulse. In the mid-IR experiments temperature is controlled with a refrigerator.

photolysis experiment an MbCO sample at a controlled temperature is photodissociated with a pulse from a laser and rebinding is monitored through optical absorption measurements. At tempertaures below about 200K we have observed rebinding at many different wavelengths as indicated in Fig. 8 and so have monitored rebinding processes at different locations in the heme pocket [7]. In the Soret and α and β bands changes in the electronic structure of the heme, mainly the pi

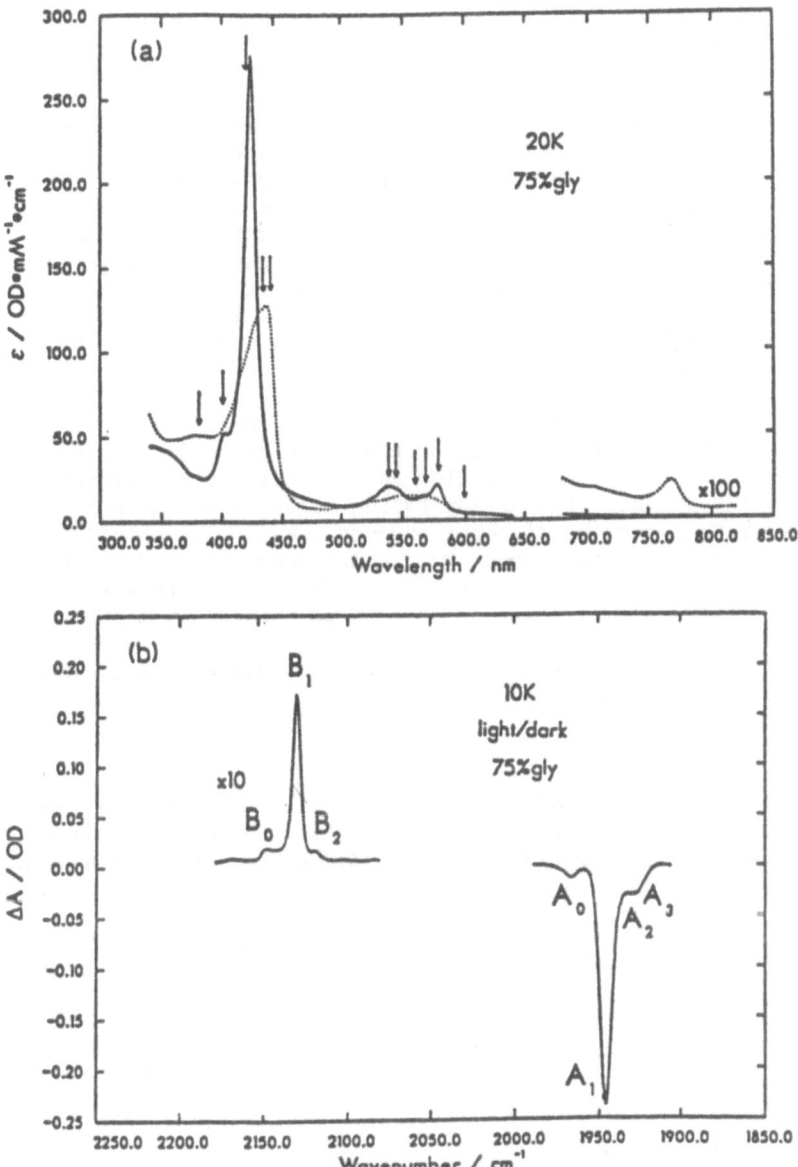

Figure 8. Spectral range where ligand rebinding is monitored (a) shows the absorption spectra of Mb (dotted line) and MbCO (solid line) in the Soret, the α and β, and the near-IR regions. Arrows indicate where we have measured rebinding in flash photolysis experiments. (b) gives the IR difference spectrum (Mb-MbCO).

electron ring, are observed. The near-IR band (band III, near 760 nm) is a charge transfer band and involves the Fe and heme [27]. The IR bands near 1950 cm^{-1} result from the stretching mode of the bound CO, and the ones near 2100 cm^{-1} result from the photodissociated CO within the heme pocket.

Four different systems are used for the flash photolysis experiments. (i) The nanosecond (ns) visible-wavelength flash photolysis system employs a 30 ns pulse from a frequency-doubled, Q-switched Nd+-glass laser (530 nm, 300 mJ). Rebinding is monitored with light from a tungsten lamp passed through a monochromator. The photomultiplier signal is digitized with a Lecroy transient digitizer from 10 ns to 300 s and a logarithmic time-base digitizer from 2 µs to 300 s. (ii) The Olis-Cary 14 spectrophotometer is interfaced to an IBM PC/AT microcomputer and monitors transient spectra from 60 to 3×10^4 s. Temperature for these two instruments is controlled by a storage cryostat. (iii) FTIR experiments are performed on a Mattson Sirius 100 FTIR spectrometer with 1 or 2 cm^{-1} resolution. Transient spectra from this instrument have a time resolution from 10 s to 5 ks. (iv) Fast kinetics in the IR region are measured on a microsecond mid-IR system in which the IR monitoring light is produced by a tuneable-diode laser and directed through the protein solution and monochromator onto a LN_2-cooled HgCdTe detector. This fast IR system allowed a time resolution from about 5 µs to 20 s. Photolysis in the kinetics experiments with the IR systems is induced by a 500 ns, 0.3 J, 590 nm pulse from a dye laser using a rhodamine 6G dye. Temperature during the IR experiments is controlled with a closed-cycle helium refrigerator.

4.2 Ligand Binding

Kinetic data for CO rebinding to the separated beta chain of adult human hemoglobin in 75% glycerol-water (v/v) at 260K is shown in Fig. 9 [28]. Rebinding is monitored in the Soret band at about 442 nm.

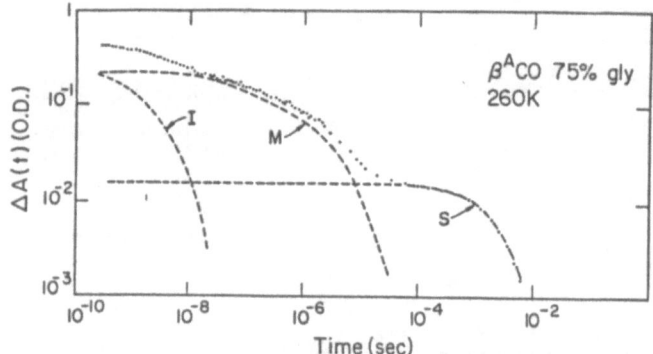

Figure 9. Rebinding of CO to the separated beta chain of human Hb (β^A) at 260K. The dots represent actual experimental data. I is the geminate rebinding from the heme pocket. M is geminate rebinding from the globin matrix, and S is the rebinding from the solvent.

Since the extinction coefficients of liganded and photodissociated protein differ significantly in the Soret region, the absorbance change yields the fraction N(t) of proteins that have not rebound a ligand at time t after photodissociation. N(t) exhibits three processes denoted by I, M, and S. These processes can be understood by referring to Figs. 4 and 5. The ligand coming from the solvent (well S) migrates through the globin matrix (well M) to the heme pocket (well B). From well B the ligand either moves back into the globin or binds covalently to the Fe (well A). If the bond between the ligand and the Fe is broken by a laser pulse the ligand moves to the heme pocket (B). Rebinding from B depends on temperature: Below about 180K the ligand remains in the heme pocket and rebinds directly to the Fe (process I). Above about 200K some ligands move into the globin matrix before returning to the pocket and rebinding (process M). Some ligands that move into the globin matrix will continue on to the solvent. Any ligand in the solvent can then enter the globin matrix and ultimately bind to the Fe (process S).

4.3 IR Spectrum of CO Stretching Bands in MbCO

The stretching bands of the bound CO in the IR are excellent probes of processes in the heme pocket [7]. We have measured the "dark" or bound-state IR spectrum of MbCO at temperatures between 15 and 300K in 75% glycerol-water, water, and solid polyvinylalcohol (PVA). A typical IR spectrum is shown in Fig. 10 for 75% glycerol-water. The IR spectra

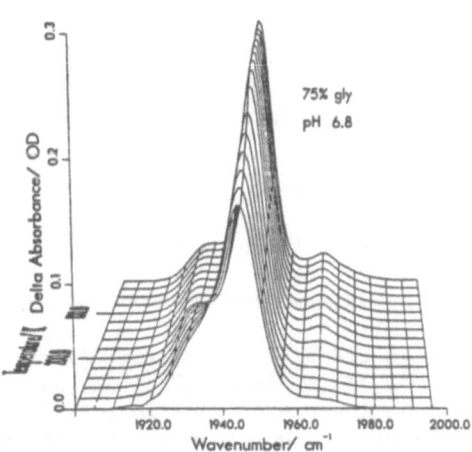

Figure 10. Temperature-dependent IR spectrum of the CO stretching bands in MbCO. Solvent: 75% glycerol-water; pH 6.8.

can be decomposed into Gaussian superpositions of Lorentzians, called Voigtians. Examples of such decompositions at 10K and 240K are presented in Fig. 11. Four different IR bands can be distinguished assuming that all bands are symmetrical [29]. We denote the bands in the

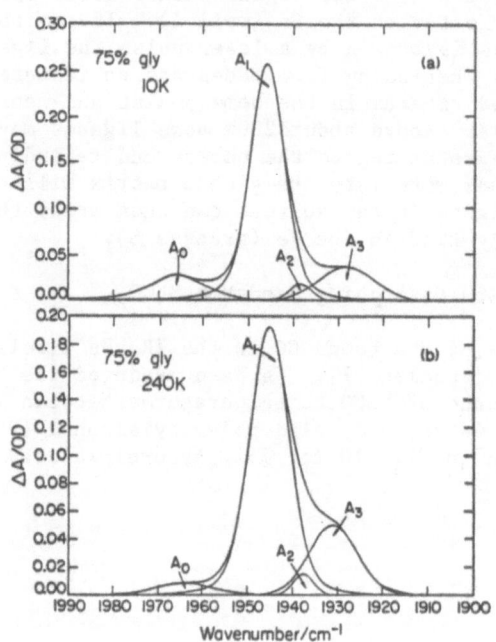

Figure 11. Voigtian decomposition of the IR spectrum of MbCO into four A substates at 10K and 240K. Solvent: 75% glycerol-water; pH 6.8.

bound-state spectrum in order of decreasing wavenumber by A_0 ($\sim$ 1966 cm^{-1}), A_1 ($\sim$ 1946 cm^{-1}), A_2 ($\sim$ 1941 cm^{-1}), and A_3 ($\sim$ 1930 cm^{-1}). The Voigtian fits result in the center position, the full width at half maximum, and the area of the four bands. Since we observe four bands in the bound-state IR spectrum we might expect four bands in the IR spectrum of the photodissociated CO at low temperature where the CO is in the heme pocket (well B). We however are able to observe only three bands as shown in Fig. 8. In fact B_2 at 2119 cm^{-1} is an intermediate that rapidly decays to B_1 at temperatures above 15 K. We observe only two B bands above 20K. This is not surprising since the

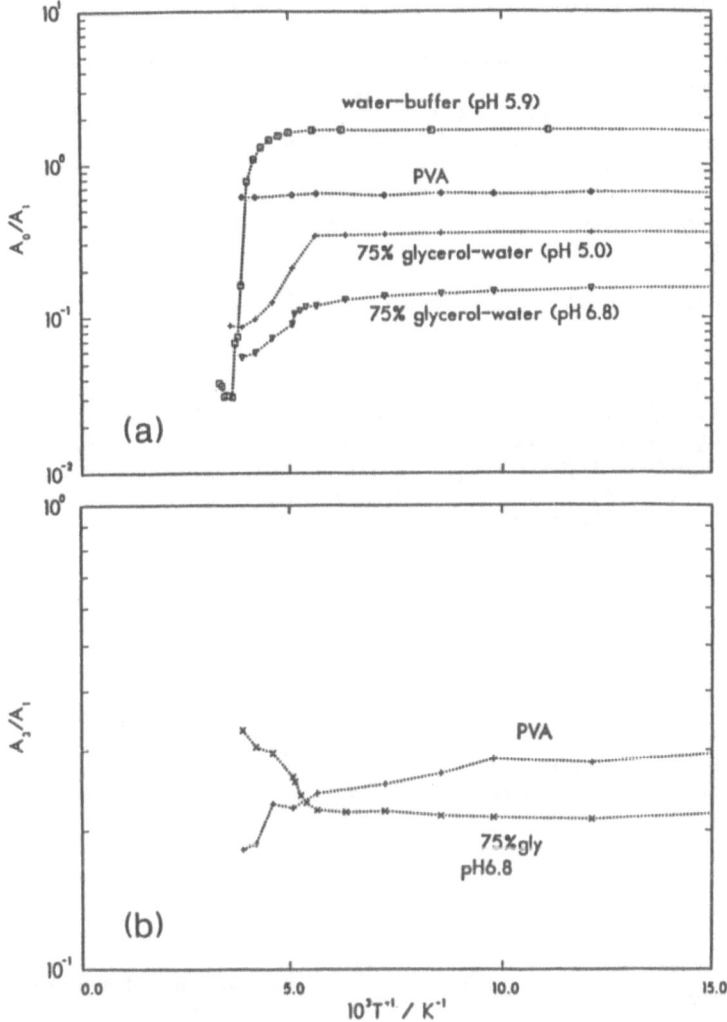

Figure 12. Plots of (a) A_0/A_1 and (b) A_3/A_1 versus 1000/T. A_i/A_1 is the ratio of the areas of the stretching bands in the bound states A_i and A_1.

integrated extinction coefficients of the B bands are small so that a
measurement sensitive enough to resolve all four B bands is difficult
[30].

The Voigtian decomposition of the A bands allows the intensity of
the individual A bands to be determined as a function of temperture.
Using the most intense band A_1 as the reference we plot in Fig. 12 the
logarithm of the ratios A_0/A_1 and A_3/A_1 as a function of 1000/T. One
feature stands out in Fig. 12: Above the temperature where the
solvent freezes (~ 180K for 75% glycerol-water) the ratios change
rapidly, below about 180K the change is much slower. We interpret this
behavior to show that the A states interconvert freely above about
180K, but that below about 180K the proteins are frozen into
particular substates.

5. LOW TEMPERATURE BINDING AND CONFORMATIONAL SUBSTATES

5.1 Low Temperature Binding -- Nonexponential Kinetics

Below about 180K in 75% glycerol-water we observe only process I in
the case of ligand rebinding to various heme proteins. We interpret
process I by assuming that the photodissociated ligand remains in the
heme pocket (B) and rebinds from there. Fig. 13 obtained by monitoring at 440 nm shows that process I is nonexponential below 180K for

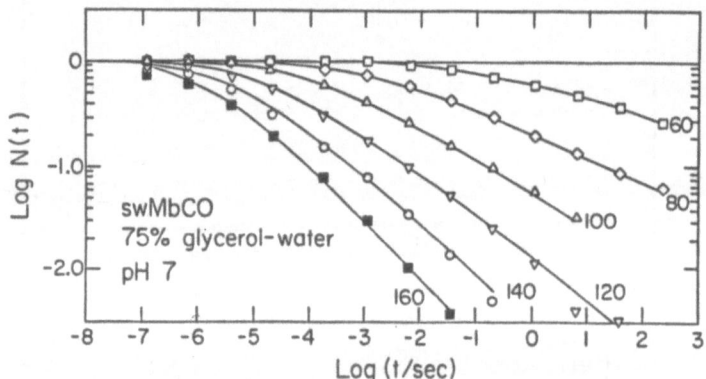

Figure 13. Rebinding Mb + CO → MbCO between 60 and 160K monitored in
the Soret at 440 nm. Solvent: 75% glycerol-water; pH 7.

the binding of CO to Mb [8,31,32]. Similar kinetics are observed for
both dioxygen and CO in a variety of monomeric heme proteins including
Mb [8], separated alpha and beta chains of normal adult Hb [28,33],
separated beta chains of mutant Hb Zurich [34], protoheme [35], and
soybean leghemoglobin [36].

We interpret the nonexponential kinetics by considering Fig. 5
which gives a schematic view of a cross section through the active
center of Mb or a separated chain of Hb. Before photolysis CO is
covalently bound to the Fe and the heme group is nearly planar. The

Fe has electronic spin zero. After photodissociation the ligand is in
the heme pocket, the heme group is domed, and the Fe is displaced by
about 0.05 nm from the mean heme plane toward the distal histidine.
The Fe then has electronic spin two [37]. If the B → A transition is
represented as a potential barrier with an activation enthalpy H_{BA},
the rate coefficient above about 40K is given by an Arrhenius relation

$$k_{BA} = A_{BA}(T/T') \exp(-H_{BA}/RT), \qquad (5)$$

where T is the kelvin temperature, R is the gas constant, and T' is an
arbitrary temperature which we usually assume to be 100K. Then A_{BA} is
the preexponential at 100K. Below 40K molecular tunneling can occur
and modify Eq. (5). We do not treat tunneling here [38].

If the activation enthalpy barrier is unique, then the fraction
N(t) of proteins unbound is a single exponential $N(t) = \exp(-k_{BA}t)$
which is not observed. The simplest explanation of the nonexponential
time dependence postulates that at low temperatures different Mb
molecules have different barriers. If $g(H_{BA})dH_{BA}$ is the probability
that a protein has activation enthalpy between H_{BA} and $H_{BA} + dH_{BA}$,
then N(t) can be expressed

$$N(t) = \int dH_{BA} g(H_{BA}) \exp(-k_{BA}t). \qquad (6)$$

If data are available over a wide range of time and tempertaure the
preexponential and distribution $g(H_{BA})$ can be determined. Fig. 14
shows several representative examples of such activation enthalpy
distributions. The distributions are temperature independent to a
good approximation.

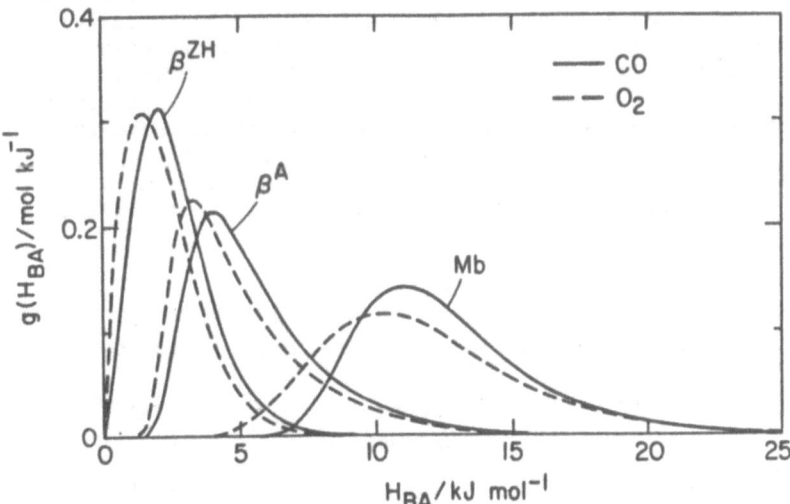

Figure 14. Examples of activation enthalpy distributions for CO and
O_2 rebinding to various monomeric heme proteins in 75% glycerol–water.
Proteins include: sperm whale Mb, separated beta chains of normal
adult Hb (β^A), and separated beta chains of mutant Hb Zurich (β^{ZH}).

5.2 Nonexponential Kinetics and Conformational Substates

The activation enthalpy distributions in Fig. 14 can arise in two
distinct ways: (i) All proteins are identical and have the full
spectrum of activation enthalpies. This situation can occur if each
protein possesses a number of distinct sites with different barriers
for the B → A transition. The sample is then an ensemble of identical
proteins and the activation enthalpy spectrum is homogeneously
broadened. (ii) Each protein is different since it exists in a
different CS with a different barrier height for the B → A transition.
The sample is then an ensemble of different proteins and the activation enthalpy spectrum is inhomogeneously broadened. The two cases
are distinguished by optical pumping experiments using either a
multiple flash or continuous illumination. The results of such
experiments first done in 1975 and repeated several times since then
show that the activation enthalpy spectrum is due mainly to inhomogeneous broadening and that at low temperature proteins are different
since each protein is in a different CS [7,8]. We use the notation
CS^1 for these substates which are responsible for the nonexponential
kinetics at low temperature.

Below about 180K in 75% glycerol-water, the individual A bands for
MbCO do not interconvert so that the individual A bands correspond to
four distinct substates. We have also performed optical pumping
experiments at low temperature showing that the B substates do not
interconvert and that each B substate rebinds to only one A substate
[7]. In particular $B_1 → A_1$ and $B_0 → A_3$. Rebinding can also be
observed by monitoring the formation of the A bands after photodissociation. Fig. 15 gives the rebinding kinetics for the individual A
bands from about 3 μs to 5 ks. The kinetics for the individual A
bands are also nonexponential over the entire timecourse and each of
the A bands has a different preexponential and activation enthalpy
distribution. Thus each A substate has its own distribution of CS^1 so
we designate the set of four substates corresponding to the A bands by
CS^0. Table 1 gives the preexponential and peak of the activation
enthalpy distribution and shows that a subtle combination of enthalpic
and entropic effects determine the binding rates of the individual A
substates.

Table 1

Barrier parameters for CO binding to A substates.

Substate	H_{peak} kJ/mol	$\log A_{BA}(100K)$ s^{-1}
A_0	10	10.8
A_1	9.5	9.3
A_3	18	9.8
Soret (440nm)	10.1	9.0

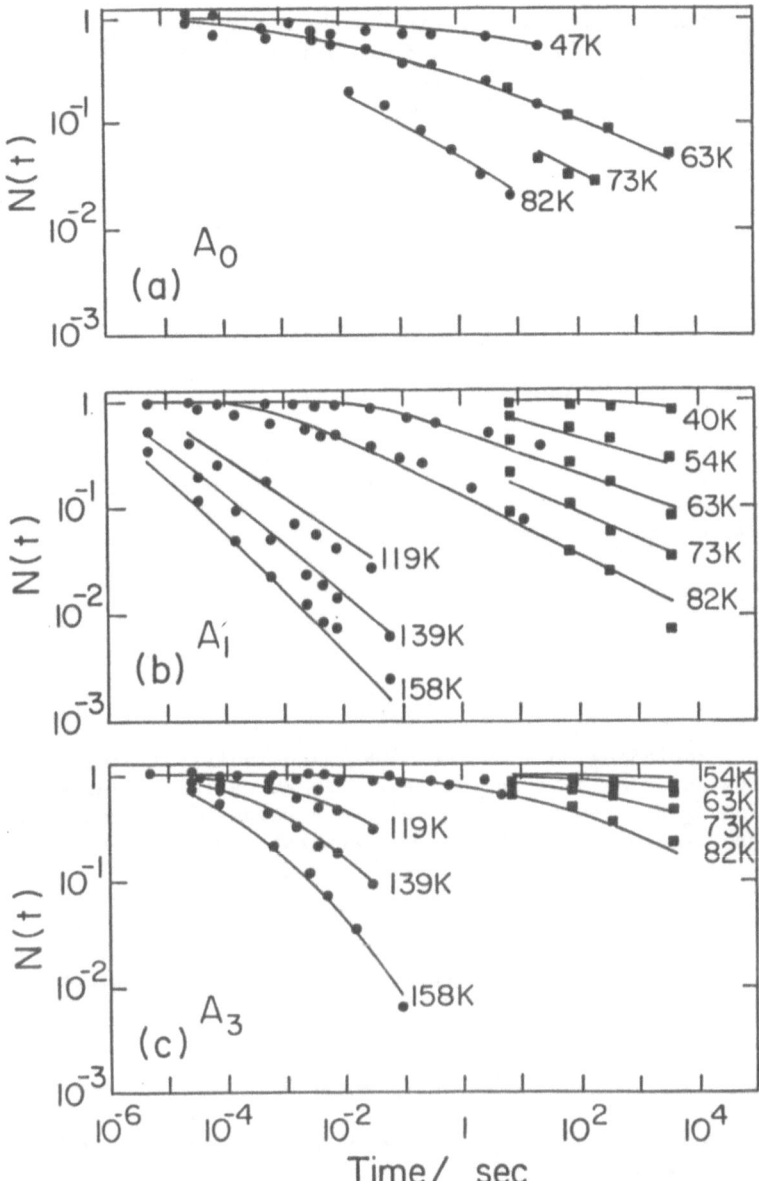

Figure 15. Rebinding kinetics to the A substates for various temperatures in 75% glycerol–water with pH 6.8. The solid lines are fits to the data using Eqs. (5), (6), and (11). ● : mid-IR flash photolysis data. ■ : FTIR data. (a) A_0, (b) A_1, and (c) A_3.

We have also monitored rebinding of CO to Mb at 40K in the Soret, the near-IR (band III, 760 nm), and the integrated A and B bands in the IR. Fig. 16 shows that all data lie on the same rebinding curve

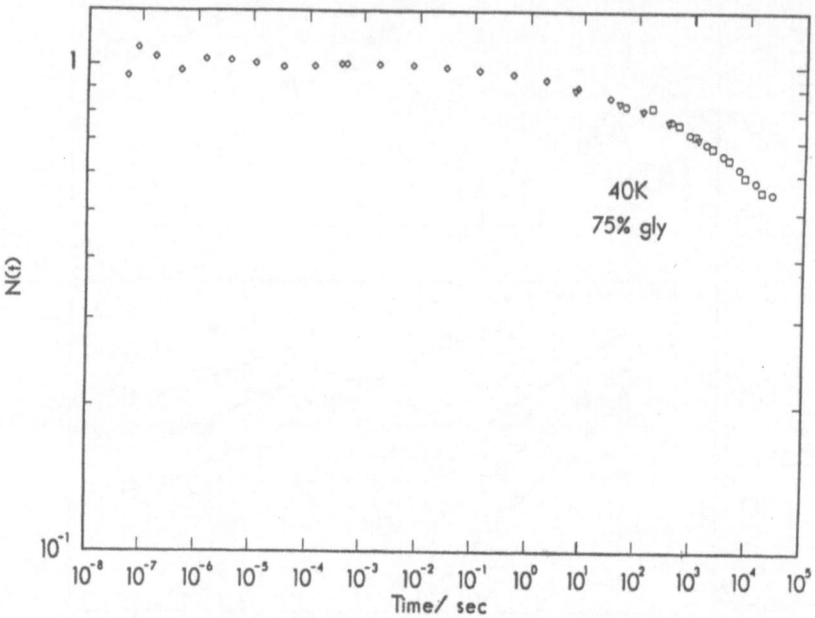

Figure 16. Rebinding Mb + CO → at 40K. The individual rebinding curves have not been matched but have been normalized independently. ◇ : 440 nm; ▽ : integrated areas of the A bands; ○ : integrated area of the Soret region, 400–450 nm; □ : integrated area of the 760 nm band.

and demonstrates that rebinding is nonexponential in time from at least 100 ns to 100 ks. Thus the concept of conformational substates is an intrinsic property of proteins over this entire timecourse (12 orders of magnitude in time) and the Soret provides information on the average rebinding.

6. MODEL OF THE ACTIVATION ENTHALPY SPECTRUM

The nonexponential rebinding kinetics discussed above establish that proteins have a <u>large number of</u> CS^1 with about equal energy. Above a characteristic temperature T_f (the freezing temperature or, perhaps, glass temperature) proteins fluctuate among the CS^1, and the distribution of protein conformational energy is given by a Boltzmann distribution. However below T_f fluctuations among the CS^1 are frozen out and the ensemble of proteins is in a nonequilibrium distribution of conformational energies which we take to be temperature-independent for reasonable experimental timescales. Thus the ensemble of proteins exhibits broken ergodicity [39] below T_f in a manner reminiscent of glasses and spin glasses. Since the conformational energy space of the protein exhibits a many-valley structure separated by large barriers there are a large number of nearly degenerate conformational substates. We take the density of conformational substates CS^1 to be a temperture-independent power law. Therefore

$$\rho(E_i) = \text{const. } (E_i - E_{i0})^\mu \qquad (7)$$

where E_i is the protein conformational energy; E_{i0} is the minimum conformational energy; and the subscript i indicates a possible dependence on the protein-ligand state. The power μ is determined from the low-temperature ligand binding kinetics. Young and Bowne first postulated a power law density of states for the protein conformational energy [40]. Gratton and collaborators have also used a power law density of states to analyze fluorescence lifetime experiments with proteins [41].

Using the density of substates in Eq. (7), the probability that a protein has conformational energy between E_i and $E_i + dE_i$ is given by

$$p(E_i, T_0) = \text{const. } (E_i - E_{i0})^\mu \exp(-E_i/RT_0) \qquad (8)$$

where $T_0 = T_f$ if $T < T_f$ and $T_0 = T$ if $T > T_f$. The constant in Eq. (8) is determined by normalization. The probability density $p(E_i, T_0)$ is temperature dependent above T_f but is temperature independent below T_f reflecting broken ergodicity owing to the existence of a large number of quasidegenerate conformational substates.

The activation enthalpy for the B → A transition can be simply related to the conformational energy of the protein using the model of Young and Bowne [40]. We do not give the details here but summarize the main points. The conformational energy E_i is written as

$$E_i = E_{i0} + \varepsilon_i \qquad (9)$$

where i = A, B, T representing the liganded, photodissociated (deoxy), and transition states, respectively, for the covalent bond formation step B → A. The substate-dependent term can be taken to be the same for the liganded and photodissociated states (state A and state B) of the protein, $\varepsilon_A = \varepsilon_B = \varepsilon > 0$, although this is not essential as pointed out in Ref. 40. The activation enthalpy H_{BA} is then given by a Bronsted relation

$$H_{BA} = H_{min} + \varepsilon/\alpha_f RT_f \qquad (10)$$

using the parameters, H_{min} and α_f, from Young and Bowne. Eq. (10) is obtained by taking the difference in enthalpy between the transition state and the heme pocket. Eqs. (8), (9), and (10) result in the following temperature-independent gamma distribution of the activation enthalpy spectrum for $T < T_f$;

$$g(H_{BA}) = \text{const. } (H_{BA} - H_{min})^\mu \exp[-\alpha_f(H_{BA} - H_{min})] \qquad (11)$$

where the peak activation enthalpy is given by $H_{min} + \mu/\alpha_f$. The fits to the rebinding kinetics in Figs. 13 and 15 are obtained using the analytic $g(H_{BA})$ in Eq. (11) [40,28]. Furthermore, the temperature-dependent distribution of activation enthalpy predicted by the model of Young and Bowne for temperatures above T_f has been shown to be in excellent agreement with experiment in the case of CO binding to separated beta chains of normal adult Hb [28].

7. HIGH TEMPERATURE BINDING AND THE A SUBSTATES

7.1 Process S -- the Solvent Process

A complete explanation of binding and relaxation processes at inter-mediate temperature is very complicated and awaits development of a detailed theory. The theory must include the protein undergoing EF and fims while interacting with the ligand. However at 300K the description of ligand binding is simpler and in fact largely inde-pendent of the details of a particular theory owing to the rapid EF and fims in the protein so that the ligand probes averaged barriers [8]. Thus the single-particle model of Fig. 5 is appropriate. The concentration-dependent binding from the solvent (process S) is exponential in time if the concentration of the ligand is much larger than that of the protein and the fraction $N(t)$ of unbound protein can be written as

$$N(t) = N_s \exp(-\lambda_s t) + \ldots \quad (12)$$

where the dots stand for any contributions to the rebinding kinetics from processes I and M. Also λ_s is a pseudo-first order rate coeffi-cient that is proportional to the ligand concentration and N_s is the fraction of photodissociated proteins whose ligands have migrated to the solvent. The association rate coefficient λ_s is given by

$$\lambda_s(c,T) = \bar{k}_{BA} P_B(c,T) N_s \quad (13)$$

where $P_B(c,T)$ is an equilibrium coefficient between the heme pocket and the solvent at ligand concentration c [26]. The term $\bar{k}_{BA}$ is the average rate coefficient for the B → A transition. Eq. (13) holds under very general conditions which include any number of wells in the globin matrix with arbitrary sequential and/or parallel pathways [42]. See Fig. 17. Analysis of the ligand binding data for many monomeric heme proteins establishes that the main regulation of the physiologi-cally significant association rate occurs at the heme.

7.2 The A Substates at High Temperatures

We have seen in Fig. 12 that the A states begin to interconvert above about 180K in 75% glycerol-water and interconvert freely above 200K. Above 200K the behavior of the ratios A_i/A_1 can be understood if the substates A_0 to A_3 have different binding enthalpies and entropies and interconvert freely. The ratios A_i/A_1 then follow a van't Hoff relation

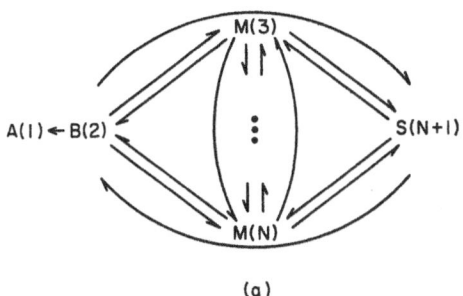

(a)

$$A(1) \leftarrow B(2) \rightleftharpoons M(3) \rightleftharpoons \cdots \rightleftharpoons M(N) \rightleftharpoons S(N+1)$$

(b)

Figure 17. (a) The generalized sequential barrier model involving N+1 wells. State A (well 1) is at the heme iron; state B (well 2) is the heme pocket; state M (wells 3 to N) is the globin matrix; and state S (well N+1) is the solvent. Covalent bond formation occurs from B to A. Parallel and/or sequential paths are allowed between any other two states including the solvent as shown. (b) A sequential barrier model with nearest neighbor transitions only.

$$A_i/A_1 = \exp[S_i/R] \, \exp[-H_i/RT] \, \varepsilon_i(T)/\varepsilon_1(T) \tag{14}$$

where S_i and H_i are the entropy and enthalpy of substates with respect to A_1 and $\varepsilon_i(T)$ is the extinction coefficient of substate A_i at temperature T. Assuming that the ratio of the extinction coefficients is unity, the relative binding enthalpies and entropies have been extracted from the data in Fig. 12. The binding parameters for two pH values are displayed in Table 2. The binding parameters for the A substates above 200K show that A_0 is the most deeply bound substate and also has the smallest number of accessible substates or volume. The substate A_3 is bound less tightly than A_1 by about 2 kJ/mol.

Table 2

Relative binding parameters of the A substates.
R is the gas constant. Solvent: 75% glycerol/water.

Substate	H_A(kJ/mol)		S_A/R	
	pH 6.8	5.	6.8	5.
A_0	−4	−8	−5	−7
A_1	0	0	0	0
A_3	2	2	0	0

128

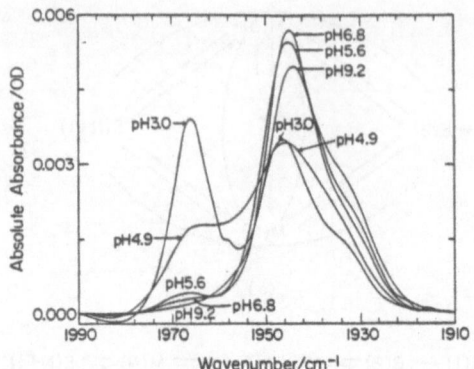

Figure 18. CO stretching bands of MbCO in 75% glycerol—water at various pH values.

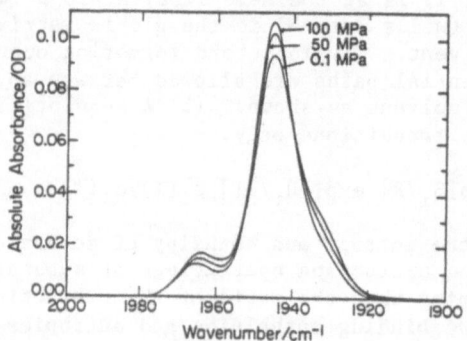

Figure 19. CO stretching bands of MbCO at various pressures in 75% glycerol—water with 0.25 M Tris-HCl buffer. (1 MPa = 9.9 bar.)

The characteristics of the CO stretching bands in the heme pocket depend strongly on the environment of the protein including solvent state (Fig. 12), pH (Fig. 18), and pressure (Fig. 19). One of the most biologically significant features of the data on the CO stretching bands is this sensitivity of the occupation of the substates in the heme pocket to external parameters since this suggests mechanisms for control of biological processes. It was found earlier that the association rate of CO binding to Mb increased with decreasing pH [26]. The IR data on the CO stretching bands provides more insight into this pH dependence: A change in pH can affect the kinetics of an individual binding path-way and also shift the system from one pathway to another. Such changes also occur on changes in pressure and solvent state. The pressure dependence suggests that the A_0 substate has the smallest volume and therefore is favored by higher pressure. This behavior is consistent with the binding entropy of the A_0 substate in Table 2. Thus IR observation of ligand binding to heme proteins is a tool well suited for the investigation of the effect of external parameters on the control of protein reactions.

8. EXPERIMENTAL STUDY OF PROTEIN RELAXATIONS

Two experiments probe the properties of fim 1 among the substates CS^1 and fim 0 among the CS^0. The CS^1 are responsible for the nonexponential binding of CO to Mb, and the CS^0 correspond to the four A substates, A_0-A_3.

(i) In a "pressure titration" experiment (Fig. 20) the sample is first cooled to state β at 120K and atmospheric pressure and the rebinding kinetics $N(t)$ in Eq. (6) are measured [43,44]. In the next

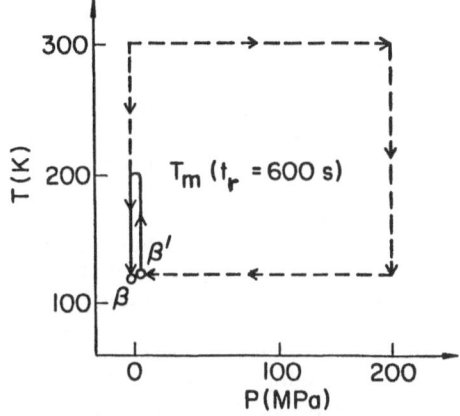

Figure 20. Nonequilibrium states of MbCO prepared by taking the two pathways shown. Kinetics are first measured at state β. The system is kept at temperature T_m for the relaxation time t_r before cooling again to 120K.

step the sample is raised to a high pressure (~ 2 kbar) at high temperature and cooled at 120K. The pressure is then released and $N(t)$ measured at state β'. The rebinding kinetics are significantly faster in state β' than in state β indicating that the two states of the sample are not in thermal equilibrium even though they are at the same temperature and pressure. Starting at state β' the sample is raised to temperature T_m, left there for 600 s (the relaxation time t_r), cooled again to 120K and $N(t)$ measured. This procedure is repeated with increasing T_m until $N(t)$ has relaxed to follow the curve corresponding to state β. The result is definitive: Up to T_m = 160K, $N(t)$ for state β' is unrelaxed, but at T_m = 200K $N(t)$ has completely relaxed to the kinetics corresponding to state β. The data are summarized in Fig. 21. Thus at 1 bar and for the relaxation time t_r = 600 s protein relaxation sets in at about 160K and extends over a range of about 40K. It should be emaphsized that the temperature region, 160K to 200K, is dependent on the relaxation time t_r. However the general features are clear -- the protein motions among the CS^1

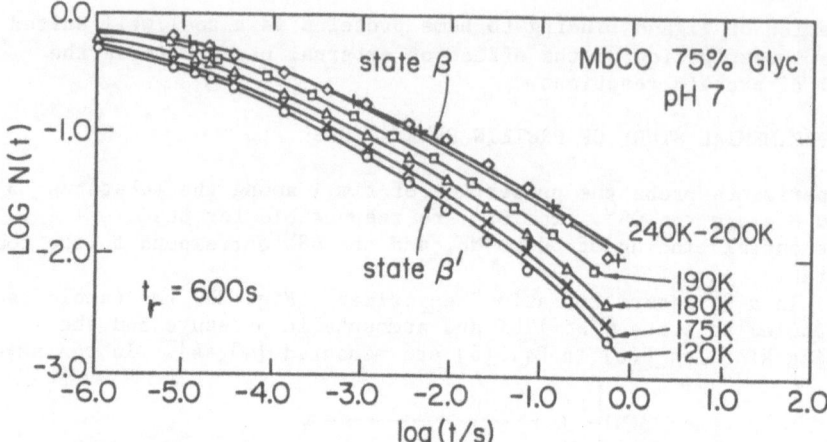

Figure 21. Pressure titration showing the onset of protein relaxation corresponding to the transition from state β' to state β at 160K for a relaxation time t_r = 600 s. Protein relaxation is over at 200K for this t_r.

and possibly the CS^0 necessary to restore thermal equilibrium are guaranteed by the pressure titration experiments.

(ii) A second experiment ("temperature-jump") involves the nonequilibrium behavior of the CO stretching bands [7]. Below about 190K the ratios A_0/A_1 and A_3/A_1 deviate from the van't Hoff line and become essentially temperture independent. This implies that the A substates are not in thermal equilibrium below about 180K in 75% glycerol-water. This conclusion is corroborated by measurements of the time dependence of the IR spectrum after the temperature is lowered from say 195K to the measurement temperature T. We denote the integrated absorbance of band A_0 at time t after the temperature is lowered by $A_0(t)$. $A_0(0)$ is the integrated absorbance at 195K, and $A_0(\infty)$ the equilibrium value at temperature T. We estimate $A_0(\infty)$ by extrapolating the van't Hoff line, obtained from 195 to 200K, to the temperature T. The fraction $n_0(T)$ of MbCO molecules not yet transferred to substate A_0 at time t after the temperature jump is given by

$$n_0(t) = [A_0(t)-A_0(\infty)]/[A_0(0)-A_0(\infty)]. \qquad (15)$$

$n_0(t)$ at 190K is shown in Fig. 22. The data indicate that the approach to equilibrium is nonexponential in time indicating a wide range of relaxation times. We have also observed the time to reach equilibrium at 180K and found it to be extremely long. The A_i substates are not in equilibrium below 180K.

The results of the pressure titration and temperature jump experiments are also consistent with the results of other experimental techniques including microwave absorption in the hydration shell [45] and Raleigh scattering of Mössbauer radiation [46].

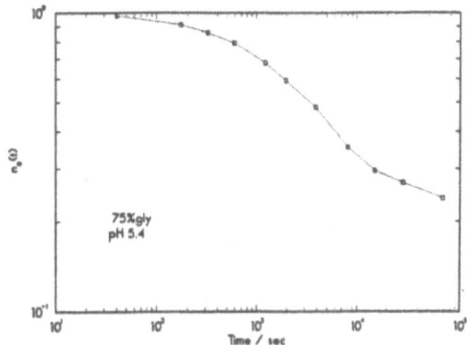

Figure 22. Approach to equilibrium of substate A_0 after a temperature "jump" from 195K to 190K. $n_0(t)$ is the fraction of MbCO molecules that have not reached equilibrium in substate A at the time t after the temperature jump.

9. PROTEINS AND GLASSES

A remarkable analogy between proteins and glasses emerges from both experimental [7,8] and theoretical studies [24]. We sketch some of the most significant similarities in this section. We have presented many of the main arguments in this review, but the complete picture is presented in the references [7,8,47].

(i) Disorder or Nonperiodicity. Amorphous solids, glasses, and spin glasses are disordered [48,49]. Proteins are also disordered in the sense that they are not spatially periodic and exhibit no reasonably obvious symmetries. The primary structure of amino acids is certainly not periodic and folding into the tertiary structure only increases the disorder.

(ii) Frustration. Spin glasses and possibly amorphous solids and glasses are frustrated systems in the sense of G. Toulouse [50]. A simple example illustrates the concept. A system of just three particles with spins are shown schematically in Fig. 23. Assume that the interaction between the atoms results in antiparallel spins having the lower energy. The system is then frustrated in the sense that the system has two states of equal energy separated by a barrier. In proteins frustration can exist owing to steric interactions. A particular amino acid residue has multiple interactions with nearby atoms and may have two or more favored states resulting from these interactions. Proteins consequently can be frustrated [24].

(iii) Energy Valleys and Conformational Substates. If a system is both disordered and frustrated as in (i) and (ii), the system no longer has a unique ground state, but possesses a quasidegenerate ground state. That is the conformational energy surface for the system possesses many energy valleys that are separated by high

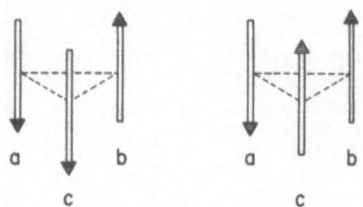

Figure 23. Example of frustration involves a system of three spins for which the interaction favors the antiparallel orientation of spins. This system has two states of equal energy separated by a barrier.

barriers [51,52]. In the case of proteins the energy valleys correspond to CS. In this regard see Fig. 24 below [7,10].

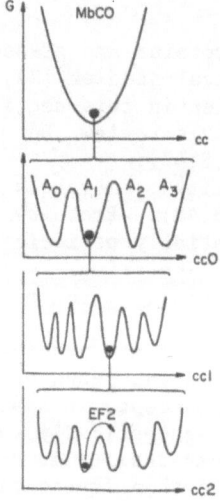

Figure 24. The hierarchical model of conformational substates in Mb illustrated by a schematic diagram of the Gibbs energy surfaces. The cci represent conformational coordinates. We show only the first three tiers of substates [7].

(iv) Replica and Substate Symmetry Breaking. In proteins different CS have different properties even though the primary structure is the same. Parisi has coined the term replica for one of the many different pure states of a spin glass with a given energy and replica–symmetry breaking to denote the fact that the replicas can have different properties [53]. In analogy we refer to the different properties of the quasidegenerate CS as "substate–symmetry breaking"[7].

(v) Temperature and Timescales. Proteins undergo fluctuations among the CS with an average relaxation rate $k_r = 1/t_r$ where t_r is the relaxation time. The response of the protein to an experimental probe depends crucially on the relative magnitudes of the relaxation time t_r and the characteristic experimental time t_{obs}. If $t_{obs} \gg t_r$ then the protein fluctuates among all the CS during the experiment and is ergodic. The structure of the CS is not significant and the multivalley conformational energy surface can be replaced by a single smooth well. If $t_{obs} \ll t_r$ each protein is frozen into a single CS and the system is nonergodic ("broken ergodicity") so the experiment samples an ensemble of different proteins. If the relaxation time depends on temperature through say an Arrhenius relation the properties of the system can exhibit dramatic temperature effects with various motions being frozen out as temperature is decreased. This situation applies also to glasses and spin glasses [51,54].

(vi) Hierarchical Structure. The experimental evidence for the proteinquake in Mb argues convincingly for the existence of a hierarchical structure in the conformational energy space of the protein. Thus we have postulated a five-tiered hierarchy of conformational substates in proteins as shown schematically in Fig. 24 which shows the conformational Gibbs energy as a function of the protein coordinate [7,10]. Experimentally the evidence for a hierarchical structure may be stronger for proteins than for amorphous solids and glasses. A wide body of theoretical knowledge however has been developing for hierarchical structures [51,53,54], and proteins may serve as laboratories to test these theories.

The concept of broken ergodicity has an interesting interpretation in terms of the hierarchical structure. In each of the tiers there are different characteristic barriers for relaxation so that the spectrum relaxation times can be very different for each of the tiers. Thus one tier can be ergodic while a higher tier may still be nonergodic with fluctuations frozen out.

(vii) Nonexponential Time Courses. The hierarchical structure in (vi) can result in nonexponential time courses in two ways — on the one hand the nonexponential rebinding kinetics for the binding of small ligands to heme proteins at low temperature results from a frozen in distribution of proteins among the CS^1 with different rebinding rates. The time dependence of the rebinding kinetics follows a power law in time with perhaps logarithmic corrections as follow naturally from the gamma distribution of activation enthalpies in Eq. (11), that is,

$$N(t) \sim (t/t_0)^{-n} [\log (t/t_0)]^{\mu}, \quad t \text{ large}, \tag{16}$$

where the $n = \alpha_f RT$. On the other hand the protein relaxation fim 0 which is illustrated in Fig. 22 also is nonexponential.

The limited data available on fim 0 in Fig. 22 can be fit by either a power law or a stretched exponential

$$n_0(t) \sim \exp[-(t/t_0)^\beta], \quad 0 < \beta < 1 \tag{17}$$

as is frequently observed in glassy systems [55].

(viii) Slaved Glass Transition [7]. As we have discussed in previous sections, Mb in 75% glycerol-water exhibits a remarkable transition between about 180K and 200K. The transition occurs in an interval of about 40K in the pressure titration experiments and about 20K in the temperature jump experiments on the A substates. Also the approach to equilibrium in the temperature jump experiments is nonexponential in time. These characteristics are similar to those observed in glass transitions which occur in a finite temperature range and which are highly nonexponential in time.

There is one property of the transition in proteins that differs significantly from that in glasses: The transition in proteins depends crucially on the solvent surrounding the protein. In 75% glycerol-water the transition in Mb is at approximately 180K where the solvent also undergoes a glass transition. In water solvent (Fig. 12) the transition is nearer to 260K which is approximately the freezing point of the buffered solvent. In solid polyvinylalcohol (PVA) no transition in MbCO is apparent up to at least 300K. The EF0, equilibrium fluctuations of the CS^0, are suppressed when the protein surrounding is frozen or solid. Thus the fluctuations among the CS^0 are coupled to the motions of the solvent and we call the transition a "slaved glass transition."

One point requires emphasis -- the slaved glass transition in proteins in not a property of the protein alone; protein and solvent together must be considered as one system. The hydration shell of the protein most likely plays an important role both in the glass transition and in the function of the protein [56].

10. SUMMARY AND OUTLOOK

Time domain studies of the binding of small ligands to simple heme proteins over about the last ten years have resulted in a picture of protein dynamics and function involving ever increasing complexity [1-5,7,10]. Models of the binding process have evolved from a one-step process as described for example by Antonini and Brunori [19] to the single particle model in which the ligand moves through a sequence of static barriers produced by the protein with control by the final barrier at the heme iron [4,8,26,44] and the proteinquake model in which the protein rearranges its structure after the Fe-CO bond is broken [4,7,10].

Studies over a narrow time range or a narrow temperature range do not allow the many different processes which occur in proteins to be disentangled or fully characterized. The existence of nonexponential, distributed behavior in time for functional processes such as ligand binding [8] and for protein relaxations [10] is a crucial experimental discovery in protein dynamics. These phenomena have in turn led to the concept of conformational substates [8] which probably have a hierarchical structure [10]. Development of these more realistic concepts and models of protein dynamics and function has come only by combining studies over wide ranges of time and temperature as well as other relevant parameters.

The hierarchical model of protein substates and motions suggests connections to amorphous solids and glasses [24,25]. Much further experimental and theoretical study remains to be done if these connections are to lead to deeper understanding of protein structure and function. The hierarchical model complements other theoretical approaches as for example molecular dynamics. Recent molecular dynamics simulations in the 300 ps time range have in fact shown evidence for conformational substates [57]. Together these theoretical models can be expected to provide an even more realistic picture of protein dynamics.

Further progress in understanding protein dynamics and function will require extension of our time domain studies in several directions: We will explore the effects of pressure more completely over wide ranges in time and temperature. The infrared wavelength region needs to be studied much more fully. We also intend to apply our experimental methods and models to modified myoglobins and more complex proteins including bacteriorhodopsin, cytochrome P450, and tetrameric hemoglobin.

ACKNOWLEDGEMENTS

I thank Professor Hans Frauenfelder for many discussions and fruitful collaboration. I thank all of the members of the biomolecular physics group at the University of Illinois at Urbana-Champaign for providing a stimulating atmosphere in which to work and discuss. It is also they who performed many of the experiments summarized here. This work was supported in part by the US National Science Foundation Grant DMB82-09619, National Institutes of Health Grant GM18051, and Office of Naval Research Grant N00014-86-K-00270.

REFERENCES

1. G. Careri P. Fasella and E. Gratton, Ann. Rev. Biophys. Bioeng. **8**, 69 (1979).
2. P. G. Debrunner and H. Frauenfelder, Ann. Rev. Phys. Chem. **33**, 283 (1982).
3. H. Frauenfelder, in Structure and Motion: Membranes, Nucleic Acids & Proteins, eds., E. Clementi, G. Corongiu, M. H. Sarma, and R. H. Sarma (Adenine Press, Guilderland, N.Y., 1985) pp. 205-217.
4. Hans Frauenfelder and Robert D. Young, Comments on Molecular & Celluar Biophysics **3**, 347 (1986).
5. Structure, Dynamics, and Function of Biomolecules, eds. A. Ehrenberg, R. Rigler, A. Graslund, and L. Nilsson (Springer, New York, 1987).

6. J. L. Martin, A. Migus, C. Poyart, Y. Lecarpentier, R. Astier and A. Antonetti, Proc Natl. Acad. Sci. USA **80**, 173 (1983).
7. A. Ansari, J. Berendzen, D. Braunstein, B. R. Cowen, H. Frauenfelder, M. K. Hong, I. E. T. Iben, J. B. Johnson, P. Ormos, T. B. Sauke, R. Scholl, A. Schulte, P. J. Steinbach, J. Vittitow, R. D. Young, Biophysical Chemistry (1987), in press.
8. R. H. Austin, K. W. Beeson, L. Eisenstein, H. Frauenfelder, and I. C. Gunsalus, Biochemistry **14**, 5355 (1975).
9. D. Beece, L. Eisenstein, H. Frauenfelder, D. Good, M. C. Marden, L. Reinisch, A. H. Reynolds, L. B. Sorensen, and K. T. Yue, Biochemistry **19**, 5147 (1980).
10. A. Ansari, J. Berendzen, S. F. Bowne, H. Frauenfelder, I. E. T. Iben, T. B. Sauke, E. Shyamsunder, and R. D. Young, Proc. Natl. Acad. Sci. USA **82**, 5000 (1985).
11. A. Cooper, Proc. Natl. Acad. Sci. USA **73**, 1740 (1976).
12. H. B. Callen, Thermodynamics, (Wiley, New York, 1962); L. D. Landau and E. M. Lifshitz, Statistical Physics, (Pergamon, London-Paris, 1980).
13. H. B. Callen and T. A. Welton, Phys. Rev. **83**, 34 (1951).
14. R. Kubo, Rep. Prog. Phys. **29**, 255 (1966).
15. M. Lax, Rev. Mod. Phys. **32**, 5 (1960); M. Suzuki, Prog. Theor. Phys. **56**, 77 (1976); F. Schlögl, Z. Phys. **B33**, 199 (1979).
16. A. Einstein, Ann. Physik **17**, 549 (1905).
17. L. Stryer, Biochemistry, Second Edition (W. H. Freeman, San Francisco, 1981), Chs. 2 and 3.
18. R. E. Dickerson and I. Geis, Hemoglobin: Structure Function, Evolution, and Pathology (Benjamin/Cummings, Menlo Park, CA, 1983).
19. E. Antonini and M. Brunori, Hemoglobin and Myoglobin in Their Reactions with Ligands (North-Holland, Amsterdam, 1971).
20. J. C. Kendrew, Science **139**, 1259 (1963).
21. H. Frauenfelder, G. A. Petsko, and D. Tsernoglou, Nature **280**, 558 (1979).
22. J. Kuriyan, S. Wilz, M. Karplus and G. A. Petsko, J. Mol. Biol. **192**, 133 (1986).
23. H. Frauenfelder, H. Hartmann, M. Karplus, I. D. Kuntz, Jr., J. Kuriyan, F. Parak, G. A. Petsko, D. Ringe, R. F. Tilton, M. L. Conolly, and N. Max, Biochemistry **26**, 254 (1987).
24. D. Stein, Proc. Natl. Acad. Sci. USA **82**, 3670 (1985).
25. Hans Frauenfelder, in Structure and Dynamics of Nucleic Acids, Proteins, and Membranes, eds. E. Clementi and S. Chin (Plenum Press, New York, 1986) pp. 169-177.
26. W. Doster, D. Beece, S. F. Bowne, E. E. Di Iorio, L. Eisenstein, H. Frauenfelder, L. Reinisch, E. Shyamsunder, K. H. Winterhalter, and K. T. Yue, Biochemistry **21**, 4831 (1982).
27. T. Iizuka, H. Yamamoto, M. Kotani, and T. Yonetani, Biochim. Biophys. Acta **371**, 126 (1974).
28. A. Ansari, E. E. Di Iorio, D. D. Dlott, H. Frauenfelder, I. E. T. Iben, P. Langer, H. Roder, T. B. Sauke, and E. Shyamsunder, Biochemistry **25**, 3139 (1986).

29. M. W. Makinen, R. A. Houtchens and W. S. Caughey, Proc. Natl. Acad. Sci. USA **76**, 6042 (1979); W. Caughey, H. Shimada, M. G. Choc, and M. P. Tucker, Proc. Natl. Acad. Sci. USA **78**, 2903 (1981).

30. J. O. Alben, D. Beece, S. F. Bowne, W. Doster, L. Eisenstein, H. Frauenfelder, D. Good, J. D. McDonald, M. C. Marden, P. P. Moh, L. Reinisch, A. H. Reynolds, E. Shyamsunder, and K. T. Yue. Proc. Natl. Acad. Sci. USA **79**, 3744 (1982).

31. R. H. Austin, K. Beeson, L. Eisenstein, H. Frauenfelder, I. C. Gunsalus, and V. P. Marshall, Science **181**, 541 (1973).

32. R. H. Austin, K. Beeson, L. Eisenstein, H. Frauenfelder, I. C. Gunsalus, and V. P. Marshall, Phys. Rev. Letters **32**, 403 (1974).

33. N. Alberding, S. S. Chan, L. Eisenstein, H. Frauenfelder, D. Good, I. C. Gunsalus, T. M. Nordlund, M. F. Perutz, A. H. Reynolds, and L. B. Sorensen, Biochemistry **17**, 43 (1978).

34. D. D. Dlott, H. Frauenflder, P. Langer, H. Roder, and E. Di Iorio, Proc. Natl. Acad. Sci. USA **80**, 6239 (1983).

35. N. Alberding, R. H. Austin, S. S. Chan, L. Eisenstein, H. Frauenfelder, I. C. Gunsalus, and T. M. Nordlund, J. Chem. Phys. **65**, 4701 (1976).

36. F. Stetzkowski, R. Banerjee, M. C. Marden, D. K. Beece, S. F. Bowne, W. Doster, L. Eisenstein, H. Frauenfelder, L. Reinisch, E. Shyamsunder, and C. Jung, J. Biol. Chem. **260**, 8803 (1985).

37. H. Roder, J. Berendzen, S. F. Bowne, H. Frauenfelder, T. B. Sauke, E. Shyamsunder, and M. B. Weissman, Proc. Natl. Acad. Sci. USA **81**, 2359 (1984).

38. N. Alberding, R. H. Austin, K. W. Beeson, S. S. Chan, L. Eisenstein, H. Frauenfelder, and T. M. Nordlund, Science **192**, 1002 (1976).

39. R. G. Palmer, Adv. Phys. **31**, 669 (1982).

40. R. D. Young and S. F. Bowne, J. Chem. Phys. **81**, 3720 (1984).

41. J. R. Alcala, E. Gratton and F. G. Prendergast, Biophys. J. **51**, 597 (1987).

42. R. D. Young, J. Chem. Phys. **80**, 554 (1984).

43. L. Eisenstein and H. Frauenfelder, in Frontiers in Biological Energetics. Vol. 1. Academic Press, New York, 1978) pp. 680-688.

44. E. Shyamsunder, Ph.D. dissertation, University of Illinois at Urbana-Champaign, 1986; R. D. Young, in Structure, Dynamics, and Function of Biomolecules, eds. A. Ehrenberg, R. Rigler, A. Graslund, and L. Nilsson (Springer, New York, 1987) pp. 39-42.

45. G. P. Singh, F. Parak, S. Hunklinger, and K. Dransfeld, Phys. Rev. Lett. **47**, 685 (1981).

46. Yu. Krupyanskii, F. Parak, D. Engelman, R. L. Mössbauer, V. I. Goldanskii and I. Suzcheliev, Z. Naturforsch. **C37**, 57 (1982).

47. Hans Frauenfelder, Ann. New York Acad. Sci., in press.

48. S. Brawer, Relaxation in Viscous Liquids and Glasses, (American Ceramic Society, Columbus, OH, 1985).

49. Heidelberg Colloquium on Spin Glasses, Lecture Notes in Physics 192, J. L. van Hemmen and I. Morgenstern, Eds., (Springer, Berlin, 1983).

50. G. Toulouse, Comm. Physics **2**, 115 (1977).

51. G. Toulouse, Helv. Phys. Acta **57**, 459 (1984).

138

52. R. Rammal, G. Toulouse and M. A. Virasoro, Rev. Mod. Phys. **58**, 765 (1986).
53. M. Mézard, G. Parisi, N. Sourlas, G. Toulouse and M. Virasoro, Phys. Rev. Lett. **52**, 1156 (1984).
54. R. G. Palmer, D. L. Stein, E. Abrahams and P. W. Anderson, Phys. Rev. Lett. **53**, 958 (1984).
55. K. Binder and A. D. Young, Rev. Mod. Phys. **58**, 801 (1986).
56. W. Doster, A. Bachleitner, R. Dunau, M. Hiebb and E. Lüscher, Biophys. J. **50**, 213 (1986).
57. R. Elber and M. Karplus, Science **235**, 318-321 (1987).

CHAPTER 9

DOMAIN MOTIONS IN PROTEINS

W. S. Bennett
Max-Planck-Institut für Molekulare Genetik
Abteilung Wittmann
Ihnestraße 73,   D-1000 Berlin 33
Federal Republic of Germany

ABSTRACT. Proteins are characterized by a number of slow processes (time constants $> 10^{-4}$ sec), which are thought to be associated with cooperative motions of large groups of atoms. Structural studies of proteins have led to the characterization of large-scale, concerted conformational changes in a number of systems. Such conformational changes are often coupled with ligand binding or covalent modification of the protein, making them amenable to experimental manipulation. It is usually possible to correlate these large-scale motions with the biological or enzymatic function of the protein. In some cases, the conformational changes appear to be coupled with perturbations of spectroscopic probes that provide some information about the time scales of the perturbation, although it is rarely possible to assign rates to the large-scale motions themselves.

1. INTRODUCTION

The study of proteins as chemical entities has a long history and was an integral part of the development of modern chemistry [41]. Interest in proteins as physical systems, and particulary in the physics of dynamic processes in proteins, is a considerably more recent development, however. An early stimulus of this interest was the observation by DeVault and Chance [31] of a temperature-independent rate for the transfer of an electron from a cytochrome to the reaction center of a photosynthetic bacterium following photobleaching at temperatures below about 100 K. One explanation suggested by DeVault and Chance for the temperature independence of this reaction was a quantum-mechanical tunneling process involving a remarkably wide barrier of 30-80 Ångstroms. Although consideration of vibrionic effects on the tunneling process indicates that the barrier width is probably considerably smaller than originally thought (about 8 Å [56]), the work of DeVault and Chance and subsequent discussions demonstrate that the study of biological systems can lead to the observation and analysis of interesting physical phenomena, in spite of the enormous chemical complexity of the sample. In this regard, it is worth noting that the observations of DeVault and Chance were made not with isolated photosynthetic components, but with suspensions of bacteria. Studies of nonexponential reaction rates in proteins at low temperatures continue to spark interesting discussions of the physics involved, as is elaborated by Young in this volume.

139

G. J. Long and F. Grandjean (eds.), The Time Domain in Surface and Structural Dynamics, 139–178.
© 1988 by Kluwer Academic Publishers.

In this chapter, we will survey a topic that comes considerably closer to fulfilling our intuitive expectation that most attempts to study dynamic processes in a biological sample, even one as comparatively simple as an isolated protein, would be hampered by the complexity of the system at the atomic level. In particular, we will examine what is known about the dynamics of concerted motions in proteins with the goal of understanding how these motions are related to the biological functions of a protein molecule. The starting point for our discussion will be the evidence accumulated in recent years that some proteins undergo reasonably well-defined conformational changes involving concerted motions of large groups of atoms [9]. One might hope that a knowledge of the major conformational states available to a protein molecule would aid in understanding the dynamics of the transitions between them. Instead, we will find that our original expectations were not far off the mark, although some progress has been made in describing the dynamic behavior of a few examples of domain motion. I hope that in the process of examining the problems involved in the study of these motions, we will be able to focus on some of the unresolved questions that are at least in principle amenable to experimental study.

2. BASICS OF PROTEIN STRUCTURE [30]

Let us start by establishing the scale of the problem. We will be concerned here with proteins that are soluable in aqueous solutions of moderate ionic strength near neutral pH. Such proteins are moderately large, linear polymers of discrete size and defined composition; their molecular weights range from about 5,000 to 100,000 daltons. A protein of average size has a molecular weight of around 30,000 daltons and consists of about 270 amino acids linked via peptide bonds into a polypeptide chain of defined sequence. This average protein molecule would contain about 2100 nonhydrogen atoms, mostly C, N, O and S. In aqueous solution, proteins are found to associate strongly with about 0.3 grams of water per gram protein, which means that our average protein molecule would associate noncovalently with about 500 water molecules. Clearly the basic system consisting of a protein molecule and associated water molecules has a large number of degrees of freedom.

Proteins can often be crystallized from aqueous solution, yielding highly hydrated crystals (typically around 50% solvent by volume). The structures of a number of crystalline proteins have been determined by X-ray diffraction and provide the experimental basis for a few broad generalizations, which we summarize here (see Creighton [30] for a more detailed discussion). Under conditions of temperature and solution compositon (i.e., pH and ionic strength) not far removed from those of its normal physiological environment, a protein molecule adopts a unique, compactly folded conformation termed its "native" structure. Soluable proteins are usually roughly spherical in shape; for this reason they are often called "globular" proteins. Our average protein, for example, would have a diameter of about 42 Å if it were spherical. The hydrophobic (aliphatic or aromatic) amino-acid side chains tend to be on the inside of the protein and the polar or ionizable side chains on the outside. The polypeptide backbone tends to be organized in short segments (up to about 15 residues) with one of a few repeating patterns of hydrogen bonds, mostly either "α-helices" or extended strands that are found side-by-side in "β-sheets." These hydrogen-bonding patterns are termed the "secondary structure" of the protein and serve to satisfy the hydrogen-bonding potential of the backbone in the hydrophobic interior of the molecule. In spite of this high degree of organization, the atoms of a folded protein molecule are about as densely packed as those in crystals of small organic molecules.

Proteins of average size or larger tend to fold into two (or a few) globular substructures called "domains" [60]. Domains are usually formed from only one or two contiguous regions of the polypeptide chain, so that they are covalently attached to one another by one or two segments of the backbone. The degree of noncovalent association between domains varies widely, however, from extensively interacting pairs that are barely distinguishable from a single domain to tethered domains that have little or no contact. We will see that proteins at both extremes can display the large-scale motions that are our main interest.

3. CRYSTAL STRUCTURES AND PROTEIN DYNAMICS

The general features of protein structure presented in the last section are derived from the results of X-ray diffraction studies, which provide a view of the protein molecule that is averaged over all of the molecules in one or more crystals over a time span of at least a few hours and more often several days. All information about the dynamics of any motion of the atoms during this time is lost. Even before the first X-ray structures of proteins were determined, however, it was clear that these molecules displayed a variety of dynamic processes with an extraordinarily broad range of rates.

The earliest evidence of this behaviour was provided by hydrogen-exchange experiments (for a recent review, see [7]). Simply put, when one measures the rate at which isotopically labeled hydrogen atoms are exchanged between a protein and its solvent, one finds that the rates of exchange at different sites are distributed over several (typically seven or more) orders of magnitude. The most slowly exchanging sites in the protein do so with half-times as long as months. For comparison, rates of hydrogen exchange from isolated amino acids and peptides are about 10^3 min^{-1}.

The rate at which a given hydrogen atom in a protein exchanges is strongly dependent on its local environment, even for relatively exposed sites on the surface of the molecule [110]. However, such environmental effects cannot account for the rates of the more slowly exchanging sites. It is generally believed that the exchange rates at these sites are limited by the rates of structural fluctuations which either transiently expose to solvent or allow the penetration of solvent molecules to sites of exchange that are in some sense "buried" inside the protein. Cooper [29] has pointed out that rather substantial fluctuations in the energy (and other thermodynamic properties) of a system the size of a protein are to be expected. Such fluctuations probably involve states with energies comparable to those accessible to the unfolded protein, since the free energy of stabilization of the folded conformation of a protein is generally only a few kcal/mol [95].

Although the nature of the fluctuations responsible for the hydrogen-exchange behavior of proteins has not been established, it is interesting that similar fluctuations occur both in solution and in the lattice of a crystalline protein. Tüchsen *et al.* [109] have shown that the overall distribution of exchange rates for the enzyme lysozyme is the same in solution and in the crystal. In addition, neutron diffraction studies, which can detect hydrogen atoms that remain unexchanged after long periods of time [65], indicate that the subset of atoms that exchange most slowly in crystals of another protein (bovine pancreatic trypsin inhibitor, BPTI) is similar, but not identical, to the group that exchanges most slowly in solution [122]. Taken together, these observations on the hydrogen-exchange behavior of crystalline lysozyme and BPTI suggest that a crystal lattice imposes constraints on the states available to a protein which may modify the exchange rates at some sites but do not produce an appreciable change in the overall spectrum of rate constants.

The success with which one can account for many of the properties of soluable proteins in terms of their crystal structures [30], together with the limited data available on hydrogen exchange from crystalline proteins, clearly suggests that the constraints imposed by a crystal lattice usually have little effect on either the average structure of a protein molecule or the range of conformations available to it. This does not mean that the electron-density distribution derived from the diffraction pattern of a protein crystal delimits the full range of conformations available to the molecule. Slower fluctuations, such as those that govern the rates of slowly exchanging hydrogen atoms, probably occur too infrequently to contribute significantly to the diffraction pattern [97]. Still, it is normally reasonable to expect that the average conformation of a protein in the crystal is a good approximation to that found in solution.

As we shall see, however, the situation is somewhat more complicated for proteins which display domain motions. Such motions are often large enough that the crystal lattice can only "trap" the molecule in a single conformation. In such cases, the question of whether a protein has a few distinct conformations or moves relatively freely within a range of possible conformations must be answered experimentally. At the level of detail provided by diffraction analysis of protein crystals, this issue can only be resolved by studying the structure of a protein in a number of different crystalline environments.

The possibility that the crystal lattice may have a significant influence on the structure of these flexibile proteins also complicates the interpretation of structural results. When a protein is found to have more than one conformation, the role (if any) that each conformation plays in the function of the protein must also be established experimentally. Since the determination of the structure of even one crystal form of a protein requires a major investment of effort, and since the crystallization of a protein is largely a matter of luck to start with, one often does not have as clear an answer to these questions as one would like. Nonetheless, we will see that a few systems have been studied in sufficient detail to establish some general trends.

4. OVERVIEW OF PROTEIN DYNAMICS

The hydrogen-exchange studies described above reveal the existence of fluctuations in proteins that are slower than the experimental manipulations required for sample preparation (several minutes). At the other end of the spectrum, one expects the fastest atomic vibrations in a protein molecule to be similar to those found in the isolated chemical groups of which it is composed [67], which have characteristic times down to about 10^{-14} sec. Contemporary discussions of the physical aspects of protein dynamics often take the review of Careri *et al.* [23] as a starting point; these authors were apparently the first to catalog various physical processes observed in proteins and comparable model systems according to their time scales. The

TABLE I.

A. Time Scale of Processes Observed in Globular Proteins

Process	Apparent Time Constant (sec)	
localized vibrations	10^{-14} –	10^{-12}
rotation of surface groups	10^{-11} –	10^{-8}
flipping of buried aromatic rings	$< 10^{-5}$ –	$> 10^{-1}$
hydrogen-exchange fluctuations	$< 10^{2}$ –	$> 10^{8}$

B. Time Scales of Related Processes

rotational relaxation of bound H_2O	10^{-11} –	10^{-9}
rotational relaxation of protein	10^{-8} –	10^{-7}
reciprocal of turnover number	10^{-8} –	10^{-1}
relaxation on binding a small ligand	10^{-4} –	10^{0}

time constants of a selection of these processes that will be of interest to us are summarized in Table I.

It should be clear that one must make rather broad generalizations in preparing a summary of this sort. The numbers listed can only be taken as illustrations of trends, not as statements of firm limits. The characteristic time of a kinetic process normally depends on temperature, for example, and the rates of the slower processes observed in proteins often vary by several orders of magnitude in the relatively narrow temperature range between the freezing point of water and the point at which the protein begins to unfold. Moreover, interpretation of observed relaxation rates in terms of the rates of structural fluctuations of the protein generally involves assumptions about the mechanism of the process.

For concreteness, we have tabulated the time constants of slower processes observed for a specific small protein, BPTI, at a fixed temperature near that found in its physiological environment. The molecular weight of BPTI is only 6,000 daltons, which has allowed individual rates of a large number of processes in the molecule to be studied by NMR methods [111]. The rates at which six-membered aromatic rings in the side chains of phenylalanine and tyrosine residues "flip" (i.e., rotate by 180°) are taken to illustrate time constants of relatively localized fluctuations. Such processes appear to require concerted movements of atoms in the immediate neighborhood of the ring over distances less than an Ångstrom [53]. The inequality symbols ($<$ or $>$) in the table indicate that the limits to the range of time constants listed are imposed by the experimental technique in this case; i.e., rings were observed that had both faster and slower flipping rates than could be resolved. As we have seen, the measurement of hydrogen-exchange rates is limited by experimental manipulations at the upper end of the rate distribution (shorter characteristic times), while the uncertainty at the lower end (longer characteristic times) results because some of the exchangeable hydrògen atoms do not measurably exchange in the course of the experiment (nearly two months [112]).

Even when one allows for substantial uncertainty in the exact numbers, a few general conclusions can be drawn from the table. First, it is clear that dynamic processes with a wide range of characteristic times can be observed for a given protein in a narrow temperature range. Second, from a comparison of parts A and B of the table, one can see that the time scales of various dynamic processes of the protein itself overlap both with the characteristic times of events in the immediate environment of the protein and with the time scales of processes associated with enzymatic catalysis (the biological function of many soluable proteins). The latter processes include the "turnover number" of an enzymatic reaction (the maximum rate of product formation [39]) and the rates commonly associated with conformational changes in enzyme molecules following the binding of a substrate or other small ligand[1] [47].

The demonstration of overlapping time domains for dynamic processes within a protein and in its immediate environment was in fact the main point of Careri et al. [23], who proposed that the coupling of such processes might be a source of free energy for the enormous rate enhancements observed for enzymatic reactions (up to 10^{10} times the rate of the uncatalyzed reaction; see Creighton [30]). The problem of explaining the catalytic efficiency of enzymes is a major stimulus to the study of both the chemistry and the physics of proteins. The specific proposal of Careri et al. is but one of a number of theories of free energy "transduction" by enzymes, in which the role of protein fluctuations is often dominant (reviewed by Welch et al. [116]). Our focus, however, will be on the more modest, if no less elusive, problem of trying to establish the time scales of some reasonably well characterized, large-scale motions in proteins. A more general review of protein dynamics, with particlar emphasis on the physical picture of proteins suggested by computational studies, was recently given by McCammon [74].

5. CHARACTERIZATION OF DOMAIN MOTIONS IN PROTEINS

An obvious explanation for the long characteristic times in Table IA is that they are associated with processes that involve cooperative motion of large groups of atoms [23]. Concerted, if not cooperative, motions have now been observed in number of proteins. The distinction made here is somewhat artificial: we take a concerted process to be one that results in movement of parts of a molecule without necessarily changing the correlations between positional fluctuations of the atoms; a cooperative process would be the opposite. From a practical point of view, it is easy to establish the existence of concerted motions, but not to characterize their cooperativity.

The large-scale, concerted motions that have been described by crystallographic studies of proteins can be divided into three classes [9]. Two of these involve essentially rigid-body movement of the globular domains of a protein molecule relative to one another, and the third is characterized by the transition of a region on the

---

[1] The relaxation times cited in the table are typical of those observed by rapid-reaction techniques [47]. Faster processes associated with changes in ligation state are observed in flash photolysis experiments on heme proteins; see the chapter by Young in this volume.

surface of a protein from a (crystallographically) disordered state to an ordered state under defined conditions. Although it is by no means a globular domain, the surface region involved in the third class of concerted motions was originally named (for reasons that will become clear later) the "activation domain." Thus concerted motions in proteins are often referred to as "domain motions;" the proteins involved are sometimes said to display "domain flexibility" or "segmental flexibility." Before turning to the dynamics of domain motions, we describe the basic features of each of the three classes. For more extensive references to the earlier literature, consult Bennett and Huber [9] or the more recent work cited.

5.1.   Flexibly Linked Domains

Domain flexibility in proteins was first recognized in immunoglobulins (antibodies), a chemically heterogeneous family of molecules found primarily in the blood of higher animals. The function of these proteins is to recognize foreign molecules (antigens) in the animal and to initiate a complicated sequence of cellular processes that constitute the immune response to the foreign substance.

Much of what is known about the atomic structure and dynamics of antibodies is based on studies of the immunoglobulin G (IgG) class, which is the most abundant group of antibody molecules found in blood. The basic structure of the IgG molecule is shown schematically in Figure 1. It is formed by four polypeptide chains, two identical "light" chains (molecular weight: 25,000 daltons) and two identical "heavy" chains (50,000 daltons). The four chains are linked by "disulfide bridges" (two cysteine residues whose sulfur atoms have reacted to form an S-S bond) and noncovalent interactions, so that the total molecular weight of the molecule is about 150,000 daltons.

Each chain is folded into a series of separate globular domains, all of which have molecular weights of about 12,000 daltons. Within each chain, the domains are connected by a single segment of the polypeptide backbone but have little other contact with one another. The light chain consists of two such domains (V_L and C_L in the figure) and heavy chain of four (V_H, $C_H 1$, $C_H 2$ and $C_H 3$). Each of the domains except $C_H 2$ interacts noncovalently with a domain from another chain, as shown in the figure, to form globular dimeric units. The chemical heterogeneity among molecules of the IgG class is localized primarily at the free end of the V_H-V_L dimer, which contains the antigen recognition site. Most of the rest of the molecule is homogeneous within an IgG sample from a given animal species, so that the chemical heterogeneity has little influence on the dynamic properties of the molecule that will be of interest to us.

The IgG molecule is easily split by proteolytic enzymes in the "hinge" region, which consists of the disulfide-linked segments of the two heavy chains between the $C_H 1$-C_L dimers to the $C_H 2$ domains. The resulting fragments of the molecule are called the antibody-combining fragment (Fab) and the crystallizing fragment (Fc). This terminology is usually also applied to the corresponding parts of the intact molecule. The names of the fragments are only of historical signficance; Fab fragments from a number of IgG molecules have also been crystallized, and have provided most of the detailed structural information we have about these molecules.

The Fab and Fc fragments of an antibody molecule are just large enough to be resolved in the electron microscope, where both appear as featureless, stubby

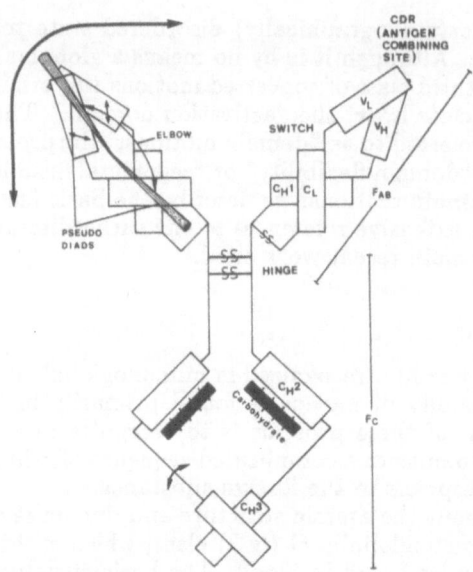

Figure 1: Schematic drawing of the immunoglobulin G molecule [58]; see text for details.

rods. A combination of hydrodynamic and electron microscopic studies of rabbit IgG [100] has shown that the angle between the Fab arms of the molecule can range from about zero to at least 140°. Statistical analysis of the frequency of occurrence of molecules in various conformations indicates that there is little barrier to motion of the Fab arms relative to one another and to the Fc region. (It is this type of motion that is usually referred to as "segmental flexibility.")

Crystallographic studies of Fab and Fc fragments have revealed that the polypeptide segments linking the dimeric globular units of each fragment are also flexible. In the Fab fragment, the segments linking the V_H-V_L and C_H1-C_L dimers are called the "switch" peptides (Fig. 1). Although the domains forming a dimer generally have different chemical compositions (except for the C_H3 dimer), the folding patterns of the domains are very similar; thus it is possible to define an approximate two-fold symmetry axis for each dimer (the "pseudo-diad" in Fig. 1). One can then characterize the relative orientation of the dimeric units of an Fab fragment by the angle between the pseudo-diads of the V_H-V_L and C_H1-C_L dimers, called the "elbow" angle; the elbow angle is taken to be 180° when the "arm" is fully extended. Elbow angles are found to range from about 115° to 180° among different Fab fragments; recent studies bearing on this question are Suh et al. [106], Amit et al. [1] and Colman et al. [27]. The range of angles observed for a given Fab in different crystal structures is considerably more limited, however (about 10° in one case and 20° in another). The orientation of domains in the Fc fragment is not as easily characterized, as the C_H2 domains do not form a compact dimer. However, comparison of different crystal structures of the Fc fragment also demonstrates the existence of limited flexibility in the polypeptide segments that link the C_H2 and

C_H3 domains.

The general conclusion derived from all of these sources of information about antibody flexibility is that the molecule consists of stable, globular units linked by flexible segments of polypeptide chain, much like beads on a string. Flexibility within the Fab and Fc regions, where the dimeric units are connected by pairs of relatively short polypeptide segments, seems to be somewhat more restricted than that found in the longer and at least partly single-stranded hinge between the Fab's and the Fc. Although the number of molecules that can be directly compared is limited, it also appears that the flexibility of the switch segments is restricted to a fairly specific joint region. This joint is formed by only one or two adjacent bonds in the polypeptide backbone. Flexibility in the hinge region, by contrast, appears to be less localized, as several highly flexible regions have been observed.

Until recently, it appeared that the globular dimeric units of the IgG molecule behaved essentially as rigid bodies. Constant (C_H1-C_L and C_H3) and variable (V_H-V_L) dimers from different IgG molecules in a number of different crystal structures (about 25 independent dimeric units) had shown no significant difference. A few recent structures have suggested that there is some variablity in the interactions between domains of the V_H-V_L dimer, however. The structure reported by Colman et al. [27] is particularly interesting, as it involves a complex between an antibody and its physiological protein antigen. It has not yet been proven whether these variant V_H-V_L dimer structures are characteristic of the IgG molecules involved or whether they result from perturbation of the normal dimer structure by antigen binding, although the latter explanation is preferred by Colman et al. This question is of interest because it may be relevant to the fundamental problem of how the antibody molecule performs its function of signaling the presence of foreign substances to other components of the immune system. Our present understanding of domain flexibility in immunoglobulins has not provided any clear-cut answers to this problem. However, the possibility of a dynamic rearrangement of domains in the V_H-V_L dimer at least suggests a way of propagating information about antigen binding to the C_H1-C_L dimer [58].

Although other proteins with flexibly linked domains have been identified (see Bennett and Huber [9] for a more extensive review), they are of little interest in the present context, as information about the dynamics of the domain movements is available only for immunoglobulins.

5.2. Hinged Domains

More restricted relative motion of globular domains has been described for a number of enzymes and for proteins involved in the recognition and transport of small metabolites. The domains of these proteins appear to have only two stable orientations. The transition from one orientation to the other, which is triggered by the binding of a specific ligand to the protein, is an essentially rigid-body rotation of the domains. We will refer to these molecules as proteins with "hinged" domains to distinguish them from the proteins described in the previous section, in which the less restricted motion of "flexibly linked" domains is observed.

The two conformations observed for proteins with hinged domains can in most cases be described as an "open" conformation, in which access to the binding site for the ligand is relatively unhindered, and a "closed" conformation, in which the bound ligand is often completely buried within the protein. This distinction is

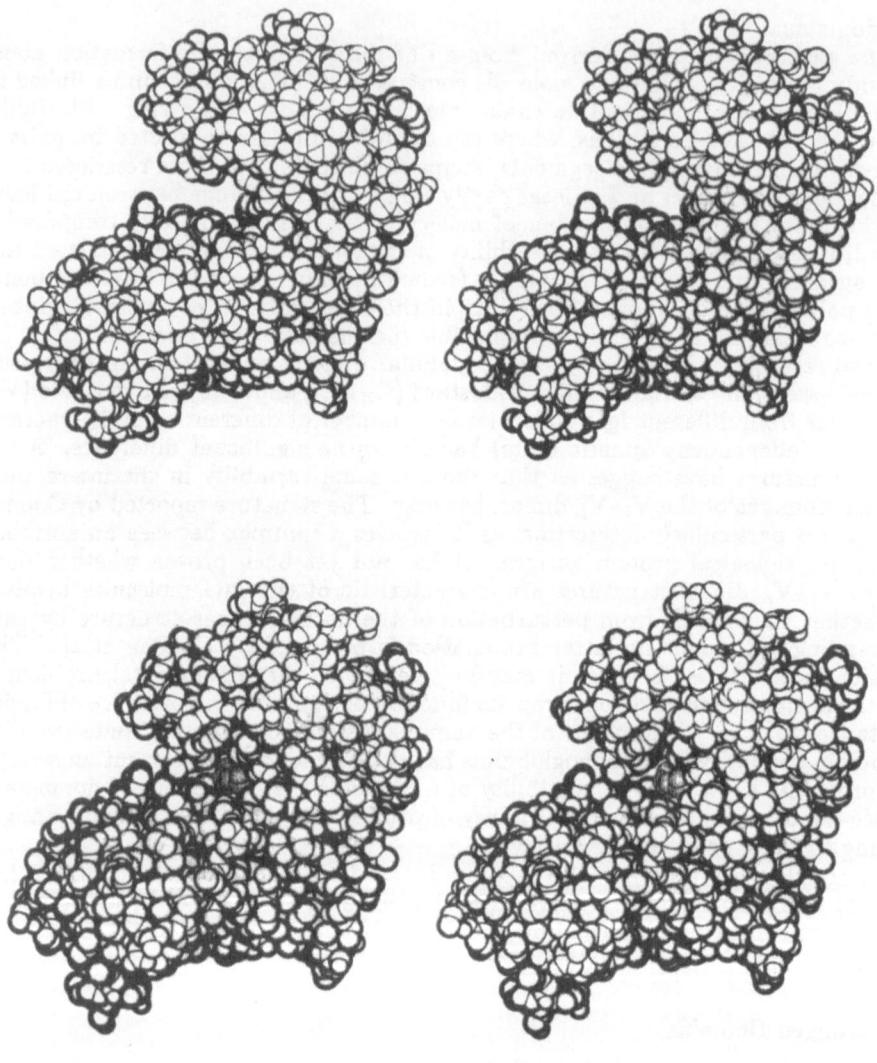

Figure 2: Stereo space-filling drawings of the unliganded hexokinase molecule (*above*) and the complex of hexokinase with glucose (*below*); the visible surface of the glucose molecule is shaded. From Ref. [103].

illustrated by space-filling drawings of the first protein for which ligand-induced domain motions of this type were established, the enzyme hexokinase (Fig. 2).

Hexokinase catalyzes the transfer of a phosphate group from a molecule of adenosine triphosphate (ATP) to the 6-hydroxyl group of glucose (or other six-carbon sugars) in the first step of glycolysis. The enzyme from yeast, on which all of the crystallographic studies have been performed [2,11], has a molecular weight of about 50,000 daltons; the smaller of the two domains accounts for about a third of the molecule and the larger for the rest. In the absence of substrate, the enzyme is in the open conformation. The binding of glucose to a site deep within the cleft between the two domains somehow causes the enzyme to switch to the closed conformation, in which the glucose molecule is almost completely buried; only the reactive 6-hydroxyl group of glucose remains accessible to solvent in the closed form. This conformational change can be described to a good approximation as a rigid-body rotation of the small domain by 12° relative to the large domain. However, since only one conformation or the other can be accommodated in a given crystal lattice, only the net result of the conformational change can be observed; whether the actual pathway taken by the small domain at all resembles a simple rotation is not known.

Hexokinase is also a good example of a protein for which the distinction between hinged and flexibly linked domains is well established. Since it is clear that the connections between the domains are flexible in both cases, it is important to determine whether the structures observed in different crystal forms are stable conformations of the protein that are not influenced by the crystal lattice, or whether the lattice has simply trapped one of a range of possible conformations of the protein, as appears to be the case in immunoglobulin crystals. This question is generally more relevant to the open conformation of a hinged protein, as the closed conformation is necessarily constrained by specific interactions between the protein and its ligand. The most convincing, if demanding, proof that the structure of a protein is not influenced by crystal-packing interactions is to determine the structure of several crystallographically independent copies of the molecule. Three independent copies of the open conformation of hexokinase have been studied at reasonably high resolution (3.5 Å or better), and low-resolution studies indicate that three more crystal forms also contain essentially the same structure [103].

The general picture of the glucose-induced conformational change in hexokinase provided by crystallographic studies is also supported by studies of the enzyme in solution. The most direct evidence that a conformational change of similar magnitude does in fact occur comes from low-angle X-ray scattering measurements [76], which show that the radius of gyration of hexokinase is reduced in the presence of glucose by about 1 Å. Although the low-angle scattering technique does not allow the conformational change in solution to be characterized in detail, the observed change in radius of gyration is the same as that predicted from the crystallographic models. Measurements of the glucose-binding properties of hexokinase in different crystal forms [118] and isotope-trapping studies of the crystalline hexokinase-glucose complex [119] are also consistent with the proposed conformational change; the latter experiments, moreover, demonstrate that the crystalline complex is catalytically competent.

Table II lists other proteins for which evidence of hinged domains has been presented. In most cases, crystal structures of both open and closed forms have

TABLE II
Proteins with Hinged Domains

protein [Ref.]	mass (daltons)			rotations		copies	
	subunit	n	domain	subunit	domain	open	closed
AATase [61,3]							
mitochondrial	45,000	2	9,000	*0°*	13°	2	3
cytosolic	45,000	2	9,000	*similar*		2	0
ABP [84]	33,000	1	15,000	–	≈18°	0	1
LIV-BP [98]	37,000	1	17,000	*similar*		1	0
ACTase [66]							
C subunit	34,000	6	15,000	10°	*13°*	3	2
R subunit	17,000	6	6,000	15°	*12°*	3	2
CSase [117]	50,000	2	12,000	*0°*	18°	1	3
GAPDH							
B. stearo. [71]	36,000	4	18,000	1°	4°	7	5
lobster [82]	36,000	4	18,000	none		8	
Hexokinase [11]	50,000	2	16,000	?	12°	6	1
LADH [28]	40,000	2	16,000	*0°*	10°	2	2
PFK [37]	33,000	4	15,000	4°	5°	4	1
PGK [94]	45,000	1	22,000	–	≈11°	2	0

Notes: See text for keys to the abbreviations used in the column labeled "protein." The column labeled "n" contains the number of subunits in the physiological oligomer of the protein. Items in italics are estimates based on published figures or qualitative analyses. Values marked as approximate (≈) are estimates derived from low-angle X-ray scattering studies in the references cited. A dash indicates that the entry is not applicable, and the question mark indicates that no information is available.

been determined, and in several cases,᠎ multiple copies of the open conformation allow the protein to be classified as one with hinged, rather than flexibly linked, domains. The list is growing rapidly; just a few years ago, hexokinase was the only protein for which there was convincing evidence on this point [9]. Most of the proteins in the list are oligomeric; i.e., they are normally found in solution as aggregates of a small number of "subunits," each of which is an independently folded polypeptide chain. For each protein in the table, we have listed the molecular mass of the subunit, the number of subunits in the oligomer (n), the mass of the small domain of the subunit, and the number of independent copies of each conformation that have been observed so far.

With the exception of phosphofructokinase (PFK) and the transport proteins, arabinose-binding protein (ABP) and "leucine, isoleucine, valine-binding protein" (LIV-BP), the domains of each protein in the list are connected by two segments of the polypeptide chain. For PFK, a third segment is found connecting a single α-helix in one domain with the other domain; i.e., the helix is near one end of the polypeptide chain and is associated with the rest of the domain by noncovalent interactions only [36]. PFK also differs from the other proteins listed in that the movement of its domains is not a rotation of one toward the other, but a twisting of one domain relative to the other around an axis roughly parallel with the long dimension of the subunit [37]. The conformational change in PFK also does not isolate the ligands of the molecule from solvent, as is the case for all of the other proteins for which the structure of the closed conformation is known. Low-angle X-ray scattering studies of PFK in solution are consistent with the crystallographic results, in that they show little difference between the two states of the enzyme for which crystal structures are available; however, the low-angle studies indicate that another ligand, ATP, induces larger changes in the structure of the tetramer [70].

The two transport proteins, ABP and LIV-BP, have similar folding patterns [98]. As is the case for PFK, the domains of these molecules are are connected by three segments of polypeptide. In ABP and LIV-BP, however, one segment is near the end of the chain and has only limited contact with one of the domains. Although only one conformation of each of the transport proteins has been established by crystallographic studies, low-angle X-ray studies on ABP have established that the binding of arabinose changes the radius of gyration by an ammount consistent with a rigid-body rotation of the domains by about 18° [84]. The differences between the unliganded structure of LIV-BP and the liganded structure of ABP are consistent with the proposed conformational change.

The classification of phosphoglycerate kinase (PGK) as a hinged enzyme is also based largely on evidence provided by low-angle X-ray scattering studies [94] but was independently suggested on structural grounds [5]. Since PGKs from two different organisms (yeast and horse) have essentially identical open conformations in independent crystal lattices, it seems likely that this assignment is correct. For the transport proteins, in contrast, only one copy of the open conformation has been observed, so the possibility remains that there is not a unique open conformation. In the absence of more direct information, it seems reasonable to regard the transport proteins as hinged, since the closed structure of ABP is similar to those of established hinged proteins in the extent to which the domains interact.

Extensive interactions between domains (e.g., Fig. 2) are in fact characteristic of hinged proteins. With the exception of the transport proteins and PGK, all of the proteins with hinged domains exhibit such interactions in both the liganded and unliganded conformations. In this regard, hinged proteins resemble smaller oligomeric proteins, such as hemoglobin, which are subject to allosteric regulation. Since allosteric regulation is characterized by the cooperative binding of a ligand to an oligomeric protein, it is normally associated with changes in the "quaternary structure" of the oligomer (i.e., in the arrangement of subunits). This is the case for hemoglobin [91], in which the subuits are relatively small (molecular weights of about 16,000 daltons) and not subdivided into globular domains. On binding approximately two molecules of O_2, one pair of subunits in the hemoglobin tetramer rotates about 14° with respect to the other; the subunits participate in numerous intramolecular interactions in both quaternary structures.

A new aspect of the structure of allosteric proteins has emerged from recent studies of proteins with larger subunits, however. Three of the hinged enzymes in Table II display cooperative behavior: aspartate carbamoyltransferase (ACTase), glyceraldehyde-3-phosphate dehydrogenase (GAPDH) from *Bacillus stearothermophilus* and PFK. In each of these proteins, conformational changes induced by ligands include alterations of both quaternary structure and tertiary structure (i.e., the arrangement of bonded atoms in space). The changes in tertiary structure all consist of relative motions of hinged, globular domains. For this reason, the table lists equivalent rotations for movements of both the subunit as a whole and of the smaller domain within the subunit.

For ACTase, the change in quaternary structure is dominant (in terms of the distances over which parts of the protein are displaced, which is not evident from the rotations listed in Table II). However, the change in quaternary structure is accompanied by substantial reorientations of domains in both the catalytic (C) subunits and the smaller regulatory (R) subunits [66]. In contrast, the subunit and domain movements observed so far in PFK are of comparable magnitude [37], while for GAPDH, domain reorientation is dominant [71].

It should be noted, however, that a relationship between the ligand-induced conformational changes (at either the subunit or domain level) and allosteric behavior remains to be established for all of these enzymes. This is particularly evident for GAPDH. Lobster GAPDH appears to undergo no large-scale conformational changes upon binding ligands, even though it displays the same cooperativity with respect to ligand binding as the enzyme from *B. stearothermophilus*. Crystal structures of the lobster enzyme in the presence and absence of the ligand are essentially the same as that found in the closed form of the bacterial enzyme [82]. If the well-studied example of hemoglobin serves as a guide, it may be expected that an explanation of the allosteric effects in these enzymes will ultimately require an understanding of the relationship of the large-scale conformational changes to relatively subtle changes in tertiary structure at the ligand-binding site and at the subunit or domain interfaces.

Localized changes in tertiary structure are in fact quite commonly observed in comparisons of the open and closed conformations of all of the proteins listed in Table II. When we describe the domain motions as "essentially" rigid-body rotations, we mean that the overall structure of the domain (in particular, the relative positions of elements of secondary structure) is maintained during the conformational change. The domain rotations are invariably accompanied by localized changes in the conformation of side chains and loops of the main chain at the surface of the domain, especially in regions where contacts with the ligand or the other domain differ (or where lattice contacts differ between crystal forms). An interesting exception to this pattern is seen in citrate synthetase (CSase), where small variations in tertiary structure are found throughout the small domain in a comparison of two closed structures [117], which were crystallized in the presence of different combinations of substrates. The source of these delocalized perturbations is not known.

Most of the domain motions described in the table have a well-defined trigger. Usually it is the binding of a single ligand, even when the protein has two or more physiological ligands. For PGK, however, the binding of two ligands (both phosphoglycerate and ATP) is required to change the radius of gyration [94]. The situation is also more complicated for enzymes subject to allosteric control by ligands other than substrates, such as ACTase and PFK. The equilibrium between the allosteric states of such enzymes (i.e., between the active and inactive conformations) can often be shifted in either direction, depending on which ligands are bound. Interestingly, all of the closed conformations of hinged enzymes described so far, including those of allosteric enzymes, have been obtained by crystallizing the proteins in the presence of substrates (or inhibitors that are structural analogs of substrates). The open conformations of PFK and ACTase have been crystallized in the presence of allosteric inhibitors, although the the open conformation of ACTase is also found in crystals of the unliganded enzyme [57].

The requirements for triggering domain closure in liver alcohol dehydrogenase (LADH) have been studied extensively and appear to be more complicated than those of the other hinged enzymes, even though LADH is not subject to allosteric control. The conformation of LADH seems to be sensitive to the environment of the catalytic Zn ion and to the nature of the alcohol substrate [24]. However, this behavior may result from ambiguities introduced by using crystal-structure analysis to assess the effects of ligand binding. In solution, the cofactor nicotinamide-adenine dinucleotide (NAD$^+$ or NADH) is sufficient to induce some conformational change [26], but crystals containing the closed conformation can usually only be grown with a ternary complex of LADH, NADH and alcohol [34]. Whether the enzyme in the binary complex of LADH and NAD$^+$ studied in solution is in the closed conformation observed in crystalline ternary complexes is not yet known.

Another interesting, if subtle, effect of crystallization is seen in crystals of aspartate aminotransferase (AATase). In animals, this enzyme is found in closely related cytoplasmic and mitochondrial variants called isozymes; the open forms of the two isozymes are quite similar. Crystal structures of both open and closed forms have been reported for the mitochondrial isozyme [61]. Although no large-scale structural differences between the subunits of the dimer are evident in crystals of the open form of either isozyme, both crystal lattices appear to impose functional asymmetry on the molecule that is not found in solution. The asymmetry appears as nonequivalent kinetics of the subunits in crystals of the mitochondrial isozyme (the enzyme is active in the crystalline form [63]) and as differences in the binding of substrate to the subunits in crystals of the cytosolic isozyme [3]. In the latter case, ligand binding produces numerous small conformational changes throughout one subunit but none in the other. These small changes are presumably the result of an attempt by the enzyme to switch to the closed conformation that is hindered by intermolecular contacts in the crystal.

Here again we encounter the problem that the localized changes in tertiary structure accompanying domain motions in hinged proteins (or quaternary changes in allosteric proteins) are generally difficult to interpret in terms of the causes or consequences of the large-scale movement. This is in part because the relative movement of globular domains (both hinged and flexibly linked) are large enough to be established with some confidence at the early stages of the crystallographic analyses of the protein, before sufficient resolution has been achieved or the models sufficiently refined to determine the significance of localized differences between two structures.

5.3. Domains Showing Disorder-order Transitions

The problem of interpreting crystallographic results is even more difficult for the third type of domain flexibility, in which the basic observation is that a part of the structure is disordered. By "disorder" we mean the absence (or poor definition) of features in the electron density distribution derived by diffraction analysis of a crystalline protein. Such disorder reflects departures from exact repetition of the distribution of scattering centers in equivalent positions of the crystal, usually due to either thermal motion of the groups involved or localized conformational variations from one molecule to another in the lattice. For proteins, a few instances are also known of seqenence "microheterogeneity" (the occurence of two or more amino acids at the same sequence postion in different molecules of a sample; e.g., Hendrickson and Teeter [54]), which can also appear as disorder at low resolution.

Localized disorder of surface residues is quite common in crystalline proteins. Charged side chains at the surface of the molecule are often disordered, especially the long, unbranched sidechains of lysine and arginine, as are residues at the ends of the polypeptide chain. Disorder at this level can rarely be associated with the function of the protein. Moreover, what appears to be localized disorder in the early stages of the crystallographic analysis of a protein sometimes proves to be an artifact resulting from errors in the experimental phases or can be resolved into relatively well-defined conformational heterogeneity as the resolution of the structure analysis or the refinement of the molecular model progresses (e.g., Smith et al. [102]).

There are also a few reports of more extensive disorder, involving large fractions of a protein molecule. For example, the entire Fc region is disordered in crystals of two intact IgG molecules with normal hinges [78,35]; in these crystals, the disorder is probably due to static departures from the lattice symmetry (see Fig. 3 of Ref. [9]). Another example is tyrosyl-tRNA syntetase, about a quarter of which (99 resides) cannot be located in the crystal [13]. The source of the disorder and whether the disordered region might form a globular domain are not known for this enzyme.

More interesting is the observation in trypsinogen (the inactive precursor, or "zymogen," of the proteolytic enzyme trypsin) of a contiguous disordered region on the surface of the molecule [38]. This region is called the "activation domain" because it is found ordered after the zymogen is converted to the active enzyme. The activation domain should not be confused with either of the two "β-barrel" folding domains of trypsin (and trypsinogen). These domains do not form distinct globular units like the flexibly linked or hinged domains we have encountered so far (see Fig. 3) and do not move significantly relative to one another upon zymogen activation or substrate binding [114]. Rather, the enzyme and its zymogen both consist of a single globular unit (with molecular weights of about 23,000 daltons); differences between the two are confined primarily to the activation domain.

The ordering of the activation domain is clearly related to the function of the enzyme. In the ordered state, the activation domain borders the active site of the enzyme and forms part of the substrate binding site. The contribution of the

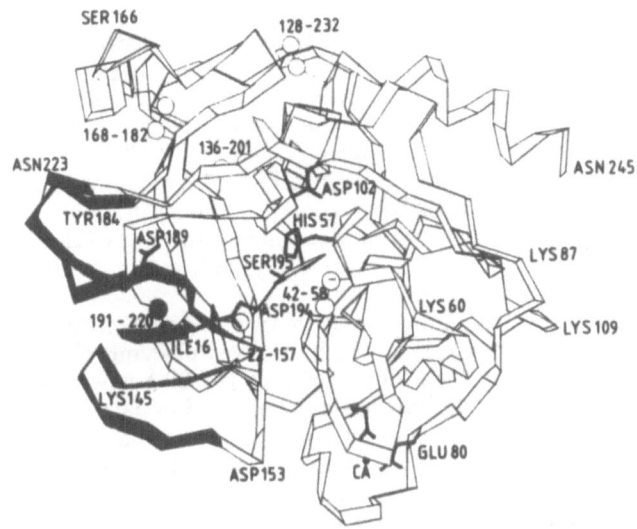

Figure 3: Schematic drawing of the folding of the polypeptide backbone of trypsin [16]; regions that are disordered in trypsinogen are shaded. Selected side chains are drawn with dark lines; sulfur atoms involved in disulfide bonds are represented by large circles.

activation domain to substrate binding is evidently critical to the biological function of the enzyme, as the reduced activity of the zymogen toward small substrates appears to be due almost entirely to its reduced affinity for them. Moreover, the conformations of catalytic residues in the zymogen and the active enzyme are very similar.

The activation domain consists of four segments of the polypeptide chain that are not involed in any regular secondary structure (Fig. 3):

1. A segment at the amino terminus of the polypeptide chain[2], residues 10 to 18.

2. The loop from residues 142 to 151 forms one edge of the domain. Parts of this loop are relatively exposed, and it is sometimes called the "autolysis loop," because it can be cleaved by other molecules of trypsin. Possibly due to partial cleavage of this sort, the conformation of the autolysis loop is relatively poorly defined at places, even in the "ordered" state.

3. The opposite edge of the domain is formed by the loop from residues 217 to 223, which forms part of the substrate binding site in the ordered state.

4. Between the latter two segments is a loop from residues 184-193, which is also part of the substrate binding site in the active enzyme.

---

[2] The amino terminus of trypsin is assigned residue number 10 for hisorical reasons.

These four segments together contain 36 residues and constitute about 15% of the molecule. The border between the ordered and disordered regions of trypsinogen is quite sharp, with the electron density of the polypeptide chain fading to background levels within one or two residues. Five of the seven junctions between ordered and disordered regions of the chain are at glycine residues. At three of these junctions, the chain on the ordered side seems to be "anchored" by aromatic residues. There are no aromatic residues within the activation domain itself. Few specific interactions are observed between the activation domain and the rest of the molecule in the ordered state, but a disulfide bridge (Cys191-Cys220) and a number of hydrogen bonds are found within the domain.

The activation domain can be ordered in two ways. One way is to perform the physiological activation reaction, which involves cleaving the amino-terminal segment of the zymogen between residues 15 and 16. The new amino terminus of the molecule, Ile16, then forms a "salt bridge" (i.e., an ion pair) with Asp194 of the middle disordered segment. These new interactions result in a concerted ordering of the activation domain, which leaves Asp194 and Ile16 almost completely buried in the protein. Under physiological conditions, this reaction is effectively irreversible, since the "activation peptide" (residues 10 to 15) diffuses away after it is cleaved off and is never present in sufficient concentrations for there to be an appreciable reverse reaction. As is the case for the hinged proteins described earlier, the crystallographic analyses of trypsin and trypsinogen on which this description is based provide only "snapshots" of the molecule in the disorderd and ordered states; details of the conformational changes required to get from one state to the other are not known.

There is, however, a second way to order the activation domain of trypsinogen, which provides some additional information about the sequence of events involved. The binding of very strong ligands has been found to shift the activation domain from the disordered to the ordered state at the expense of some binding energy [15,17]. The binding of BPTI alone shifts the internal segments of the activation domain to a conformation very similar to that observed in trypsin, except that the autolysis loop is less well ordered than in the active enzyme. In this binary complex, the amino-terminal segment remains disordered, and the pocket next to Asp194, normally occupied by Ile16 and Val17, is formed but filled with solvent. The ternary complex of trypsinogen, BPTI and a dipeptide of sequence Isoleucine-Valine (Ile-Val), however, is indistinguishable from the trypsin-BPTI binary complex, except that residue 18, which is still attached to the intact amino-terminal segment, remains disordered. The Ile-Val dipeptide occupies the pocket next to Asp194, and the autolysis loop is as ordered as in trypsin.

Considerable interest has focused on the question of whether the disorder observed in the activation domain of trypsinogen is dynamic or static in nature (and earlier on whether it was real or artifact; see Ref. [9]). From the observation of disorder, it is clear that the polypeptide segments of the activation domain can adopt a variety of conformations in the zymogen. The conceptual distinction that one would like to make between static and dynamic disorder centers in its simplest form on the question of whether or not there are transitions between these conformations on the time scale of a diffraction experiment at physiological temperatures.

Earlier evidence on this point seemed to indicate static disorder (negligible transitions), since attempts to "freeze out" motion in the activation domain at temperatures down to about 100 K produced little change in the structure; only a few residues in the amino-terminal segment were appreciably better ordered at low temperature [113]. In addition, crystallographic studies on a chemically modified trypsinogen (Hg-trypsinogen), in which an Hg atom was inserted into the Cys191-Cys220 disulfide bond, revealed that atomic displacements in the activation domain are so large that even the electron-dense Hg atom is disordered. This also seems easier to reconcile with static disorder, as motions with an r.m.s. displacement of at least 1.6 Å would be required to account for a disordered Hg atom[3]. More recently, a novel application of the perturbed angular correlation (PAC) technique to Hg-trypsinogen has revealed motions of the Hg nucleus with a correlation time of about 11 nsec, which are not present when the activation domain is stabilized by strong ligands [21]. Whether these motions are local to the Cys191-Cys220 disulfide or are an indication of widespread motion within the activation domain is not yet clear. The PAC experiments are discussed in more detail by Butz in this volume.

Another interesting feature of the activation domain that may be relevant to its dynamics is that it is apparently quite "soft" (in the sense of being easily deformed) when forced into the active (ordered) state by ligand binding. It was a matter of considerable luck that the activation domain participates in no intermolecular contacts in crystals of trypsinogen or its complexes with BPTI or BPTI and Ile-Val. This is not the case for the crystalline complex of trypsinogen and another protein inhibitor (pancreatic secretory trypsin inhibitor; PSTI), however, where intermolecular contacts appear to distort the activation domain away from its trypsin-like conformation [18].

A similar situation is seen in the comparison of chymotrypsin, another protease that is closely related to trypsin, with its zymogen chymotrypsinogen. The overall folding patterns of chymotrypsin and trypsin are similar, except for small variations in some surface loops. Both are activated by a proteolytic cleavage that allows Ile16 to form an internal salt bridge with Asp194. Earlier comparisons of unrefined models of chymotrypsin and chymotrypsinogen indicated that the conformation of the region corresponding to the activation domain is different in the zymogen, but not disordered as in trypsinogen [124]. However, parts of this region may be involved in contacts with other molecules in the chymotrypsinogen crystals used for these studies [9]. More recently, refined models of two crystallographically independent copies of chymotrypsinogen from another crystal form have been reported by Wang et al. [114]. In one copy, the region corresponding to the activation domain of trypsinogen is not involved in intermolecular contacts. This region is mostly ordered in both models, but the conformations found in the two new models differ from one another and from the conformations observed in chymotrypsin and in the earlier chymotrypsinogen model.

Thus the activation domain of chymotrypsinogen is also soft, but is apparently less flexible than the activation domain of trypsinogen. Other properties of the two zymogens are consistent with this idea [16]. Although there are many chemical differences between trypsinogen and chymotrypsinogen that might alter the flexibility of their activation domains, Wang et al. suggest that the covalent attachment of the amino terminal segment of chymotrypsinogen to the ordered region of the molecule may be important. Unlike the corresponding part of trypsinogen,

---

[3] One must anticipate some limitation to the flexibility of the activation domain; if the energy barrier for achieving the ordered conformation were small, even the relatively weakly binding natural substrates would be able to induce the active conformation, and there would be little difference in the activities of the zymogen and the "activated" enzyme.

the amino terminal segment of chymotrypsinogen has a well-defined conformation, presumably because its position is fixed by a disulfide bond that is absent in trypsinogen. This segment may serve to stablilize the conformations of other parts of the activation region, with which it makes contact.

Other examples in which parts of a protein are found in both ordered and disordered states in different crystal forms are confined mainly to relatively short, disorderd surface loops that are ordered by ligand binding; such features have been observed in several proteins involved in nucleic acid binding, for example [9,51]. Of particular interest, however, is the enzyme phospholipase, which exists in an "active" form that hydrolyzes phospholipids in large aggregates, such as micelles or membranes, and an "inactive" form (here called the "proenzyme") that is inactive toward large substrates but reactive toward monomeric phosphoglycerides. A loop of 14 residues and a segment ten residues long at the amino terminus are disordered in the proenzyme of phospholipase A_2 from bovine pancreas. The proenzyme is activated by proteolytic cleavage of seven residues from the amino terminal segment, which allows the internal loop to adopt the conformation found in the active enzyme [32]. Recent studies of the phospholipase from snake venom suggest it may have a similar mechanism of activation [20]. Although the disordered regions of phospholipase are small, they constitute a substantial fraction (20%) of the molecule, whose molecular weight is only about 14,000 daltons, well within the range accessible to computer simulations of protein dynamics.

Proteolytic processing of the precursor form of a protein to its active form is a widespread mechanism of biological control. The extent to which such processing is associated with structural transitions such as those described for trypsin, chymotrypsin or phospholipase is unknown, although it is clear that in a few instances, considerably more extensive rearrangements of the polypeptide chain must be involved. For both the hemagglutinin from influenza virus (a trimeric protein with a total molecular weight of about 225,000 daltons; crystal structure by Wilson *et al.* [121]) and the human α_1-proteinase inhibitor (a single chain of 53,000 daltons; Loebermann *et al.* [73]), only the structure of the processed form of the molecule is known. In each case, however, the ends of the polypeptide chain that were joined in the precursor are separated by a large distance (20 and 60 Å, respectively) in the processed protein, suggesting that a substantial rearrangement of the polypeptide chain accompanies processing. Unfortunately, both the details and the dynamics of the conformational changes involved in these interesting examples of protelytic activation remain to be determined.

6. DYNAMICS OF DOMAIN MOTIONS IN PROTEINS

Establishing the existence of domain flexibility is relatively simple compared to obtaining experimental information about the dynamics of the motions involved. The very features of domain motions that make them easy to detect, namely the size of the regions that move and the distances over which they move, are clearly a major source of difficulty. Most techniques that allow the study of dynamic processes at the molecular level provide information about the motion or local environment of a spectroscopic probe. When dealing with proteins one usually has either a small number of probes, whose dynamics must somehow be extrapolated to a model of motions of the protein, or a large number of probes whose signals must be resolved or for which only an average signal can be measured. As a consequence, there is little direct information about the dynamics of domain motions of proteins *per se*.

Another obvious complication to studies of the dynamics of domain motions is the variety of motions inherent to proteins, which form a background against which the events of interest must be resolved. In this regard, one might expect that domain motions involving a small number of defined conformational states would be the easiest to study, but it turns out that this is not the case.

6.1. Dynamics of Flexibly Linked Domains

The domain motions whose dynamics are best understood are those of the flexibly linked domains of immunoglobulins. This is due largely to the landmark work of Yguerabide *et al.* [125], who realized that a specifically labeled population of antibody molecules could be obtained by raising an immune response against the label itself, in their case a dansyl fluorophore. They were also the first to study this problem with time-resolved (rather than steady-state) fluorescence depolarization techniques.

In this experiment, the key simplification is the preparation of a molecule with a single probe, in contrast to the multiple labels that were introduced by earlier chemical modification techniques. This is done by immunizing an animal (usually a rabbit) with an antigen consisting of the dansyl group conjugated to a macromolecular carrier. When immunoglobulins are isolated from the animal, one typically finds that about 5 to 10% of the (heterogeneous) sample will recognize the dansyl group attached to a small polar molecule (to increase its soluability) and bind it with high affinity.

Depolarization of the fluorescence from these specifically labeled samples following excitation by a polarized nanosecond light pulse was interpreted in terms of a two-exponential process. The faster component, ϕ_S, corresponded to a rotational correlation time of 33 nsec and the slower, ϕ_L, to one of 168 nsec. In contrast, depolarizaton of fluorescence from the isolated Fab fragments of these antibodies could be described as a single-exponential process with a correlation time of 33 nsec. Thus Yguerabide *et al.* assigned ϕ_S of the intact immunoglobulin molecules to intramolecular motion of the Fab arms and assumed that ϕ_L corresponded to global tumbling of the molecule.

The basic conclusion that antibodies in solution exhibit intramolecular motion on the nanosecond time scale is now well established, although improvements in technique have led to somewhat smaller correlation times and different interpretations of the molecular details of the motions. A good summary of current techniques of sample preparation and data processing is given by Hanson *et al.* [48]. These authors also used antibodies from rabbits, but took care to remove high molecular-weight aggregates from the samples and to correct the decay of polarization anisotropy for instrumental factors, such as the width of the exciting pulse. The data were again analyzed in terms of a two-component decay, but with correlation times of 14-17 nsec for ϕ_S and 80-100 nsec for ϕ_L, where the range given shows the variation between two different samples from different rabbits. The samples were well characterized by hydrodynamic techniques, which allowed an estimate of the rotational correlation time of a sphere with the same sedimentation properties to be made (about 167 nsec). This estimate and other arguments led Hanson *et al.* to conclude that the slower component of the decay was too fast to be assigned to global tumbling of the molecule. Instead they suggested that ϕ_L corresponds to a waving movement of the Fab arms and that ϕ_S corresponds either to twisting of the Fab arms or to flexibility in the elbow regions.

The interpretation of the slower correlation time for these rabbit antibodies in terms of intramolecular motions has recently been supported by study of the same samples after "anchoring" the Fc segment by binding to protein A, a bacterial protein that interacts specifically with the Fc region of immunoglobulins. Since protein A is polyvalent, Hanson et al. [49] were able to prepare and characterize antibody-protein A aggregates of various sizes, which had estimated "equivalent-sphere" global rotational correlation times up to a microsecond. The observed depolarization, interpreted in terms of a two-exponential process, had ϕ_S components of about 15 nsec for all complexes, while ϕ_L increased modestly to 140-160 nsec for the larger complexes. Hanson [50], in an analysis of these results, argues that the small increase in the slower component is probably due to steric effects, which would suggest that the hinge segments of these antibodies allow nearly unrestricted motion of the Fab arms.

Somewhat different conclusions were reached by Reidler et al. [96] in a simialr study of monoclonal mouse antibodies. These authors were unable to resolve the anisotropy decay curve into two unique components, so only an average rotational correlation time, $\langle\phi\rangle$, was reported. For one antibody molecule studied in detail, $\langle\phi\rangle$ was 84 nsec; the correlation time for the isolated Fab fragment of this molecule was found to be 32 nsec. Anchoring the Fc with protein A or an anti-Fc antibody increased $\langle\phi\rangle$ for the intact antibody to 140-150 nsec. Reidler et al. used calculations based on theoretical models of the hydrodynamic behavior of segmented molecules to argue that the observed increase indicated substantially restricted rotation of the Fab arms of this molecule.

Although the samples studied by Reidler et al. were not as well characterized hydrodynamically as those of Hanson et al., the preparative procedures used should have removed any (unwanted) aggregates. Hanson [50] suggests that the differences in the conclusions of the two groups may have resulted because mouse antibodies (or at least the ones studied by Reidler et al.; monoclonal antibody samples are chemically homogeneous) are less flexible than rabbit antibodies. He also presents an interesting analysis of the limitations of the fluorescnce depolarization technique due to the relatively short fluorescence lifetime of the dansyl probe (about 25 nsec); the time scale of these measurements is evidently too short to resolve the slower components of the motion. This limitation explains why one does not see separate components for global tumbling of the molecule in the anisotropy decay curve and may explain why a unique decomposition of the ansotropy is difficult to obtain in some cases. Other fluorophores with somewhat longer fluorescence lifetimes have been used in similar experiments on immunoglobulins, but it has not yet been possible to resolve the depolarization into more than two components (reviewed briefly by Hanson et al. [48]).

In addition to these experimental problems, it is clear that a major stumbling block to the interpretation of the fluorescence depolarization data already available is the difficulty of predicting the hydrodynamic behavior of macromolecules of known three-dimensional structure, especially when the molecule is composed of flexible segments. As we have seen, the interpretations above are based on simple Stokes-law arguments or approximate analytical models of the behavior of flexibly linked bodies with simple shapes. Although the development of exact analytical expressions for the hydrodynamic behavior of macromolecules remains a topic of considerable interest (e.g., Wegener [115]), Monte Carlo simulations of the depolarization of fluorescence from flexibly linked macromolecules [52] indicate that even for simple systems, such as two flexibly linked rods, the depolarization displays a complex dependence on a number of factors that precludes a simple interpretation.

An extension to traditional hydrodynamic models that is of interest in view of the complex shape of the molecular surface of a protein (e.g., Fig. 2) is the representation of an arbitrary body by a number of small spheres (or "beads"), whose hydrodynamic properties are readily calculated [42]. This approach has been used to calculate translational friction coefficients for a number of rigid proteins with encouraging results [108]; the theoretical results differ systematically from the observed values but can be corrected with reasonable assumptions about the distribution of bound solvent molecules. A basic problem with the application of continuum hydrodynamics to bodies the size of an average protein has yet to be resolved, however. As pointed out by Kuntz and Kauzmann [69], it is not clear that the boundary conditions normally applied to solve the equations of motion in continuum hydrodynamics are meaningful for a body whose diameter is only about a factor of ten greater than that of a water molecule.

6.2. Dynamics of Disorder-order Transitions

While the detailed interpretation of the nanosecond motions of flexibly linked domains remains open to discussion, it is at least clear what is moving and what the time scale of the motions is. This is not the case for either of the other two types of domain flexibility. The disorder-order transition of trypsinogen, for example, appears to be associated with a slow relaxation process, but it is not clear whether the spectroscopic probes use to follow the process are associated with the activation domain itself or reflect a coupled process elsewhere in the protein.

Initial studies of the rate of the transition were performed on a derivative of the zymogen in which a p-guanidinobenzoate group (pGB) is covalently attached to a reactive residue in the active site. The pGB derivatives of trypsin and trypsinogen have characteristic circular dichroism spectra, and the addition of Ile-Val dipeptide to pGB-trypsinogen shifts its spectrum to one that is indistinguishable from that of pGB-trypsin [15], indicating that this combination of ligands can shift the activation domain into the ordered state (at least in the vicinity of the active site). Nolte and Neumann [85] studied the relaxation kinetics of this system with temperature-jump techniques and observed slow relaxation times of about 0.5 sec. From the concentration dependence of the process, they concluded that the binding of Ile-Val was rapid and that the slow process observed was a substrate-induced change in the protein, presumably the crystallographically characterized disorder-order transition; the rate of the transition (in the disorder to order direction) was about 26 sec^{-1}. The relaxation rates were determined in this case by monitoring changes in UV absorption at 280 nm. Since both the protein and the bound pGB group have absorption maxima near this wavelength and similar molar extinctions, it was not clear whether the absorption changes in these experiments reflected events in the immediate vicinity of the active site or elsewhere in the protein. As the changes observed were only a few tenths of a percent of the total absorption of the system, the events responsible might well have been highly localized.

Very recent experiments by Ascenzi *et al.* [4] are quite interesting in this regard. These authors have studied the binding of Ile-Val to binary complexes of trypsinogen with p-GB, BPTI and PSTI by stopped-flow techniques, monitoring changes in the UV absorption of the complexes over a wide range of wavelengths. Remarkably, the rate of the slow process (again assuming rapid dipeptide binding followed by a slow change in the protein) was essentially constant (25-33 sec^{-1}), regardless of which binary complex was studied and where in the range of 240-320 nm the absorption change was monitored.

Although it is clear from the results of Ascenzi *et al.* that the slow rate is not associated with localized events at the active site, some question remains as to whether the absorption changes are due to the reorganization of the activation domain itself or to more subtle changes elsewhere in the protein. Since the the activation domain is already almost fully ordered in the crystal structure of the binary BPTI-trypsinogen complex, the rate-limiting step leading to the absorption change either occurs after this stage of the ordering process or is related to events outside the activation domain. NMR studies [90] have demonstrated the existence of conformational changes outside the activation domain that are coupled to the ordering process but not resolved in the crystal structures.

The transition of chymotrypsin from an inactive conformation at high pH to the active conformation near neutral pH is similarly slow. The high-pH form of the enzyme is apparently inactive because it does not bind subsrates. The inactive conformation is thought to result from deprotonation of the amino terminus, which disrupts the salt link between Ile16 and Asp194; thus the process observed is probably related to the transition of the activation domain from a chymotrypsinogen-like soft state to the chymotrypsin state. Fersht and Requena [40] studied the rate of this process by stopped-flow experiments, in which chymotrypsin in the inactive conformation (at high pH) was subjected to a rapid pH shift. The formation of the active conformation was followed by monitoring changes in the UV absorption of the dye proflavin, which binds only to the active enzyme. This process, which is clearly associated with changes at the active site that allow the dye to bind, has a rate (inactive to active) of about 3 sec^{-1}.

In considering these kinetic data, one should keep in mind that the transitions studied must involve a large number of elementary processes (bond isomerizations, etc.), presumably with a wide range of rates. As we have already seen, the existence of nanosecond motions in the disordered activation domain of trypsinogen has been established by PAC experiments. More detailed studies with this technique [22] suggest that motions of the Hg atom in Hg-trypsinogen involve transitions among a few preferred conformations separated by modest activation enthalpies of about 8 kcal/mol. Such intrinsic motions of the disordered state might well be related to the earliest events in the transition to order, since the simplest effect of the formation of a salt link between Ile16 (or Ile-Val) and Asp194 would be to shift the equilibrium of preferred conformations of residues in its vicinity, which could eventually lead to rearrangement of the entire activation domain.

The main obstacles to a more detailed understanding of the activation of these zymogens is clearly a lack of information about both the structure(s) and dynamics of the various regions of the activation domain at all stages of the ordering process. The studies by Butz and colleagues are an important first step, but are limited by the nature of the probe to providing highly localized information. Without additional information from other regions of the molecule at different stages of the process, it is difficult to see how one can begin to understand how formation of the Ile16-Asp194 salt bridge initiates the sequence of events that leads to an ordered activation domain. Against this background, it is clear that the specific question suggested by the kinetic studies outlined above, namely the nature of the rate-limiting slow step, is but one of many that remain to be answered.

6.3. Dynamics of Hinged Domains

In contrast to the proteins with disordered domains, the nature of the motions in hinged proteins is reasonably well defined, but the dynamics of these motions is not. There are, to be sure, numerous kinetic studies of ligand binding to proteins now known to have hinged domains, some which display multi-step binding reactions. However, in no case can the slower step(s) of the reaction be unambiguously associated with the domain movements.

Discussions of protein "isomerizations" linked with ligand binding often start with Hammes's summary of early results (prior to 1970) of the application of rapid-reaction techniques to proteins (e.g., Hammes and Schimmel [47]). His tabulation of the time constants for processes associated with conformational changes in proteins demonstrates that values in the range of 1 to 10^{-4} sec are not uncommon. Some of these classic examples of proteins that undergo slow conformational isomerizations upon ligand binding have subsequently been shown to have hinged domains (e.g., AATase, GAPDH and LADH), and one (chymotrypsin) undergoes another type of large-scale rearrangement. Others, however, show only localized conformational changes upon ligand binding. The best studied example in the latter group (not in the early lists) is hen egg-white lysozyme, a small enzyme (about 15,000 daltons) that hydrolyzes the β-(1-4)-glycosidic linkage of oligosaccharides.

A number of stopped-flow and temperature-jump studies of the binding of oligosaccharides to lysozyme [6] have resolved three slow processes in the protein following a rapid bimolecular association. The fastest process, with a forward rate constant of 200 sec^{-1}, is associated with nonproductive binding, while the slower processes (14 sec^{-1} and 0.5 sec^{-1}) precede the rate-limiting step of the reaction following productive substrate binding. The turnover number of lysozyme is about 0.04 sec^{-1} under the conditions used for these studies. Both of the structural probes monitored, changes in protein fluorescence and proton uptake by acidic groups on the protein, can be associated with residues at the active site. Moreover, crystallographic (e.g., Kelly et al. [62]) and NMR [89] studies of ligand binding to lysozyme both indicate that changes in the conformation of the protein are limited to residues at or near the active site. Thus while it is often "tempting" to associate slow kinetic processes with domain motions (for enzymes in which both are observed), such correlations are unfounded unless one can somehow rule out the possiblity that the processes are associated with more localized structural changes.

It is probably no coincidence that the classic examples of proteins in which slow isomerizations are observed by fast-reaction methods almost all bind relatively large, aromatic ligands, such as NAD^+, that serve as chromophores with which the binding reaction can be monitored. These probes necessarily reflect changes in the immediate vicinity of their binding site. In contrast, proteins in which the motion of hinged domains is linked to the binding of smaller, less hydrophobic ligands often do not exhibit multi-step kinetics. Proteins in this group include two for which the existence of the domain movements in solution is best established (by low-angle X-ray scattering), hexokinase and ABP. In both cases the available evidence indirectly suggests that the domain motions are relatively fast.

Temperature-jump studies of the rate of glucose binding to hexokinase, for example, revealed a single relaxation process under all conditions studied [55]. However, since a conformational change of the protein was expected on other grounds, and since the apparent rate constant for the bimolecular reaction, about 2×10^6 $M^{-1}sec^{-1}$, was considerably slower than the theoretical diffusion-controlled limit for molecules of this size (about 10^9 $M^{-1}sec^{-1}$), the data were interpreted in terms of a substrate-induced isomerization of the protein. The isomerization was assumed to be either too fast to resolve with the instrumentation available or not linked to changes in protein fluorescence, the optical property of the system that was monitored. With this assumption, a lower limit of 10^3-10^4 sec^{-1} (depending on the isozyme studied) could be placed on the rate of the conformational change. Similar observations indicative of a simple bimolecular association reaction have been made for the binding of ligand to a number of bacterial transport proteins, again monitored by changes in protein fluorescence known to be linked with ligand binding [79]. For ABP, the apparent bimolecular rate constant was about 2×10^7 $M^{-1}sec^{-1}$, still slower than the theoretical diffusion limit. In this case, the authors suggested that the rate of formation of the closed conformation could be as high as 2×10^8 sec^{-1} in the limit of very slow dissociation of the ligand from the open conformation.

On the other hand, it is reasonable to expect that some of the slower rates measured by relaxation methods can also be associated with domain motions. For example, pressure-jump studies of LADH [26] provide kinetic information comparable to that described above for hexokinase and ABP. In the presence of NAD^+, a slow process with reciprocal relaxation times of 60 to 300 sec^{-1} (depending on NAD^+ concentration) was observed, which could be interpreted in terms of an isomerization of the protein with a rate of about 300 sec^{-1}. Interestingly, a faster process (reciprocal relaxation times of 2500 to 7000 sec^{-1}, depending on pH) was observed when the enzyme alone was subjected to a pressure jump. As was the case in the studies of hexokinase and the transport proteins, the conformational change was monitored by changes in protein fluorescence in these experiments. Since the group most likely to be the fluorophore in LADH is a tryptophan residue located between the two domains [123], it is at least as plausible as in the former examples that the rates measured for LADH in these studies can be associated with domain motions.

If accepted at face value, these experiments seem to indicate that transitions from the open to the closed conformations of hinged proteins occur with a wide range of rates, from the "slow isomerizations" detected by rapid-reaction techniques to rates with time constants approaching the rotational correlation times of flexibly linked globular domains. It should be emphasized, however, that in no case can one unambiguously assign a rate to the crystallographically characterized motions of proteins known to have hinged domains.

6.4. Coupling of Motions on Different Time Scales

For enzymes with hinged domains, the rate of the overall domain motion is an obvious focus of interest in the dynamics of these molecules. With the exception of PFK (and possibly PGK), the substrates of the enzymes in Table II are sufficiently buried within the protein in the closed conformation that the molecule almost certainly must cycle between open and closed conformations at least once per turnover to allow reasonable rates of substrate binding and product release. Here again, however, one must expect that the overall domain motions will be coupled with changes in other, more localized motions of the protein displaying a wide range of time constants. This expectation is based on thermodynamic properties of proteins outlined in the next paragraph and is confirmed by experiment for a few hinged proteins.

It is frequently observed that dynamic processes involving proteins (e.g., ligand binding or unfolding) are accompanied by large changes in heat capacity. Sturtevant [105] has suggested that changes in the low-frequency vibrational modes of the protein should make a major contribution to these changes in heat capacity. That ligand binding can induce changes in the dynamics of a protein is well established. A recent detailed study of lysozyme [46], for example, combined hydrogen-exchange data with the results of neutron diffraction experiments to identify regions of the molecule whose exchange rates are perturbed by substrate binding.

Similar observations have been made for some hinged proteins, including changes in the hydrogen-exchange behavior of AATase upon substrate binding [93]. Since the degree to which various ligands modify the extent of hydrogen exchange is not strictly correlated with the degree to which they tend to induce the transition to the closed conformation, the domain motions apparently are not required for ligands to change the hydrogen-exchange characteristics of this enzyme. However, the widespread structural variations in the small domain that accompany domain closure in citrate synthetase suggest a mechanism by which domain motions *per se* could also be linked to changes in the fluctuations of a protein.

Given the expectation that ligand binding to proteins (and the accompanying large-scale motions of hinged proteins) would be linked to changes in their low-frequency vibrational modes, it was somewhat surprising to find that the binding of glucose to hexokinase is accompanied by no net change in heat capacity or enthalphy, especially since the enthalpy of unfolding of the enzyme increases substantially in the presence of glucose [107]. This behavior probably results from compensatory changes in the heat capacity of different components of the system, since attempts to observe the vibrational spectrum of the protein directly by inelastic neutron scattering [59] suggest that glucose binding shifts the spectrum in the direction of fewer vibrational modes or smaller amplitudes over essentially the full range of frequencies experimentally accessible (10 to 500 cm^{-1}).

Unfortunately, detailed study of low-frequency vibrations in proteins and their possible relationships to domain motions is hampered by a number of experimental factors. Distinct features in the IR or Raman spectra of proteins can often be observed only in the solid state (crystals or dehydrated films), which is usually interpreted to mean that the low-frequency vibrations of proteins are damped in solution [92]. Moreover, at the resolution attainable with current techniques, vibrational spectra of proteins tend to show only a few features, which are frequently broad peaks near 25 cm^{-1} and 75 cm^{-1} [87,72]. These peaks do not seem to correlate with protein size (i.e., domain structure). Some attempts have been made to correlate one peak or the other with the occurence of different types of secondary structure, but there appears to be no generally accepted interpretation. Finally, the particularly interesting region of the vibrational spectrum of a protein below about 10 cm^{-1} (see below) is difficult to resolve with current Raman or inelastic neutron scattering techniques due to overlap between elastic and inelastic scattering.

6.5. Computational Studies of Domain Motions

Even the fastest rates associated with domain motions in proteins (ignoring coupling considerations for the moment) indicate that these processes are much too slow to be studied by molecular dynamics simulations. Such calculations are limited by computing resources to a few hundred picoseconds, even for small proteins. However, several attempts have been made to study domain motions using other computational methods.

The first approach to the study of large-scale motions in proteins by computer simulation was introduced by McCammon *et al.* [75], who attempted to map the potential function of a proposed "hinge-bending mode" in lysozyme. To accomplish this, the location of a rotation axis between the two domains of the molecule was assumed, and a series of small rotations of one domain relative to the other was applied to an atomic model of the protein derived from the X-ray model by minimization of its potential energy. At each step the minimization procedure was repeated to relax as much of the strain introduced by previous rotations as possible (a procedure sometimes called "adiabatic mapping" [74]). The result of these calculations was a nearly harmonic effective potential function, from which a vibrational frequency of about 4 cm^{-1} could be estimated, ignoring any damping or other hydrodynamic effects. Attempts to confirm this prediction using inelastic neutron scattering [8] have been unsuccessful; as we have noted, however, observations in this frequency range are difficult.

Similar studies have been made on two hinged proteins, ABP [77] and LADH [28]. Although the effective potential functions are not as smoothly harmonic as that found for lysozyme (probably because larger rotational steps were taken to reduce computing time for these much larger molecules), the results are qualitatively the same in each case; a potential function with a single minimum near the orientation of the starting model is obtained. This result is quite reasonable for lysozyme, which seems to have only one stable conformation, and it could be correct for ABP, which may not have a unique open conformation.

However, it is troubling that the simulation of the domain motion of LADH, which started from the open conformation, did not find a minimum in the potential function near the orientation of the closed conformation. Analysis of the components of the effective potential function by Colonna-Cesari *et al.* suggests that the relaxation of strain may be incomplete[4], which may have influenced the form of the potential function. Alternatively, the absence of a second minimum could simply have resulted because the cofactor NAD^+ was omitted from the calculations; this is the interpretation preferred by Colonna-Cesari *et al.* Under the circumstances, however, one must wonder if the complementary simulaton, starting from the closed structure, would also show a single minimum at the new starting orientation. The answer to this question should be known in the near future, if Colonna-Cesari *et al.* pursue the series of studies outlined in their paper. Until a more convincing demonstration of the validity of this approach is available, one must view the results of such calculations on large proteins with some skepticism.

Another computational approach that has been attempted is the direct calculation of the normal modes of a protein. Although the rigorous normal mode analysis applied successfully to small oligopeptides has so far been attempted for only the

---

[4] Only parts of the structure were allowed to relax in order to reduce computing time.

smallest of proteins (glucagon, essentially a single α-helix of 29 residues [67]), two recent treatments of proteins as large as lysozyme by methods involving quite different simplifications of the problem are encouraging. Levitt *et al.* [72] employ a general formulation of the normal-modes problem, which allows the usual simplification of performing the calculation in terms of the natural variables of the system (the unrestrained torsion angles of the protein in this case); this procedure was applied to several small proteins, including lysozyme. Brooks and Karplus [19] use an iterative approach designed to converge on the normal mode(s) closest to an initial guess at the motion of the system (in this case the mode proposed by McCammon *et al.* for lysozyme); for these calculations, the problem was formulated in terms of the Cartesian coordinates of the molecule, which involves a larger number of parameters. Both studies employed the simplified empirical potential functions normally used to estimate the energy of a protein molecule [74].

The encouraging thing is that both of these methods arrive at essentially the same set of two normal modes, which are the modes of lowest frequency obtained by Levitt *et al.* Neither of these modes corresponds to the simple rotation proposed by McCammon *et al.*, but one of them results in similar displacements of atoms near the active site. The frequency predicted for the lowest-frequency mode is about the same in each case, in the range of 3-4 cm^{-1}.

The only difficulty with these results is that it is not entirely clear that lysozyme actually undergoes deformations of this type. The idea that the domains of lysozyme might move relative to one another as separate units is based simply on its bilobal structure. It is an elementary result of solid-state physics that the classical normal modes of a monatomic crystal are the same as those of an elastic body of the same shape in the low-frequency limit. Since one can easily imagine that an elastic model of lysozyme would preferentially bend in the narrow connecting region between the lobes, it seems reasonable to suppose that the molecule would have low-frequency vibrational modes corresponding to rotations around a hinge located in the connecting region.

Unfortunately, an increasing number of structural studies of lysosyme have failed to detect concerted displacements of the domains that might correspond to hinged motions. The significance of structural studies in this regard is that even if such motions are damped (and thus stochastic rather than oscillatory) in solution, the principle modes of motion, if correct, should still correspond to preferred deformations of the molecule. As we have already mentioned, however, substrate binding to lysozyme appears to cause only local structural changes. Comparisons of the structure of the molecule in a number of different crystal lattices, including one produced by a temperature-dependent space group transformation of the original lysozyme structure, have also failed to detect any concerted motion of the domains (e.g., Berthou *et al.* [12]), as have crystallographic studies of the protein at high pressure [68].

A direct attempt to analyze the Debye-Waller factors of lysozyme in terms of concerted displacements of the atoms [104] found that the distribution of displacements was best described in terms of libration of the whole molecule or radial "breathing" motions, rather than movement of the domains around either of two guesses at a bending axis, including the one proposed by McCammon *et al.* Moreover, studies of the viscoelastic properties of several crystal forms of lysozyme [81] and an analysis of the X-ray diffuse scattering from one crystal form [33] suggest no preferential displacements in directions that would correspond to domain motions. The latter two experiments were both interpreted in terms of rigid-body displacements of the whole molecule in the crystal lattice.

None of these experimental results can be taken as evidence that there is not a hinge-like mode in lysozyme, since the maximum r.m.s. displacements of about 0.3 Å cited for the lowest-frequency mode of Levitt *et al.* are just above the error level of a well-determined protein structure. While one might hope to be able to detect simple concerted motions at this level, it is easy to imagine that such small differences could be obscured by the larger localized displacements often found near the active site and near packing contacts. It is unfortunate that none of the more sophisticated experimental techniques mentioned above (i.e., none except the comparison of structures in different lattices and in the presence and absence of substrates) has yet been applied to a protein that is known to have hinged domains. Such studies should help to establish the extent to which domain motions restricted by a crystal lattice can be detected with these methods.

The question of the reliability of a computational procedure when applied to the study of large-scale motions in proteins is obviously one of considerable interest. Since it is unlikely that we will ever be able to experimentally characterize the various motions of a protein in enough detail to describe the sequence of events on different time scales that lead to a ligand-induced domain rotation or disorder-order transition, computational methods of some sort would seem to be our best hope in the long run of understanding such processes. Although there is still considerable work to be done in the experimental characterization of domain motions, it also seems reasonable to expect that computational studies will turn increasingly to the larger proteins which are known to exhibit domain'flexibility. As we have seen, a few attempts in this direction have already appeared [77,28], and there is no obvious reason why the normal-mode approach of Levitt *et al.* [72] could not be applied to larger proteins, especially in view of the substantial increases in computing power that have recently become available and that can be expected in the near future.

As pointed out by Levitt *et al.*, normal-mode analysis offers an attractive approach to the study of triggered processes, such as the ligand-induced conformational changes of hinged proteins or the disorder-order transition of trypsinogen, since it is reasonable to suppose that the earliest steps of such processes correspond to motions which can be described by some combination of the lowest modes of the initial structure. Moreover, recent successes in explaining some features of the rates of reactions involving proteins in terms of a quasiharmonic formalism [14,44] suggest a connection between the normal modes that we can now (one hopes) begin to calculate and the coupling of motions on different time scales (and between the protein and its environment) that we would like to understand.

7. CORRELATION OF DOMAIN MOTIONS WITH FUNCTION

As was pointed out earlier, the relationship of domain flexibility to the function of a protein must be established experimentally; evidence bearing on this question for each class of domain motions has been reviewed in detail elsewhere [9]. In this section, we focus on the more limited question of how the dynamics of domain motions in proteins are related to function.

The principle function of flexibly linked domains appears to be that of allowing the protein to interact with macromolecular ligands in a variety of orientations.

Such interactions often serve as signals in complicated cellular processes that are much slower than the binding and recognition events *per se*, so one would not necessarily expect that the dynamics of these domain motions would play a significant role in the function of the protein. However, Oi *et al.* [86] have found that the segmental flexibility of a group of monoclonal anti-dansyl mouse antibodies, as assessed by fluorescence depolarization studies on the nanosecond time scale, is correlated with the ability of the antibodies to initiate complement fixation (a relatively rapid assay of antibody function). These observations are consistent with earlier evidence that IgG molecules from which the hinge region is missing have a substantially reduced ability to interact with other components of the immune system [64]. The details of these interactions, and thus the influence that segmental motions might have on them, are unknown.

Similarly, it is not immediately obvious that the kinetics of the transition of trypsinogen or chymotrypsinogen from the inactive (disordered or soft) state to the active state are important to the function of these molecules. In the simple picture of zymogen activation normally presented, this reaction would only need to occur once in the lifetime of the protein. However, the results of thermodynamic studies of both enzymes [15,40] indicate that the equilibrium between active and inactive conformations in the "activated" enzymes is such that any given molecule must spend a significant fraction of its time (1-10%) in the inactive conformation. Thus the rate of the transition between the two conformations could well be of more importance than the simple "once in a lifetime" picture of zymogen activation would suggest. It is probably no coincidence that the rates of the transition are greater than the turnover numbers for most substrates.

The situation is more complex for hinged enzymes, as the detailed role of domain flexibility seems not to be the same in all cases. Hexokinase, for example, is a classic example of "induced fit," a phrase that enzymologists usually employ to describe a ligand-induced conformational change that brings critical reactive groups on the protein into optimal positions for catalysis. The existence of an induced-fit process of this type has been established unambiguously for hexokinase and citrate synthetase by the observation that small molecules related in structure to the substrate that triggers the domain motion can stimulate side reactions. In LADH, however, the domain motions produce no major changes in the active site, so an induced fit process in the enzymological sense probably does not occur.

Moreover, there is no common chemical step among the reactions catalyzed by the enzymes listed in Table II that would suggest a common role for the hinged domains. For hexokinase there is probably a functional requirement for the ligand-induced conformational change; it seems to prevent the enzyme from functioning as an ATPase (in effect, transferring a phosphate group to water) when a suitable sugar substrate is not present. Since a water molecule is chemically and structurally similar to the reactive 6-hydroxyl of glucose, and since water is always present in high concentrations in the physiological environment of the enzyme, it is likely that hexokinase would exihibit substantial ATPase activity if it were in the active conformation all of the time. However, it is not clear that water would serve as a particularly good substrate for any of the other hinged enzymes, since even the substrates of the other phosphate-transfer enzymes are in higher oxidation states than water.

Another possible functional role of the ligand-induced conformational change for all of the enzymes listed in the table except PFK and possibly PGK is to isolate the active site from solvent. It is often assumed that this isolation promotes catalysis by providing an apolar environment in which the reaction can proceed. Whether this assumption is true or not, the enzyme presumably must adopt the open conformation at least once per catalytic cycle to allow product release and substrate entry. If so, the importance of the kinetics of the domain motions involved is obvious, as noted earlier. It is interesting in this regard that for most of the enzymes listed in the table, the conformational change *per se* is not the rate-limiting step of the reaction. From the observation that the turnover numbers of these enzymes vary widely from a few per sec for LADH to about 6000 sec^{-1} for citrate synthetase, it would appear that selective pressures can adapt the the structure of the protein as needed to prevent the dynamics of domain motions from limiting the rate of the reaction.

8. FUTURE PROSPECTS

In this chapter I have presented an overview of the types of large-scale, concerted motions that are observed in proteins and have summarized such information as is available about the time scales on which the motions occur. At each stage, I have attempted to focus on the experimental or theoretical problems that seem to be hindering an understanding of the dynamic behavior of these systems at a level which should be attainable with current techniques.

Beyond these immediate problems, however, one can see that two major obstacles to a clearer picture of the dynamics of domain motions have come up repeatedly in our discussion. One is simply that both experimental and theoretical studies of these motions have only just begun. What little we can say on this topic comes from a few key experiments that provide explicit dynamic information about the molecules studied. Although a number of physical techniques have been adapted to the study of protein dynamics, they have rarely been applied to more than one protein (if any) that is known to undergo large-scale domain motions. One hopes that as the number of systems available for study and the nature of the questions that they pose become more widely known, a broader database will be developed from which a more coherent picture of domain dynamics can emerge.

One extension of existing techniques that appears to hold considerable promise for the study of protein dynamics (at all levels of structure) are recent developments in rapid data-collection techniques at high-intensity synchrotron X-ray sources (e.g., Moffat *et al.* [80]). There is considerable optimism that these efforts will lead to the possibility of performing time-resolved X-ray diffraction studies. It is easy to see that a time-resolved study of a relatively slow process, such as the ordering of the activation domain of trypsinogen, could considerably clarify our picture of the sequence of events involved. Although trypsin and trypsinogen can be crystallized in isomorphous lattices, it remains to be seen whether a practical way of inducing the transition between the ordered and disordered states in the crystal can be found.

The second major obstacle that has appeared repeatedly is that we have detailed information about only the "before" and "after" states of triggered domain motions. We have essentially no idea how the molecule gets from one state to the other, even though we usually have a pretty good idea of what it takes experimentally to trigger the change. As indicated earlier, this question will probably have to be addressed by computational techniques, but it is not yet clear that the computational approaches currently available can be applied to the large proteins that display domain motions with any degree of confidence.

In spite of these difficulties, the time is ripe for the development of more precise models of the sequence of events involved in such conformational changes. The current state of the art in molecular biology (meant here in the "gene technology" sense) makes it possible to create proteins modified essentially at will, often in quantities suitable for physical studies. Most of these "directed mutagenesis" experiments so far have been designed to answer questions of enzyme mechanism or genetic control, but a few attempts to study the dynamic behavior of proteins by this approach have already been reported. The first was a study of the functional role of a surface loop known to move into the active site region of lactate dehydrogenase upon substrate binding (see Grau et al. [45] for structural details and Parker et al. [88] for kinetics). Clarke et al. [25] showed by substitution of glutamine for arginine at one position in this loop that the arginine (and thus the loop movement) has an important role in the catalytic reaction of the enzyme. Wilson et al. [120] made a direct attempt to perturb the dynamics of the postulated domain motions of PGK by changing a histidine residue in the hinge region to glutamine. This subsitution had no clear-cut effect on the activity of the enzyme (or, presumably, the domain motions), which forced the reappraisal of a proposed role for the histidine residue in triggering the conformational change.

It is clear that if more detailed models for the sequence of events leading from substrate binding to domain closure in hinged proteins could be devised, the tools to test them are available in principle (if a test involving the change of a few key residues can be devised) and an interest in performing the experiments exists in practice. With this thought in mind, I would like to draw attention to a feature of hinged enzymes that recent crystallographic studies have made clear is reasonably common and which may have a bearing on the question of how substrate binding triggers the conformational change.

To place these observations in perspective, it is useful to recall some of the current models for this process. The earliest idea was simply a literal interpretation of the thermodynamic model for an induced-fit process. The equilibrium between open and closed conformations is assumed to favor the open form; interactions between substrate and protein shift the equilibrium toward the closed form. Most attempts to define a mechanism for ligand-induced domain motions are directed towards identifying ligand-protein interactions that might serve as the "attractive force" in this model.

The observation that the glucose molecule is deeply buried inside the closed conformation of hexokinase (Fig. 2) later led to the proposal of a "solvent displacement" model of the trigger mechanism for this enzyme [10]. The argument is simply that the formation of the closed conformation of the protein in the absence of substrate is highly unfavorable because it would involve the formation of an empty (save for six or so water molecules) cavity in the middle of the protein. This solvation effect is proposed to be the dominant factor in shifting the conformational equilibrium far toward the open state in the absence of glucose. In this model, the main role of the sugar in triggering the conformational change is to displace solvent from the cavity, which allows the protein to adopt its (presumably preferred) closed conformation, assuming that the sugar has polar (hydroxyl) groups in the right places to specifically compensate for hydrogen bonds that are lost when water is displaced from the cavity.

However, neither of these models accounts for subsequent observations that the ligand whose binding triggers the conformational change of a hinged protein almost always interacts directly with at least one of the hinge segments connecting the domains. This trend was first noted for hexokinase, LADH and citrate synthetase by Bennett and Huber [9] and is particularly striking for the well-defined hinge point at His274 of the latter enzyme (see Wiegand et al. [117]). Although no distinct hinge point is found in hexokinase, glucose is observed to interact directly with Asp189, the last residue of an α-helix that is immediately adjacent to a four-residue connecting segment between the domains of this enzyme [2].

A number of recent studies of hinged enzymes are consistent with the idea that specific interactions between the triggering ligand and one of the connecting segments are a common feature of these proteins. Cys174, a protein ligand of the catalytic Zn atom of LADH, is in one of the hinge segments identified by Colonna-Cesari et al. [28], for example; as mentioned earlier, the conformation of crystalline LADH is sensitive to the environment of this Zn atom. The refined structure of liganded B. stearothermophilus GAPDH [101] verifies earlier observations that NAD^+ forms a hydrogen bond with Asn313; this residue is also near the end of an α-helix that is directly adjacent to a connecting segment. The allosteric substrate of PFK, fructose-6-phosphate, is hydrogen bonded to Asp127, a residue in the middle of an irregular loop between the domains of this enzyme [36]

The refined structure of the liganded form of ACTase [66] reveals a slightly different situation. In this case, the triggering ligand appears to form a hydrogen bond with His134, a residue near one end of α-helix H5, which is normally assigned to the "polar" domain of the catalytic subunit. Although this helix is adjacent to a domain boundary, His134 is at the end of the helix distal to the connecting segment. However, the domain boundaries of ACTase (and most other hinged proteins) were somewhat arbitrarily chosen before the domain motions had been described. Helix H5 is positioned between the domains and might also be regarded as part of the connecting segment, rather than a part of the polar domain (see Fig. 8 of Krause et al. [66]).

Crystals of a closed conformation of PGK have not been reported, and the closed structure of AATase has not yet been presented in enough detail to assess whether ligands that trigger the conformational change interact with any of the hinge segments. The only clear exception to this trend so far is found in the transport proteins. Although it appears from backbone drawings of ABP (the closed form) that the sugar binding site is near the connecting strands (e.g., Gilliland and Quiocho [43]), the detailed interactions reported by Newcomer et al. [83] do not include direct interactions with any of these strands. Moreover, studies of ligand binding to LIV-BP (the open form) in the crystal indicate that the initial binding site is well removed from the hinge region. One should also mention that NAD^+ forms the same hydrogen bond with Asn313 of lobster GAPDH, which does not seem to undergo a large-scale conformational change upon ligand binding [82], that is observed in the closed conformation of the enzyme from B. stearothermophilus.

It is not at all clear what (if any) significance the direct contacts between triggering ligands and hinge segments described above might have. It is interesting that residues in the connecting segments involved in specific contacts with ligand do not themselves move much during the conformational change in any of the enzymes for which such interactions have been established. This is just another way of saying that these residues are near the hinge, or rotation axis (defined as the line that is not moved by the rotation), even for the proteins that do not seem to have a well-defined hinge in the polypeptide backbone. However, it is not obvious why this should be an important factor in either of the triggering models described above. Alternatively, in the context of our earlier focus on dynamics, one might speculate that contacts with hinge segments somehow perturb the (hypothetical) vibrational coupling between the domains; but it is not obvious why the connecting segments would be more important in this regard than the extensive noncovalent contacts that are characteristic of hinged proteins. Perhaps computational studies of hinged proteins will eventually move such considerations out of the realm of speculation.

In the meantime, it would be interesting to see what effect perturbation of these direct ligand-hinge contacts by site-directed mutagenesis has on the domain motions of the protein. Such experiments may be difficult to design, since one would prefer to modify the system in such a way as to avoid large changes in the affinity of the protein for the ligand. For this reason, it is quite encouraging that an experiment much along these lines has already been performed successfully on LADH by chemically removing the reactive Zn atom [99]. This modified form of LADH is the only one described so far for which the binary complex of enzyme and NADH can be crystallized in the triclinic crystal form (closed conformation); normally a ternary complex of enzyme, NADH and substrate (or inhibitor) is required to obtain these crystals. Since the sensitivity of the conformational equilibrium of LADH to changes in the environment of this Zn atom is seen mainly in crystallization experiments, one cannot conclude too much from this result alone; it would obviously be of interest to know if removal of the active-site Zn atom affects the dynamic properties of the enzyme in solution as well.

In view of the common occurence of ligand-hinge contacts outlined above, the intriguing results obtained by Brändén and his colleagues in their extensive studies of ligand binding to LADH should serve as a stimulus to investigations of the role of ligand-hinge interactions in other systems. As the proposals of Watson and his colleagues [120] for PGK illustrate, the connecting segments of hinged proteins are of obvious interest when one is concerned with the question of how the conformational change is triggered. One can hope that the observations of ligand-hinge contacts summarized above will help to focus attention on regions of the connecting segments of different proteins that may have similar roles in this process.

I thank Drs. W. Bode and R. Huber for permission to reproduce Fig. 3 and for many useful discussions, K. Epp for assistance with the figures, and Prof. H. G. Wittmann for support.

References

[1] A. G. Amit, R. A. Mariuzza, S. E. V. Phillips and R. J. Poljak, *Science*, **1986**, *233*, 747-753.
[2] C. M. Anderson, R. E. Stenkamp, R. C. McDonald and T. A. Steitz, *J. Mol. Biol.*, **1978**, *123*, 207-219.
[3] A. Arnone, P. D. Briley, P. H. Rogers, C. C. Hyde, C. M. Metzler and D. E. Metzler, in *Molecular Structure and biological Activity*, J. F. Griffin and W. L. Duax (Eds.), Elsevier, New York, 57-77, 1982.
[4] P. Ascenzi, G. Amiconi, M. Bolognesi, E. Menegatti and M. Guarneri, *J. Mol. Biol.*, **1987**, *194*, 751-754.

[5] R. D. Banks, C. C. F. Blake, P. R. Evans, R. Haser, D. W. Rice, G. W. Hardy, M. Merrett and A. W. Phillips, *Nature*, **1979**, *279*, 773-777.

[6] S. K. Banerjee, E. Holler, G. P. Hess and J. A. Rupley, *J. Biol. Chem.*, **1975**, *250*, 4355-4367.

[7] A. D. Barksdale and A. Rosenberg, *Meth. Biochem. Anal.*, **1982**, *28*, 1-113.

[8] H. Bartunik, P. Jollès, J. Berthou and A. J. Dianoux, *Biopolymers*, **1982**, *21*, 43-50.

[9] W. S. Bennett and R. Huber, *Crit. Rev. Biochem.*, **1984**, *15*, 291-384.

[10] W. S. Bennett and T. A. Steitz, *Proc. Natl. Acad. Sci. USA*, **1978**, *75*, 4848-4852.

[11] W. S. Bennett and T. A. Steitz, *J. Mol. Biol.*, **1980**, *140*, 211-230.

[12] J. Berthou, A. Lifchitz, P. Artymiuk and P. Jollès, *Proc. R. Soc. Lond.*, **1983**, *B217*, 471-489.

[13] T. N. Bhat, D. M. Blow, P. Brick and J. Nyborg, *J. Mol. Biol.*, **1982**, *158*, 699-709.

[14] W. Bialek and R. F. Goldstein, *Biophys. J.*, **1985**, *48*, 1027-1044.

[15] W. Bode, *J. Mol. Biol.*, **1979**, *127*, 357-374.

[16] W. Bode and R. Huber, in *Molecular and Cellular Basis of Digestion*, P. Desnuelle, H. Sjöström and O. Norén,   (Eds.), Elsevier, Amsterdam, 213-234, 1986.

[17] W. Bode, P. Schwager and R. Huber, *J. Mol. Biol.*, **1978**, *118*, 99-112.

[18] M. Bolognesi, G. Gatti, E. Menegatti, M. Guarneri, M. Marquart, E. Papamokos and R. Huber, *J. Mol. Biol.*, **1982**, *162*, 839-868.

[19] B. Brooks and M. Karplus, *Proc. Natl. Acad. Sci. USA*, **1985**, *82*, 4995-4999.

[20] S. Brunie, J. Bolin, D. Gewirth and P. B. Sigler, *J. Biol. Chem.*, **1985**, *260*, 9742-9749.

[21] T. Butz, A. Lerf and R. Huber, *Phys. Rev. Lett.*, **1982**, *48*, 890-893.

[22] T. Butz, A. Lerf and R. Huber, *Hyperfine Interactons*, **1983**, *15*, 869-880.

[23] G. Careri, P. Fasella and E. Gratton, *Crit. Rev. Biochem.*, **1975**, *3*, 141-164.

[24] E. S. Cedergren-Zeppezauer, I. Andersson, S. Ottonello and E. Bignetti, *Biochemistry*, **1985**, *24*, 4000-4010.

[25] A. R. Clarke, D. B. Wigley, W. N. Chia, D. Barstow, T. Atkinson and J. J. Holbrook, *Nature*, **1986**, *324*, 699-702.

[26] J. H. Coates, M. J. Hardman, J. D. Shore and H. Gutfreund, *FEBS Lett.*, **1977**, *84*, 25-28.

[27] P. M. Colman, W. G. Laver, J. N. Varghese, A. T. Baker, P. A. Tulloch, G. M. Air and R. G. Webster, *Nature*, **1987**, *326*, 358-363.

[28] F. Colonna-Cesari, D. Perahia, M. Karplus, H. Eklund, C.-I. Brändén and O. Tapia, *J. Biol. Chem.*, **1986**, *261*, 15273-15280.

[29] A. Cooper, *Prog. Biophys. molec. Biol.*, **1984**, *44*, 181-214.

[30] T. E. Creighton, *Proteins. Structures and Molecular Principles*, Freeman & Co., New York, 1984.

[31] D. DeVault and B. Chance, *Biophys. J.*, **1966**, *6*, 825-847.

[32] B. W. Dijkstra, G. J. H. van Nes, K. H. Kalk, N. P. Brandenburg, W. G. J. Hol. and J. Drenth, *Acta Cryst.*, **1982**, *B38*, 793-799.

[33] J. Doucet and J. P. Benoit, *Nature*, **1987**, *325*, 643-646.

[34] H. Eklund, J.-P. Samama, L. Wallén, C.-I. Brändén, Å. Åkeson and T. A. Jones, *J. Mol. Biol.*, **1981**, *146*, 561-587.

[35] K. R. Ely, P. M. Colman, E. E. Abola, A. C. Hess, D. S. Peabody, D. M. Parr, G. E. Connell, C. A. Laschinger and A. B. Edmundson, *Biochemistry*, **1978**, *17*, 820-823.

[36] P. R. Evans, G. W. Farrants and P. J. Hudson, *Phil. Trans. R. Soc. Lond.*, **1981**, *B293*, 53-62.

[37] P. R. Evans, G. W. Farrants and M. C. Lawrence, *J. Mol. Biol.*, **1986**, *191*, 713-720.

[38] H. Fehlhammer, W. Bode and R. Huber, *J. Mol. Biol.*, **1977**, *111*, 415-438.

[39] A. Fersht, *Enzyme Structure and Mechanism*, Freeman & Co., San Francisco, 84-133, 1977.

[40] A. R. Fersht and Y. Requena, *J. Mol. Biol.*, **1971**, *60*, 279-290.

[41] J. S. Fruton, *Molecules and Life*, Wiley & Sons, New York, 87-179, 1972.

[42] J. García de la Torre and V. A. Bloomfield, *Quart. Rev. Biophys.*, **1981**, *14*, 81-139.

[43] G. L. Gilliland and F. A. Quiocho, *J. Mol. Biol*, **1981**, *146*, 341-362.

[44] R.F. Goldstein and W. Bialek, *Comments Mol. Cell. Biophys.*, **1986**, *4*.

[45] U. M. Grau, W. E. Trommer and M. G. Rossmann, *J. Mol. Biol.*, **1981**, *151*, 289-307.

[46] R. B. Gregory, A. Dinh and A. Rosenberg, *J. Biol. Chem.*, **1986**, *261*, 13963-13968.

[47] G. G. Hammes and P. R. Schimmel, in *The Enzymes*, **1970**, *II*, P. D. Boyer (Ed.), Academic Press, New York, 67-114.

[48] D. C. Hanson, J. Yguerabide and V. N. Schumaker, *Biochemistry*, **1981**, *20*, 6842-6852.

[49] D. C. Hanson, J. Yguerabide and V. N. Schumaker, *Mol. Immunol.*, **1985**, *22*, 237-244.

[50] D. C. Hanson, *Mol. Immunol.*, **1985**, *22*, 245-250.

[51] S. C. Harrison, in *The Microbe 1984: Part I Viruses*, B. W. J. Mahy and J. R. Pattison (Eds.), Cambridge Universiy Press, 29-73, 1984.

[52] S. C. Harvey and H. C. Cheung, *Biopolymers*, **1980**, *19*, 913-930.

[53] R. Hetzel, K. Wüthrich, J. Deisenhofer and R. Huber, *Biophys. Struct. Mech.*, **1976**, *2*, 159-180.

[54] W. A. Hendrickson and M. M. Teeter, *Nature*, **1981**, *290*, 107-113.

[55] J. G. Hoggett and G. L. Kellett, *Eur. J. Biochem.*, **1976**, *68*, 347-353.

[56] J. J. Hopfield, *Proc. Natl. Acad. Sci. USA*, **1974**, *71*, 3640-3644.

[57] R. B. Honzatko, H. L. Monaco and W. N. Lipscomb, *Proc. Natl. Acad. Sci. USA*, **1979**, *76*, 5105-5109.

[58] R. Huber and W. S. Bennett, *Nature*, **1987**, *326*, 334-335.

176

[59] B. Jacrot, S. Cusack, A. J. Dianoux and D. M. Engelman, *Nature*, **1982**, *300*, 84-86.

[60] J. Janin and S. J. Wodak, *Prog. Biophys. molec. Biol.*, **1983**, *42*, 21-78.

[61] J. N. Jansonius, G. Eichele, G. C. Ford, J. F. Kirsch, D. Picot, C. Thaller, M. G. Vincent, H. Gehring and P. Christen, *Biochem. Soc. Trans.*, **1984**, *12*, 424-427.

[62] J. A. Kelly, A. R. Sielecki, B. D. Sykes, M. N. G. James and D. C. Phillips, *Nature*, **1979**, *282*, 875-878.

[63] H. Kirsten, H. Gehring and P. Christen, *Proc. Natl. Acad. Sci. USA*, **1983**, *80*, 1807-1810.

[64] M. Klein, N. Haeffner-Cavaillon, D. E. Isenman, C. Rivat, M. A. Navia, D. R. Davies and K. J. Dorrington, *Proc. Natl. Acad. Sci. USA*, **1981**, *78*, 524-529.

[65] A. A. Kossiakoff, *Ann. Rev. Biochem.*, **1985**, *54*, 1195-1227.

[66] K. L. Krause, K. W. Volz and W. N. Lipscomb, *J. Mol. Biol.*, **1987**, *193*, 527-553.

[67] S. Krimm and J. Bandekar, *Adv. Prot. Chem.*, **1986**, *38*, 181-364.

[68] C. E. Kundrot and F. M. Richards, *J. Mol. Biol.*, **1987**, *193*, 157-170.

[69] I. D. Kuntz and W. Kauzmann, *Adv. Prot. Chem.*, **1974**, *28*, 239-345.

[70] M. Laurent, M. N. Tijane, C. Roucous, F. J. Seydoux and A. Tardieu, *J. Biol. Chem.*, **1984**, *259*, 3124-3126.

[71] A. G. W. Leslie and A. J. Wonacott, *J. Mol. Biol.*, **1984**, *178*, 743-772.

[72] M. Levitt, C. Sander and P. S. Stern, *J. Mol. Biol.*, **1985**, *181*, 423-447.

[73] H. Loebermann, R. Tokuoka, J. Deisenhofer and R. Huber, *J. Mol. Biol.*, **1984**, *177*, 531-556.

[74] J. A. McCammon, *Rep. Prog. Phys.*, **1984**, *47*, 1-46.

[75] J. A. McCammon, B. R. Gelin, M. Karplus and P. G. Woylnes, *Nature*, **1976**, *262*, 325-326.

[76] R. C. McDonald, T. A. Steitz and D. M. Engelman, *Biochemistry*, **1979**, *18*, 338-342.

[77] B. Mao and J. A. McCammon, *J. Biol. Chem.*, **1983**, *258*, 12543-12547.

[78] M. Marquart, J. Deisenhofer, R. Huber and W. Palm, *J. Mol. Biol.*, **1980**, *141*, 369-391.

[79] D. M. Miller, J. S. Olson, J. W. Pflugrath and F. A. Quiocho, *J. Biol. Chem.*, **1983**, *258*, 13665-13627.

[80] K. Moffat, D. Bilderback, W. Schildkamp and K. Volz, *Nucl. Instr. and Meth.*, **1986**, *A246*, 627-635.

[81] V. N. Morozov and T. Ya. Morozova, *J. theor. Biol.*, **1986**, *121*, 73-88.

[82] M. R. N. Murthy, R. M. Garavito, J. E. Johnson and M. G. Rossmann, *J. Mol. Biol.*, **1980**, *138*, 859-872.

[83] M. E. Newcomer, G. L. Gilliland and F. A. Quiocho, *J. Biol. Chem.*, **1981**, *256*, 13213-13217.

[84] M. E. Newcomer, B. A. Lewis and F. A. Quiocho, *J. Biol. Chem.*, **1981**, *256*, 13218-13222.

[85] H. J. Nolte and E. Neumann, *Biophys. Chem.*, **1979**, *10*, 253-260.

[86] V. T. Oi, T. M. Vuong, R. Hardy, J. Reidler, J. Dangl, L. A. Herzenberg and L. Stryer, *Nature*, **1984**, *307*, 136-140.

[87] P. C. Painter, L. E. Mosher and C. Rhoads, *Biopolymers*, **1982**, *21*, 1469-1472.

[88] D. M. Parker, D. Jeckel and J. J. Holbrook, *Biochem. J.*, **1982**, *201*, 465-471.

[89] S. J. Perkins, L. N. Johnson, D. C. Phillips and R. A. Dwek, *Biochem. J.*, **1981**, *193*, 553-572.

[90] S. J. Perkins and K. Wüthrich, *J. Mol. Biol.*, **1980**, *138*, 43-64.

[91] M. F. Perutz, *Ann. Rev. Biochem.*, **1979**, *48*, 327-386.

[92] W. L. Peticolas, *Meth. Enzymol.*, **1979**, *61*, 425-458.

[93] K. Pfister, E. Sandmeier, W. Berchtold and P. Christen, *J. Biol. Chem.*, **1985**, *260*, 11414-11421.

[94] C. A. Pickover, D. B. McKay, D. M. Engelman and T. A. Steitz, *J. Biol. Chem.*, **1979**, *254*, 11323-11329.

[95] P. L. Privalov, *Adv. Prot. Chem.*, **1979**, *33*, 167-241.

[96] J. Reidler, V. T. Oi, W. Carlsen, T. M. Vuong, I. Pecht, L. A. Herzenberg and L. Stryer, *J. Mol. Biol.*, **1982**, *158*, 739-746.

[97] D. Ringe and G. A. Petsko, *Prog. Biophys. molec. Biol.*, **1985**, *45*, 197-235.

[98] M. A. Saper and F. A. Quiocho, *J. Biol. Chem.*, **1983**, *258*, 11057-11062.

[99] G. Schneider, H. Eklund, E. Cedergren-Zeppezauer and M. Zeppezauer, *EMBO J.*, **1983**, *2*, 685-689.

[100] V. N. Shumaker, G. W. Seegan, C. A. Smith, S. K. Ma, J. D. Rodwell and M. F. Shumaker, *Mol. Immunol.*, **1980**, *17*, 413-423.

[101] T. Skarzynski, P. C. E. Moody and A. J. Wonacott, *J. Mol. Biol.*, **1987**, *193*, 171-187.

[102] J. L. Smith, W. A. Hendrickson, R. B. Honzatko and S. Sheriff, *Biochemistry*, **1986**, *25*, 5018-5027.

[103] T. A. Steitz, M. Shoham and W. S. Bennett, *Phil. Trans. R. Soc. Lond.*, **1981**, *B293*, 43-52.

[104] M. J. E. Sternberg, D. E. P. Grace and D. C. Phillips, *J. Mol. Biol.*, **1979**, *130*, 231-253.

[105] J. M. Sturtevant, *Proc. Natl. Acad. Sci. USA*, **1977**, *74*, 2236-2240.

[106] S. W. Suh, T. N. Bhat, M. A. Navia, G. H. Cohen, D. N. Rao, S. Rudikoff and D. R. Davies, *Proteins*, **1986**, *1*, 74-80.

[107] K. Takahashi, J. L. Casey and J. M. Sturtevant, *Biochemistry*, **1981**, *20*, 4693-4697.

[108] D. C. Teller, E. Swanson and C. de Haen, *Meth. Enzymol.*, **1979**, *61*, 103-124.

[109] E. Tüchsen, A. Hvidt and M. Ottesen, *Biochemie*, **1980**, *62*, 563-566.

[110] E. Tüchsen and C. Woodward, *J. Mol. Biol.*, **1985**, *185*, 405-419.

[111] G. Wagner, *Quart. Rev. Biophys.*, **1983**, *16*, 1-57.

[112] G. Wagner and K. Wüthrich, *J. Mol. Biol.*, **1982**, *160*, 343-361.

[113] J. Walter, W. Steigemann, T. P. Singh, H. Bartunik, W. Bode and R. Huber, *Acta Cryst.*, **1982**, *B38*, 1462-1472.

[114] D. Wang, W. Bode and R. Huber, *J. Mol. Biol.*, **1985**, *185*, 595-624.

[115] W. A. Wegener, *Biopolymers*, **1986**, *25*, 627-637.

[116] G. R. Welch, G. Somogyi and S. Damjanovich, *Prog. Biophys. molec. Biol.*, **1982**, *39*, 109-146.

[117] G. Wiegand, S. J. Remington, J. Deisenhofer and R. Huber, *J. Mol. Biol.*, **1984**, *174*, 205-219.

[118] K. D. Wilkinson and I. A. Rose, *J. Biol. Chem.*, **1980**, *255*, 7569-7574.

[119] K. D. Wilkinson and I. A. Rose, *J. Biol. Chem.*, **1981**, *256*, 9890-9894.

[120] C. A. B. Wilson, N. Hardman, L. A. Fothergill-Gilmore, S. J. Gamblin and H. C. Watson, *Biochem. J.*, **1987**, *241*, 609-617.

[121] I. A. Wilson, J. J. Skehel and D. C. Wiley, *Nature*, **1981**, *289*, 366-373.

[122] A. Wlodawer, J. Walter, R. Huber and L. Sjölin, *J. Mol. Biol.*, **1984**, *180*, 301-329.

[123] J. K. Wolfe, C. F. Weidig, H. R. Halvorson, J. D. Shore, D. M. Parker and J. J. Holbrook, *J. Biol. Chem.*, **1977**, *252*, 433-436.

[124] H. T. Wright, *J. Mol. Biol.*, **1973**, *79*, 1-11.

[125] J. Yguerabide, H. F. Epstein and L. Stryer, *J. Mol. Biol.*, **1970**, *51*, 573-590.

CHAPTER 10

QUASI-ELASTIC AND INELASTIC NEUTRON SCATTERING

A. J. Dianoux
Institut Laue-Langevin
Avenue des Martyrs, 156 X
38042 Grenoble Cedex
France

ABSTRACT. The method of quasi-elastic and inelastic neutron scattering
is outlined and its usefulness for studying molecular motions in condensed matter is stressed. After a short introduction to the theory of
neutron scattering, some details are given on quasi-elastic scattering
and its application to the study of various non-periodic motions. High
resolution inelastic scattering which permits to reveal quantum aspects
of motions at low temperature is described next. Finally, some account
of inelastic scattering is given with emphasis to the derivation of the
vibrational density of states for disordered systems.

1. NEUTRON SCATTERING STUDIES OF MOLECULAR MOTIONS

1.1. General

It is outside the scope of these two lectures to present a comprehensive
review of the technique of neutron scattering applied to the problem of
molecular motion. Since this School is concentrating upon the time
domain, I will let aside Neutron Diffraction which gives only structural
information, although some slow dynamical processes can be studied, as
for example, the dynamics of intercalation.

Neutron Scattering covers nowadays a very wide dynamical range from
about 10^{-14}s to 10^{-6}s corresponding to energy transfers from 100 meV to 1
neV (1 meV = 1.52 x 10^{12} rad/s). This last limit corresponds to the new
technique of neutron spin-echo which is described by C. Zeyen in these
proceedings.

This presentation will be focussed on quasi-elastic and inelastic
scattering on non-oriented samples (powders, glasses, liquids). I will
not touch upon the whole field of lattice dynamics (phonon dispersion
curves) which is done on single crystals.

The normal approach for the study of molecular motions is to perform
spectroscopic experiments. The idea is to prepare a probe in a well
specified state (for example, a monochromatic neutron beam) and to make
it interact with the system under study. The interaction with the
degrees of freedom of the system will change the state of the probe (for
example, change the energy of the incident neutron beam). One can relate
this change to some correlation functions which describe the dynamical

179

G. J. Long and F. Grandjean (eds.), The Time Domain in Surface and Structural Dynamics, 179–212.

180

state of the system. Depending upon the technique one can have access to
dynamical quantities (e.g. : NMR), to spatial information (e.g. : X-Ray
or Neutron diffraction), or to both (quasi-elastic and inelastic neutron
scattering).

1.2. Principle of a spectroscopic measurement

Consider figure 1 which is a sketch of a spectroscopic experiment.

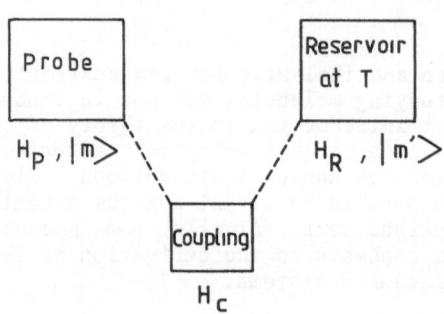

Figure 1. Sketch of a spectroscopic experiment

The system under study, at thermal equilibrium T, constitutes the
reservoir R describes by the Hamiltonian H_R. The probe, which is
described by the Hamiltonian H_p, interacts with the reservoir via a
coupling described by the Hamiltonian H_C. We are following here the
presentation by Volino [1] to which we refer the reader for details.
    When the interaction is switched on, the probe changes its state $|m\rangle$
to a final state $|n\rangle$. In the linear approximation (H_C small compared to
H_p or H_R), this change is characterized by a probability per unit time
W_{nm}. The principle of a spectroscopic experiment is to measure a
quantity proportional to W_{nm} as a function of $|n\rangle$ or $|m\rangle$. Since W_{nm} is a
function of the dynamical variables of the reservoir R this measurement
will yield information about the molecular motion of our sytem.
    Using the Fermi golden rule, one obtains [1] :

$$W_{nm} = \frac{2\pi}{\hbar^2} C_{\bar{H}_C \bar{H}_C}(\omega) \qquad (1)$$

where $C_{\bar{H}_C \bar{H}_C}(\omega)$ is the Fourier transform (spectral density) of the auto
correlation function of $\bar{H}_C$. $\bar{H}_C$ is the average of the coupling
Hamiltonian H_C between the initial and final state of the probe :

$$H_C = \langle n|H_C|m\rangle. \qquad (2)$$

For a classical system, one writes

$$C_{\bar{H}_C \bar{H}_C}(t) = \langle \bar{H}_C(o) . \bar{H}_C(t)\rangle \qquad (3)$$

where $\langle...\rangle$ is a thermal average and the operators in $\bar{H}_C$ are replaced
by their classical expressions.

The relationship between W_{nm} and a measurable quantity depends upon the type of experiment. This has to be established for each particular case.

1.3. Introduction to Neutron Scattering

We give here only a very brief overview. Interested readers are encouraged to consult the excellent textbook by Marshall and Lovesey [2].

Thermal neutrons are produced in the moderator of a reactor. They have a Maxwellian distribution of velocities such that their average kinetic energy is determined by the temperature of the moderator ($\bar{E} \simeq 26$ meV at room temperature). The associated wavelength is $\lambda = 1.8$ Å. One sees immediately that neutrons have the right length and energy scale to study excitations in condensed matter.

$$(\text{Recall} : 1 \text{ meV} \sim 8 \text{ cm}^{-1})$$

Neutron-nuclei interaction is very short range ($\sim 10^{-12}$ cm compared to the size of an atom $\sim 10^{-8}$ cm). It can be written as the <u>Fermi pseudo-potential</u>

$$V(\vec{r}) = \frac{2\pi\hbar^2}{m_n} b_i \; \delta(\vec{r} - \vec{r}_i) \tag{4}$$

where m_n is the neutron mass. b_i is the scattering amplitude of the nuclei i : it is independent of the neutron energy (in the thermal and sub-thermal range). For an assembly of nuclei, having different isotopes and spins, one defines the scattering lengths :

$$- \text{coherent} \quad b_i^{coh} = \langle b_i \rangle \tag{5}$$
$$- \text{incoherent} \quad b_i^{inc} = [\langle b_i^2 \rangle - \langle b_i \rangle^2]^{1/2}$$

The cross-sections are defined by :

$$\sigma_{coh}^i = 4\pi \; b_i^2 \; {}^{coh}$$
$$\sigma_{inc}^i = 4\pi \; b_i^2 \; {}^{inc} \tag{6}$$

We list below these cross-sections in the case of Hydrogen and Deuterium where the differences are enormous.

	spin	σ_{coh}	σ_{inc}	
H	1/2	1.76	79.7	<u>Table 1</u>
D	1	5.6	2.0	

The unit is in barn (1 barn = 10^{-24} cm^2). H has a very big incoherent scattering cross-section which is more than twenty times bigger than for other nuclei. Dynamical studies are mostly done by using <u>incoherent</u> scatterers.

Besides the nuclear interaction with the nuclei, the neutrons interact with the nuclear and electronic spins via dipole-dipole interaction. This permits studies of magnetic structures or magnetic

182

excitations (spin waves).

Using the Fermi pseudo-potential (4), the coupling Hamiltonian for an assembly of nuclei i is given by :

$$H_c = \frac{2\pi\hbar^2}{m_n} \sum_i b_i \; \delta(\vec{r} - \vec{r}_i) \qquad (7)$$

The neutron states before and after scattering are defined by the wave-vectors $\vec{k}_o$ and $\vec{k}$. Using Eq. 2, one obtains :

$$\tilde{H}_c = \langle k|H_c|k_o\rangle \; \alpha \sum_i b_i \; \exp{(\vec{q}.\vec{r}_i)} \qquad (8)$$

with $\vec{q} = \vec{k} - \vec{k}_o$ the momentum transfer. The correlation function (Eq. 3) is directly :

$$C_{\tilde{H}_c\tilde{H}_c}(t) \; \alpha \sum_{i,j} \langle b_i b_j \; \exp i \; \vec{q}. \; [\vec{r}_i(t) - \vec{r}_j(0)]\rangle \qquad (9)$$

The scattered intensity in a solid angle $d\Omega$ and an energy window $d\omega$ is given by :

$$I = I_o \; N \; \frac{k}{k_o} \; S(\vec{q},\omega) \; d\Omega d\omega \qquad (10)$$

for an incident neutron flux I_o on N scattering centres.

The scattering law, $S(\vec{q},\omega)$ is the Fourier transform of the correlation function (Eq. 9) (often called the intermediate scattering law)

$$S(\vec{q},\omega) = \frac{1}{2\pi} \int_{-\infty}^{+\infty} C(t) \; e^{-i\omega t} \; dt \qquad (11)$$

A complete derivation is given in reference [1].

2. QUASI-ELASTIC NEUTRON SCATTERING

2.1. General

We will restrict ourselves to incoherent scattering. The correlation function called the intermediate scattering law, writes :

$$I_s(\vec{q},t) = \frac{1}{N} \sum_i b_i^{2\,\text{inc}} \langle \exp i \; \vec{q} \; [\vec{r}_i(t) - \vec{r}_i(0)]\rangle \qquad (12)$$

Writing :

$$\vec{r} = \vec{d} + \vec{\rho} + \vec{u}, \qquad (13)$$

where $\vec{d}$ specifies the centre of mass position, $\vec{\rho}$ defines the position of the scattering centre with respect to the c.o.m and $\vec{u}$ stands for a small vibrational displacement around the average position. For uncoupled motions, we can write with obvious notations :

$$I_s(Q,t) = I_s^{\text{trans}} . \; I_s^{\text{rot}} . \; I_s^{\text{vib}} \qquad (14)$$

By Fourier transform, one obtains :

$$S_s(\vec{q},\omega) = S_s^{trans} \otimes S_s^{rot} \otimes S_s^{vib} \tag{15}$$

where $\otimes$ stands for the convolution product.

- Vibrational part : When we restrict to the quasi-elastic region, the characteristic times we are interested in are much longer than the reciprocal of the vibrational frequencies ($\tau_c \gg \frac{1}{\omega_{vib}}$). I_s^{vib} is thus a simple Debye-Waller factor :

$$S_s^{vib} = I_s^{vib} = e^{-Q^2 \langle u^2 \rangle} \tag{16}$$

where $\langle u^2 \rangle$ is a mean square amplitude of vibrations.

- Translational part : for the simple translational diffusion model characterized by the self-diffusion coefficient D_t, valid at low Q-values (Q a $\ll$ 1, where a is a molecular radius) one has :

$$I_s^{trans}(Q,t) = e^{-D_t Q^2 |t|} \tag{17}$$

which gives by Fourier transform

$$S_s^{trans}(Q,\omega) = \frac{1}{\pi} \frac{D_t Q^2}{(D_t Q^2)^2 + \omega^2}$$

The full width at half maximum (FWHM) is given by $\Delta E = 2D_t Q^2$ (fig. 2)

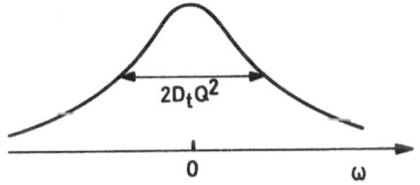

Figure 2. Theoretical incoherent neutron quasi-elastic scattering spectrum for isotropic translational diffusion

- Rotational part : In the case of the rotational diffusion on a sphere of radius ρ, one obtains [3],

$$S_s^{rot}(Q,\omega) = j_0^2(Q\rho) \delta(\omega) + \sum_{\ell=1}^{\infty} (2\ell+1) j_\ell^2(Q\rho) \frac{1}{\pi} \frac{\ell(\ell+1)D_r}{[\ell(\ell+1)D_r]^2 + \omega^2} \tag{18}$$

Using Eqs. 11 and 12 it is easy to show that :

$$I_s(\vec{q},0) = \int_{-\infty}^{+\infty} S_s(\vec{q},\omega)\, d\omega \equiv 1 \tag{18}$$

184

Eq. 18 must verify this, which gives the sum rule for spherical Bessel functions :

$$\sum_{\ell=0}^{\infty} (2\ell+1) \; j_\ell^{\;2}(Q\rho) \equiv 1 \tag{19}$$

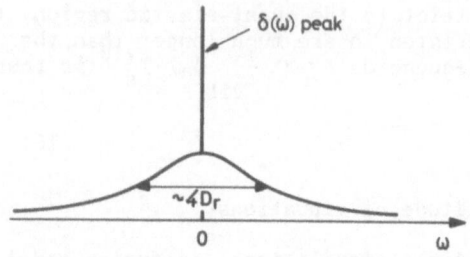

Fig. 3. Theoretical incoherent neutron quasi-elastic scattering spectrum for a purely rotational diffusion motion

On figure 3, we have sketched the scattering law Eq. 18. In an actual experiment, the elastic peak reproduces the shape of the resolution function.

2.2. Elastic Incoherent Structure Factor (EISF)

In the preceding paragraph, we have seen an example of a scattering law with a $\delta(\omega)$ peak, whose intensity is a function of $\vec{Q}$. This is quite general for any bounded motion : for $t \to \infty$, $I_s(Q,t)$ does not decay to zero since there will be always a finite probability to find the particle inside this volume. Since for ergodic systems one has :
$\vec{r}(t=0) \equiv \vec{r}(t \to \infty)$, this constant value is given by :

$$A_o(\vec{Q}) = |<e^{i\vec{Q}.\vec{r}}>|^2 = |\int p(r) \; e^{i\vec{Q}.\vec{r}} d^3r|^2 \tag{20}$$

for one particle moving inside a bounded volume; $\vec{r}$ is the position vector of this particle with probability $p(\vec{r})$. It is called the EISF. The scattering law can thus be written as :

$$S_s^{\;bounded}(\vec{Q},\omega) = A_o(\vec{Q}) \; \delta(\omega) + \sum_n \text{broadened terms} \tag{21}$$

Experimentally, the EISF is given by :

$$A_o(Q) = \frac{I_e}{I_e+I_q} \tag{22}$$

This is a model independent quantity.

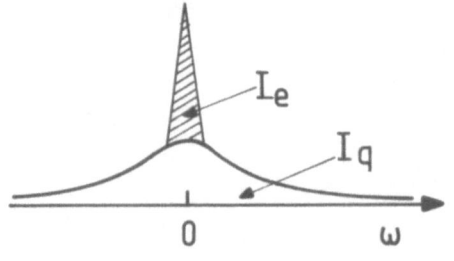

Figure 4. Experimental determination of the Elastic Incoherent Structure Factor (EISF)

An excellent introduction to quasi-elastic neutron scattering is found in reference [4], while an up to date and extensive presentation of the method can be found in reference [5].

2.3. Selected Examples

I present below three examples where quasi-elastic neutron scattering has been used to study molecular motions in solids, adsorbed phases or liquids. A recent review on plastic crystals can be found in Ref. [6].

2.3.1. Reorientations in pivalic acid [7]

Pivalic acid $(CH_3)_3$ CCOOH (or trimethylacetic acid) undergoes a solid-solid transition at T = 280 K between a low temperature, triclinic phase and an orientationally disordered cubic phase. In this phase (a plastic phase) the molecules are associated in non-polar dimer units as evidenced from dielectric relaxation measurements (Fig. 5). As sketched on the figure at least three kinds of reorientations can occur :

   i) rotation of the methyl groups about their threefold axis
  ii) rotation of the t-butyl group about the central c-c bond of the molecule
 iii) whole-molecule tumbling.

In addition over-damped librations can give a quasi elastic signal.

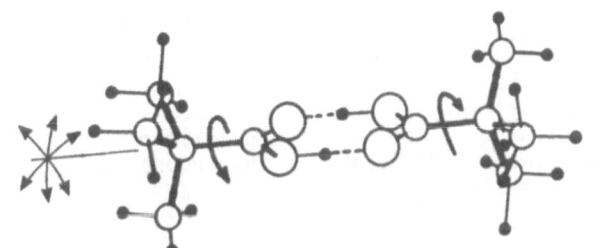

Figure 5. Sketch of a dimeric unit of pivalic acid

Partially deurated samples have been used to investigate in a precise way the motions of the different parts of the molecule : $(CH_3)_3$ CCOOD (noted D_1), $(CH_3)_3$ CCOOH (noted D_0), $(CD_3)_3$ CCOOH (noted D_9) and $(CD_3)_3$ CCOOD (noted D_{10}).

   Typical examples of time-of-flight spectra obtained with the D_1 compound in its plastic phase are shown on figure 6. The abscissa scale is the time-of-flight τ, where

$$\tau(\mu s/m) = 252.8 \times \lambda(\text{Å}).$$

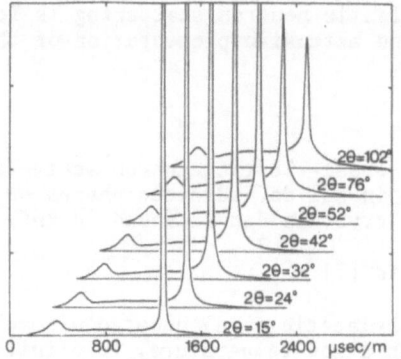

Figure 6. Time-of-flight
spectra of pivalic acid in its
plastic phase at T = 303 K,
for different scattering angles
The incident wavelength is
λ = 5.1 Å

One sees clearly a sharp elastic line sitting above a rather wide
quasi-elastic component whose intensity increases with the scattering
angle 2θ. Furthermore one should note that in this case, the inelastic
and quasi-elastic parts are not well separated, leading to difficulties
to extract a reliable EISF. Nevertheless, by taking into account in an
approximate way this inelastic intensity, one can extract an experimental
EISF by using a non-linear least-square fitting procedure [8,9]. The
result is shown in figure 7 for the D_1 compound, at different
temperatures in the plastic phase and using different incident
wavelengths. It is consistent with a reorientation of the t-butyl group
by 60° jumps. However, with the incident wavelength of 5.9 Å (which
gives the best resolution) the points lie systematically slightly below
the theoretical B_6 curve : this could be due to the effect of a slower
rotational motion of the mehtyl groups around their C_3 axis. The
determination of the EISF has also been done in the liquid phase for the
D_0 compound (figure 8). One can note that for Q = 0, the EISF tends to
1, indicating that on the instrument time scale, no translational
broadening is visible. The experimental EISF lies on the theoretical
curve predicted for whole-dimer tumbling. Therefore it appears that in
the liquid phase, strong correlations still exist between pivalic
molecules.
    Now that the dynamical model has been established, one can fit all
angles simultaneously to obtain the correlation time of t-butyl motion.
An example of such a fit is shown on figure 9 and the inverse of the
correlation time is drawn on figure 10.
    Using the two compounds D_9 and D_{10}, the authors were able to obtain
the scattering of the acid proton alone. This is shown on figure 11.
Here again the extraction of the EISF gives immediately an answer
concerning the motion of this acid proton (figure 12) : it consists of
180° jumps of the carboxylic group. Of course a proton exchange
mechanism cannot be ruled out, but if this motion exists, it occurs on a
slower time scale than the 180° jumps of the carboxylic group whose
correlation time is of the order of 1ps.

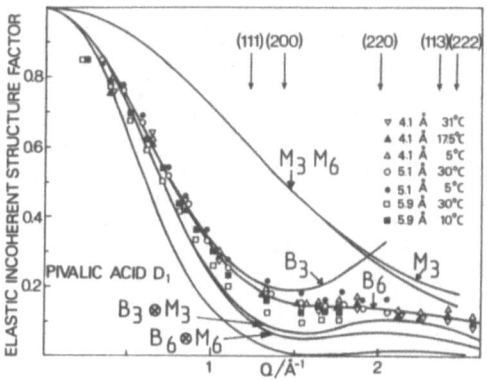

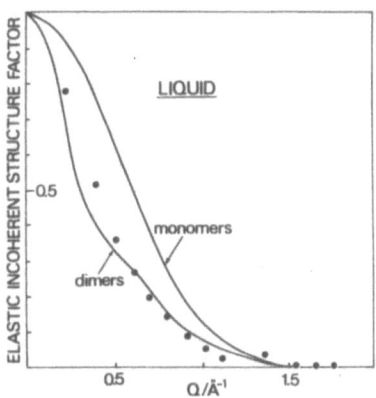

Figure 7. EISF as a function of the momentum transfer Q, for the D_1 compound. Different combinations of rotations for methyl (M_3 = 120° jumps, M_6 = 60° jumps) or t-butyl group (B_3 = 120° jumps, B_6 = 60° jumps) are considered. Squares, circles and triangles correspond to incident wavelengths of 5.9, 5.1 and 4.1 Å respectively. Bragg peak positions are indicated by vertical arrows.

Figure 8. EISF for the D_0 compound in the liquid state at T = 330°K. Full curves are theoretical ones for models based upon an overall tumbling of monomers or dimer units.

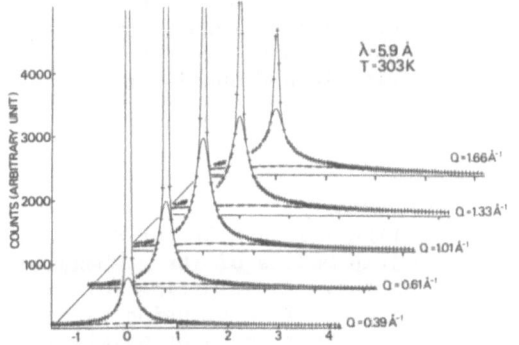

Figure 9. Example of fits with the model based on uniaxial rotation of the t-butyl group and taking into account the underlying inelastic intensity by the scattering of an overdamped oscillator. The dashed line separates this contribution The separation between elastic and quasi-elastic scattering is also indicated.

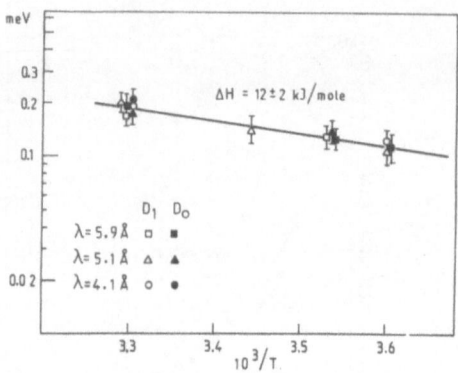

Figure 10. Reciprocal of the correlation time for t-butyl motion in pivalic acid versus reciprocal temperature. Open symbols represent results of refinements with D_1 compound, full symbols correspond to D_0. Three incident wave-lengths are reported.

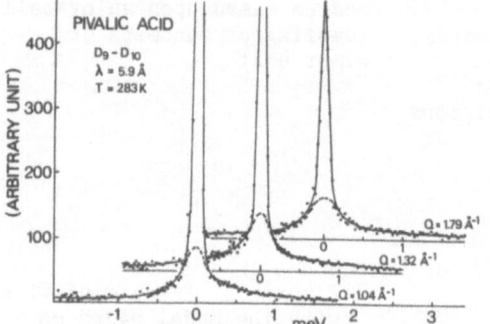

Figure 11. Quasi-elastic spectra due to the scattering by the acid protons of pivalic acid. They are obtained by subtracting the spectra measured with the D_{10} compound from the corresponding spectra recorded with the D_9 compound, at a temperature of 283°K and using a wavelength of 5.9 Å. The full line is the result of the refinement of the model based upon 180° jumps of the carboxylic group. The dotted line shows the separation between elastic and quasi-elastic scattering obtained by the fit.

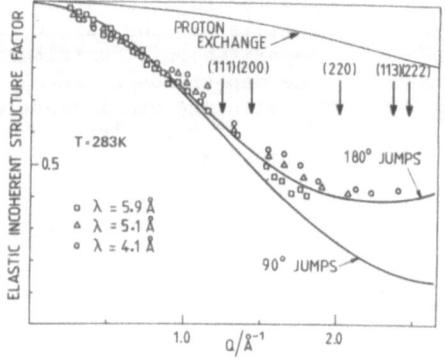

Figure 12. EISF for different jump models of the carboxylic group. The experimental points are obtained by using three different incident wavelengths. Vertical arrows indicate the position of the Bragg peaks.

2.3.2. Water mobility in an ionic membrane

This experiment is an excellent example of the power of quasi-elastic neutron scattering to provide <u>both</u> dynamical and structural information [10]. Water mobility in an acid Nafion ® membrane containing 15% water by weight has been studied by using different resolutions ranging from 18.5 to 9 µeV FWHM. An example of a spectrum is given on figure 13. One sees that it consists of a sharp peak superimposed on a broad component having about 100 µeV FWHM.

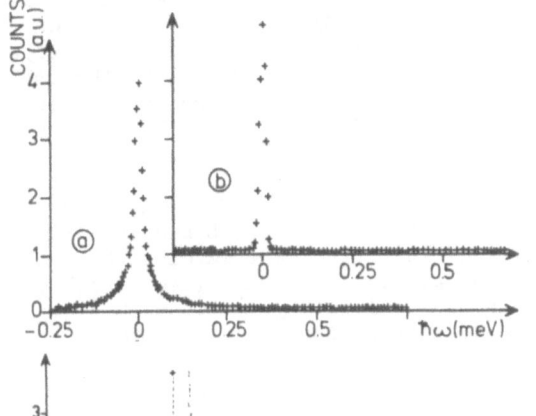

Figure 13. Examples of neutron quasielastic spectra of acid Nafion ® membrane for Q = 0.59 Å⁻¹. (a) Membrane, (b) resolution function obtained with a vanadium sample (energy resolution 19.5 µeV FWHM temperature ca. 25°C).

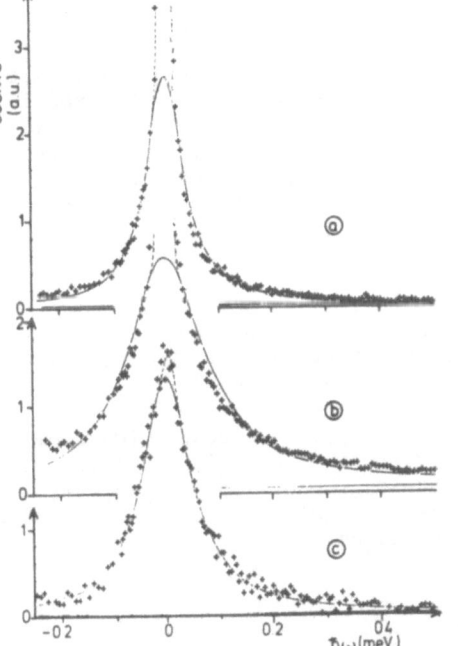

Figure 14. Broadened component of neutron quasielastic spectra : experimental points and best-fit Lorentzian line : (a) Nafion membrane containing 15 wt % H₂O,Q = 0.59 Å⁻¹; (b) the same but Q = 1.05 Å⁻¹; (c) pure bulk water at 28°C, Q = 0.59 Å⁻¹. The small narrow component in (c) comes from the quartz sample holder. Note the systematic deviation from the Lorentzian shape in (a) and (b).

By studying carefully the shape of this broadened component, one finds that it deviates systematically from a single Lorentzian (Fig. 14). Furthermore, the width of the best fitted Lorentzian is practically constant at low Q and then increases with Q. This is shown on Figure 15.

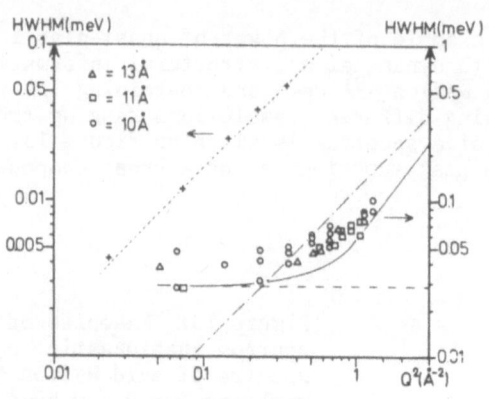

Figure 15. Half-width at half-maximum of broadened component of neutron quasielastic spectra obtained from acid Nafion membrane containing 15 wt% water. The points are the widths of the best-fit Lorentzian curves : (0) from spectra obtained with incident wavelength 10 Å (□) 11 Å, (Δ) 13 Å. The full line is the theoretical width predicted by the model with diffusion in a sphere with $D = 1.8 \times 10^{-5}$ cm^2/s and a = 4.25 Å. The two theoretical asymptotes for $Q \to 0$ and $Q \to \infty$ are also shown. (+) half-width at half-maximum of the best-fit Lorentzian lines to spectra obtained from bulk water at 28°C (incident wavelength 10 Å). The straight line passing through the points (+) is the theoretical width predicted by the simple self-diffusion model with $D_t = 2.5 \times 10^{-5}$ cm^2/s. Note the different vertical scales for the Nafion® and bulk-water samples.

This behaviour is consistent with a model of diffusion inside a sphere [11]. We cannot report here the mathematical details, but this model depends upon two parameters : a self-diffusion coefficient D and the radius of the sphere a. By applying this model to the measured spectra, the values of D and a have been extracted and reported on figure 16. One sees that the water diffuses nearly as fast as in pure water but in a volume of less than 10 Å in diameter.

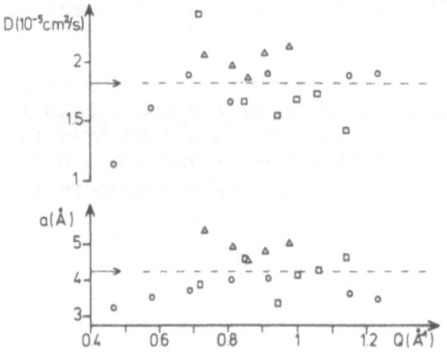

Figure 16. Values of diffusion coefficient D and radii of spheres a obtained by fitting the theoretical scattering law to all the quasi-elastic spectra with the central part excluded : (0) from spectra obtained with incident wavelength 10 Å, (□) 11 Å, (Δ) 13 Å. The mean values of D and a are also indicated.

The long-range diffusion coefficient of water has been measured by tracer diffusion using HTO. The result is $D_t = 1.6 \times 10^{-6}$ cm^2/s at 25°C. This is more than one order of magnitude smaller than the short range diffusion coefficient. Using this value, the spectra obtained with the best resolution (9 μeV FWHM) could be fitted satisfactorily (Fig. 17b).

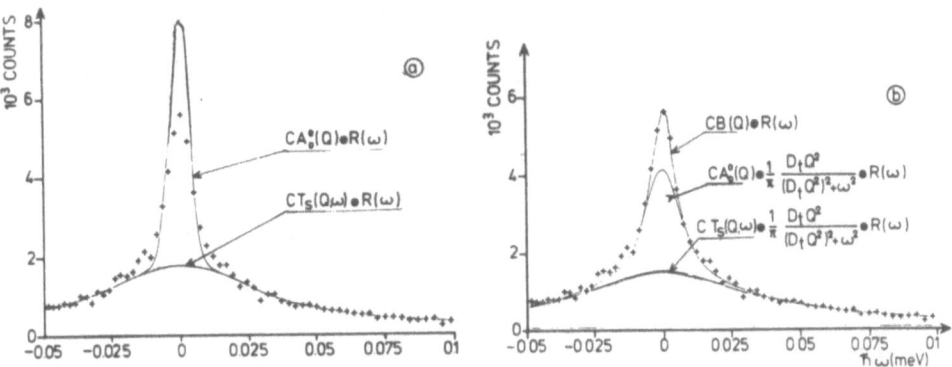

Figure 17. Quasielastic spectrum for Q = 0.635 Å^{-1} obtained with an incident wavelength of 13 Å, and best-fit curves using D = 1.8 x 10^{-5} cm^2/s, a = 4.25 Å (a) or D = 1.8 x 10^{-5} cm^2/s, a = 4.25 Å, and D_t = 1.6 x 10^{-6} cm^2/s (b). The separation between the various components of the theoretical spectra is also indicated. In (a) the best fit gives B(Q)=0.

2.3.3. Supercooled water

This study has been done on pure water, but in the supercooled range down to -20°C [12]. This is achieved by using an array of capillaries having an inner diameter of 0.3 mm. It is essential to take into account the rotational motion of the water molecules to correctly describe the lineshape. For simplicity, it has been assumed that the protons have an isotropic rotational diffusive motion around the center of mass of the molecule and characterized by a rotational diffusion coefficient D_r [3]. One can define a characteristic relaxation time $\tau_1 = 1/6D_r$. Having determined D_r it is possible to extract the translational width Γ_s versus the momentum transfer Q. This is done on figure 18 for different temperatures. This width can be fitted with the Random Jump Diffusion model [13] :

$$\Gamma_s = \frac{DQ^2}{1+DQ^2\tau_0} \qquad (23)$$

where D is the diffusion coefficient and τ_0 is a residence time.

On figure 19 are reported the two characteristic times τ_0 and τ_1. One can see that the residence time τ_0 defined by Eq. 23 increases dramatically where the temperature decreases. On the contrary, the rotational correlation time τ_1 increases less rapidly and follows an

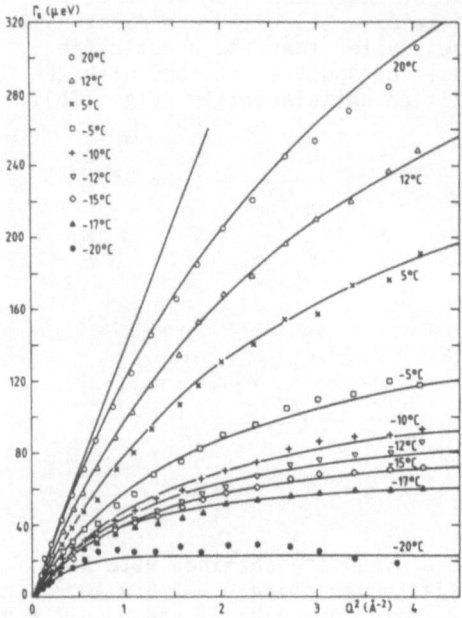

Figure 18. Linewidth Γ of the translational component of the spectrum vs Q^2. Note that a one-Lorentzian fit, as commonly done, gives much larger linewidths with a different Q dependence. —— : best fit using Eq. (23). The straight line gives the self-diffusion constant at 20°C.

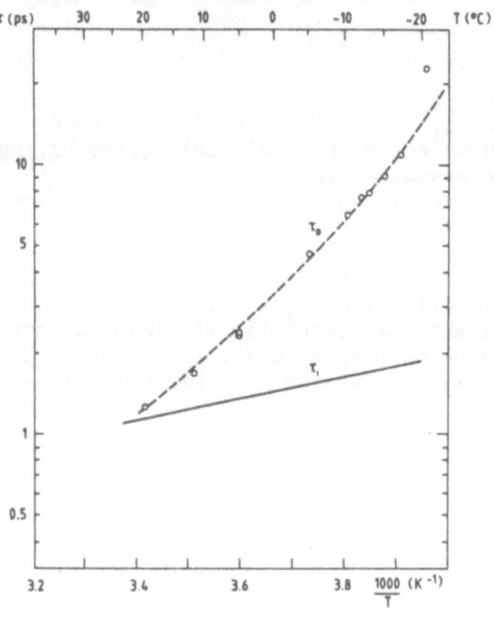

Figure 19. Residence time τ_0 of the jump diffusion and an Arrhenius plot of the relaxation time τ_1.

Arrhenius behaviour. The behaviour of τ_0 with temperature is similar to many other relaxation times and transport properties of water. Another interesting feature is the nearly constant value with temperature of the vibrational amplitude $\langle u^2 \rangle^{1/2} \sim 0.5$ Å. This value corresponds to an amplitude of vibration around 25°, when according to Molecular Dynamics simulations, the hydrogen bond is broken. More on this subject can be found in the proceedings of a Workshop which was held at the ILL in April 1984 [14].

3. HIGH RESOLUTION INELASTIC SCATTERING

3.1. Tunneling motions

At low temperature (T < 5K) reorientation of molecular groups can be properly described as a quantum mechanical process. In the neutron spectra this quantum tunneling appears as inelastic lines due to the splitting of the torsional ground state. The determination of this splitting is a very sensitive probe of the height and shape of the potential barrier. At higher temperature, (T ~ 20K) a quantum relaxation process occurs caused by the interaction with librational excitations and phonons. At still higher temperature the reorientations become thermally activated and can be treated classically, as it has been done in the preceding paragraph. The existence of translational tunneling is much more rare phenomenon; we will give an example of hydrogen tunneling in a metal. One of the simplest case of tunneling motion is the reorientation of methyl group in solids. A recent review on these tunneling motions can be found in reference [15], while recent experimental and theoretical results were presented in a Workshop held at the ILL [16].

3.2. Methyl groups dynamics

The energy levels of the one-dimensional rotator are given by the eigen-values of the Schrödinger equation :

$$[- B \frac{\partial^2}{\partial \theta^2} + V(\theta)] \Psi(\theta)] = E\Psi(\theta) \qquad (24)$$

where $B = \frac{\hbar^2}{2I}$ is the rotational constant (I is the momentum of inertia). The rotational potential $V(\theta)$ is described by a Fourier expansion, taking into account the basic threefold symmetry of a methyl group :

$$V(\theta) = \sum_n \frac{V_{3n}}{2} [1 + \cos n\theta] \qquad (25)$$

Usually it is sufficient to truncate this expansion at the second order. More complicated potentials must be considered when there is a coupling between the tunneling motion of two methyl groups. According to the C_3 symmetry, two types of rotator eigenfunctions can be found : A state having a nuclear spin 3/2 and E states with nuclear spin 1/2. The energy difference between these two states is called the tunnel splitting. The energy levels and associated wavefunctions are calculated by diagonaliza-tion of the Hamiltonian (24) using a suitable basis of free rotator wave

194

functions [15]. Two kinds of excitations can be distinguished as can be seen on figure 20 :
i)   the torsional (or librational) transitions, with energies in the meV range ($\hbar\omega_\tau$). They describe oscillations of the methyl group in one minimum of the potential.
ii) the tunneling transitions with energies in the µeV range ($\hbar\omega_t$). Each torsional state n is split into two tunneling states A_n and E_n owing to the overlap of the wavefunctions in the adjacent wells of the potential.

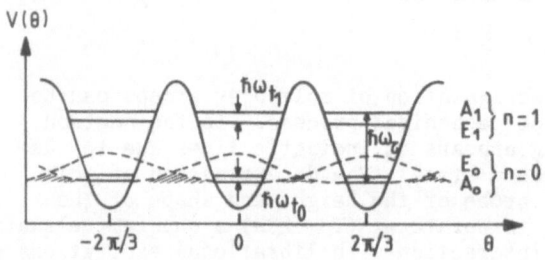

Figure 20. Tunnel splittings for a threefold symmetric potential.
$\hbar\omega_{t0}$: tunnel splitting of the groundstate
$\hbar\omega_{t1}$: tunnel splitting of the first excited state
$\hbar\omega_\tau$: torsional splitting
The dashed regions represent the overlap of wavefunctions in neighbouring potential wells.

A review on methyl groups reorientation in molecular crystals is found in ref. [17] while recent results are presented in [18]. I describe below two examples of such studies for isolated and coupled methyl groups.

3.2.1. Simple tunneling motion in toluene [19]

The authors have recorded the spectra for perdeuterated toluene $C_6D_5CD_3$ and for the deuterated compound with a protonated methyl group $C_6D_5CH_3$, using two different backscattering spectrometers at the ILL, with respective resolutions of 0.4 and 7 µeV. The results are reported on figure 21. They are a good illustration of the huge isotopic effect of the tunneling transitions : the tunneling frequency is about 21 times smaller for the $-CD_3$ compound than for the $-CH_3$ one. A hindering potential of about 20 meV is determined, with a predominant V_3 term (Eq. 25).

3.2.2. Coupled tunneling motion in lithium acetate [20]

Coupling between different molecular groups, leading to collective rotational motion, can be important in many cases where groups are located close together. It thus can be expected that cooperative phenomena will be observable for low hindering barriers and short distances between groups. One of this system is lithium acetate dihydrate (LIAC) $CH_3COOLi-2H_2O$ where X-ray results have shown that the methyl groups can be considered as one-dimensional rotors grouped in pair with their planes of rotation separated by 2.5 Å along the C_3 axes.

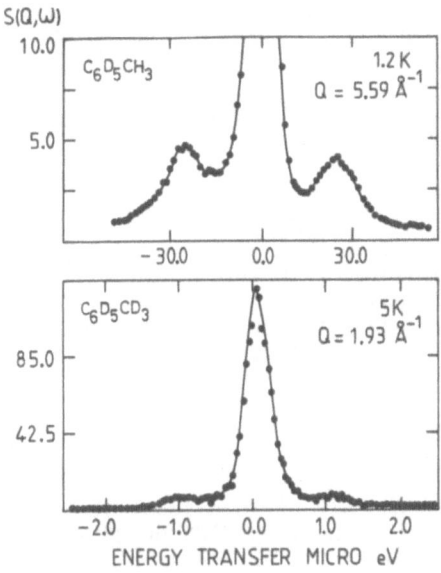

Figure 21. Tunneling spectra of toluene at low temperature. Upper curve : protonated methyl group. Lower curve : deuterated methyl group.

It is easy to show, using Eq. 24, that low barrier corresponds to high tunnel splitting. This is a case in LIAC where NMR [21] and Neutron Inelastic Scattering [22] agree with a tunnel splitting of 250 μeV. Figure 22b shows the tunneling spectrum of LIAC measured at 2.6°K with

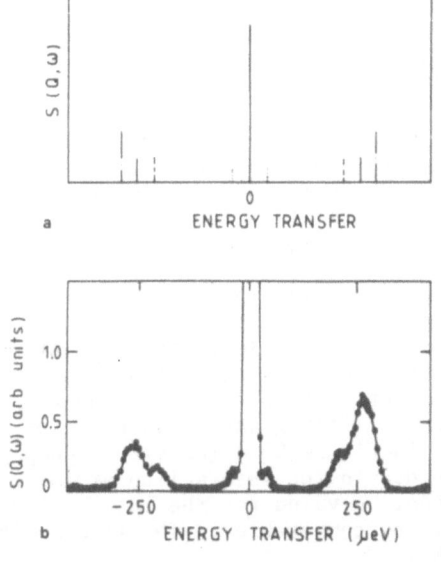

Figure 22 a) : Theoretical tunneling spectrum for a pair of coupled methyl groups for a coupling strength much higher than the tunnel splitting.
b) Tunneling spectrum of LIAC at 2.6 K measured with a high resolution time-of-flight spectrometer (resolution FWHM : 20 μeV). The average scattering angle is 100°. Nota : in this figure, ω > 0 corresponds to neutron energy loss.

a resolution (FWHM) of 20 μeV. It is seen that the main peak around 250 μeV is clearly splitted into two peaks with one of them having twice the intensity of the other. Furthermore a much weaker inelastic peak appears at 40 μeV at the base of the elastic peak.

The authors have solved the Schrödinger equation (Eq. 24) for a pair of coupled rotors, writting the coupling term as :

$$W(\theta, \phi) = \frac{W}{2} [1 - \cos 3(\theta - \phi)] \qquad (26)$$

The theoretical resulting spectrum is shown in Figure 22a which is in very good agreement with the measured spectrum. The data were fitted with the theoretical scattering law convoluted with the resolution function of the spectrometer. The continuous line in Fig. 22b shows that the theoretical scattering law described quantitatively the data with a strength of the coupling term (Eq. 26) being about twice as strong as the threefold hindering barrier for the single particle motion (8.6 meV compared to 3.9 meV for the V_3 term).

3.3. Hydrogen tunneling in a metal [23]

In the preceding section we have seen an example of rotational tunneling. Defects in lattices with variable concentrations (e.g. CN^- or OH^- in alkali halides) constitute another class of tunneling centers. At low temperature the tunnel spectrum consists of well defined excitations. With increasing concentration, interaction between the defects lead to asymmetric pocket states (like for two-level systems in amorphous materials) and the excitations become smeared out in energy.

Hydrogen tunneling in $Nb(OH)_x$ falls into that category and has been studied recently by Neutron Scattering [23]. The thermal destruction of a tunneling state, above say 10 K is generally attributed to interactions with phonons. If the tunnel system is in a metallic environment, there is an additional relaxation path originating from the interaction with electrons. Nb is a superconductor with a transition temperature of $T_c = 9.2°$ K and the OH defect can be considered as a model system to study interaction of a tunnel state with conduction electrons.

Figure 23 presents the concentration dependence of the inelastic scattering measured at 1.5° K. The ordinates are plotted as a percentage of the elastic intensity. One sees clearly the effect of interactions between the tunnel systems as the concentration is increased, starting with $x = 1.5 \times 10^{-4}$. The full line is the result of a fit to a scattering law for a two-site model, with a distribution of matrix element and of the energy difference between the two sites (due to elastic defect-defect interaction).

Figure 24 shows the temperature dependence for the sample with $x = 2.2 \times 10^{-3}$. Up to 5 K, only the intensities for neutron energy loss and energy gain vary in accordance with the detailed balance factor. At higher temperatures, the lineshape of the inelastic scattering changes and the separation from the elastic scattering becomes less pronounced : this comes from a decrease of the energy splitting and an increase of the damping. These effects are summarized in Figure 25. The strong damping is attributed to a Korringa relaxation due to the interaction of the tunnel systems with conduction electrons. A value for the interaction strength has been obtained and shown to be comparable to the interaction of the electrons with the phonons in Nb.

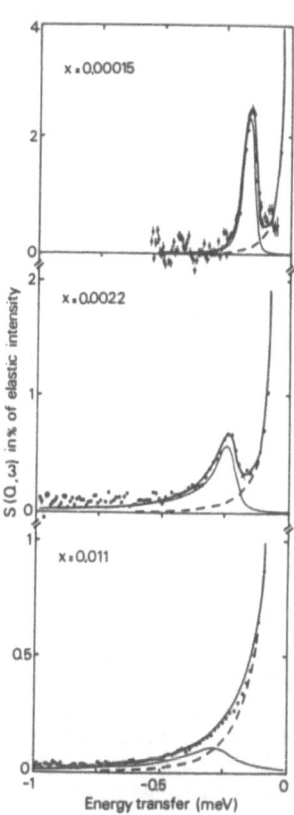

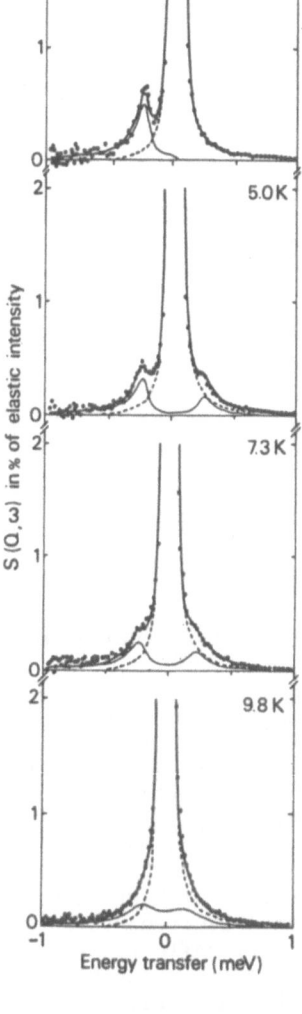

Figure 23. Concentration
dependence of the inelastic
scattering scattering for $N_b(OH)_x$
at 1.5 K.
Note the different scales of the
ordinates.

Figure 24. Temperature dependence
of the inelastic scattering
for $N_b(OH)_{0.0022}$.

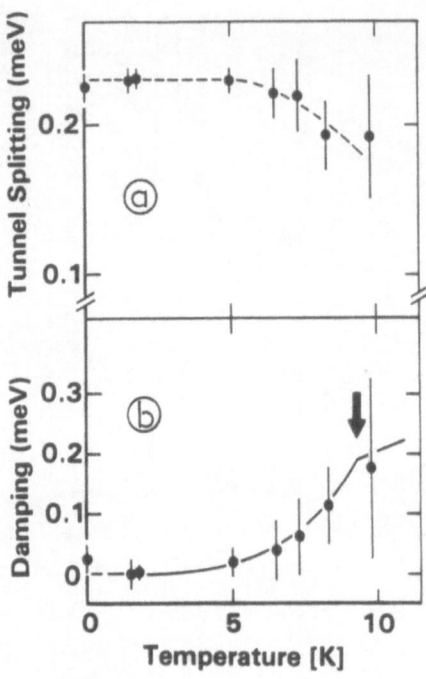

Figure 25. Temperature dependence of (a) : the tunnel splitting and of (b) : the damping, for $Nb(OH)_{0.0022}$. The arrow marks the superconducting transition temperature for pure Nb.

3.4. Elementary excitations in liquid ^{3}He [24]

Neutron inelastic scattering offers a unique method to examine directly the elementary excitations in liquid ^{3}He on a microscopic scale, for wave vectors of order 1 Å^{-1}. Through their spin dependent interaction with the ^{3}He nuclei, neutrons probe the nuclear spin fluctuations in addition to the density fluctuations which are probed by the spin averaged interaction.

The experimental difficulty comes from the high absorption cross section of the ^{3}He nuclei, which lead to the probability for a neutron to be scattered without being absorbed of only 2 x 10^{-4}. This imposes very stringent requirements on the sample cell : it should cause minimal scattering, while being thick enough to sustain pressures in excess of 2MP$_a$. Similarly the energy independent background of the spectrometer had to be small (less than 43 counts/hour/detector).

Energy distributions at constant Q were generated by regrouping the constant scattering angle TOF-data into wave vectors bins of 0.1 Å^{-1} in the range 0.3 to 2 Å^{-1} and into energy bins of 0.02 meV, resulting in 18 constant-Q representations of the dynamical structure factor $S(Q,\omega)$ on an absolute scale :

$$S(Q,\omega) = S_{coh}(Q,\omega) + \frac{\sigma_{inc}}{\sigma_{coh}} S_{inc}(Q,\omega) \qquad (27)$$

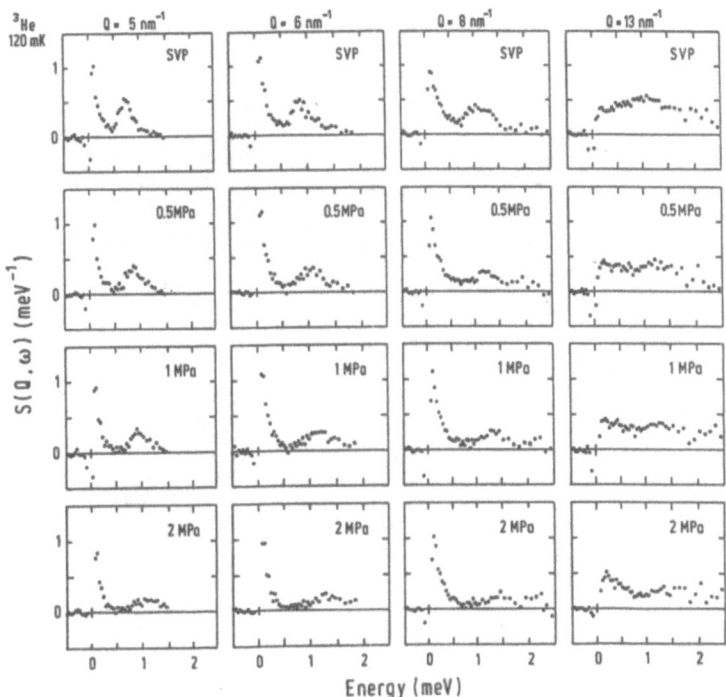

Figure 26. Dynamical structure factor of liquid ^{3}He at 0.12 K for pressures and momentum transfers as shown.

The experimentally determined scattering function is shown in Figure 26. It should be noted that the large elastic peak coming the sample cell scattering obscures the signal for energies below 0.1 meV. At low Q, the low energy "paramagnon" scattering is largely independent of pressure, while at larger wave vectors the spectra show pronounced softening of the single-pair excitations with increasing density. This latter effect is a manifestation of the increase in the effective quasiparticle mass.

The density dependence of the zero sound mode (collective density fluctuation mode) is even more distinct. Figure 27 shows the frequencies and intrinsic widths of this mode at the four pressures measured.

The frequency of the mode increases, the width increases and the spectral weight decreases as the pressure is raised. These results should serve as a useful test of theories on excitations in neutral Fermi liquids. They are in qualitative agreement with extensions of the Landau Fermi liquid theory to finite wave vectors.

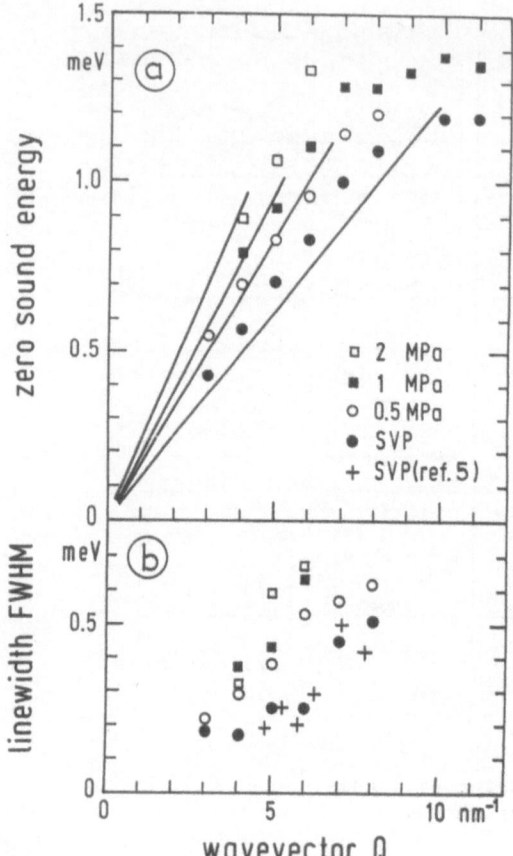

Figure 27. a) : Energy and b) : linewidth (FWHM) of the zero sound mode for different pressures and at T = 0.12 K. The solid lines in a) are derived from the ultrasonic sound velocities at these pressures and temperature.

4. Inelastic Scattering

4.1. General

Because of the large incoherent scattering cross-section and the large amplitude of vibration of the hydrogen nucleus, Incoherent Inelastic Neutron Scattering of an hydrogeneous material is dominated by those normal modes of vibration which involve hydrogen motion. It is thus possible to observe normal modes which are infrared or Raman inactive or which are associated with very small dipole moment or polarizability changes. Another situation where Neutron Scattering is favourable is for the study of complexes which fluoresce or are unstable in a laser beam.

Neutron Scattering has a distinct advantage over all the other spectroscopic techniques : the spectra result from a <u>direct coupling</u> with the <u>positions</u> of the scattering particles (see Eq. 9). By making use of the <u>selective deuteration</u> method, it is possible to assign vibrational bands

unambiguously. The low frequency librational and torsional modes are
easily seen with Neutron Scattering. This arises because these modes are
associated with large amplitude of vibration and give rise to intense
bands in the neutron spectra. This contrast with the results obtained
using optical (infrared or Raman) spectroscopy where torsions and librations are either formally forbidden by the optical selection rule or
produce only very small changes in the dipole moment and/or polarizability of a molecule.

I will not present here the whole field of molecular spectroscopy
with neutrons. Interested readers can consult the textbook by Boutin and
Yip [25] and an excellent review which has appeared recently [26]. In
the remaining of this section, I will concentrate on the generalized
frequency distribution and its derivation for disordered systems.

4.2. Generalized Frequency Distribution

I present here a very brief account of the derivation leading to the
generalized frequency distribution. A full derivation can be found in
ref. [25]. It is straightforward to show that for isolated or coupled
harmonic oscillators, the intermediate incoherent scattering law (Eq. 12)
takes the form :

$$I_{inc}(Q,t) = e^{-\frac{1}{2} Q^2 \gamma(t)} \tag{28}$$

where $\gamma(t)$ is called the width function. For an harmonic oscillator, in
the classical limit ($\hbar \omega_o \ll k_B T$) one has :

$$\gamma(t) = \frac{2 k_B T}{M \omega_o^2} (1 - \cos \omega_o t) \tag{29}$$

$\gamma(t)$ is related directly to the mean square displacement

$$\gamma(t) = \frac{1}{3} \langle r^2(t) \rangle \tag{30}$$

In the Gaussian approximation, one writes for any scattering law :

$$S_{inc}(Q,\omega) = \frac{1}{2\pi} \int_{-\infty}^{\infty} dt \ e^{i\omega t} e^{-\frac{1}{2} Q^2 \gamma(t)} \tag{31}$$

one can show [25] that $\gamma(t)$ is related to the Fourier transform of the
frequency distribution $g(\omega)$. A practical way to extract this quantity is
the extrapolation procedure due to Egelstaff and Schöfield [27].

Putting $\alpha = \frac{\hbar^2 Q^2}{2Mk_B T}$ and $\beta = \frac{\hbar\omega}{k_B T}$

the frequency distribution is obtained as :

$$g(\beta) = 2\beta \sinh(\beta/2) \lim_{\alpha \to o} \frac{S(Q,\omega)}{Q^2} \tag{32}$$

This expression is nearly equal to :

$$g(\omega) \sim \omega^2 \lim_{Q \to o} \frac{S(Q,\omega)}{Q^2} \qquad (33)$$

An example of a vibrational density of states is the one predicted by the Debye model :

$$g(\omega) = \frac{3}{\omega_D^3} \omega^2 \qquad \text{for} \qquad \omega \leqslant \omega_D$$

$$= 0 \qquad \text{for} \qquad \omega > \omega_D \qquad 34)$$

This generalized frequency distribution can be compared to the one obtained by Raman scattering. It will be shown that while in the neutron case one obtains a quantity which can be directly compared to Molecular Dynamics simulations, in the Raman case, the low frequency part is quite insensitive to the whole molecule vibrations.

4.2.1. Inelastic scattering of pivalic acid

It has been mentioned previously, that in pivalic acid, the inelastic part of the spectrum overlaps with the quasi-elastic part. The authors have extracted the generalized frequency distribution from the spectra measured with a time-of-flight spectrometer [7], but without doing the extrapolation to Q going to zero implied by Eq. 33. It is noted $P(\alpha,\beta)$. Figure 28 gives the results for the D_1 compound below and above the plastic phase transition temperature and for the D_{10} compound in the plastic phase. One notices that at high energy transfers ($\omega > 40$ meV) the resolution degrades rapidly : this comes from the low energy of the incident neutrons (~ 5 meV). There is a marked change of the spectra in the low frequency range (0 - 20 meV) at the phase transition : all the peaks disappear and the general shape is that of a single broad band. Furthermore this part of the spectrum is shifted to lower frequencies.

In a phenomenological approach, the authors have analyzed the low frequency part of the spectra obtained in the plastic phase at each scattering angle, in terms of a single overdamped oscillator, whose scattering law is given by :

$$S(Q,\omega) = e^{-Q^2 \langle u^2 \rangle} \times \left\{ \delta(\omega) + \frac{Q^2 \langle u^2 \rangle}{e^{\frac{\hbar\omega}{kT}}-1} \frac{1}{\pi} \frac{\omega\gamma}{(\omega^2-\omega_0^2)^2 + \omega^2\gamma^2} \right\} \qquad (35)$$

$\langle u^2 \rangle$ is the mean square amplitude of the displacement of the scattering centre in the oscillatory motion. It is related to the mean square angular libration of the molecule $\langle \theta^2 \rangle$ by :

$$\langle u^2 \rangle = \frac{I}{M} \langle \theta^2 \rangle \qquad (36)$$

where I and M are the moment of inertia and the mass of the molecule respectively. Figure 29 shows the scattering law, Eq. 35, for different values of the damping coefficient. When the damping increases, the shape of the scattering function gradually changes from a two peak stucture towards a single central peak.

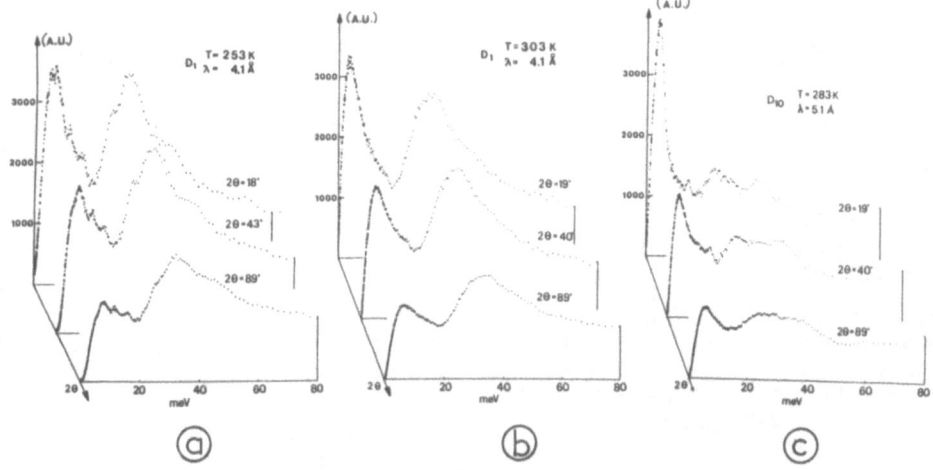

Figure 28. Experimental values of the function $P(\alpha, \beta)$ for pivalic acid as a function of the energy transfer, for three scattering angles.
(a) : pivalic acid D_1 in the low temperature phase
(b) : pivalic acid D_1 in the plastic phase
(c) : pivalic acid D_{10} in the plastic phase.

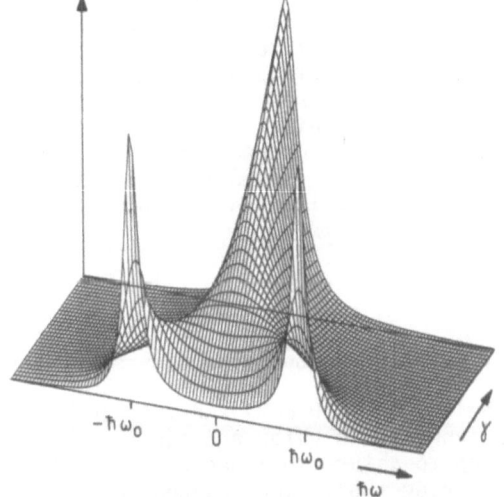

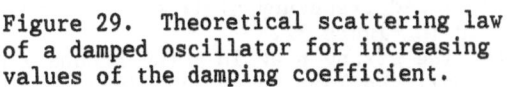

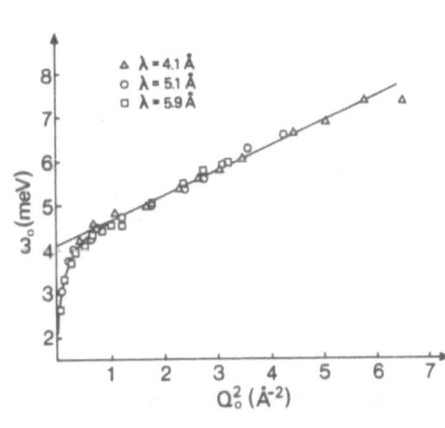

Figure 29. Theoretical scattering law of a damped oscillator for increasing values of the damping coefficient.

Figure 30. Frequency of the overdamped oscillator in the phenomenological description of the low-frequency part of the inelastic spectra obtained from refinements to experimental data of pivalic acid D_1.

204

Figure 30 presents for each scattering angle the frequency ω_0 of the overdamped oscillator. The values obtained with the three incident energies were found to coincide perfectly. From these experimentally determined values, the authors could evaluate the contribution of the inelastic scattering to the quasi-elastic part of the spectra. Figure 9 shows such a fit for all scattering angles simultaneously. The amount of inelastic scattering is controlled by $\langle u^2 \rangle$ since from Eq. 35 :

$$\frac{I^{inel}(Q)}{I^{quasi}(Q)+I^{inel}(Q)} = \frac{Q^2\langle u^2 \rangle}{1+Q^2\langle u^2 \rangle} \simeq 1 - e^{-Q^2\langle u^2 \rangle} \qquad (37)$$

The second equality holds because Eq. 35 has been obtained using a one-phonon expansion. The value of $\langle u^2 \rangle$ is found to be nearly independent of temperature in the plastic phase and equal to 0.42 Å. Using Eq. 36, this corresponds to an angular amplitude of oscillation $\langle \theta^2 \rangle^{1/2} \simeq 8°$ of the dimer unit. In Figure 31 is reported the ratio of the inelastic intensity over the whole intensity obtained with the least square fitting procedure used to obtain the EISF (see Fig. 7). The full line is the variation calculated with Eq. 36, and using the value $\langle u^2 \rangle = 0.42$ Å^2. The agreement is quite acceptable taking into account that an appreciable multiphonon contribution exists at the higher Q values.

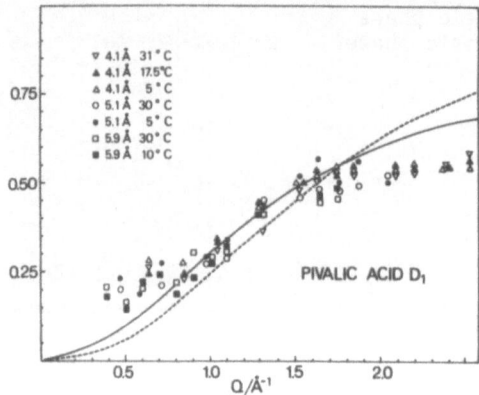

Figure 31. Ratio of the intensity arising from librations over the whole scattered intensity as a function of the momentum transfer Q. The full line is obtained from Eq. 35 for an overdamped oscillator, while the dashed line corresponds to fluctuations of the dimer axis.

4.2.2. Low- and very low-frequency dynamics in ZnCl$_2$ aqueous solutions

A comprehensive investigation on the low ($\omega < 50$ meV) and very low (0.02 $< \omega < 5$ meV) frequency vibrational dynamics in water and aqueous solutions of ZnCl$_2$ (for concentrations up to saturation) has been done by TOF neutron spectroscopy and Raman scattering [28]. The reduced depolarized Raman intensity $g^R(\omega)$ and the generalized frequency distribution deduced from neutron data $g^N(\omega)$ are compared in Figure 32.

One can see that for pure water the strong peak in $g^N(\omega)$ around 7 meV (~ 60 cm^{-1}) is only apparent as a small shoulder in $g^R(\omega)$. In a very drastic approximation [29] one can write :

$$g^R(\omega) = C^R(\omega) \cdot g(\omega) \qquad (38)$$

where $C^R(\omega)$ is the electron-vibration coupling function. It can be obtained by forming the ratio $g^R(\omega)/g^N(\omega)$ and is shown in Figure 33 for pure water and three concentrations of $ZnCl_2$.

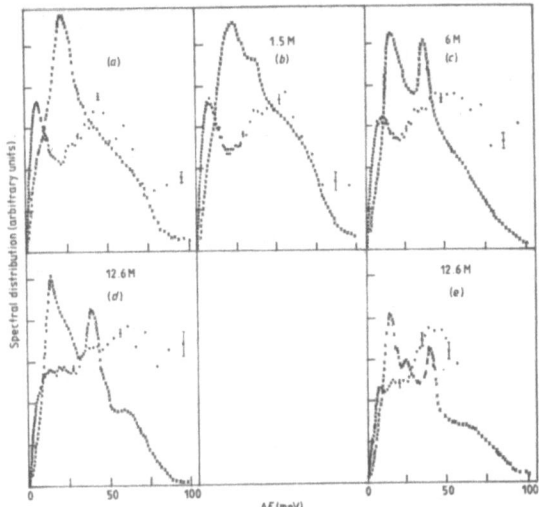

Figure 32. Reduced depolarised Raman intensity $g^R(\omega)$ (crosses) and generalized frequency distribution $g^N(\omega)$ (dots) for pure D_2O (a) and solutions of $ZnCl_2$ in D_2O (b-d). The reduced spectra for the saturated $ZnCl_2$ solution in H_2O are also shown for comparison (e).

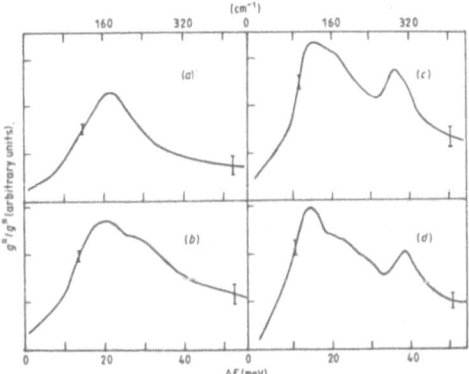

Figure 33. Concentration dependence of the experimentally obtained Raman coupling function $C^R(\omega) = g^R(\omega)/g^N(\omega)$: pure water (a), 3 M $ZnCl_2$ in H_2O (b), 6 M $ZnCl_2$ in H_2O (c) and 12.6 M $ZnCl_2$ in H_2O (d).

The peak in $g(\omega)$ around 7 meV has been found in MD simulations [30], [31] and shown to arise from oscillations in the centre-of-mass velocity correlation function. Furthermore it has been shown that this translational motion become spectrally active due to interaction-induced dipoles [31]. One sees on figure 32 that this peak broadens and shifts to higher frequency with increasing $ZnCl_2$ concentration. These data should be of some value to derive realistic potentials to be used in the MD simulations of aqueous ionic solutions. The very-low frequency part of $g^N(\omega)$ and $g^R(\omega)$ are reported in figure 34 for H_2O and D_2O. The observed behaviour of $g^N(\omega)$ is clearly far from a typical ω^2 Debye dependence which can be expected at such low frequencies. There is a big contribution of other low frequency modes like structural relaxation modes observed in all amorphous substances. By using a model for the Raman coupling function for acoustic modes, a correlation length varying from 3.5 Å for pure water to 7.5 Å for the saturated solution has been determined. This is a quantitative confirmation of the local structural change with increasing solute concentration and of the connected increase of the spatial corelation of the locally ordered patches.

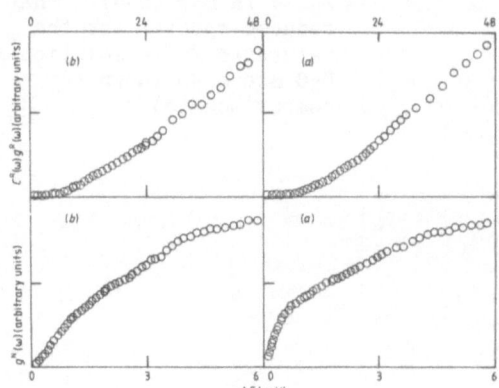

Figure 34. Very-low frequency spectrum of the generalized distribution $g^N(\omega)$ and of the reduced Raman intensity after subtraction of the quasielastic contribution for H_2O (a) and D_2O (b).

4.2.3. Fractal dynamics of amorphous materials

A currently fashionable explanation for the form of the density of states of amorphous materials is based on "fractons" which are excitations on a fractal network [32]. The theoretical predictions are that for a percolating network the nature of the thermal vibrations would change at a particular frequency. Below that frequency, the vibrations would be phonons which could propagate and $g(\omega)$ would vary as ω^2 as in the Debye model (Eq. 34). However, above that frequency (whose wavelength would correspond to the length of some characteristic structural unit of the material), $g(\omega)$ would increase more rapidly than ω^2 and this region would correspond to localized vibrational modes which were called "fractons".

Inelastic neutron scattering on epoxy resins and on the amorphous and crystalline monomer basis of these resins [33,34] seems to support that proposal as can be seen on Figure 35. For the crystalline material $g(\omega)$ varies quite nicely as ω^2 up to about 3 meV beyond which there is some structure which is typical of the density of states of crystals (Fig. 35a). The slight deviation at the lowest energy is probably due to the contamination of the wings of the huge elastic peak coming from Bragg scattering. The amorphous material (Fig. 35b) shows a breakaway from ω^2 dependence to a higher power at an energy of about 1.2 meV. Even if one does not accept the phonon-fracton hypothesis, the difference between the two curves is unequivocal.

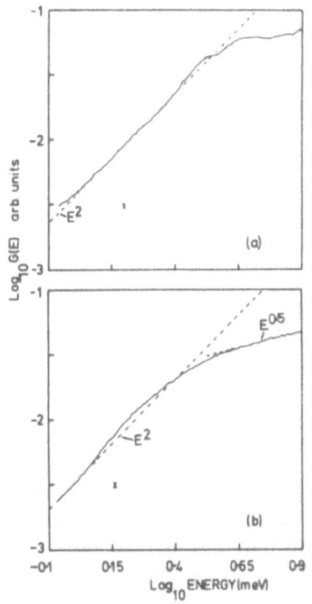

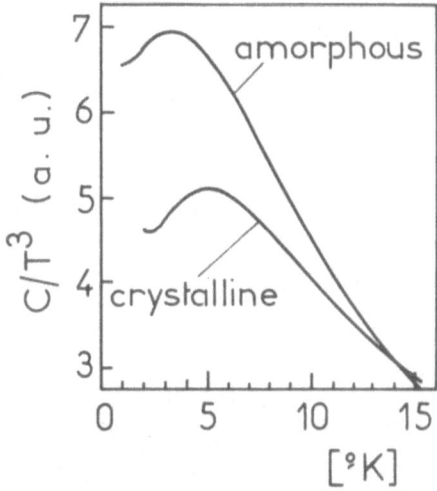

Figure 35. Dependence of the density G(E) on the energy E for (a) : crystalline and (b) : amorphous monomer basis DGEBA. The dashed line shows the slope for an E^2 dependence. The vertical markers indicate probable error bars.

Figure 36. Specific heat C, plotted as C/T^3 versus T, calculated from the density of state curves of Fig. 35, for amorphous and crystalline DGEBA. The curves have been normalized at 14K, where the measured specific heats are found equal.

Specific-heat (C) data of amorphous materials shows that C/T^3, instead of being constant as one would expect for a Debye solid, exhibit a maximum in the liquid helium range. Using the data of Fig. 35, one can calculate C/T^3 for the two sets of sample (For the crystalline material, the Bragg tail at the lowest energy has been subtracted for this calculation). The results are shown on Figure 36; the two curves have been normalized at 14K, where the specific heat of the two samples has been found equal. In each case C/T^3 has a peak, but it is much reduced and shifted to higher temperature for the crystalline material. This work demonstrates that it is very difficult, if not impossible, to draw conclusions about details of the density of states curve from specific-heat data alone.

4.3. Low frequency modes in vitreous silica

In the preceding section, the measurements were done on hydrogeneous samples. The predominantly incoherent scattering permits to extract the vibrational density of states from the neutron data, but any analysis of the character of these vibrational modes is impossible. On the contrary with a purely coherent scatterer, one can do this analysis from an exami nation of the inelastic structure factor. This has been done in the cas of vitreous silica [35].

An approximation for the one-phonon scattering cross-section for amorphous materials is given by [36] :

$$\frac{d^2\sigma^{(1)}}{d\Omega d\omega} = \frac{k}{k_o} \frac{3\hbar N}{2} \frac{g(\omega)}{\omega} n_B I^{(1)}(Q) \qquad (39)$$

where N is the number of atoms and n_B the Bose factor : $1/(e^{\frac{\hbar\omega}{kT}} -1)$. The inelastic structure factor $I^{(1)}(Q)$ is given by :

$$I^{(1)}(Q) = \left| \sum_j b_j e^{-W_j} e^{i\vec{Q}.\vec{R}_j} (\vec{Q}.\vec{e}_j)/M_j^{1/2} \right|^2 \qquad (40)$$

where $\vec{R}_j$ is the position vector, M_j the mass, and $\vec{e}_j$ the displacement vector of atom j.

On the assumption that the Q dependence of $I^{(1)}(Q)$ is the same for all modes in the low frequency range considered, $g(\omega)$ can be obtained by scaling the inelastic intensities to give a common curve, shown on Figur 37. The variation with Q of this inelastic scattering provides detailed

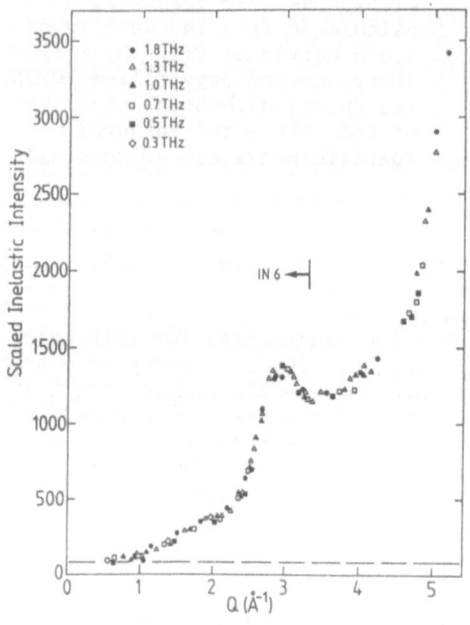

Figure 37. Inelastic intensities of vitreous silica scaled to a common curve. The dashed line is the estimated multiple scattering contribution, obtained from the low Q par

information on the nature of the vibrational modes. In particular for sound waves, $I^1(Q)$ should be proportional to $Q^2 S_0(Q)$ [36], where $S_0(Q)$ is the elastic structure factor. It is seen on Figure 37 that for $Q = 1.6$ Å^{-1} where $S_0(Q)$ has a strong peak, there is only a small shoulder. This is an indication that sound waves in this Q range are only a small part of the vibrational excitations. A detailed analysis of $I^{(1)}(Q)$ is able to explain the peak at $Q = 2.6$ Å^{-1} and the sharp rise around $Q = 5$ Å^{-1} : this is assigned to relative rotations of five tetrahedra [35].

The density of states, whose scaling is nearly independent of the model used to fit $I^{(1)}(Q)$, is shown in Figure 38. One can see that it is much higher than the Debye density of states calculated with a mean sound velocity of 4.1 km/s, and agrees quite well with the one deduced from the specific heat.

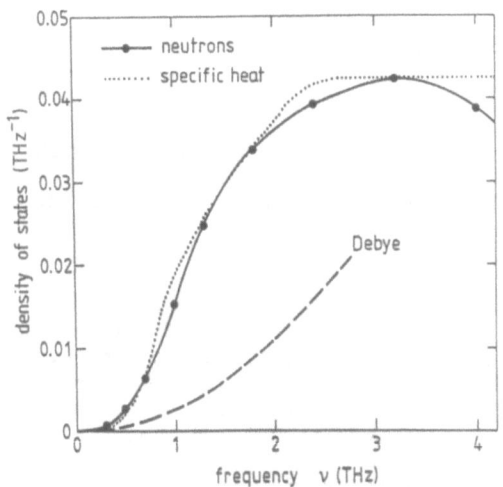

Figure 38. Normalized density of states derived from the scaling factor used to bring the inelastic intensities to the common curve of Fig. 37. The dashed line is the sound wave contribution while the dotted line is derived from the measured specific heat.

Figure 39 compares the heat capacity calculated from the density of states of Fig. 38 to the measured data. The agrement is again quite good and the peak at 11K indicates that the density of states increases first more rapidly than ω^2 and then less rapidly as can be seen on Fig. 38.

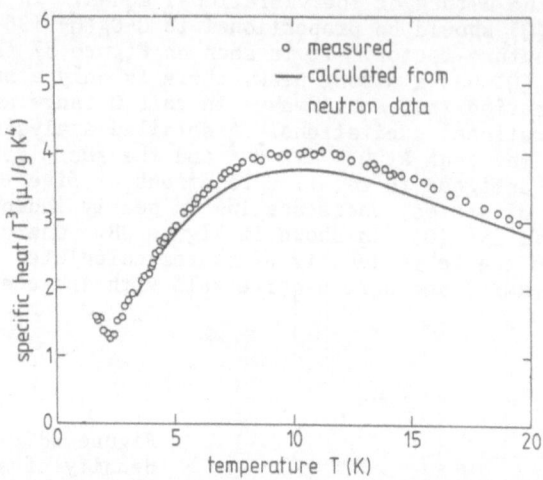

Figure 39. Specific heat C of vitreous silica plotted as C/T^3 versus temperature T. The points are experimental data and the line is calculated from the density of states obtained by neutron scattering.

5. CONCLUSIONS

Neutron scattering is certainly the most powerful spectroscopic techni-ques to study motions in condensed matter. However, due to its relati-vely low flux and its high cost which make it available only at selected places around the world, it should be used in conjunction with the other spectroscopic techniques, to provide the information which are not obtainable by them.
      Neutron Scattering (quasi-elastic or inelastic scattering) measures directly a correlation function of the displacement vector of the particle. Different kinds of motion can be separated through the momentum transfer variation of the spectrum. It is very well adapted to the study of the dynamics of hydrogeneous samples due to the high value of the incoherent cross section of the hydrogen. Quantum motions are directly visible in the neutron spectra at low temperature. These spectra provide excellent data to test theories of quantum tunneling in solids or elementary excitations in quantum fluids.

REFERENCES

[1] F. VOLINO, 'Local dynamics in polyatomic fluids' in "Microscopic Structure and Dynamics of Liquids", J. DUPUY and A.J. DIANOUX Eds. NATO-ASI series B33 - Plenum Press (1978)

[2] W. MARSHALL and S.W. LOVESEY, "Theory of Thermal Neutron Scattering", O.U.P. (1971);
S.W. LOVESEY, "Theory of Neutron Scattering from Condensed Matter", O.U.P. (1984)

[3] V.F. SEARS, Can. J. Phys. 44, 1279 (1966); 44, 1299 (1966)

[4] T. SPRINGER, "Quasi-elastic Neutron Scattering for the Investigation of Diffusive Motions in Solids and Liquids", Springer Tract in Modern Physics, Vol. 64, Springer Verlag (1972)

[5] M. BEE and C. POINSIGNON, "Application of Quasi-elastic Neutron Scattering to Solid State Chemistry, Biology and Material Science", Adam & Hilger (1988)

[6] M. BEE, J. Chimie-Physique 82, 205 (1985)

[7] M. BEE, W. LONGUEVILLE, J.P. AMOUREUX, and R. FOURET, J. Physique 47, 305 (1986)

[8] F. VOLINO, A.J. DIANOUX and H. HERVET, J. Physique C37, 55 (1976)

[9] R.E. LECHNER, J.P. AMOUREUX, M. BEE and R. FOURET, Comm. Phys. 2, 207 (1977)

[10] F. VOLINO, M. PINERI, A.J. DIANOUX and A. DE GEYER, J. Pol. Science, Poly. Phys. Ed., 20, 481 (1982)

[11] F. VOLINO and A.J. DIANOUX Mol. Phys. 41, 271 (1980)

[12] J. TEIXEIRA, M.C. BELLISSENT-FUNEL, S.H. CHEN and A.J. DIANOUX, Phys. Rev. A31, 1913 (1985)

[13] See e.g. P.A. EGELSTAFF "An introduction to the liquid state", Academic Press (1967)

[14] Workshop on Water, "Structure and dynamics of water and aqueous solutions : anomalies and their possible implications in biology", J. Physique C7 (1984)

[15] W. PRESS, "Single-particle rotations in molecular crystals", Springer Tract in Modern Physics, Vol. 92, Springer Verlag (1981)

[16] A. HEIDEMANN et al, Eds., "Quantum aspects of molecular motions in solids", Springer Proc. Phys. 17 (1987), Springer Verlag

[17] D. CAVAGNAT, J. Chimie-Physique 82, 239 (1985)

[18] M. PRAGER, J. STANISLAWSKI and W. HAUSLER, J. Chem. Phys. 86, 2563 (1987)

[19] D. CAVAGNAT et al, J. Physique 45, 97 (1984)

[20] S. CLOUGH et al, Z. Phys. B55, 1 (1984)

[21] P.S. ALLEN and P. BRANSON, J. Phys. C, 11, L121 (1978)

[22] S. CLOUGH, A. HEIDEMANN and N. PALEY, J. Phys. C 13, 4009 (1980)

[23] A. MAGERL, A.J. DIANOUX, H. WIPF, K. NEUMAIER and I.S. ANDERSON, Phys. Rev. Lett. 56, 159 (1986);
H. WIPF, D. STEINBINDER, K. NEUMAIER, P. GUTSMIEDL, A. MAGERL and A.J. DIANOUX, ref. 16, p. 153

[24] R. SCHERM, K. GUCKELSBERGER, B. FAK, K. SKÖLD, A.J. DIANOUX, H. GODFRIN and W.G. STIRLING, Phys. Rev. Lett. 59, 217 (1987)

[25] H. BOUTIN and S. YIP, "Molecular spectroscopy with neutrons", MIT Press (1968)

[26] J. HOWARD and T.C. WADDINGTON, "Molecular spectroscopy with neutrons", in 'Advances in infrared and Raman spectroscopy", Vol. 7, p. 86, Heyden (1980)

[27] P.A. EGELSTAFF and P. SCHOFIELD, Nucl. Sci. Eng. **12**, 260 (1962)

[28] G. MAISANO, P. MIGLIARDO, M.P. FONTANA, M.C. BELLISSENT-FUNEL and A.J. DIANOUX, J. Phys. C **18**, 1115 (1985)

[29] R. SHUKER and R.W. GAMON, Phys. Rev. Lett. **25**, 222 (1970)

[30] P. BOPP in "The Physics and Chemistry of aqueous ionic solutions", M.C. BELLISSENT-FUNEL and G.W. NEILSON, Eds, NATO-ASI, **C205**, p. 217, Reidel (1987)

[31] P.A. MADDEN and R.W. IMPEY, Chem. Phys. Lett. **123**, 502 (1986)

[32] B. DERRIDA, R. ORBACH and K.W. YU, Phys. Rev. B **29**, 6645 (1984)

[33] H.M. ROSENBERG, Phys. Rev. Lett. **54**, 704 (1985)

[34] A.J. DIANOUX, J.N. PAGE and H.M. ROSENBERG, Phys. Rev. Lett. **58**, 886 (1987)

[35] U. BUCHENAU, N. NÜCKER and A.J. DIANOUX, Phys. Rev. Lett., **53**, 2316 (1984) and errata, **56**, 539 (1986); U. BUCHENAU, M. PRAGER, N. NÜCKER, A.J. DIANOUX, N. AHMAD and W.A. PHILLIPS, Phys. Rev. B **34**, 5665 (1986)

[36] U. BUCHENAU, Z. Phys. B **58**, 181 (1985)

CHAPTER 11

PROBING STRUCTURAL DYNAMICS BY VERY HIGH RESOLUTION
NEUTRON THREE AXIS SPIN ECHO SPECTROSCOPY

Claude M.E. ZEYEN
INSTITUT LAUE-LANGEVIN 156X
F-38042 GRENOBLE-CEDEX

ABSTRACT

Adding the neutron spin-echo (NSE) method to the classical and widely used three-axis spectroscopy technique (TAS),(TAS+NSE=TASSE) it is possible to boost its resolution characteristics by several orders of magnitude. This not only means that TAS measurements will then be feasible with a very significantly enhanced energy resolution, but also that new types of high resolution quasielastic and inelastic experiments will be made possible. The domain of large momentum and large energy transfers at relative resolutions of better than say 1% is at the present time uncovered by any technique. TASSE is not just another "trick", the resolution profit of which one has to pay by much reduced intensity, but it is a fundamentally different approach to inelastic spectroscopy featuring significantly better neutron economy and an accessible time domain for solid or liquid state dynamics ranging from 10^{-14} to 10^{-6} seconds combined with momentum transfers of 10^{-2} to 10 Å^{-1} a unique feature of neutron scattering. The study of lifetimes of collective excitations in solids as a function of their frequency and momentum will thus be possible. Lifetimes of higher energy optical excitations, energy widths of critical scattering or central components in the vicinity of phase transformations could be fully investigated.

After a general introduction to NSE we will describe the design calculation of the optimized TASSE and give its potential capabilities. An outline of unique applications in a wide range of time domains of structural and magnetic dynamics this novel technique allows will then be given.

1. INTRODUCTION

Neutron triple axis spectroscopy (TAS) has for a long time been the only technique giving direct access to the determination of dispersion relations in single crystal materials[1]. Exceptionally, for short lived excitations, the energy distribution of inelastic peaks could also be analyzed, hence yielding the

G. J. Long and F. Grandjean (eds.), The Time Domain in Surface and Structural Dynamics, 213–232.
© 1988 by Kluwer Academic Publishers.

lifetime of the collective excitation under study[2]. Most commonly though, the anharmonic mechanisms responsible for this finite lifetime do not produce sufficient energy broadening to make the effect measurable. TAS has limited relative energy resolution $\partial E/E > 1\%$ essentially due to intensity limitations of available neutron sources. Therefore the domains of large momentum and large energy transfer are only covered by relatively low resolution methods. For example for any excitation above a few meV the best resolution is of the order of say .05 meV. It is in order to be able to perform large energy (> 4 meV) and/or momentum (> 3 Å^{-1}) transfer neutron spectroscopy at relative energy resolutions up to 10^{-6} that we developed the combined technique described below. We will first recall the principles of neutron Larmor precession, in particular of spin echo (NSE) [3] and then introduce the improvements to this technique by optimal precession magnet design which will allow it to reach its intrinsic limits in inelastic neutron scattering.

2. NEUTRON LARMOR SPIN PRECESSION SPECTROSCOPY

Neutron Larmor precession spectroscopy techniques are based on the condition that for all neutrons in a beam, having the same energy E, the amount of precession performed while traversing a magnetic field region is identical. The energy of thermal neutrons being much larger than the Zeeman energy, the classical picture of Larmor precession is applicable and the equation of motion of the magnetic moment $\underline{m}$ of a neutron traversing a magnetic field induction $\underline{B}$ is:

$$\partial \underline{m}/\partial t = \gamma\, \underline{m} \times \underline{\dot{B}} \qquad\qquad (2.1)$$

v_z being the constant neutron velocity along a beam axis z and γ the neutron gyromagnetic constant ($\gamma = 1.8\ 10^8$ radian/sec. Tesla), the time and z dependence can be exchanged so that the z variation of $\underline{m}$ reads:

$$\partial \underline{m}/\partial z = \gamma/\, v_z\ (\underline{m} \times \underline{B}) \qquad\qquad (2.2)$$

Initiating the precession with $\underline{m}$ perpendicular to $\underline{B}$ for all neutron velocities, which corresponds to the experimental situation, we require that this situation remains along the entire precession path. In other words the spatial variations of $\underline{B}$ have to satisfy

$$\partial(\underline{m} \cdot \underline{B})/\partial t = 0$$

and we look for the conditions imposed on $\underline{B}$ and the expression for the total amount of Larmor precession performed in a given field distribution.

Clearly the field $\underline{B}$ has to keep the same direction or vary very slowly as experienced by the moving neutron. Thence the amount of precession performed by a given neutron while traversing a time independent magnetic field region will be a measure of its velocity or energy. The total amount of Larmor turns N accumulated over a length L across $\underline{B}$ will be proportional to the line integral along the path of the neutron of the modulus of $\underline{B}$:

$$N + (\gamma / E) \int^{L} |\underline{B}| \, dl \qquad\qquad (2.3)$$

If the relation between the neutron energy is to be unique, the $|\underline{B}|$ line integrals have to be exactly identical for all possible neutron trajectories within the beam, whether they are parallel or inclined with respect to the beam axis (within the angular beam divergence). If the field does not strictly point in the same direction everywhere along the integration path L, we will have to check the proportionality between N and $\int^{L} |\underline{B}| \, dl$ by solving the exact equation of motion (2.1).

Larmor precession was introduced in neutron spectroscopy by Mezei [3] making use of an echo technique analogous to the one proposed by Hahn [4]. A short review of various Larmor precession techniques used was recently given by Rekveldt [5]. The NSE principle was first applied to weakly inelastic small angle scattering for which long collimating distances are needed, hence favouring the use of long solenoids which naturally present relatively homogeneous line integrals. Using cold neutrons of wavelengths in the range of 5 to 9 Å, energy resolutions for quasielastic scattering of the order of nano eV were obtained [3]. If the same ideas are to be applied to thermal neutron spectroscopy adapted to high energy excitations and large momentum transfers [6] the magnet homogeneity question becomes more difficult to solve. This led us naturally to a systematic study of the best field shapes for neutron Larmor precession methods. These recent results will be summarized in chapters 3,4 and 5.

The **operation of NSE spectrometers** can shortly be idealized as follows. Before and after the sample of a polarized neutron spectrometer with polarization analysis, two identical cylinder precession magnets are positioned with their (z) axis along the neutron beam. Before the first magnet the precession is started by flipping the neutron spins into a plane perpendicular to z , direction $(1, 0, 0)$. They will start precessing around the field oriented along z and after traveling a time t the spin vector of a neutron of energy E_i will point in the direction:
$(\cos \omega t, \sin \omega t, 0)$
where ω is the Larmor frequency. At the end of the first magnet B_i (length L) they will have turned by a total angle $\omega t = 2\pi N_i$ given by (2.3).

At the sample position, again using a flipper, the spins are turned from the $(\cos 2\pi N_i, \sin 2\pi N_i, 0)$ into the $(\cos 2\pi N_i, - \sin 2\pi N_i, 0)$ direction. The second precession field brings about N_f precessions for neutrons of energy E_f, $(E_f - E_i)$ is the energy change of the neutron upon scattering, so that the direction of the spin is

$(\cos 2\pi[N_i - N_f], - \sin 2\pi[N_i - N_f], 0)$ which flipped back to z by a final flipper yields
$(0, \sin 2\pi[N_i - N_f], \cos 2\pi[N_i - N_f])$.

The final spin analyzer only transmits neutrons along the $(0, 0, 1)$ direction resulting in a detector signal proportional to $\{1 + \cos 2\pi[N_i - N_f]\}/2$.

We can see from this that the measured neutron intensity is directly related to $N_i - N_f$ and therefore $E_i - E_f$ which is precisely the neutron energy change related to an excitation of energy hv in the sample (h is the Planck constant, v the frequency of the excitation) via energy conservation $hv = E_i - E_f$. Whereas TAS determines E_i and E_f individually to deliver hv as a difference, NSE gives direct access to $E_i - E_f$. Further details can be found in references [7] and [3]. This NSE relation holds for a broad neutron incident spectrum provided the Larmor phase angle $\varphi = 2\pi N$ for all individual neutrons in the beam is uniquely related to their energy. If this condition is not fulfilled, neutrons which are scattered in the same manner but travel through different sections of the magnets, may end-up with different Larmor phase angles. This effect blurs the final spin polarization and appears similar to an effect of finite resolution. It is classically corrected-for by measuring a reference polarization [8] with the spectrometer in the same (or similar if the experiment concerns inelastic scattering) configuration with the sample replaced by a reference elastic scatterer. For inelastic scattering experiments though, for which the two magnets are run at significantly different fields, this correction method may be unreliable [6]. We therefore looked for magnet designs [9] the homogeneity of which is good enough to allow echo polarizations to be measured without need for corrections.

Another resolution limitation of NSE arises from the fact that the relationship between the inelasticity and $N_i - N_f$ is in fact nonlinear and high order terms may only be neglected if the spectral distributions of in- and outgoing beams are not too wide and if the maximum amount of precession is not too high. For practical purposes typical in neutron inelastic scattering, one has to restrict N to approximately 10^4. A precession magnet for inelastic NSE should therefore guarantee a trajectory-independent number of turns up to $N = 10^4$, such that for a given energy, the deviation in φ is not seen by the final spin analyzer. Beam polarizations are standardly measured to within 1%, which imposes a 10^{-6} relative line integral homogeneity to the precession field. As shown in [9] and shortly reviewed in chapters 3,4, 5 this is at least conceptually possible. Magnets designed this way are under construction.

That precession magnets for neutron beam experiments should rather have cylinder symmetry stems from the fact that the higher symmetry [10] favours homogeneity and to a certain extend reduces stray fields, provided the magnet diameter is kept small. We will therefore place ourselves in the case of cylinder magnets and solve the problem: $\int^L |\underline{B}| \, dl = \text{constant}$

for all neutrons in the beam of given size and angular divergence. This equation has analytical solutions, one of which is a continuous field shape B(z) along the cylinder axis particularly well suited to our problem, we will call this optimal field shape OFS. Even in OFS magnets though there are residual inhomogeneities, proportional to the length of the field distribution, and further corrections are usually needed.

As we shall see, OFS is easily implemented in practice, while other more exotic field shape solutions of the above variational problem, may be technically more involved.

To reduce the radial field inhomogeneity of solenoids, in-beam spiral coils have been introduced by Mezei [11]. Those line integral variations originating from the angular beam divergence though are not corrected by this technique. Neutrons traversing the magnet at an angle to its axis have longer pathlengths, an effect which becomes significant if large beam divergencies are useful and shorter magnets are used for example for three-axis spectrometers with spin-echo (TASSE). In this paper we also derive the theory of optimal correction coils compensating for magnet inhomogeneity terms as well as for pathlength effects. Such coils are calculated for the optimal field shape magnets, but we show that they can profitably be applied to other magnets too, such as existing solenoids for example. For OFS magnets these correction coils carry minimum current and very significantly improved homogeneities are obtained. For other magnets the improvements achieved with such correcting current distributions depend on the specific field shape and can be envisaged if their intrinsic homogeneity is reasonable. The limits here are given by the ratio of the correction coil field to the local main field. Under the strong field approximation used here, corrections operate as a superposed axial field and are effective for small values of the above ratio only.

Other conditions precession magnets have to fulfil are good stability in time and a linear field-current relationship. For inelastic neutron scattering experiments where the two magnets need to be operated at very different field values, superconducting magnets used in the persisting current mode, have the advantage of better time stability than independently powered classical magnets. For quasielastic work, both (classical) magnets can be powered in series by one supply which solves the stability problem. For line integrals of more than .5 Tesla Meter and not too long magnets, superconducting coils seem to be favourable.

Reduced lateral stray field levels to avoid magnetic cross-talk is a crucial point since a change of field in one magnet should not be seen in the beam area of the other-one, nor should the fringe field of one magnet reduce the homogeneity of the other one. This is of particular importance for scattering experiments done at large and variable angles, because the magnets may stand quite close to each other. Magnetic shielding is not really possible because it would require the use of magnetic materials which in turn would induce nonlinearities in the current-to-magnetic-induction relation of the magnets. The only free parameter to reduce stray fields is thus the diameter of the cylinder coil which we will keep as small as permitted by the beam size.

3. OPTIMAL PRECESSION MAGNETS

OFS magnets described here are useful for all spatial Larmor precession techniques. We start calculating the optimal field shape in cylinder geometry followed by the determination of the current distributions required to obtain the desired fieldshape. In-beam current distributions are needed to remove at the same time the residual magnet inhomogeneities and the pathlength differences due to finite beam divergence.

For lifetime determinations on dispersive excitations, we need to be able to rotate the lines of constant spin echo phase in energy / momentum transfer space. This can be done using special field gradient coils which can locally adapt the above NSE lines to the curvature of the dispersion surface [12].

Finally the overall response of a Larmor precession spectrometer equipped with optimized magnets must be calculated by solving the spin equation of motion across the complete apparatus numerically. This is needed to verify that field wiggles due to finite conductor dimensions and fringe fields of ancillary devices such as flippers do not perturb precession or destroy the homogeneity of the main coils. The result is a feasibility demonstration of the above optimization calculations, leading towards an optimal use of Larmor precession.

We start from a rotation symmetric magnetic field induction $\underline{B}(r,z)$ the rotation axis being z, and r the radial coordinate. Furthermore we suppose no currents flowing for r<R. Using Maxwell's equations the line integral of the modulus of $\underline{B}$ over the length L can be written to lowest order (sufficient for the present purpose) as a series expansion:

$$\int^L |\underline{B}|\, dz = \int^L B_z(0,z)\, dz + 1/8\, r^2 \int^L \{(\partial B_z(0,z)/\partial z)^2 / B_z(0,z)\}\, dz - 1/4\, r^2 \int^L \{\partial^2 B_z(0,z)/\partial z^2\}\, dz \qquad (3.1)$$

in the vicinity of the z axis for r<<R.

For less symmetric fields similar expressions can be worked-out, but the coefficients of the r^2 terms are always larger. For example for a transverse field distribution having a symmetry plane the r^2 coefficients are four times larger.

We now look for a symmetric solution on the z axis $B_z(o,z) > 0$ in the finite interval $-1/2L< z < 1/2L$, such that $B_z \approx 0$ and $\partial B_z/\partial z = 0$ at the boundary points z = -1/2L and z = 1/2L. These vanishing field and zero gradient conditions favour better flipper operation at the points where precession is started and stopped. They also guarantee minimum magnetic interaction between the various components of the spectrometer, so that the field in the sample and flipper areas can be controlled independently of the precession fields. The optimal field shape will be the solution(s) of minimizing

$$1/8\, r^2 \int^L \{(\partial B_z(0,z)/\partial z)^2 / B_z(0,z)\}\, dz - 1/4\, r^2 \int^L \{\partial^2 B_z(0,z)/\partial z^2\}\, dz \qquad (3.2)$$

Note that for continuous $B_z(0,z)$ functions over this interval the second term integrates to zero because of the boundary conditions. Hence there is no solution which satisfies (3.2) $\equiv 0$ exactly since the first term always is positive and $\neq 0$.

Let us search for a continuous solution first, minimizing the first term of (3.2). Suppose that $f(z) = B_z(0, z)$ is the OFS, then for any small perturbation $h(z)$ to $f(z)$ such that:

$$h(-1/2L) = h(1/2L) = 0 = \int^L h(z)\, dz \qquad (3.3)$$

the deviation (' denotes the z derivative)

$$\int^L \{(f+h')^2/(f+h)\}\, dz - \int^L (f')^2/f\, dz =$$

$$\int^L \{2h'(f'/f) - h(f'/f)^2\}\, dz + \text{order } (h^2) \text{ terms} \qquad (3.4)$$

must be of order h^2 . This implies that for any $h(z)$ satisfying (3.3):

$$\int^L 2h'(f'/f)\, dz - \int^L h(f'/f)^2\, dz$$

$$= 2h\,(f'/f)\Big|_{-1/2L}^{1/2L} - \int^L h\,\{2(f'/f)' + (f'/f)^2\}\, dz$$

$$= -\int^L h\,\{2(f'/f)' + (f'/f)^2\}\, dz = 0$$

Remembering the third condition of (3.3) this implies:

$$2(f'/f)' + (f'/f)^2 = 0, \qquad \text{or} \pm A^2 \qquad (3.5)$$

with A a positive constant. The $+A^2$ solution leads to an unphysical solution containing exponential functions. Let us develop the $-A^2$ possibility:

$$2(f'/f)' + (f'/f)^2 = -A^2 \qquad (3.6)$$

can be solved easily by setting g = f'/f, which yields
g = A tan 1/2 A (C - z), C being an integration constant, leading to

$$f(z) = D \cos^2 1/2\, A\, (C - z) \qquad \text{D a constant} \qquad (3.7)$$

This is a physical solution which we rewrite, taking the boundary conditions into account and setting

$$B_z(0,0) = B_0$$

$$B_z(0, z) = B_0 \cos^2 \pi z/L = B_0/2\,(1 + \cos 2\pi z/L) \qquad \text{(OFS)} \qquad (3.8)$$

The integrals in (3.1) then write:

$$\int^L B_z dz = 1/2\, B_0 L \text{ and}$$

$$\int^L \{(\partial B_z/\partial z)^2 / B_z\}\, dz = 1/2\, B_0 L\, (2\pi/L)^2$$

which leads to the **inhomogeneity formula** for the $\cos^2$ fieldshape

$$\eta_{OFS} = \{\int^L |\underline{B}|(r,z)dz - \int^L B_z(0,z)dz\} / \int^L B_z(0,z)dz = 1/2(\pi r/L)^2 \qquad (3.9)$$

It can easily be verified that this fieldshape gives indeed better homogeneity than for example other standard bell-shaped curves.

The optimal $\cos^2$ field shape can be reproduced in practice by simple current distributions exhibiting line integral inhomogeneities for parallel trajectories which agree with the theoretical inhomogeneity expression (3.9). The solution to this inverse problem is described in the next section. By means of a numerical approximation our double aim will be to reproduce the theoretical OFS as accurately as possible using small diameters to minimize fringe fields. This is possible with magnets which have a length of about .8 L, L being the length of the field distribution.

Since the theoretical inhomogeneity corresponds to the shape of a field it is independent of the diameter of any current distribution. The important consequence is that we do not have to use solenoids of large diameters to make them homogeneous, but instead use the smallest diameters compatible with the beam size thus considerably reducing costs.

Magnets producing an optimal field shape (OFS) can also be made shorter than simple solenoids at comparable homogeneity. Because in OFS magnets the inhomogeneities are less localized than in solenoids, their corrections are easier to implement. Unlike simple solenoids, for which the homogeneity increases with increasing length **and** diameter, the OFS homogeneity improves faster with length and does not depend on its diameter which can therefore be made small enough to reduce lateral stray fields to such a low level that it does not influence the field in the other magnet.

An approximate formula for the line integral homogeneity of simple solenoids was given by Mezei [13]. The relative line integral inhomogeneity due to the radial field increase as one moves away from the axis is given by $\eta_{SOL} \cong r^2/2DL_S$, with r the beam radius, D the solenoid diameter and L_S its length. If we set $L_S = .8 L$ and $D = .08$ Meter, we can compare with the homogeneity formula (3.9) and form the equal length and beam diameter ratio $\eta_{SOL}/\eta_{OFS} \cong 2L_S$. For reasonable η values L_S has to be larger than 1 Meter so that OFS is at least two times better than a solenoid of same diameter.

For $\eta_{OFS} = 5.10^{-4}$ (beam diameter .04 Meter) a length of 1.7 M is required whereas a solenoid needs to have a length of 5 Meters to reach this homogeneity.

Let us now come back to the first solution of (3.5) which after setting $g = f'/f$, leads to the differential equation:

$2g' + g^2 = 0$ the solution of which is:

$g = (k + 1/2z)^{-1}$ (k is a constant) leading to:

$f = C(k + 1/2z)^2$ (C a constant)

$$B_z(0,z) = B_0(1 - 2|z|/L)^2 \tag{3.10}$$

This solution has a finite derivative at $z = 0$, and since $\partial^2 B_z/\partial z^2 \neq 0$ $\forall z$ the second term of (3.2) is nonzero and exactly compensates the first term, so that for solution (3.10) equation (3.2) gives a minimum of exactly 0, that is perfect line integral homogeneity for parallel trajectories The corresponding field shape is discontinuous at $z = 0$, it gives perfect homogeneity for all neutron paths parallel to the z axis and this even for very short magnets. The limitations here possibly arise from too fast field variations at $z = 0$ for. The practical implementation too is more involved than for the smoother $\cos^2$ shape. Such rapidly varying field shapes needs high current densities and probably an in-beam current sheet to reproduce the discontinuity at $z = 0$ accurately enough. This is not really a problem since a few in-beam correction coils are required anyhow to remove the pathlength differences due to finite angular divergence. Calculations have nevertheless convinced us that the implementation of fieldshape (3.10) is worth further investigation the result of which will be described elsewhere.

4. CURRENT DISTRIBUTION FOR OFS MAGNETS:

To solve the inverse problem of finding the current distributions which reproduce OFS along the beam axis z $B_z(0, z) = B_0 \cos^2 \pi z/L$ in the best way we start with a number N of superposed and concentric solenoid layers of radii r_i and decreasing lengths l_i, i varying from 1 to N, all carrying the same current density. The initial guess for the set of N values of l_i is taken in such a way that the shape of the current distribution is close to the field shape to be produced. The r_i are calculated for given wire diameters and insulator thicknesses, starting from R the radius of the innermost layer. The optimum set of l_i values for the lengths of the superposed coils are then determined numerically by a least square routine. Typically 10 superposed solenoids are sufficient to properly reproduce the $\cos^2$ shape.

The next step is to look for the best inner diameter of such a composite coil system. In order to reduce the stray field we require the smallest possible diameter. Since the homogeneity is not directly related to the magnet diameter, the latter can be kept as small as allowed by the neutron beam diameter (typically .04 Meter). If the coils are made superconducting sufficient space for cryostat walls and thermal insulation has to be left between beam and coils. The smallest practical superconducting coil diameter with a room temperature bore for the above beam diameter seems to be of the order of .08 Meter.

But in fact there is a more fundamental lower limit for the inner solenoid diameter. Since the overall current distribution of the N solenoids necessarily varies stepwise along z, there will be wiggles in the radial component of the off-axis field at positions corresponding to $z = l_i$. The effect of these current density steps increases as the neutron passes closer to the wires and for given current density steps, there will be a minimum value for the innermost diameter of the magnet. Relation (2.3) may therefore not hold anymore if the field variations as seen by the neutron become comparable to the Larmor frequency.

To evaluate the depolarization effect of such field vector wiggles precisely one has to integrate the equation of motion of the individual spin (2.1) across the exact field distribution calculated exactly from the current distribution. We performed such calculations numerically for the above superconducting magnet (I = 20 Amps, wire diameter .0002 Meter) and for various neutron trajectories and came to the conclusion that the practical limit of .08 M for the inner coil diameter can only be used if the number of superposed solenoids is made larger than the initial value of 10 needed to obtain a sufficiently good field shape on the axis. With 20 layers the OFS approximation is only slightly better but the wiggles of the radial field component at the outer beam radius are sufficiently weak not to be seen by the neutron spins. In fact with 20 layers there are no real wiggles at the $|z| = l_i$ values but rather inflexion points of the radial component distribution.

We also checked that the field interaction of OFS magnets with such a diameter is indeed vanishingly small for all spectrometer settings, the fringe fields do not perturb the homogeneities.

Table 1 shows numerical line integral calculations for various neutron paths across the magnet, including the reference integral along the magnet axis. It can be seen that the inhomogeneity for parallel trajectories corresponds indeed to the value calculated using the OFS inhomogeneity formula (3.9). For classical, necessarily thicker conductors, detailed design calculations are required from case to case in order to determine the minimum magnet diameter.

OFS magnet homogeneities depend on their lengths only, for practical lengths η_{OFS} is limited to ~10^{-4}. Standard angular beam divergencies introduce pathlength inhomogeneities of the same order so that further corrections are required.

5. IN-BEAM CORRECTION COILS:

The only way to further improve the irrotational OFS solution described above is to introduce currents in the neutron beam. Fortunately most conductors are rather transparent to neutrons. Rewriting Maxwell's equations including an unknown current distribution J(r,z) and generalizing the analytical path integrals expression to oblique trajectories we will minimize the line integral variation again.

We will derive the formalism of in-beam corrections for general cylinder field shapes, having in mind their application to OFS magnets and classical solenoids, and why not to an improved correction of already existing solenoids.

We consider current distributions applied to field regions where the main field $\underline{B}$ is essentially oriented along the cylinder axis, i.e. $|B_r| \ll |B_z|$, so that $B \sim B_z$, and significantly stronger than the one generated by the correction coil field induction $\underline{B}^c$ (strong field limit). The total field $\underline{B}^t$ can then be written as

$$|\underline{B}^t| = (B_z + B_z^c)^2 + (B_r + B_r^c)^2$$

$$= |B| + B_z^c + \{ B_r B_r^c + 1/2 (B_r^{c\,2} + B_z^{c\,2})\}/|B| \tag{5.1}$$

$B_z^c(r,z)$ is exactly the term proportional to the correction current density we wish to have in order to correct for the line integral variations of $|B|$. To obtain this simple field superposition, we minimize the third term which first requires that for the main field, $|B_r| \ll |B_z|$ which will limit the application of corrections to the central high field part of the magnet. More quantitatively, we impose that the radial to longitudinal component field ratio $|B_r|/|B_z|$ should be small enough not to produce field integral errors larger than 10^{-6}. For a $\cos^2$ shape field for example, the residual integral error is $1/2(\pi r/L)^2$ so that the relative integral error will have to be less than $2(L/\pi r)^2 \cdot 10^{-6}$. This sets an upper limit to the ratio $(B_r/B_z)^2$. From this limit we can deduce the maximum central z range, $|z| < L_c$ for the application of in-beam correction: $B_r/B_z < [2(L/\pi r)^2 10^{-6}]^{1/2}$ (for $\cos^2$ shape fields) \tag{5.2}

For other fields shapes this condition has to be adapted.

Second, B_r^c and B_z^c being both proportional to J (r,z), the correction current distribution we are looking for, we need to minimize $\int^L J^2(r,z) / |B|^2$ dz (5.3) which completes the minimization of the third term of (5.1) and establishes the framework of the strong field approximation.

The uncorrected field modulus is of the form (compare 3.1) $|B(r,z)| = B_z(0,z) + r^2 f(z)$ which defines f(z). Hence the radial part of the correction current must be of order r^2, i.e. J $(r,z) = r^2 J'(z)$ and we need to find the longitudinal (z) shape of this current distribution $j(r,z) = \partial J(r,z)/\partial r = r J'(z)$. We express that the Larmor precession angle φ should be the same for all trajectories and depend on the neutron energy only.

$$\varphi + \gamma/v_z \int^L_{} |B^t| dz = \gamma/v_z \int^B_A \{ B_z(0,z) + r^2(z) f(z) + B_z^c(r,z) \} \, dz$$

for a general trajectory connecting point A(x_A, y_A, -L/2) to B(x_B, y_B, L/2). We calculate $\int^B_A B_z^c(r,z)$ dz using Maxwell's relation $\nabla \times \underline{B} = \mu_0 \underline{J}$ since the trajectories now traverse current density $\underline{J}$. Using $x(z) = x_A + v_x/v_z \, z$ and $y(z) = y_A + v_y/v_z \, z$, we calculate a loop integral along a general trajectory as given on the figure.

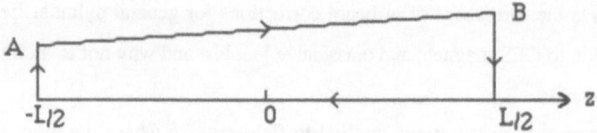

The current term reads:

$$\mu_0 \int_A^B \int_o^{r(z)} r\, J'(z)\, dr\, dz = \mu_0/2 \int_A^B r^2(z)\, J'(z)\, dz \qquad (5.3)$$

This yields an expression for $\int_A^B B_z{}^C(r,z)\, dz$ leading to an explicit formula for the Larmor phase angle along a general trajectory across the magnet:

$$\phi + \gamma/v_z \int_A^B \{ C(z) + r^2(z)\, [\, f(z) - \mu_0/2 J'(z)]\, \}\, dz \qquad (5.4)$$

taking the z components of the main coil together with the constant contribution of the correction coils

$$C(z) = B_z(0,z) + 1/L \int_A^B B_z{}^C(0,z) dz.$$

$J'(z)$ is the z - dependant part of the correction current distribution we are looking for. Note that r is a function of z now. For **parallel trajectories** $\partial r/\partial z = 0$ and the solution of our problem is given by:

$$\int_A^B [\, f(z) - \mu_0/2 J'(z)]\, dz = 0 \qquad (5.5)$$

We may then replace $J'(z)$ by a delta function at any value z_0 within the limits imposed by the strong field condition (4.2). The correction is then given by $j(r) = r J'(z_0)$ with:

$$J'(z_0) = \delta(z-z_0)/\mu_0 \int_A^B f(z)\, dz\, , \quad |z_0| < L_C \text{ clearly depending on the shape of the precession field.}$$

This is similar to the correction coils introduced by Mezei[11], except that here we draw the conclusion that they should be positioned inside the magnet for the sake of the strong field approximation.

To obtain $J'(z)$ for **oblique trajectories** too, we substitute

$$r^2(z) = x_A{}^2 + x_B{}^2 + \{(v_x{}^2 + v_y{}^2)/v_z{}^2\} z^2 \text{ in (5.4) which gives, remembering that all functions}$$

of z have to be even and substituting for $E + v_x{}^2 + v_y{}^2 + v_z{}^2$

$$\phi + \gamma/\sqrt{E} \, (\int_0^B \{ 2C(z) - z^2(v_x{}^2 + v_y{}^2)/v_z{}^2 \, [\, \mu_0 J(z) - 2f(z)]\, \}\, dz \,)(1 - (v_x{}^2 + v_y{}^2)/v_z{}^2)^{-1}$$

$$(5.6)$$

This expression is of the form $\varphi = (A - D\alpha)/(1 - \alpha)$, with $\alpha = (v_x^2+v_y^2)/v_z^2$. We want φ to depend on the neutron energy E only, not on the direction under which it travels. Expressing therefore the independence of φ with respect to α leads to the condition $D/A = 1/1 = 1$ or explicitly $\int_0^B \{ 2C(z) \, dz =$

$\int_0^B [2f(z) - \mu_o J(z)] \} \, dz$ 　　　　　　　(5.7)

$\int^L J^2(r,z) / |\underline{B}|^2 \, dz$ minimal (4.3) and conditions (4.5) and (4.7) now describe our variational problem completely. Using a similar perturbation technique than for the analytical OFS solution above we obtain J'(z). The general form of the solution is :

$$J'(z) = 1/\mu_o(X \, z^2 - Y) \, C^2(z)$$

X and Y are constants. The general shape of J'(z) for the case of an OFS magnet of dimensions L = 1.6 Meter and D = .08 Meter, can be seen in figure 1.

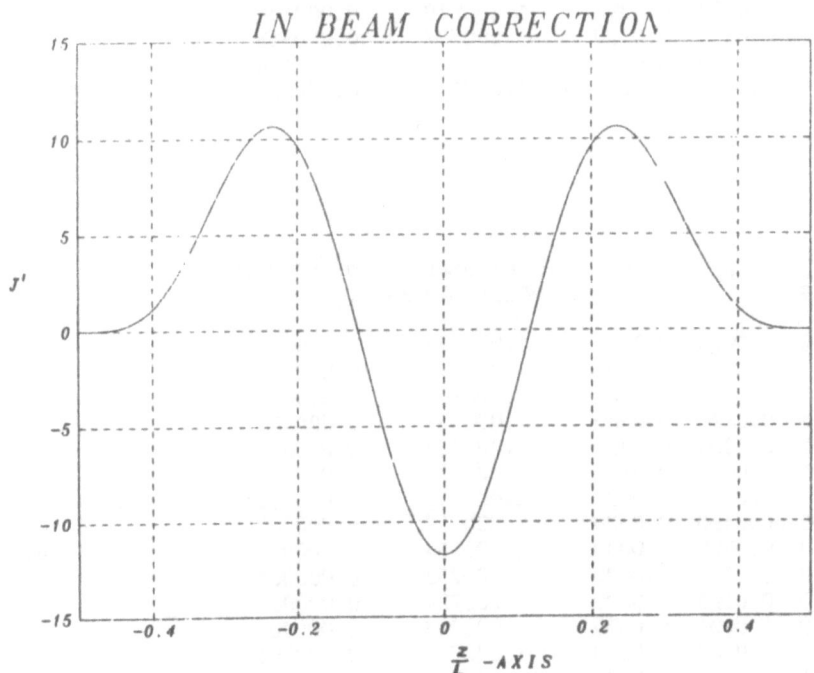

Fig. 1

Table 1. EXACT LINE INTEGRALS FOR OFS MAGNET

Starting point of neutrons (X_A, Y_A = 0, Z_A = 0)
End point $\qquad$ (X_B, Y_B, Z_B = 2M)

X_A METERS	X_B	Y_B	LINE INTEGRAL TESLA-METER	DIFFERENCE	
0.0000	0.0000	0.0000	1.000224	REFERENCE NEUTRON	
0.0095	0.0000	0.0000	1.000263	0.000039	
0.0095	0.0095	0.0000	1.000327	0.000103	
0.0095	0.0190	0.0000	1.000471	0.000247	OFS alone
0.0095	0.0000	0.0095	1.000302	0.000078	
0.0095	0.0000	0.0190	1.000419	0.000195	no in-beam
0.0095	-0.0095	0.0000	1.000276	0.000052	corrections
0.0095	-0.0190	0.0000	1.000368	0.000144	(optimized
0.0190	0.0000	0.0000	1.000380	0.000156	cylinder
0.0190	0.0095	0.0000	1.000471	0.000247	magnet)
0.0190	0.0190	0.0000	1.000640	0.000416	
0.0190	0.0000	0.0095	1.000419	0.000195	
0.0190	0.0000	0.0190	1.000537	0.000313	
0.0190	-0.0095	0.0000	1.000368	0.000144	
0.0190	-0.0190	0.0000	1.000434	0.000210	

X_A METERS	X_B	Y_B	LINE INTEGRAL TESLA-METER	DIFFERENCE	
0.0000	0.0000	0.0000	1.000732	REFERENCE NEUTRON	
0.0095	0.0000	0.0000	1.000732	0.000000	
0.0095	0.0095	0.0000	1.000732	0.000000	
0.0095	0.0190	0.0000	1.000733	0.000001	
0.0095	0.0000	0.0095	1.000732	0.000000	
0.0095	0.0000	0.0190	1.000733	0.000001	OFS
0.0095	-0.0095	0.0000	1.000732	0.000000	
0.0095	-0.0190	0.0000	1.000732	0.000000	with in-beam
0.0190	0.0000	0.0000	1.000733	0.000001	corrections
0.0190	0.0095	0.0000	1.000733	0.000001	
0.0190	0.0190	0.0000	1.000734	0.000002	
0.0190	0.0000	0.0095	1.000733	0.000001	
0.0190	0.0000	0.0190	1.000733	0.000001	
0.0190	-0.0095	0.0000	1.000732	0.000000	
0.0190	-0.0190	0.0000	1.000732	0.000001	

Practically the continuous solution J(r,z) has to be discretisized to be implemented inside the magnet. This means that we will dispose along the magnet axis, and within $|z| < L_c$ given by the strong field limit, a small number of identical spiral coils representing a current density $j(r) = a\,r$ at positions calculated to reproduce the function $J'(z)$ best. The calculation of these optimal positions for the spirals is performed numerically along the formalism described above, but in a discrete manner taking a given number (here 8) of spiral sheets.

The solution of this is an optimal current distribution which reduces the remaining **OFS** inhomogeneity to 2.10^{-6} for all neutron trajectories as can be seen in table 1.

For a **solenoid** of similar dimensions, $L_S = 1.6$ Meter, $D = .08$ Meter, $J'(z)$ is given in figure 2. The current distribution shows stronger peaks near $z/L = \pm .4$ The quality of the uncorrected and corrected solenoid (with a discretization technique as described above, 8 spirals) is described in table 2. It should be noted that this result cannot be improved by further increasing the number of spiral coils. The final corrected homogeneity obtained is ten times worse than for OFS. Therefore OFS magnets not only yield better intrinsic homogeneity, but owing to their ideally distributed inhomogeneities, they can also be corrected much more efficiently to reach final homogeneities which surpass similar corrected solenoids by one order of magnitude.

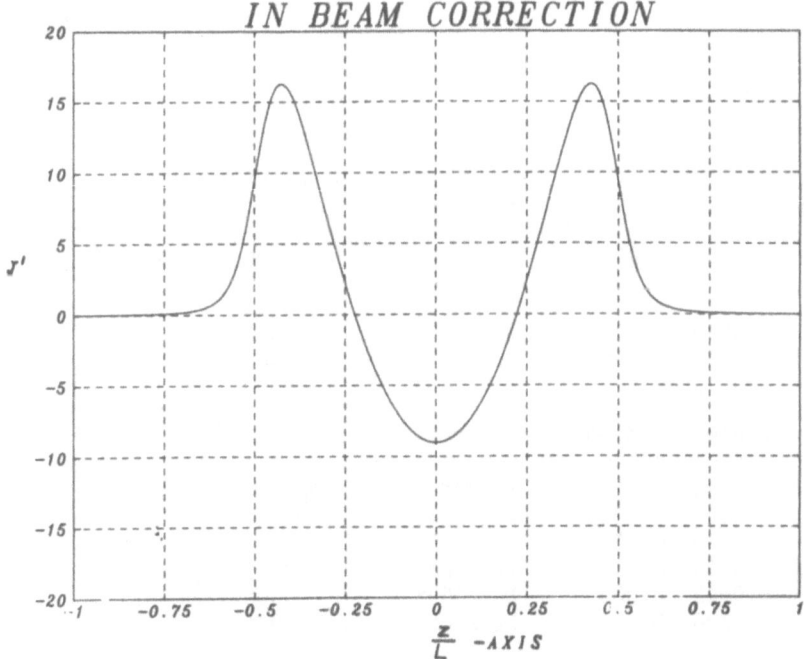

Fig. 2.

Table 2. EXACT LINE INTEGRALS FOR SOLENOID MAGNET

Starting point of neutrons (X_A, $Y_A = 0$, $Z_A = 0$)
End point (X_B, Y_B, $Z_B = 2M$)

X_A METERS	X_B	Y_B	LINE INTEGRAL TESLA–METER	DIFFERENCE	
0.0000	0.0000	0.0000	1.17125	REFERENCE NEUTRON	
0.0095	0.0000	0.0000	1.17139	0.00014	
0.0095	0.0095	0.0000	1.17158	0.00033	
0.0095	0.0190	0.0000	1.17207	0.00082	Solenoid alone
0.0095	0.0000	0.0095	1.17153	0.00028	
0.0095	0.0000	0.0190	1.17195	0.00070	no in-beam
0.0095	−0.0095	0.0000	1.17147	0.00022	corrections
0.0095	−0.0190	0.0000	1.17184	0.00059	
0.0190	0.0000	0.0000	1.17182	0.00057	
0.0190	0.0095	0.0000	1.17207	0.00082	
0.0190	0.0190	0.0000	1.17263	0.^0138	
0.0190	0.0000	0.0095	1.17195	0.00070	
0.0190	0.0000	0.0190	1.17239	0.00114	
0.0190	−0.0095	0.0000	1.17184	0.00059	
0.0190	−0.0190	0.0000	1.17214	0.00089	

X_A METERS	X_B	Y_B	LINE INTEGRAL TESLA–METER	DIFFERENCE	
0.0000	0.0000	0.0000	1.173650	REFERENCE NEUTRON	
0.0095	0.0000	0.0000	1.173652	0.000002	
0.0095	0.0095	0.0000	1.173654	0.000004	
0.0095	0.0190	0.0000	1.173658	0.000008	
0.0095	0.0000	0.0095	1.173653	0.000003	
0.0095	0.0000	0.0190	1.173659	0.000009	Solenoid
0.0095	−0.0095	0.0000	1.173658	0.000008	
0.0095	−0.0190	0.0000	1.173656	0.000006	with optimized
0.0190	0.0000	0.0000	1.173653	0.000003	in-beam corrections
0.0190	0.0095	0.0000	1.173657	0.000007	
0.0190	0.0190	0.0000	1.173670	0.000020	
0.0190	0.0000	0.0095	1.173654	0.000004	
0.0190	0.0000	0.0190	1.173663	0.000013	
0.0190	−0.0095	0.0000	1.173655	0.000005	
0.0190	−0.0190	0.0000	1.173655	0.000005	

6. THREE AXIS SPECTROSCOPY WITH SPIN ECHO

By adding the neutron spin-echo (NSE) method to the classical three-axis spectroscopy technique (TAS), (TAS+SE=TASSE) the TAS transmission function is used to isolate the inelastic event to be studied, the actual measurement is then performed with a much improved accuracy using the NSE part. Theoretical possibilities of this method have been discussed [7] [14]. A test spin echo spectrometer has been built [6] demonstrating the need for better magnets and that the proposed tilted magnet scheme for phonon focussing is not practical. Although the effect of focussing could be observed, it was in practice accompanied by higher order effects which introduced uncorrectable errors. In reference 12 a new technique for linear and nonlinear focussing (local matching of slope and curvature of dispersion surface) will be discussed.

Spin Echo spectrometers not operated with optimized magnets, will collect data marred by depolarization effects which require corrections based on reference polarization measurements. For quasi-elastic scattering these measurements are performed on reference samples, they are time consuming but possible, at least as long as the sample scattering function does not exhibit too strong local (Bragg peaks etc). For inelastic spectroscopy such resolution determinations are impossible for imperfect magnets. This particular remark is based on experience with a test TASSE [6] and has been the principal motivation for the mathematical study of optimal Larmor precession magnets.

The price one has to pay for the impressively improved measuring accuracy of NSE is relatively modest and twofold. First NSE makes use of the neutron spin and thus requires the use of polarized neutron beams. Second well defined magnetic field distributions have to be added which results in a somewhat more complex device.

The reduction of neutron intensity accompanying the use of polarized beams has been shown to be of the order of a factor of five [6] in favourable cases. In fact the introduction of NSE has already motivated significant progress in polarizing devices and it is anticipated that further improvements are to be expected. As to the complexity of the apparatus, it is still comparable to, if not less involved than, those of certain optical techniques such as for example C.A.R.S. [15] used in nonlinear phonon dynamics experiments.

7. APPLICATIONS

Before assessing the scientific applications of TASSE and passing to an enumeration of what can presently be thought-of as typical applications, we note that in the past, several new techniques (for example backscattering and small-angle scattering) have been implemented without proper forecast of the vast fields of applications they opened-up consecutively. In other words the present attempt to foresee the future uses of TASSE is necessarily incomplete and might prove to be partially wrong.

Any method allowing significantly more accurate measurements of physical quantities, has to be suspected of some usefulness.

Because incoherent scattering is mostly spin-flip, NSE is more advantageous for coherent scattering effects, although the unique capability of polarization analysis to separate between the two scattering types will prove to be very useful. TASSE will therefore be complementary to other high-resolution techniques such as backscattering [16] which is essentially limited to incoherent quasielastic and small energy transfer inelastic scattering at moderate momentum transfer. TASSE's energy and momentum transfer capabilities are those of classical TAS, in other words they are given by the incident neutron energy used.

To maximize the dynamical range and the momentum transfer possibilities we use thermal neutron beams for TASSE. Indeed with the resolution potential of TASSE, it is not necessary to lower the incident neutron energy to obtain a sufficient energy resolution.

We can now enumerate a series of classes of high resolution experiments to be envisaged with TASSE. Certain types of such measurements have already been performed successfully using the first TASSE test facility [6]:

7.1. Large momentum-transfer very-high resolution quasielastic scattering (with automatic coherent/incoherent separation for example in molecular dynamics, slow diffusion processes, relaxation mechanisms, dynamics of disordered systems, spin glasses,amorphous materials, liquids).

7.2. Separation of elastic / inelastic scattering at very high resolution (Diffuse scattering, Central peaks, Huang scattering), even short wavelengths as used in crystallography are possible.

7.3. Phonon energy-shifts (dependence on external parameters for ex. Grüneisen parameters).

7.4. Linewidth of any collective excitation, (anharmonicities, phonon-phonon, magnon-phonon, electron-phonon interactions, impurity effects, nonlinear phonon dynamics, dynamics of phonon distributions out of equilibrium etc). The study of nonlinear phonon dynamics in solids as a function of their frequency and momentum is not generally possible for the moment. For acoustical phonons the question of the frequency dependence of the lifetime, due to anharmonic decay, is a much discussed problem. Theories describing the possible mechanisms of anharmonic decay for these phonons are actively worked-on [15]. At present no experimental technique is able to give lifetimes of acoustical phonons for all wave-vectors throughout the Brillouin zone.

7.5. Interacting modes, crossing effects. Here the slope sensitivity of the method can help to single-out particular excitations.

7.6. Phonon focussing, phonon lifetimes at any point of the Brillouin-zone and for any slope or curvature of the dispersion surface. In general very high accuracy in measuring detailed local properties of $S(Q,\omega)$ both in Q and ω and including curvature effects, for example close to phonon anomalies. This question is further developed in 12.

7.7. High Energy Phonon and Magnon Studies at High Resolution and chemical spectroscopy, the use of optimized superconducting precession magnets allows even for high neutron energies (such as available on pulsed spallation sources) relative energy resolutions $\Delta E/E$ of 10^{-4} to 10^{-6}. For these neutron sources an even greater dynamical range can be obtained.

7.8. Phase transition studies: soft modes and central peaks, critical slowing-down of relaxational fluctuations.

The contribution of TAS to the understanding of the mechanisms of displacive (soft-modes) or order-disorder phase transformations in solid materials has been altogether very important [1]. However the presently available energy resolutions do not generally allow the investigation of what really occurs very close to the transition point at least in those cases where the corresponding modes are not too heavily overdamped.

TASSE should be useful to solve the central component problem near displacive and order-disorder phase transformations.

8. CONCLUSION

The high energy resolution, the dynamical range of six or seven orders of magnitude of Spin-echo combined with the unique property of neutron three-axis spectroscopy (TASSE) to obtain inelastic data at virtually any useful momentum transfer in the solid and liquid states are the highlights of the above described spectroscopy technique. In the study of dispersive excitation lifetimes the theoretical possibility to perform generalized phonon focussing [12], that is local matching of both the slope and the curvature of the dispersion surface will also be essential.

It is the progress in the design of Larmor precession magnets summarized in chapters 3, 4 and 5 which provides the combined three axis and spin echo spectroscopic technique with sufficiently homogeneous line integrals over large beam cross sections and short distances to enable real high resolution inelastic spectroscopy. The delicate, time consuming and sometimes impossible correction procedures necessary with simple solenoid precession magnets can be avoided. Stray fields are kept at a low nondisturbing levels.

The described superconducting magnets can easily be made strong enough to reach similarly high relative energy resolutions for the harder neutron spectra of spallation sources giving access to higher excitation energies and increased dynamical range.

REFERENCES

[11 Dorner B, 1982, Springer Tracts in Modern Physics **93**, Springer, Berlin.
[2] Shapiro S M, Shirane G, and Axe J D, 1975, Phys. Rev. **B12**, 4899.
[3] Mezei F, 1972, Z. Phys. **255**, 146 and Proceedings of a workshop on neutron spin echo, 1982, Lecture notes on physics **128**, Springer, Berlin.
[4] Hahn E L, 1950, Phys. Rev. **80**, 580.
[5] Rekveldt Th, 1987, Nucl. Instr. and Meth., to be published.
[6] Zeyen C M E, 1982, A.I.P. conference **89**, 101-110, Faber J. ed.
[7] Pynn R, 1978, J. Phys. E **11**, 1133.
[8] Mezei F, 1982, Proceedings of a workshop on neutron spin echo, Lecture notes on physics **128**, Springer, Berlin, p.10.
[9] Zeyen C M E et al., 1987, J. Phys. E, to be published.
[10] Hayter J and Pynn R, 1980, On the production of Homogeneous field integrals for neutron spin echo, ILL internal scientific report 81HA5.
[11] Mezei F, 1982, Proceedings of a workshop on neutron spin echo, Lecture notes on physics **128**, Springer, Berlin, p.183.
[12] Zeyen C M E et al., 1987, to be published.
[13] Mezei F, 1982, Proceedings of a workshop on neutron spin echo, Lecture notes on physics **128**, Springer, Berlin, p.184.
[14] Pynn R, 1982, Proceedings of a workshop on neutron spin echo, Lecture notes on physics **128**, Springer, Berlin, p.159.
[15] Bron W, 1985, Nonequilibrium Phonon Dynamics **124** NATO ASI, pp.43, 59, 129.
[16] Heidemann T, Proceedings of a workshop on neutron spin echo, Lecture notes on physics **128**, Springer, Berlin, p.122.

Acknowledgements:

The magnet calculations have been done as a NATO sponsored (grant 86/0217) collaboration with Peter Rem, Robin Hartmann, Nico van den Hijligenberg and Louis van de Klundert from Twente University, Enschede, The Netherlands

CHAPTER 12

DYNAMIC PROCESSES AS FOLLOWED BY ELECTRON PARAMAGNETIC RESONANCE

J.P. Launay
Laboratoire de Chimie des Métaux de Transition,
Université Pierre et Marie Curie,
4 place Jussieu, 75252 Paris Cedex 05.

The use of Electron Paramagnetic Resonance (EPR) for the study of dynamic processes has received increasing attention in recent years. In fact, time dependent processes and their influence on EPR line shapes can be treated in close analogy with the case of NMR. In a qualitative way, it can be said that the observed spectra depend upon the time scale of the dynamic process under consideration, with respect to some characteristic time scale of the method which is related to a difference in resonance frequencies for localized states. However due to the large shifts in frequencies arising from g factor anisotropies or hyperfine splittings, the spectral shapes can become quite complex and the notion of a fixed time scale vanishes upon further scrutiny. This is because each process leading to a shift in resonance frequencies can lead to a different "EPR time scale". In addition, lineshapes are strongly dependent upon relaxation rates which themselves can be modulated by a dynamic process. This extends considerably the range of "time scales" which can be probed by EPR.

In the following, we shall first recall some basic principles of EPR and the reader is invited to consult references such as [1,2] for more details.

1. BASIC PRINCIPLES OF EPR

When a free electron is subject to a magnetic field H, the energy of the two Zeeman sublevels can be written as $\pm 1/2\, g\, \beta\, H$, where β is the Bohr magneton ($9.274.10^{-21}$ erg. G^{-1}) and g the Landé factor (2.0023). The resonance condition

$$h\upsilon = g\,\beta\,H \tag{1}$$

corresponds to electromagnetic radiation in the radiofrequency domain, i.e. $\nu = 9.5 \cdot 10^9$ Hz (X band) for $H = 3400$ G or $\nu = 35.10^9$ Hz (Q Band) for $H = 12\,500$ G. In an actual substance, equation (1) is used to define an experimental g factor which can be different from 2.0023 when orbital angular momentum is involved. (Incidentally, this is why the terminology Electron Spin Resonance should be avoided, because no pure electron spin resonance without any orbital contribution exists). The resonance condition (1) can also be obtained by a semi classical argument in which the magnetic moment is assimilated to a small vector precessing around H. The resonance frequency is then simply the precession (Larmor) frequency. Both approaches are equivalent and one or the other will be used indistinctly in the following.

G. J. Long and F. Grandjean (eds.), The Time Domain in Surface and Structural Dynamics, 233–249.
© 1988 by Kluwer Academic Publishers.

In a magnetically dilute solid, the position of the resonance lines depends upon the orientation. This can be described by an anisotropic spin Hamiltonian :

$$\hat{H} = \beta \vec{H} \, \vec{g} \, \hat{S} \tag{2}$$

where $\vec{H}$ and $\vec{S}$ are vectors ang g a tensor-like quantity[3]. For an orthorhombic system, taking the principal axes of g as coordinate axes :

$$\hat{H} = \beta \, (g_x \, H_x \, \hat{S}_x + g_y \, H_y \, \hat{S}_y + g_z \, H_z \, \hat{S}_z) \tag{3}$$

Thus the position of the resonance line is characterized by g_x, g_y, g_z respectively, according to the orientation of the magnetic field. A frequent case is the axial symmetry i.e. $g_x = g_y$ leading to the simple expression

$$g^2 = g_\perp^{\,2} \sin^2 \theta + g_{//}^{\,2} \cos^2 \theta \tag{4}$$

where θ is the angle between the magnetic field and the z $(//)$ axis. In a polycrystalline sample, g can vary between $g_\perp$ and $g_{//}$ and since it is the first derivative of the absorption spectrum which is recorded, a characteristic feature is observed (Fig.1). As we shall see later, this spectrum will be considerably altered by any molecular motion involving a change in H orientation (tumbling or pseudotumbling).

When the unpaired electron is localized near a nucleus with a non zero magnetic moment, hyperfine splitting occurs. It can be considered as resulting from the interaction of the electronic magnetic moment with the local field generated by the magnetic nucleus. This interaction is described by the anisotropic additional term in the spin Hamiltonian :

$$\hat{H} = \mathbf{A} \, \hat{S} \, \hat{I} \tag{5}$$

When H is parallel to z, it frequently reduces to

$$\hat{H} = A \, \hat{S}_z \, \hat{I}_z \tag{6}$$

assuming that the field is strong enough to align $\vec{S}$ and $\vec{I}$ along z. Thus (2I+1) lines will result, their spacing being just A, usually expressed in Gauss. As for g, A is a tensor like quantity[4]. Consequently, the position of the hyperfine lines are dependent upon the nature and number of the nuclei with which the unpaired electron interacts, as well as the orientation of the magnetic field. Obviously the hyperfine structure is particularly sensitive to dynamic delocalization processes.

Finally two important quantities must be introduced. These are the spin-lattice (longitudinal) relaxation time T_1 and the spin-spin (transversal) relaxation time T_2. They enter in the phenomenological Bloch equations describing the dynamic behaviour of the macroscopic magnetization :

$$\frac{dM_z}{dt} = -\frac{M_z - M_0}{T_1}$$

$$\frac{dM_x}{dt} = \omega_0 M_y - \frac{M_x}{T_2} \tag{7}$$

$$\frac{dM_y}{dt} = -\omega_0 M_x - \frac{M_y}{T_2}$$

$1/T_1$ has the dimension of a rate constant and characterizes the return of nuclear magnetization to an equilibrium value in the direction of the applied field, following a perturbation. This is achieved through energy exchange between the spin population and the surrounding (lattice). T_2 is related to the decay of the transverse macroscopic magnetization and requires more insight. If the unpaired electrons of the sample experience slightly different effective magnetic fields due to spatial inhomogeneities, the different magnetic moments will precess at different Larmor frequencies and the resulting coherence loss averages rapidly the transverse magnetization to zero. But this coherence loss can also be caused by time dependent effects when the static plus time average magnetic field can be considered to be the same at each dipole, but the instantaneous magnetic field is not. This will correspond to inhomogeneous and homogeneous broadenings respectively.

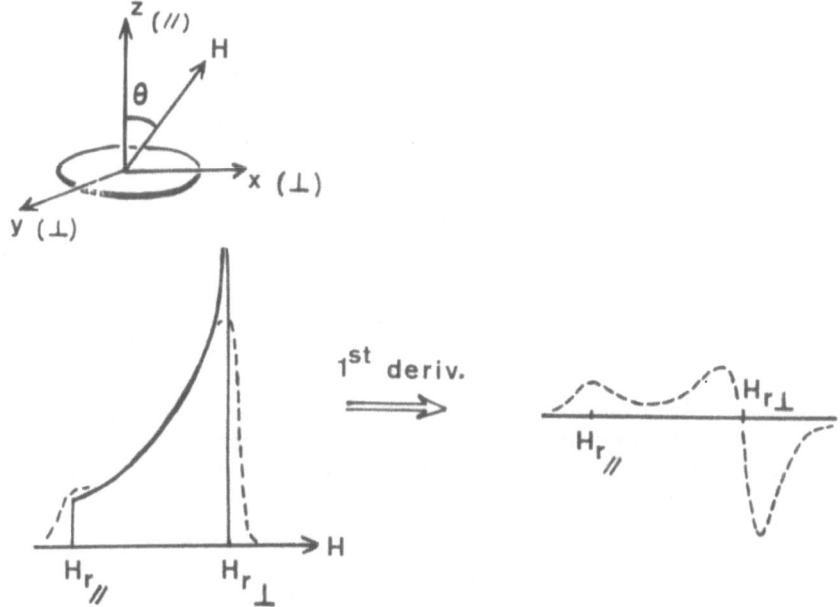

Figure 1. Anisotropic line shape for a polycrystalline axially symmetric system.

236

The standard derivation of the lineshape through Bloch equations consists in adding the effect of the radiofrequency (rf) field H_1, giving :

$$\frac{dM_z}{dt} = -\frac{M_z - M_0}{T_1} + \gamma(-M_x H_1 \sin \omega t - M_y H_1 \cos \omega t)$$

$$\frac{dM_x}{dt} = \gamma H_0 M_y - \frac{M_x}{T_2} + \gamma M_z H_1 \sin \omega t \tag{8}$$

$$\frac{dM_y}{dt} = -\gamma H_0 M_x - \frac{M_y}{T_2} + \gamma M_z H_1 \cos \omega t$$

where γ is the magnetogyric ratio, i.e. $-g\, e/2mc$. These equations are then written in the "rotating frame" turning around z at the Larmor frequency (Fig. 2).

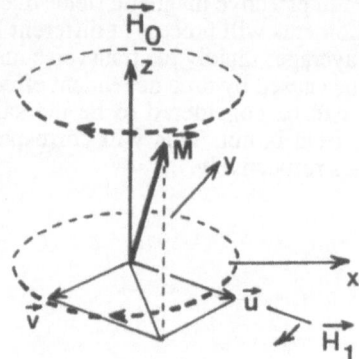

Figure 2. The rotating frame.

After this transformation, one has:

$$\frac{du}{dt} + \frac{u}{T_2} + (\omega_0 - \omega)v = 0$$

$$\frac{dv}{dt} + \frac{v}{T_2} - (\omega_0 - \omega)u + \gamma H_1 M_z = 0 \tag{9}$$

Here u is the in-phase component of the transverse magnetization while v (out of phase) is perpendicular to u. The last two equations can be condensed into a simple form, by putting $M = u + iv$ (complex magnetization) :

$$\frac{d\overline{M}}{dt} + \frac{\overline{M}}{T_2} - i(\omega_0 - \omega)\overline{M} = i\gamma H_1 M_z \tag{10}$$

One then solves (10) for stationary states and assuming $M_z = M_0$. It can be demonstrated that the absorption lineshape is given by the imaginary part of $\overline{M}$, i.e. v. This yields for the line shape:

$$\frac{\omega \gamma H_1^2 M_0 T_2}{1 + T_2^2 (\omega_0 - \omega)^2} \tag{11}$$

which is a Lorentzian function with half width at half height :

$$\Gamma = 1/T_2 \tag{12}$$

A consequence of this result is that T_2 can be considered as the average duration of the rf wavetrain which is absorbed [5].

In actual systems, however, the linewidth depends upon a number of other processes, including T_1. Thus the linewidth is used to define an experimental T_2 and a frequently used expression is, in the absence of saturation :

$$(1/T_2)_{exp} = 1/T_2 + 1/2T_1$$

where T_2 includes both homogeneous and inhomogeneous broadenings while T_1 is due to spin lattice relaxation. This expression shows that a short T_1 can also lead to a broadening by lowering the lifetime of the states (lifetime broadening).

2. EPR AND TIME-DEPENDENT PROCESSES

2.1. Coalescence of lines

In the presence of a dynamic process, modified Bloch equations can be used to predict the new lineshapes. The simplest case is for instance an electron jump between two sites A and B, accompanied by a jump in resonance frequencies from ω_a to ω_b. The jump process will be described by the same first order rate constant k in both directions. We thus assume that the two sites are equally probable and that $M_{0a} = M_{0b}$ (same equilibrium magnetization). Then the Bloch equations describing the complex magnetizations of A and B sites can be written:

$$\frac{d\overline{M_a}}{dt} + \frac{\overline{M_a}}{T_2} - i\,(\omega_a - \omega)\,\overline{M_a} = -i\,\gamma\,H_1\,M_{0a} + k\,(\overline{M_b} - \overline{M_a})$$

$$\tag{13}$$

$$\frac{d\overline{M_b}}{dt} + \frac{\overline{M_b}}{T_2} - i\,(\omega_b - \omega)\,\overline{M_b} = -i\,\gamma\,H_1\,M_{0b} + k\,(\overline{M_a} - \overline{M_b})$$

Assuming $1/T_2$ negligible, i.e. an infinitely sharp line in the absence of exchange :

$$\frac{dM_a}{dt} - i\,(\omega_a - \omega)\,\overline{M}_a + k\,(\overline{M}_a - \overline{M}_b) = -\frac{1}{2}\,i\,\gamma\,H_1\,M_0$$

$$(14)$$

$$\frac{dM_b}{dt} - i\,(\omega_b - \omega)\,\overline{M}_b + k\,(\overline{M}_b - \overline{M}_a) = -\frac{1}{2}\,i\,\gamma\,H_1\,M_0$$

Solving these equations for stationary states and taking the imaginary part of $\overline{M}_A + \overline{M}_B$, one gets for the line shape:

$$\frac{1}{2}\,\gamma\,H_1\,M_0 \; \frac{k\,(\omega_a - \omega_b)^2}{(\omega - \omega_a)^2\,(\omega - \omega_b)^2 + 4\,k^2\,[\omega - 1/2\,(\omega_a + \omega_b)]^2} \tag{15}$$

This equation describes the very fundamental process called "coalescence" (see Fig.3):
- When the exchange rate is small, i.e. $k \ll |\,\omega_a - \omega_b\,|$, two lines are observed near the Larmor frequencies ω_a and ω_b. Thus the spectrum corresponds to a static mixture of A and B.
- When k increases, each line begins to broaden and exhibits a Lorentzian lineshape with width $\Gamma = k$. At the same time the centers of the lines begin to move towards each other. Thus exchange leads to an additional broadening of the individual signals. At this stage this effect is equivalent to a decrease in T_2.

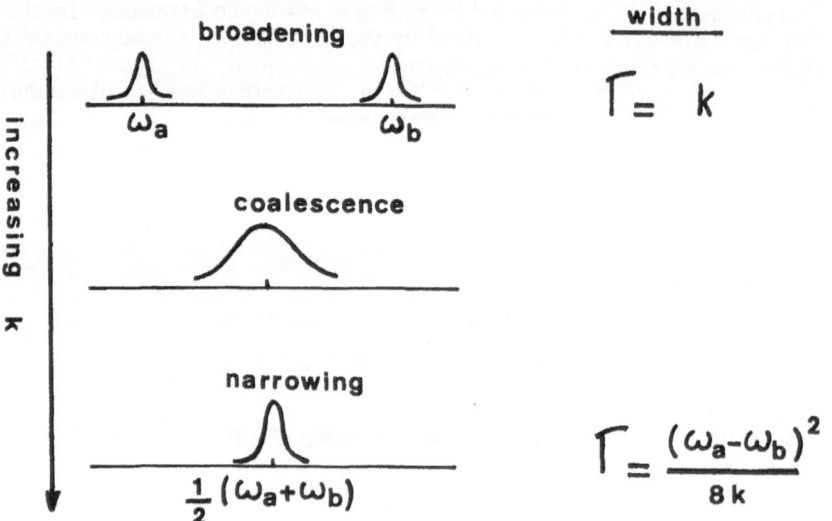

Figure 3. The coalescence process.

- When k is of the order of $| \omega_a - \omega_b |$, the two broad lines coalesce in a single line at the average frequency $(\omega_a + \omega_b)/2$.
- When k is greater than $| \omega_a - \omega_b |$, the line begins to <u>narrow</u>. Once again, a Lorentzian line is obtained but its width is now given by :

$$\Gamma = | \omega_a - \omega_b |^2 / 8\, k \tag{16}$$

This is an exchange <u>narrowing</u>.

The coalescence process is very general and can be used to define a time scale which is of the order of $1/| \omega_a - \omega_b |$. However, in practice one has generally more than two lines involved and the populations are not equal. Thus a more systematic formalism is necessary. We shall follow exactly the same method as for NMR[6].

For sake of generality we call the fractional populations of the sites P_a and P_b. Then, for the steady state, eq(14) can be written :

$$P_a\, i\, \gamma\, H_1\, M_0 = -i\, (\omega - \omega_a)\, \overline{M_a} - k\, \overline{M_a} + k\, \overline{M_b} \tag{17}$$

$$P_b\, i\, \gamma\, H_1\, M_0 = k\, \overline{M_a} - i\, (\omega - \omega_b)\, \overline{M_b} - k\, \overline{M_b}$$

or in matrix form :

$$i\omega_R\, M_0 \begin{pmatrix} P_a \\ P_b \end{pmatrix} = \begin{bmatrix} -i(\omega - \omega_a) - k & k \\ k & -i(\omega - \omega_b) - k \end{bmatrix} \begin{pmatrix} \overline{M_a} \\ \overline{M_b} \end{pmatrix} \tag{18}$$

Retaining the relation between the lineshape and the imaginary part of $\overline{M}_a + \overline{M}_b$, it is easy to show after simple matrix manipulation that

$$I_{(\omega)} = \omega_R\, M_0\, \mathrm{Re}(P_a, P_b)\, [A]^{-1} \begin{pmatrix} 1 \\ 1 \end{pmatrix} \tag{19}$$

where ω_R is the Larmor frequency and Re means " the real part of". Thus any complex system can be described by the A matrix which can be easily constructed by writing :

$$A = \Omega + D$$

where

$$\Omega_{ij} = -\delta_{ij}[i(\omega - \omega_i) - 1/T_{2(i)}] \tag{20}$$

$$D_{ij} = p_{ij}/\tau_i \qquad (i \neq j)$$

$$D_{ii} = -1/\tau_i$$

In these expressions ω_i and T_2 (i) are the resonance frequency and transverse relaxation time of site i, τ_i is the mean lifetime of site i, and p_{ij} the fraction of transfers from site i which terminate in site j. Once the **A** matrix is written, the lineshape can be computed after matrix inversion for each point of the resonance line.

Finally it should be recalled that eq (19) has been obtained in a semi-classical model, i.e. using a precessing vector model. In the quantum model, the lineshape is computed as the Fourier transform of the correlation function of transverse magnetization[7], but the final result is exactly the same as eq (19).

2.2. The modulation of relaxation times

As we have seen, in the absence of exchange, the linewidth depends generally on T_2 and also in some cases on T_1. Both kinds of relaxations are caused by time-dependent magnetic or electric fields at the electron, and these fields in turn come from the random thermal motions which are present in any form of matter. An essential condition for a relaxation process is that the molecular motion has a suitable time scale. Thus for instance interactions which change sign at a rate much faster than the resonance frequency (ca 10^{10} Hz), such as molecular vibrations (10^{13} Hz) are not effective.

An important notion for a given process is thus its correlation time. Considering a random variable of the time f(t), the autocorrelation function can be defined as the following average :

$$G(t, t+\tau) = \overline{f(t) \cdot f^*(t+\tau)} \tag{21}$$

It happens frequently that $G(t, t+\tau)$ depends only on the time difference τ, i.e. $G(t, t+\tau) = G(\tau)$ and that $G(\tau)$ can be represented by an exponential law

$$G(\tau) = G(0) \exp - \tau/\tau_c \tag{22}$$

This defines a "correlation time " τ_c which is a measure of the average time necessary for the system to forget its initial value, i.e. to achieve the independence between f(t) and f(t+τ). Conversely, for times $t \ll \tau_c$, the system can be considered as not having had enough time to evolve. Returning now to the coupling between dynamic processes and relaxation times, it can be shown that the interactions which cause transition between α and β spin states and fluctuate at the Larmor frequency are particularly efficient for "flipping" the spins and thus produce spin lattice (T_1) broadening. Usually T_1 is minimum and contributes most to the linewidth when the correlation time of the dynamic process equals the period of spin precession.

Another influence is provided by random forces which modulate the spin energy levels at low frequencies without directly causing transitions between them.

They can accelerate the loss of coherence between the individual magnetic moments and thus produce a T_2 broadening. These two effects can add their contributions to the one of chemical exchange. A particulary important example is the case of electron hopping between an occupied and an empty state. Here the correlation time is simply the average time of residence of the unpaired electron on a given site. As we shall see later, this process can give rise to the usual coalescence of lines when the rate of electron hopping begins comparable to the characteristic time scale of the spectrum. But in addition, a strong influence on the linewidth may occur by an indirect effect, i. e. the modulation of the spin lattice relaxation time by electron hopping. The theoretical treatment is rather complex but can be summarized as follows[8] :

When electron hopping occurs, a given spin experiences a fluctuating internal magnetic field H(t). These fluctuations are caused by the differences in chemical environment, or orientations between the different sites, but in addition they contain also <u>non local</u> <u>contributions</u> due to spin orbit coupling. Thus a transition from site i to site j has a finite probability $W_{i\uparrow \to j\downarrow}$ of being accompanied by a spin flip. The effective spin lifetime τ_s can then be shown to be related to the hopping lifetime τ_h by :

$$\frac{1}{\tau_s} = \left(\frac{(H_i^x)^2 + (H_i^y)^2}{\omega_0^2 + (\tau_h)^{-2}} + C \right) \frac{1}{\tau_h} \tag{23}$$

The first part of the parenthesis corresponds to local field fluctuation and its effect is maximum when $\omega_0 = 1/\tau_h$, i. e. the hopping rate equals the Larmor frequency, as in usual T_1 processes. But the second term C introduces a new effect, namely the spin flip rate (hence the linewidth) tends to be proportional to the hopping rate, <u>without narrowing</u> <u>for high hopping rates</u>. Thus a considerable broadening of the the lines may occur. Unfortunately there is no simple way to evaluate C. In principle it could be calculated by

$$C = \left\{ \frac{<i| \, H_{SO}| \, j>}{<i| \, e^2/\epsilon r| \, j>} \right\}^2 \tag{24}$$

where i and j are the sites between which electron transfer occurs and H_{SO} is the spin-orbit operator; but this expression is not easy to use for quantitative estimations. To help fixing ideas, in amorphous silicon, the C constant could be around 10^{-2}, which means that an average of 100 spin hops would be necessary for one spin flip to occur.

2.3. The EPR time scales

At the present time we can summarize the previous discussion by giving approximate values of the EPR time scales. As we have seen, there are at least three time scales to consider:

242

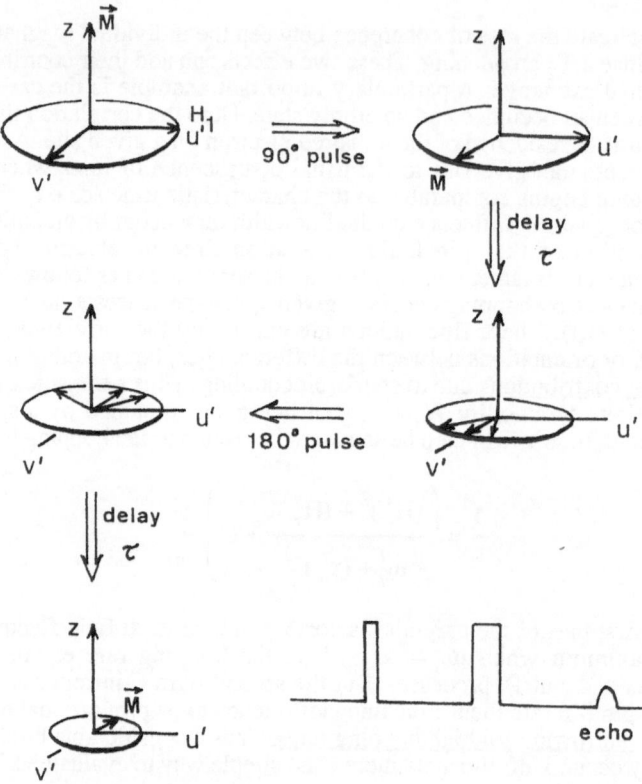

Figure 4. Principle of Electron Spin Echo. U' and V' are rotating axes. During the first delay time, the magnetic moments corresponding to different parts of the sample spread in the equatorial plane, but they are refocussed during the second delay.

- A time scale based on coalescence of lines which have been separated by g factor anisotropies. For systems with non degenerate ground states, the separation between the corresponding lines can vary from a few gauss to a few hundred gauss in X band. This corresponds to a time scale between 10^{-7} and 10^{-9} sec.
- A time scale based on coalescence of hyperfine lines. Their separation being typically a few tens of Gauss, the corresponding time scale is around 10^{-8} sec. Incidentally, this is the most frequently cited figure.
- A time scale based on the modulation of relaxation times, particularly T_1. This time scale can be much shorter than the other two and can give information on dynamic processes with characteristic rates near or even above the Larmor frequency.

It should be also mentioned that the advent of pulsed methods, and particularly Electron Spin Echo, allows the direct and selective measurement of relaxation times. This is realized through the application of intense and carefully timed rf pulses that cause the magnetic moments to precess around H_1 in addition to H_0. A simple sequence is $\pi/2$, τ, π giving rise to an Electron Spin Echo at time 2 τ, from which the transverse relaxation time T_2 can be extracted (Fig.4). The interest is that the measured T_2 is the "true" relaxation time, freed from inhomogeneity effects. Other sequences can yield T_1. These methods have not been very much used in chemistry at the present time, but they offer great potentialities and two examples will be described below.

3. EXAMPLES OF DYNAMIC PROCESSES FOLLOWED BY EPR

The EPR technique has been used for studying a great variety of dynamic processes. Before dealing with some specific examples, we shall recall some basic notions about random processes.

3.1. Generalities on Markov processes[11]

Let us consider a system represented by a random variable X_t. This random variable can take different values $s_1, s_2,...$ and we assume that the process is discrete, i.e. one can define time steps $t_1, t_2, ...$

Markov processes are a special class of stochastic processes satisfying the following property :

$$P\{X_t = s \,|\, X_1 = s_1, X_2 = s_2, ... X_{t-1} = s_{t-1}\} = P\{X_t = s \,|\, X_{t-1} = s_{t-1}\}$$

In this notation, the first bracket represents the probability of the system to be in configuration s at time t knowing that it has been in configurations $s_1, s_2, ...$ at previous times $t_1, t_2, ...$. The Markov property simply states that this probability depends only on the last configuration X_{t-1} and not on the more ancient history of the system. Thus Markov processes have no memory. An example is provided by thermal electron transfer between different sites. The average residence time (correlation time) is much longer than the transit time. Thus if the system is probed at a given instant, almost all the mobile electrons will be found in localized states and practically none will be found "in flight". Owing to this "long" residence time, the electron forgets where it comes from between two consecutive hops. A number of thermal reactions in which a system can undergo interconversions between several forms separated by an energy barrier are in fact Markov processes.

3.2. Molecular motions

The simplest case of molecular motion is the so called tumbling process in solution. If we take for instance a di-t-butylnitroxide radical, the coupling of the unpaired electron with ^{14}N (I=1) gives 3 lines in fluid solution[12]. In the solid state (frozen solution), there are parallel and perpendicular features associated with each of these lines, since the g and A tensors are axial. For temperatures where molecular rotation is slow, the spectrum exhibits a particular shape because the tumbling can average certain M_I lines more than others (see Fig.5). This is a combined effect of the g and A anisotropies and it gives an unsymmetrical spectrum exhibiting unequal linewidths with respect to the center. The case of vanadyl can be explained in the same way and the 8 hyperfine lines exhibit a characteristic envelope.(Fig. 6)

From the shape of the spectrum and particularly the dependence of the linewidth on M_I it is possible to compute the correlation time for molecular rotation. The theoretical treatment was first tackled by H. McConnell[13] and then expanded in a rigorous way by D. Kivelson[14]. The linewidth can be expressed as

$$\Gamma = \pi \sqrt{3}\,(a_1 + a_2\,M_I + a_3\,M_I^2)$$

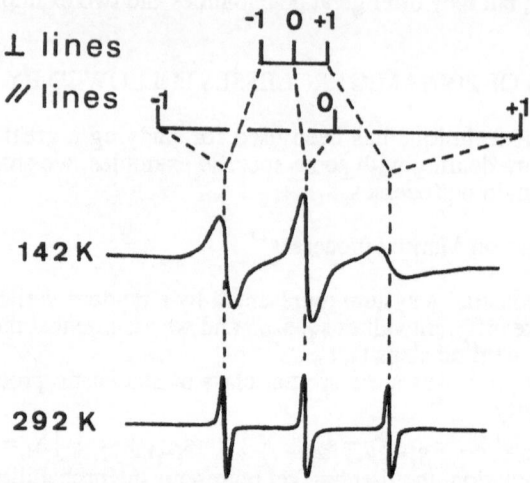

Figure 5. Di-t-butylnitroxide in ethanol at 142 K (viscous solution with slow tumbling) and at 292 K (fluid solution). Adapted from ref[12].

in which

$$a_1 = [(7/45)(\Delta\gamma . H_0)^2 + 63\, b^2/16\,]\, \tau_c + K$$

$$a_2 = (7/15)\, b\, \Delta\gamma . H_0\, \tau_c$$

$$a_3 = (b^2/10\,) . \tau_c$$

with

$$\Delta\gamma = \beta\, (g_{//} - g_\perp)\, /\hbar \quad \text{and} \quad b = 2\, (A_{//} - A_\perp)\, /3$$

Thus an experimental determination of a_1, a_2, a_3 allows the determination of τ_c, remarking that if the anisotropic parameters $g_{//}$, $g_\perp$, $A_{//}$, $A_\perp$ are known, the only other unknown is the residual linewidth K.

Another example occurs with paramagnetic ion pairs such as sodium naphthalenide in mixtures of THF and diethylether[15]. Here there is an hyperfine structure due to the coupling with Na^+ ions ($I = 3/2$) which are associated with the radical anion. The temperature evolution shows that there is a fast equilibrium between two different ion pairs, one having a large (2.2 G) and the other a small (0.8 G) ^{23}Na hyperfine splitting. The differential broadening of the $M_I = \pm 3/2$ lines with respect to the $M_I = \pm 1/2$ lines has been used to obtain a value of the rate constant for interconversion in the limit of fast exchange.

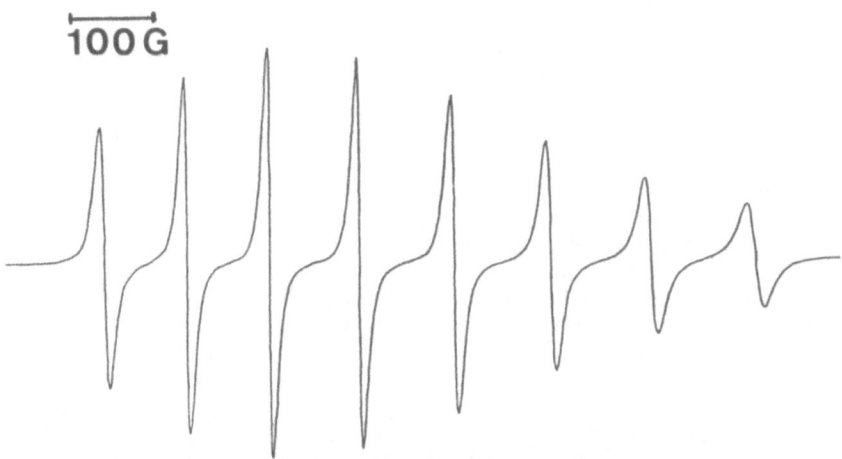

Figure 6. EPR spectrum of the $[VO(nta)(H_2O)]^-$ ion in water at room temperature, from ref[28]. nta = nitrilotriacetate.

3.3. Electron motions

Since the EPR method directly probes the behaviour of unpaired electrons, it is particularly well adapted to follow electron motions. Of particular interest is the case of intramolecular evolution in which the different forms differ only by the localization of a mobile electron.

One of the first examples studied by EPR was the case of multisite organic radicals of general formula :

$$O_2N-\langle O \rangle-X-\langle O \rangle-NO_2^{\bullet-}$$

where the monoradical is generated electrochemically[16] from the parent neutral dinitro compound. Incidentally, due to the proximity of reduction waves, the dianon is also generated but fortunately it is EPR silent . Thus one has a symmetrical system in which an unpaired electron can exchange between two chemically equivalent sites. If the electron was perfectly localized on a site, hyperfine coupling with ^{14}N ($I = 1$) and with the pairs of protons ortho and meta to the NO_2 group would give a 27 line spectrum. If on the contrary perfect delocalization occured, a 125 line spectrum would occur. For intermediate cases, a complex spectrum arises since the coalescence condition can be verified for certain sets of lines but not for others. Thus the pattern of hyperfine lines is very sensitive to the electron transfer rate. Computer simulation allowed a precise determination of the electron transfer rate for X = O and X = S. The values were found around 10^6 sec^{-1} at 25°C and the method was sensitive enough to show the influence of the solvent on the rate. Analogous experiments have been recently reported on dianthrylalkanes radical anions[17].

In a related domain, EPR has been used to study the electronic structure of the so-called bacteriochlorophyll special pair cation (Bchl$^+$.Bchl) in which a paramagnetic "hole" is shared between two bacteriochlorophyll molecules. The problem was to determine whether the hole is localized or delocalized in the dimer. Conventional EPR is not adapted, as a result of too large a linewidth and the occurrence of inhomogeneous broadening. But the technique of Electron Spin Echo allows us to extract <u>homogeneous</u> linewidths[18]. The corresponding phase memory times are in the microsecond domain but their variation upon isotopic substitution gives the contribution of the electron exchange to the homogeneous linewidth. Using the results for fast exchange, (eq 16) it is then possible to show that the correlation time for electron hopping is less than 7 picosec. With such a time scale for electron transfer, it is most likely that the hole in the special pair is completely delocalized. The advantage of Electron Spin Echo is that very small linewidth variations can be measured, thus expanding the time scale of EPR from the nanosecond to the picosecond domain.

In the field of Molecular Inorganic Chemistry, electron motions have been extensively studied in the recent years by two groups, in Pierre et Marie Curie University (Paris) and Georgetown University (Washington). The compounds are partially reduced, i.e. mixed valence, polymetallic oxoanions of Molybdenum and Tungsten (Fig. 7). Thus for instance, in the $[Mo_6O_{19}]^-$ ion, which is formed by the association of 6 MoO_6 octahedra sharing edges, it is possible to introduce just one additional electron by electrochemistry. The resulting $[Mo_6O_{19}]^-$ species exhibit at low temperature (77 K, frozen solution) a spectrum with an anisotropic 6 line hyperfine structure showing that the unpaired electron is localized on one molybdenum atom (for ^{95,97}Mo, abundance 25%, I = 5/2). The central line, due to non magnetic isotopes shows the resolution between $g_\perp$ and $g_{//}$. Broadening effects are thus very small and the hopping rate can be estimated to be < ca 5.10^7 sec^{-1}.

When the temperature is raised, the hyperfine structure disappears suddenly near 117 K, showing that the rate of electron exchange is now near 10^8 sec^{-1}. Then the central line begins to increase continuously with a purely Lorentzian line shape above 200K. The linewidth can be greater than the hyperfine parameter, showing that it is actually a T_1 process, i.e. a modulation of the spin-lattice relaxation time induced by hopping. The measured peak to peak width can be expressed as

$$\Delta H_{pp(T)} = \Delta H_{pp(o)} + \delta H_{pp(T)}$$

i.e. as the sum of a temperature independent width $\Delta H_{pp(o)}$ (due to dipolar interactions, nonresolved hyperfine splittings and ligand field heterogeneities) and a temperature dependent width $\delta H_{pp(T)}$ due to lifetime broadening. This width is assumed to be dominated by T_1, and thus to be proportional to the hopping rate υ_h. Unfortunately the absolute value of υ_h cannot be measured since the C constant in eq. (23) is unknown, but an Arrhenius plot $\ln \delta H_{pp}$ vs $1/T$ gives the activation energy for hopping, which is found to be 0.155eV[20]. Interestingly, the same experiment performed with $[XMo_{12}O_{40}]^{n-}$ anions, for which, in addition to edge sharing octahedra, one finds <u>corner sharing octahedra</u> gives a much lower activation energy

(0.035 to 0.045 eV). It is thus tempting to relate this low activation energy to an easier thermal process between corner sharing octahedra. This is nicely supported by the behaviour of β [SiMo$_3$W$_9$O$_{40}$]$^-$ in which only corner sharing modes are present for the MoO$_6$ octahedra and which gives also low activation energies [20].

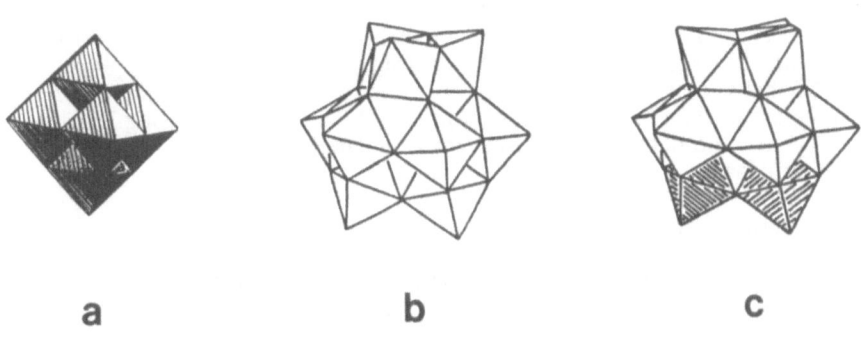

Figure 7. Structures of several polyoxoanions:
(a) [M$_6$O$_{19}$]$^{n-}$ (b) [XM$_{12}$O$_{40}$]$^{n-}$ (c) β [SiMo$_3$W$_9$O$_{40}$]$^{n-}$
In (c), the MoO$_6$ octahedra are shaded.

    The use of vanadium substituted compounds offers additional possibilities to measure the amount of delocalization, since ^{51}V (I = 7/2, abundance 100%) usually yields nicely resolved hyperfine structures. Thus in [SiW$_9$V$_3$O$_{40}$]$^{n-}$ where vanadium octahedra are corner shared, the delocalization is much greater than in [P$_2$W$_{15}$V$_3$O$_{62}$]$^{n-}$ where the octahedra are edge shared. This is shown by the "trapped" 77K ESR of the phosphate compound [22] whereas the silicate exhibits an unresolved structure even at 10K [21,23]. In addition, for [P$_2$W$_{16}$V$_2$O$_{62}$]$^{n-}$ or [P$_2$W$_{15}$V$_3$O$_{62}$]$^{n-}$, the passage from an 8-line spectrum at low temperature to a 15-line (V$_2$) or 22-line (V$_3$) spectrum at room temperature clearly demonstrates that the occurence of thermally activated hopping drives the system from the slow to the fast exchange limit. In the slow exchange limit, the line broadening allows an absolute measurement of the rate and of the corresponding activation energy which is found between 0.006 and 0.022 eV for the three-vanadium system [22]. The [P$_2$W$_{15}$V$_3$O$_{62}$]$^{n-}$ ion exhibits furthermore a spectacular behaviour upon protonation of one of the oxygen bridges. This is interpreted by a considerable reduction of the electron transfer rate across the protonated bridge which becomes suddenly "insulating".

    Temperature dependent line broadening has also been used to study intramolecular electron hopping in reduced Ruthenium complexes [27] of the type Ru (bpz)$_3^+$ (bpz = bipyrazine). In these compounds the extra electron introduced by reduction is localized on one of the bipyrazine ligand and can hop from one ligand to the other above ca 200 K.

In the field of one-dimensional conductors, EPR has been used to monitor spin mobility. An interesting consequence of the one-dimensionality is that the decay law for transverse magnetization is predicted to be $\exp[-(\alpha t)^{3/2}]$ instead of an exponential law as for tridimensional systems[24]. This gives a peculiar motional narrowed EPR line shape which is the Fourier transform of $\exp[-(\alpha t)^{3/2}]$ and exhibits a non-Lorentzian profile. The experimental observation of this effect is quite difficult in materials like polyacetylene because a large part of the commonly reported linewidth comes from a small amount of oxygen trapped localized spins, the width of which is not motional narrowed. However careful measurements performed on vacuum-sealed samples combined with an analysis of the EPR line shape up to 100 times the linewidth allows the extraction of the contribution due to diffusive spins[25]. Finally, the 1D behaviour has been observed directly in the time domain for organic conductors of the type $[(\text{fluoranthene})_2]^+ X^-$ ($X = PF_6$, AsF_6) using Electron Spin Echo[26] and the decay law of transverse magnetization followed effectively the theoretical law.

AKNOWLEDGEMENTS

Helpful discussions with the researchers from Laboratoire de Spectrochimie du Solide are gratefully aknowledged, in particular J.Livage, C.Sanchez, M.Henry,and F.Taulelle.

REFERENCES

1   J.E. Wertz and J.R. Bolton, Electron Spin Resonance, M^{C}Graw Hill, New York, 1972
2   A. Abragam and B. Bleaney, Résonance Paramagnétique Electronique des Ions de Transition, Presses Univ. France, 1971
3   J.E. Wertz and J.R. Bolton, op. cit. p135
4   J.E. Wertz and J.R. Bolton, op. cit. p145
5   A. Abragam and B. Bleaney, op. cit. p526
6   C.S. Johnson and C.G. Moreland. J. Chem. Ed. 50,477 (1973)
7   A. Abragam. Les principes du Magnétisme Nucléaire. Presses Univ. France 1961, Chap. X.
8   B. Movaghar and L. Schweitzer. Phys. Stat. Sol(b) 80, 491(1977)
9   B. Movaghar, L. Schweitzer and H. Overhof. Phil. Mag. B 37, 683 (1978)
10   S.I. Weissman . Ann. Rev. Phys. Chem. 33, 301 (1982)
11   G. Baumann and G.S. Easton J. Theoret. Biol. 93, 785, (1981)
12   O.H. Griffith, D.W. Cornell and H.M. M^{C}Connell. J. Chem. Phys. 43,2909(1965).
13   H.M. M^{C}Connell. J. Chem. Phys. 25, 709 (1956)
14   D. Kivelson. J. Chem. Phys. 33, 1094 (1960)
15   N. Hirota. J. Phys. Chem. 71, 127 (1967)
16   J.E. Harriman and A.H. Maki. J. Chem. Phys. 39,778(1963)
17   W. Huber and K. Müllen. Acc. Chem. Res. 19, 300 (1986)
18   M.K. Bowman and J.R. Norris. J. Am. Chem. Soc. 104, 1512 (1982)
19   M. Che, M. Fournier and J.P. Launay. J. Chem. Phys. 71,1954(1979)
20   C. Sanchez, J. Livage, J.P. Launay, M. Fournier and Y. Jeannin.

J. Am. Chem. Soc. 104, 3194 (1982)

21   M.M. Mossoba, C.J. O'Connor, M.T. Pope, E. Sinn, G. Hervé, and A. Tézé
J. Am. Chem. Soc. 102 , 6864 (1980)

22   S.P. Harmalker, M.A. Leparulo, and M.T. Pope,   J. Am. Chem. Soc. 105 ,
4286 , (1983)

23   M.T. Pope Private communication

24   M.J. Hennessy , C.D. Mc Elwee and P.M. Richards   Phys.Rev. B
7, 930, (1973)

25   K. Holczer, J.P. Boucher, F. Devreux, and M. Nechtschein   Phys. Rev. B
23, 1051 (1981)

26   J. Sigg, Th. Prisner, K.P. Dinse, H.Brunner, D. Schweitzer and K.H. Hausser
Phys. Rev. B 27, 5366 (1983)

27   J. N. Gex, M. K. De Armond, and K. W. Hanck, J. Phys. Chem. 91, 251 (1987)

28   M. Daoudi, Thèse 3e cycle, Univ. P. et M. Curie, Paris, 1983

CHAPTER 13

THE TIME DOMAIN IN INTRAMOLECULAR ELECTRON TRANSFER
REACTIONS

J. P. Launay
Laboratoire de Chimie des métaux de Transition,
Université Pierre et Marie Curie, 4 Place Jussieu,
75252 Paris Cedex 05 France

In another chapter of this book, the application of Electron Paramagnetic Resonance (EPR) to the dynamic study of intramolecular electron transfer has been described. In fact electron transfers can be studied by a range of spectroscopic methods of which EPR is only one example. Of course each method has to be selected according to its particular time scale. Before dealing with specific examples, it is thus necessary to briefly discuss the range of electron transfer time scales. In this chapter, the case of polymetallic mixed valence systems will be mainly considered, but of course intramolecular electron transfer is not limited to inorganic species.

1. RANGE OF INTRAMOLECULAR ELECTRON TRANSFER TIME SCALES

Intramolecular electron transfer occurs spontaneously in mixed valence compounds. In these systems, it is well established that the rate depends upon the competition between the electronic interaction between metal sites (favoring delocalization) and the electron phonon coupling, i.e. polaron formation (favoring localization). For strong electronic coupling or weak electron phonon coupling, the system exists in a delocalized state (class III in Robin-Day's classification[1]) which can be described by a perfect resonance mixture of the localized forms. In this case, there is no real dynamic process and these compounds will not be considered here. Conversely for weak electronic coupling or strong electron phonon interaction, the electron transfer becomes a real dynamic process because the electron motion becomes slow enough to couple with atomic motions. This represents class II systems in which electron transfer occurs by a thermally activated process.

The range of electron transfer time scales, (broadly defined as the reciprocal of the first order rate constant) is nevertheless enormous. A lower limit can be estimated near 10^{-12} sec, i.e. when it approaches the period of a molecular vibration. There is no theoretical upper limit; in the related case of intermolecular electron transfer, very low bimolecular rates are observed, for spin-forbidden reactions of the $Co^{2+/3+}$ couple[2] $(k \sim 10^{-6} \text{ mole}^{-1}. \text{l. sec}^{-1})$. This corresponds to half life times as long as several years. Thus there is no universal experimental method to study electron transfer processes and most of the time it is necessary to use a combination of techniques.

G. J. Long and F. Grandjean (eds.), The Time Domain in Surface and Structural Dynamics, 251–259.
© 1988 by Kluwer Academic Publishers.

252

2. METHODS BASED ON A STATIC SPECTRUM

Here we shall consider the methods by which information on the dynamics of
electron transfer are obtained by recording a static spectrum. By "static" we mean that
the spectrum is recorded in a time which is considerably longer than the characteristic
time of the dynamic process. But of course the shape of the spectrum is relevant to the
dynamics. This is what we have seen in the case of EPR. Several such methods are
schematically displayed on Figure 1, and some particular points will be briefly
discussed.

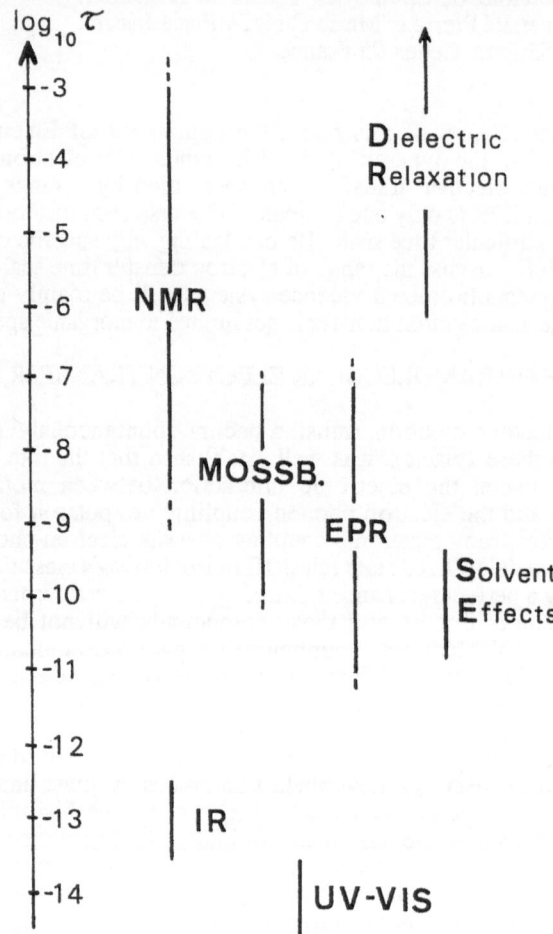

Figure 1. Characteristic times (in seconds, log scale) of several spectroscopic
techniques used to study electron transfer.

In the case of NMR, the time scales are defined in the same way as for EPR,
i.e. by reference to a chemical shift difference, or to coupling constants, or to

relaxation times[3]. For the first two processes, the time scales are in the range 10^{-4} - 10^{-5} sec, but for the last one it can reach 10^{-9} sec. An example of application to electron transfer reactions was provided some years ago by Drago, Hendrickson et al, who investigated the Creutz-Taube ion, using the displacement in ^1H chemical shifts of the pyrazine protons upon a change in Ruthenium oxidation state[4]. The occurence of a single averaged signal led to the conclusion that the electron transfer rate was "fast" in the NMR time scale (ca 10^{-4} sec. in this case).

An interesting technique, which has not been very much used so far, is dielectric relaxation. In fact it is probably the simplest conceivable method to study an electron transfer, since it is based on the dynamic behaviour of a sample submitted to an alternating electric field. The time scale is directly given by the reciprocal of the angular frequency w of the applied field and can be varied between 10^4 and 10^{-6} sec. This method has been applied to the mixed valence solid $K_3(MnO_4)_2$ and the electron transfer rate could be correlated with d.c. conductivity [5]. Unfortunately any process giving rise to a dipolar relaxation, such as polar group rotation, ion motion, etc.. can interfere and therefore the method is presently applied only to simple systems.

The use of Mossbauer spectroscopy to study dynamic processes is described elsewhere in this book (see the contributions by F. Grandjean and by S. Morup) and the theory exhibits striking mathematical analogies with the case of magnetic resonances. In the field of molecular mixed valence compounds, one of the earliest studies was performed by Brown et al on trinuclear Iron acetates [6]. They observed the coalescence of lines as function of temperature, exactly as in the case of NMR or EPR, and the detailed spectral fits yielded the rate and activation energy of electron transfer. Since then, considerable work has been performed on these systems by Hendrickson et al [7]. Using variable temperature Mossbauer spectra, they have shown that the rate of intramolecular electron transfer is in fact, and somewhat surprisingly, governed by the motion of solvate molecules. Mossbauer spectroscopy has also been used to study mixed valence organometallic "Electron Reservoirs"[26].

Finally a special mention should be given to electronic spectroscopy (UV-Vis-near IR). This method cannot be compared with the preceding ones because the energies involved are considerably above the thermal activation energy for hopping. Thus probing the system actually causes the electron transfer to occur by excitation of the so called intervalence band. In addition, the concept of a timescale is somewhat misleading in this case. On the grounds of the Franck-Condon Principle it is frequently stated that the duration of the electronic transition is "fast" (ca 10^{-15} sec). But in fact the derivation of the Franck-Condon Principle does not require any statement concerning the duration of the electronic transition! [24] Moreover the definition of the duration of the interaction between the molecule and the excitation radiation raises considerable conceptual difficulties [24]. However information on the spontaneous electron transfer can be indirectly obtained by at least two methods:

-One is the detailed study of the intervalence band by models such as Piepho-Krausz-Schatz's vibronic coupling model [8].This allows the determination of two quantities : an electron phonon coupling constant λ, and an electronic coupling parameter ε. From these, it is then possible to calculate the rate of thermal electron transfer [9].

- A second method is the study of solvent effects on intervalence transition. This gives qualitative information by the following argument: If the electron transfer is very fast, the solvent molecules have no time to reorientate and the band position is insensitive to the solvent. On the contrary, if the electron transfer is very slow, there is a component in the band energy which is due to solvent dipolar relaxation and the band position exhibits a characteristic dependence as a function of $(1/n^2 - 1/\varepsilon)$ [10]. The relevant time scale is the solvent reorientation correlation time, i.e. ca 10^{-10} sec.

3. TIME RESOLVED DYNAMIC EXPERIMENTS

In time resolved experiments, the electron transfer is triggered at time $t = 0$ by an external perturbation and the return to equilibrium is followed in the time domain by measuring an appropriate physical property. Thus the first difficulty is to trigger electron transfer, i.e. to find a physical or chemical perturbation which is accompanied by an electron migration.

3.1. How to trigger electron transfer ?

One of the simplest means, at least in the conceptual sense, is to use flash photolysis, because if the excitation is performed on an intramolecular charge transfer band, it results in an actual electron migration. In the case of binuclear ligand bridged systems, the first example was reported by Creutz, Netzel, Sutin et al[11] on the unsymmetrical dimer $[(NH_3)_5 Ru^{II} \, pyz \, Ru^{III} \, (edta)]^+$ (see Fig. 2). In the ground state, the Ruthenium atom linked to the NH_3 ligands is in the +II state and the Ruthenium atom linked to edta is in the +III state. To observe intramolecular electron transfer, it is necessary to generate the "redox isomer" in which the oxidation states have been interchanged. This was attempted by flash photolysis followed by absorption studies in the picosecond domain. Curiously the direct excitation on the intervalence band did not give the expected transient and it was necessary to perform an indirect excitation, using the metal to ligand charge transfer absorption, but finally the conversion of the "redox isomer" to the ground state was observed. Unfortunately this kind of experiment has not been generalized to other ligands due to several problems : sensitivy of $Ru(NH_3)_5 L^{2+}$ moieties to multiphoton induced reactions, and stringent requirements for the electrochemical and spectral characteristics [12].

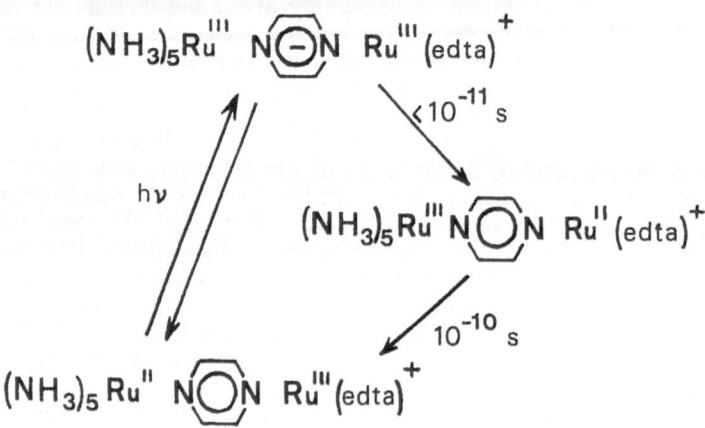

Figure 2. Flash photolysis of an asymmetrical mixed valence Ruthenium dimer, from ref [11].

In the field of biochemistry, spectacular results have been recently obtained with modifield electron transfer proteins. They are prepared by attaching a redox site $Ru(NH_3)_5$ on an histidine residue of a cytochrome for instance. Then the photochemical reduction of Ru^{III} to Ru^{II} is triggered through excitation of a nearby $Ru(bipy)_3^{2+}$ sensitizer (see Fig.3). This gives an out-of-equilibrium species and finally the intramolecular reaction occurs through the protein between the outer Ru site and the inner heme site[13,15]. Typical rates are in the range 1-100 sec^{-1}.

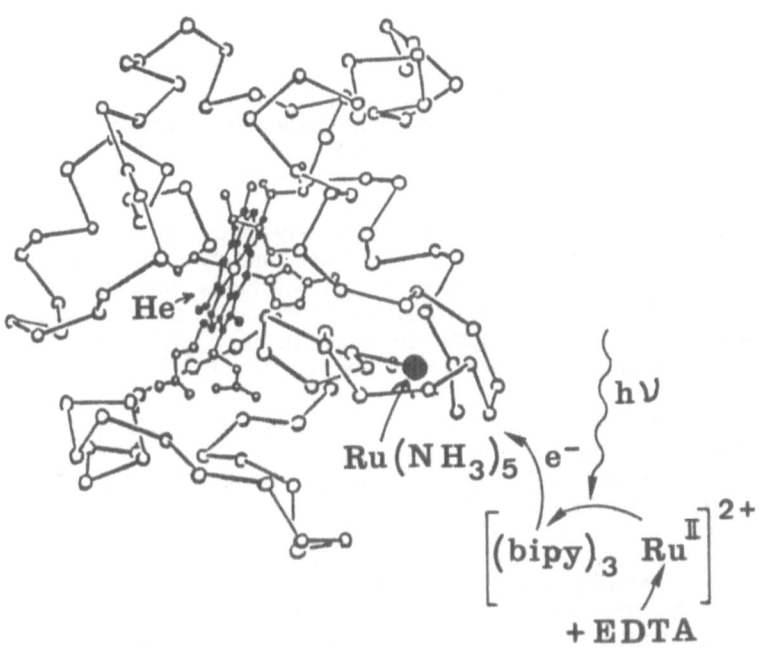

Figure 3. Electron transfer in a modified cytochrome[13,15]. He designates the Heme group.

It should be also mentioned that it is possible to make an electron transfer occur in the excited state. This is realized in binuclear coordination complexes containing two Osmium(II) ions linked by bipyridylethane . The photochemical excitation generates a mixed valence state and electron transfer is followed by the study of luminescence decay[14].

Another possibility to trigger electron transfer is through pulse radiolysis. The effect of the ionizing radiation is to generate either hydrated electrons or strongly reducing radicals. In some cases, it is possible to have this reducing agent reacting preferentially on one of the two possible sites of a molecule. This gives a non equilibrium species which then undergoes spontaneous electron transfer[15]. The key point is to have a site which is thermodynamically easy to reduce but kinetically very slow. This is possible when one of the sites is Cobalt, using the well known sluggishness of the spin forbidden electron exchange between Cobalt (II) and (III).

Finally it is even possible to trigger the intramolecular electron transfer by purely chemical means, i.e. in a test tube experiment ! This was made possible by the mastery of ligand substitution and redox reactions by H. Taube, A. Haim and others. Thus a binuclear Ruthenium-Cobalt complex which is reduced by 1 equivalent Eu^{2+} can give first a Ru^{II}-Co^{III} species which then turns slowly into Ru^{III}-Co^{II} (half life several minutes), and the reaction is easily followed in a conventional spectrophotometer[16].

3.2. How to follow the time evolution ?

Usually the time evolution is followed by optical means since they give an instantaneous picture of the system composition. Furthermore, the change in electronic structure accompanying electron transfer gives rise to important spectral modifications. Two kinds of measurements are usually performed, i.e. absorption or emission.

For absorption measurements immediately following the perturbation, it is necessary to use the techniques of laser flash photolysis in the nanosecond or picosecond domain[17]. A simplified experimental setup is schematized on Figure 4 . The exciting laser beam is divided in two parts, one of them being used for the excitation, and the other for analysis, after focussing on a D_2O cell to generate a white light continuum. The delay between excitation and probe is adjusted by a change in the optical path.

Luminescence measurements are also convenient to monitor electron transfer. Of particular interest is the case of dual luminescent systems in which two excited states have a specific emission. This is the case with molecules exhibiting the so called twisted Internal Charge Transfer effect (TICT)[18]. In these systems there are two excited states which can appear by excitation on a charge transfer band, a planar and weakly polar one, and a twisted and strongly polar one. Since these two forms differ by the amont of electron transfer between the donor and acceptor group, the dual luminescence effect provides a unique opportunity to follow electron transfer by monitoring independently and with a high sensitivity the initial and final states. Time resolved luminescence studies are currently being performed to follow the formation and decay of these species, and the importance of temperature and viscosity effects on the twisting motion[25].

4. CONCLUSION : TOWARDS MOLECULAR ELECTRONICS

Electron motions in mixed valence complexes or in any two-site donor-acceptor molecules could in principle be used to transmit a signal. Of course the detection of such a signal for an isolated molecule raises formidable problems. However there are interesting results which herald this possibility. Thus the excitation of certain bichromophoric organic donor-acceptor systems gives rise to a charge-separated excited state which emits a characteristic luminescence[19]. This can be considered as a signal indicating that the electron transfer has occurred. The transmission of electrons through large molecules is only beginning to be systematically studied. There is a reduction in the rate when the distance increases, but in some cases, this reduction may be very small or even nonexistent. This is predicted by theoretical calculations[20]

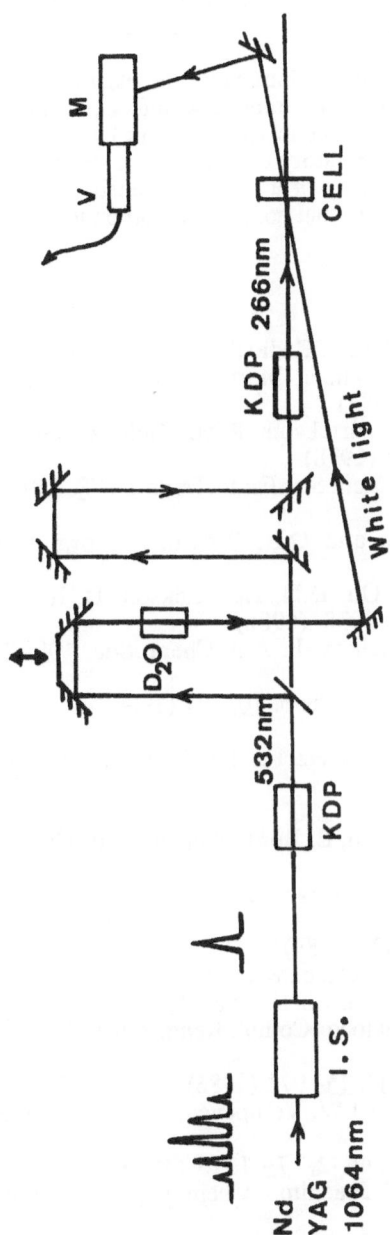

Figure 4 . Picosecond absorption system (schematic)
KDP : Non linear crystal for frequency doubling

I.S. : Impulsion Selector
V : Vidicon tube
M : Monochromator

and also by preliminary results[21] on systems linked by conjugated double bonds. Large conjugated systems with electron transfer groups at each end are now available and have been shown to be incorporable in membranes to act as "molecular wires"[22].

Finally an important challenge is the control of intramolecular electron transfer, i.e. the realization of molecules in which the electron transfer would occur through a bridging ligand with the possibility to allow or quench this transfer by an external perturbation[23]. Work is in progress in our Laboratory to devise such systems from bridging ligands with either a structural mobility or a chemical sensitivity. These systems are presently studied in solution, but the final goal is to incorporate them in a polymeric structure.

REFERENCES

1.   M.B. Robin and P. Day   Adv. Inorg. Chem. Radiochem. 10, 248 (1967)
2.   A. Hammershoi, D. Geselowitz and H.Taube. Inorg. Chem. 23, 979 (1984)
3.   R.G. Bryant. J. Chem. Ed. 60, 933 (1983)
4.   B.C. Bunker, R.S. Drago, D.N. Hendrickson, R.M. Richman and S.L. Kessel. J. Am. Chem. Soc. 100, 3805 (1978)
5.   D.R. Rosseinsky and J.S. Tonge. J. Chem.Soc.Farad.Trans 1, 78, 3595 (1982)
6.   J.T. Wrobleski, C.T. Dziobkowski, and D.B. Brown.   Inorg. Chem. 20, 684 (1981) and 20, 671 (1981).
7.   S.E. Woehler, R.J. Wittebort, S.M. Oh, D.N. Hendrickson, D. Innis and C.E. Strouse. J. Am. Chem. Soc. 108, 2938 (1986)
8.   S.B. Piepho, E.R. Krausz and P.N. Schatz. J.. Am. Chem. Soc. 100, 2996 (1978)
9.   F. Babonneau and J. Livage. New Journ. Chem. 10, 191 (1986)
10.   C. Creutz. Prog. Inorg. Chem. 30, 1 (1983)
11   C. Creutz, P. Kroger, T. Matsubara, T.L. Netzel and N. Sutin. J. Am. Chem. Soc 101, 5442 (1979)
12.   C. Creutz, private communication
13.   D.G. Nocera, J.R. Winkler, K.M. Yocom, E. Bordignon and H.B. Gray J. Am. Chem. Soc. 106, 5145 (1984)
14.   K.S. Schanze, G.A. Neyhart and T.J. Meyer. J. Phys. Chem. 90, 2182 (1986)
15.   S.S. Isied. Prog. Inorg. Chem. 32 , 443 (1984)
16.   S. K. S. Zawacky and H. Taube, J. Am. Chem. Soc. 103, 3379 (1981) J. J. Jwo, P. L. Gaus and A. Haim, J. Am. Chem. Soc. 101, 6189 (1979)
17.   See for instance D. Doizi and J.C. Mialocq Compt. Rend. Acad. Sci. Ser. II 297 , 109 (1983)
18.   W. Rettig Angew. Chem. Int. Ed. Engl. 25, 971 (1986)
19.   P. Pasman, G.F. Mes, N.W. Koper and J.W. Verhoeven, J. Am. Chem. Soc. 107, 5839 (1985)
20.   S. Larsson J. Chem. Soc. Farad. Trans. 2, 79,1375 (1983); N.S. Hush, Coord. Chem. Rev. 64, 135 (1985); C. Joachim,   Chem. Phys. (1987) in the press
21.   S. Woitellier, C.W. Spangler, J.P. Launay. Work in progress
22.   T. S. Arrhenius, M. Blanchard-Desce, M. Dvolaitzky, J. M. Lehn, and J.

Malthete Proc. Nat. Acad. Sci. USA <u>83</u>, 5355 (1986)

23. J. P. Launay, S. Woitellier, M. Sowinska, M. Tourrel, and C. Joachim
Proceed. 3rd Int. Symp. on "Molecular" Electronic Devices,
ed. F. L. Carter and H. Wohltjen, North Holland in press

24 S. E. Schwartz   J. Chem. Ed. <u>50</u>, 608 (1973)

25 F. Heisel and J. A. Miehé   Chem. Phys. Lett. <u>128</u>, 323 (1986)

26 M. H. Desbois, D. Astruc, J. Guillin, J. P. Mariot, and F. Varret J. Am.
Chem. Soc. <u>107</u>, 5280 (1985)

CHAPTER 14

DYNAMICS OF INTERIONIC ELECTRON TRANSFER

Peter Day
Oxford University, Inorganic Chemistry Laboratory
South Parks Road, Oxford OX1 3QR, England

ABSTRACT. Mechanisms and timescales are described for the transfer of electrons between metal ions (oxidation-reduction). Diffusion of the oxidant and reductant together and the formation of a binuclear intermediate are considered. Criteria are given for the presence or absence of bridging ligands in the latter, and the rate of adiabatic electron transfer is evaluated in terms of a transfer integral and vibrational-electronic coupling. Methods are given for the calculation of the transfer integral from the energies and intensities of metal-to-ligand and ligand-to-metal charge transfer transitions involving the bridging ligand.

1. RATES OF ELECTRON TRANSFER

The most elementary interionic charge transfer reaction between two chemical species A and B is the so-called 'complementary' one, i.e. having 1:1 chemical stoichiometry (1). When a one electron oxidant like Fe^{3+} meets a two-electron reductant like Sn^{2+} the reaction proceeds in steps which are more complicated, and will not be treated here. For the complementary reactions we have

$$(A^+) + (B) \rightarrow (A) + (B^+) \tag{1}$$

and the rate of the transfer is $k[A^+][B]$, where the square brackets denote concentrations. The round brackets in eq.(1) indicate distinct chemical moieties with translational freedom, not just the metal ions between which the electron is being transferred. For instance A and B could be entire cytochrome c proteins with central Fe in the +2 and +3 oxidation states.

In fluid solution the process of eq.(1) is broken down into steps, each with its own rate constant k_n as follows:

$$(A^+) + (B) \underset{k_{-1}}{\overset{k_1}{\rightleftarrows}} (A^+ \cdot B) \underset{k_{-2}}{\overset{k_2}{\rightleftarrows}} (A \cdot B^+) \underset{k_{-3}}{\overset{k_3}{\rightleftarrows}} (A) + (B) \tag{2}$$

Sutin (2) has labelled these four successive states as the initial (i),

261

precursor (p), successor (s) and final (f). In the precursor, the
reactants have diffused together to form a dimeric unit with the
correct orientation for the electron transfer, that takes place in step
2, followed by diffusion of the products away from each other. Thus,
as in eq.(1) the brackets denote translationally independent entities.
Although step 2 is the one of greatest interest to us, the overall rate
constant k is determined by all three, and imposing steady state
conditions on the intermediate processes we have

$$k = \frac{k_1 k_2 k_3}{k_2 k_3 + k_3 k_{-1} + k_{-1} k_{-2}} \tag{3}$$

or, put in another way

$$\frac{1}{k} = \frac{1}{k_1} + \frac{k_{-1}}{k_1 k_2} + \frac{k_{-1} k_{-2}}{k_1 k_2 k_3} \equiv \frac{1}{k_1} + \frac{1}{k_2*} + \frac{1}{k_3*} \tag{4}$$

where the last three terms represent the rate constants for the
processes $i \rightarrow p$, $i \rightarrow s$ and $i \rightarrow f$. Any of these steps could be so much
faster than the others that it becomes rate-determining, when of course
$k \sim k_1$, k_2 or k_3.

A particularly important category of electron transfer reactions
occurs when A and B are complexes of the same element. This is known
as mixed valency (3), and then the p and s states are equivalent. The
rate constant k_2 is then determined by nuclear rearrangements within
the $(A.B)^+$ dimer and by an electronic metal element which will be
discussed below.

2. DIFFUSION OF REACTANTS

If the electron transfer step itself is very fast the diffusion of the
reactant together (k_1) is the rate determining step. Theories of
diffusion of ions in solution (Debye equation (4)) allow one to
calculate k_1 as a function of the charge product $z_{(A^+)} z_{(B)}$. The
$p \rightarrow i$ dissociation rate constant k_{-1} is obtained from k_1 if one knows
the equilibrium constant K_1 for $i \overset{\rightarrow}{\leftarrow} p$. The latter is calculated from
the Fuoss equation (5) assuming that the reactants are considered as
charged spheres. For example, for complexes with mean ionic diameters
of 4Å in a polar solvent, (water) k_1, k_{-1} and K_1 range from 2.1, 12.2
and -10.0 (in log units) to 11.4, 2.8 and 8.5 for charge products of
respectively +12 and -12. Clearly, therefore, the rates $k_{\pm 1}$ are
extremely sensitive to the charges of the reactants. Some examples of
outer sphere electron transfer rates measured (6,7) for pairs of
oxidants and reductants of different charge products are shown in Table
1, compared with the predictions if the reactions were controlled by
$k_1(i)$. The orders of magnitude are correct, and the trend is roughly
in the observed sense.

When the reaction is diffusion limited the activation energy for i → f will be determined by that of the electron transfer step 2 so

$$k \sim k_1 \exp(-\Delta G_2*/RT) \tag{5}$$

TABLE 1. Some diffusion controlled outer sphere reactions

A^+	B	$z_{A^+} z_B$	Rate const. k/m^{-1}s^{-1} obs.	calc.
$IrCl_6^{2-}$	$Fe(Me_2phen)_3^{2+}$	−4	1.10^9	7.10^9
$Mo(CN)_8^{3-}$	$Os(bipy)_3^{2+}$	−6	2.10^9	1.10^{10}
$Fe(Me_2phen)_3^{3+}$	$IrCl_6^{3-}$	−9	1.10^9	2.10^{10}
$Os(bipy)_3^{3+}$	$Mo(CN)_8^{4-}$	−12	4.10^9	2.10^{10}

3. INNER AND OUTER SPHERE REACTIONS

When the reactants have diffused together, there are two limiting possibilities for the structure of the dimeric precursor complex distinguished many years ago by Taube (8). If there is a continuous pathway of chemical bonds from the oxidant metal ion to the reductant the mechanism is called 'inner sphere' but if (for example in the pairs of complexes in Table 1) the coordination spheres around the two metal ions remain intact throughout the electron transfer process it is called 'outer sphere'. In the latter case the electron's trajectory involves a 'jump' across a non-bonded contact between the two complexes.

There are numerous experimental criteria for deciding whether a given reaction is inner or outer sphere. To create an inner sphere transition state, the incoming complex has to displace a ligand bonded to the metal ion in the other complex. One piece of evidence would therefore be the existence of a correlation between the rate and also the activation energy for electron transfer and chemical substitution. Such a correlation has been found for the aquo V^{2+} ion and a wide range of oxidants (9). On the other hand, one knows equally well that the rates of ligand substitution of the complexes listed in Table 1 are several orders of magnitude lower than the electron transfer rates especially bearing in mind that the latter rate constants are themselves lower limits, because the overall process is diffusion controlled. Consequently the electron transfers have to be outer sphere.

The most direct possible evidence for an inner sphere mechanism would be to isolate the binuclear intermediate ($A^+.B$). This has been done for a series of complexes of Co(III) and Ru(II), both of which are kinetically inert to substitution. Reacting the two octahedral low spin d^6 complexes

$$[(NH_3)_5CoOOC\langle O \rangle NH]^{3+} + \underline{trans} -[ClRu(NH_3)_4H_2O]^+ \tag{6}$$
$$(NH_3)_5CoOOC\langle O \rangle N-Ru(NH_3)_4H_2O]^{4+}$$

The trick, exploited by Taube and others (10), is to use a bifunctional ligand that coordinates simultaneously to both metal ions. Conversely, there are cases where the structure of the ligands rules out bridging, as, for example, in the case of NH_3, which only has one electron-donating lone pair.

A much less direct, but nevertheless quite useful argument is based on the idea of linear free energy relationships ΔG^* or $\log k$ for a series of related oxidants for a given reductant or vice versa. Strong deviations from linearity for any member of the series are taken to betoken a change of mechanism. For example, using $V(H_2O)_6^{2+}$ to reduce $Co(NH_3)X^{2+}$ (X=F,Cl,Br,I), the last three fall on an excellent straight line, while F lies far off the plot, and is assigned an inner sphere mechanism (11). Correspondingly, $Ru(NH_3)_6^{2+}$ reducing $Co(NH_3)_5X^{2+}$ gives a straight line for X=Cl,Br,I very similar to $V(H_2O)_6^{2+}$ but $Fe(H_2O)_6^{2+}$ as reducing agent yields quite a different plot, and is therefore thought to be an inner sphere reaction (12).

For reactions that are definitely authenticated as outer sphere pronounced catalysis is observed by ions which do not participate directly in the electron transfer process, but which serve to bring two complexes of similar charge together. Thus the exchange between MnO_4^- and MnO_4^{2-} ($z_A z_B = +2$) is enhanced by cations in the order $K^+ < Cs^+ \ll Co(NH_3)_6^{3+}$ (13) and that between $Fe(CN)_6^{4-}$ and $Fe(CN)_6^{3-}$ ($z_A z_B = +12$) by cations $(C_n H_{2n+1})_4 N^+$ in the order $n=5 < 4 < 3 < 2 < 1$ (14). When the complexes are not isotropic, orientation effects on the electron transfer rate could also be anticipated, most spectacularly in the case of redox-active proteins like cytochrome c, where the Fe porphyrin complex that acts as the site of electron transfer is both anisotropic itself (i.e. planar), and buried on one side of the protein shell. Assuming that the redox site occupies about 3% of the surface area of the protein molecule, a rate constant diminution of $(0.03)^2 \sim 10^{-3}$ from that of a spherical centre could be anticipated, in reasonable agreement with the observed difference between cytochrome c (II/III) $(5.10^4 M^{-1} s^{-1})$ and a near spherical low spin Fe(II/III) system like the phenanthroline $(10^5 M^{-1} s^{-1})$.

4. VIBRONIC COUPLING AND ELECTRON TRANSFER

As the electron is transferred from A to B on passing from the precursor to the successor complex, the metal-ligand bond lengths expand around A and contract around B. Potential energy surfaces of $(A.B^+)$ and $(A^+.B)$ are shown in Figure 1, assuming harmonic coupling, in terms of a configuration coordinate q that is the antisymmetric combination of the totally symmetric vibrational coordinates of the two sites A and B. Of course, if A=B (i.e. we have mixed valency) the two curves are identical, $E_0 = E_1$, and merely displaced along q. The system may evolve from p to s by two routes; either a vertical Franck-Condon transition in which the electron is transferred 'instantaneously' from B to A, conserving the equilibrium bond lengths of A^+ and B, followed by relaxation to the equilibrium bond lengths of A and B^+, or thermal excitation of vibrations in (A^+,B), leading to equal bond lengths

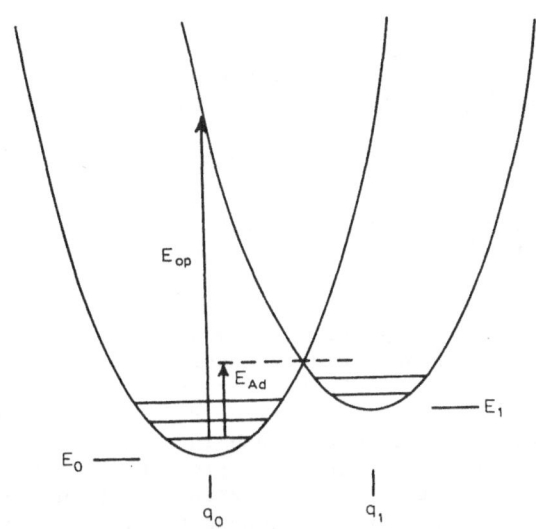

Figure 1. Vibrational potential energy surfaces for $p(E_0)$ and $s(E_1)$ complexes.

around A and B, so the electron can resonate back and forth between them. The former is an optical transition, resulting in an 'intervalence' absorption band with peak energy E_{op} and the latter a thermal (adiabatic) transition E_{Ad}. However, neither can occur if the two surfaces are truly orthogonal and there must exist a non-zero electronic matrix element V_{ps} that mixes them together.

The so-called 'transfer' integral V_{ps} is a function of both electronic and vibrational coordinates and invoking the Born-Oppenheimer approximation the wavefunctions of the p and s surfaces are

$$\Psi_{pi} = \psi_p \chi_{oi}; \quad \Psi_{sj} = \psi_s \chi_{sj} \tag{7}$$

where the ψ and χ are the electronic and vibrational wavefunctions. Then

$$V_{ps}^{ij} = \langle \psi_p \chi_{pi} | \quad | \psi_s \chi_{sj} \rangle = V_{ps} \langle \chi_{pi} | \chi_{sj} \rangle \tag{8}$$

The adiabatic transition rate from Ψ_{pi} to any vibrational level χ_{sj} of Ψ_s given by the Fermi golden rule as

$$W_{ps}^{ij} = (2\pi/h) \cdot V_{ps}^2 \langle \chi_{pi} | \chi_{sj} \rangle^2 \cdot \rho(E_{pi} = E_{sj}) \tag{9}$$

where p is the energy density of vibrational states when the energy of E_{pi} equals E_{sj}. The total adiabatic transition probability W_{ps} is obtained from eq.(9) by making a thermal average over all the χ_{pi} that overlap with χ_{sj}. Making the 'semiclassical' approximation for the vibrational overlap integrals finally gives

$$W_{ps} = (2\pi/h)V_{ps}^2[\pi/kT(E_{op}-E_s)]\exp(-E_{Ad}/kT) \tag{10}$$

For mixed valency binuclear complexes $(A^+.A)$ it is possible to estimate V_{ps} from the oscillator strength of the intervalence transition. A particularly interesting set of such complexes is the one synthesised by Stein et al. (15) in which two $Ru(NH_3)_3$ moeities are bridged by disulphido-spiropyrans. These are non-conjugated bifunctional molecules whose rigidity ensures that the metal ions are fixed at known separations. Relevant data on the complexes are in Table 2. As increasing numbers of spiropyran rings are interpolated between the S-donors, the Ru-Ru distance increases to the point where it is comparable to that found in redox protein complexes like the cytochrome c cited in section 2. The intensity of the intervalence band falls as the Ru-Ru distance increases but it remains detectable, and the transfer integral V_{sp} can be estimated from it. Finally, from V_{sp} the rate constant W_{ps} is calculated from eq.(10).

TABLE 2. Mixed valency $(NH_3)_5Ru(L)Ru(NH_3)_5$ dimers

L	R_{MM}	Intervalence band		Transfer integral cm^{-1}	Adiabatic rate constant s^{-1}
		λ_{max}	ε		
S⬡S	11.3Å	910nm	43	136	8.10^8
S⬡⬡S	14.4	808	9	55	5.10^6
S⬡⬡⬡S	17.6	690	2.3	25	$3.5.10^4$

5. MICROSCOPIC MODEL OF ELECTRON TRANSFER

In the last section we used the observed intensity of the intervalence absorption band to estimate V_{ps} but this does not yield any insight into the microscopic mechanism of the electron transfer. We now outline such a model, based on a perturbation approach to the orbitals of any atoms or molecules that lie in between the two metal centres (16,17).

Consider a simplified orbital arrangement, with two electrons in an orbital centred on the reduced ion A, one in an orbital on the oxidized ion B, and a closed shell atom or molecule having both filled and empty orbitals lying between them. Ignoring vibronic effects the zero-order wavefunctions of the ground state and the intervalence charge transfer state are respectively

$$\Psi_0 = |\phi_A \bar{\phi}_A \phi_L \bar{\phi}_L \phi_B|$$

$$\Psi_1 = |\phi_A \bar{\phi}_B \phi_L \bar{\phi}_L \phi_B| \tag{11}$$

Additionally, and crucially for the model, there are also 'local' charge transfer states involving the excitation of an electron from ϕ_A to the empty $\phi_L{}^*$ or from the filled ϕ_L to ϕ_B:

$$\Psi_2 = |\overset{*}{\phi}_L \bar{\phi}_A \phi_L \bar{\phi}_L \phi_B|$$

$$\Psi_3 = |\phi_A \bar{\phi}_A \phi_L \bar{\phi}_B \phi_B| \tag{12}$$

As noted already, there must be a transfer integral V_{01} mixing Ψ_0 and Ψ_1 if electron transfer is to take place, either optically or thermally. As a result of the mixing we have

$$\Psi_p = (1-\alpha^2)^{\frac{1}{2}}\Psi_0 + \alpha\Psi_1$$

$$\Psi_s = \beta\Psi_0 + (1-\beta^2)^{\frac{1}{2}}\Psi_1 \tag{13}$$

However, if the atomic orbitals of A and B are far apart $V_{01} \sim \langle\phi_A|\phi_B\rangle$
is zero and hence we have to invoke the orbitals of the bridging ligand, i.e. use second order perturbation to mix Ψ_0 and Ψ_1 via Ψ_2 and Ψ_3 using matrix elements V_{02}, V_{03}. The latter are larger than V_{01} because ϕ_A and $\phi_L{}^*$, or ϕ_L and ϕ_B, are in contact and have significant overlap. The valence delocalization coefficient' α, β are given by

$$\alpha = \sum_{n=2,3} (V_{0n} V_{1n})/(E_1 - E_n)(E_n - E_0)$$

$$\beta = -\sum_{n=2,3} (V_{01} V_{0n})/(E_1 - E_0)(E_n - E_1) \tag{14}$$

These expression are very suitable for application to conjugated bridging ligands like 4,4'-bipyridyl, where the sums are taken over all the filled and empty π orbitals. It is convenient to use experimental data as far as possible to estimate the E's and V's. Thus the energies E_n of the local and intervalence charge transfer states can be obtained from the spectra of model single valence and mixed valence compounds, while the V come from observed intensities of metal-ligand charge transfer absorption bands, or alternatively by calculating metal-ligand overlap integrals. The model can be validated by comparing the observed intensity of the intervalence absorption with that calculated from the following expression:

$$M_{ps} = \langle\psi_p|er|\psi_s\rangle = (1/2)e(\alpha+\beta)R \tag{15}$$

where M_{ps} is the transition electric dipole moment and R the distance between the two metal centres over which the electron is transferred.

The model decribed here was investigated first in its application to Fe^{3+}, Fe^{2+} dimers (16). More recently Richardson and Taube (18) described an extensive series of Ru^{3+},Ru^{2+} dimers with large bifunctional bridging ligands that permitted tuning of R over a wide range from 6-14Å. They used the pertubation model to calculate intervalence band intensities for nine such complexes, and the agreement with observation is indicated in Figure 2, which also contains the earlier calculations of the Fe dimers. Overall the correlation is very good, covering two orders of magnitude of intensity, and give good hope of ab initio calculation of transfer integrals. For instance, for 2,6-dicyanonaphthalene bridging two Ru the value of R is 14Å and the observed and calculated intervalence band intensities are respectively 0.16 and 0.10eÅ, yielding a transfer integral V_{ps} of about $150cm^{-1}$ and W_{ps} of 10^8s^{-1}, quite comparable to the values found in Table 2.

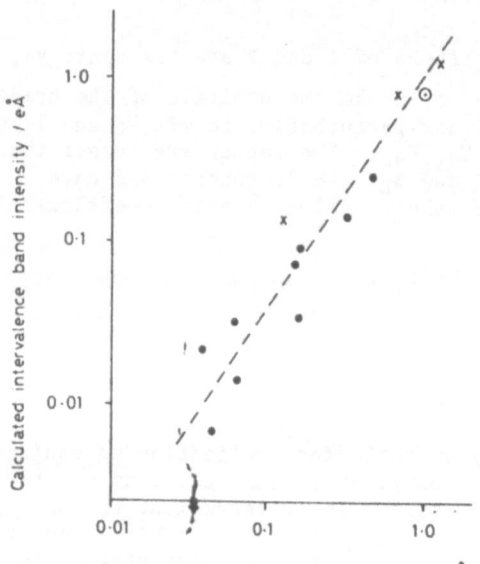

Figure 2. Observed and calculated intervalence band intensities for dimeric complexes with different bridging ligands, (x:Fe(II)-L-Fe(III), ref.(16); o Ru(II)-L-Ru(III), ref.(18).

6. CONCLUSION

Electron transfer rates in solution depend on the rate of diffusion of the reactants together and the actual transfer of the electron within the collision complex or binuclear intermediate. Inner and outer sehere reaction pathways in the latter are distinguished by experimental criteria and the adiabatic rate constant is estimated from the transfer integral. In turn the latter can either be derived experimentally from the intensity of the intervalence charge transfer absorption spectrum in the visible or near infrared, or alternatively calculated from a knowledge of metal-to-ligand and ligand-to-metal charge transfer states of an intervening mediating ligand. Finally, it is worth noting that binuclear mixed valency complexes are now being synthesised with intermetallic spacings comparable to those found in redox proteins. Theories like those applied to the simple models will be useful in understanding biological electron transfer.

REFERENCES

1. R.D. Cannon, Electron Transfer Reactions, London, Butterworths, 1980.
2. N. Sutin, Acc. Chem. Res. 1 225 (1968).
3. M.B. Robin and P. Day, Adv. Inorg. Chem. & Radiochem. 10 247 (1967).
4. P. Debye, Trans. Electrochem. Soc. 82 265 (1942).
5. R.M. Fuoss, J. Amer. Chem. Soc. 80 5059 (1958).
6. J. Halpern, R.J. Legare and R. Lumry, J. Amer. Chem. Soc. 85 680 (1963).
7. R.J. Campion, N. Purdie and N. Sutin, J. Amer. Chem. Soc. 85 3528 (1963).
8. H. Taube, J. Amer. Chem. Soc. 77 4481 (1955).
9. B. Grossman and A. Halm, J. Amer. Chem. Soc. 93 6490 (1971).
10. S.S. Isied and H. Taube. J. Amer. Chem. Soc. 95 8198 (1973).
11. M.R. Hughes, R.S. Taylor and A.G. Sykes, J.C.S. Dalton Trans. 6730 (1973).
12. P.R. Guenther and R.G. Lauk, J. Amer. Chem. Soc. 91 3769 (1969).
13. J.C. Sheppard and A.C. Wahl, Inorg. Chem. 6 672 (1967).
14. R.J. Campion, C.P. Deck, P. King and A.C. Wahl, Inorg. Chem., 6 672 (1967).
15. C.A. Stein, N.A. Lewis and G. Seitz, J. Amer. Chem. Soc. 104 2596 (1982).
16. B. Mayoh and P. Day, J.C.S. Dalton Trans. 846 (1974).
17. P. Day, Int. Rev. Phys. Chem. 1 149 (1981).
18. D. Richardson and H. Taube, J. Amer. Chem. Soc. 105 40 (1983); idem, Coord. Chem. Rev. 60 107 (1984).

CHAPTER 15

MÖSSBAUER SPECTROSCOPY AND ITS APPLICATION TO STUDIES OF TIME-DEPENDENT PHENOMENA

Steen Mørup
Laboratory of Applied Physics II
Technical University of Denmark
DK-2800 Lyngby, Denmark

1. INTRODUCTION

In 1957, Rudolf L. Mössbauer discovered /1,2/ that gamma radiation can be emitted and resonance absorbed in a recoil-free way in solids, i.e. without any loss of energy to the lattice. Because of the extremely well-defined energy of this gamma radiation, the effect (the Mössbauer effect) makes possible to study static as well as dynamic hyperfine interactions, and has developed into a new kind of spectroscopy, Mössbauer spectroscopy. As such it is now widely used in many branches of science, e.g. solid state physics, chemistry, biochemistry, metallurgy, geology, archeology, catalysis, etc.

In this chapter a brief introduction to Mössbauer spectroscopy will be given. More detailed discussions of fundamental aspects and applications of Mössbauer spectroscopy can be found in a number of books and review articles /3-11/.

2. BASIC FEATURES OF MÖSSBAUER SPECTROSCOPY

In a classical model for γ-emission, a nucleus in an excited state may be considered a damped oscillator emitting electromagnetic radiation. The electric field of the gamma radiation is then given by:

$$\epsilon(t) = \epsilon_0 \exp\left(i \frac{E_\gamma t}{\hbar} - \frac{t}{2\tau_n}\right) \tag{1}$$

where τ_n is the mean lifetime of the excited nucleus, E_γ is the nuclear transition energy, and $\hbar$ is Planck's constant h divided by 2π.

The energy distribution of the emitted radiation is given by:

$$I(\omega) \propto \left| \int_0^\infty \epsilon(t)\, e^{-i\omega t} dt \right|^2 \propto \frac{1}{(E-E_\gamma)^2 + (\Gamma_{nat}/2)2^2}; \tag{2}$$

271

G. J. Long and F. Grandjean (eds.), The Time Domain in Surface and Structural Dynamics, 271–286.
© 1988 by Kluwer Academic Publishers.

272

where $\Gamma_{nat} = \hbar/\tau_n$, and $E = \hbar\omega$.

Eq. (2) represents a Lorentzian line centered at E_γ with full width at half maximum (FWHM) equal to Γ_{nat}.

The energy dependence of the resonant absorption cross-section $\sigma(E)$ can be expressed as:

$$\sigma(E) = \sigma_o \frac{(\Gamma_{nat}/2)^2}{(E-E_\gamma)^2 + (\Gamma_{nat}/2)^2} \qquad (3)$$

where

$$\sigma_o = 2\pi\lambda^2 \frac{2I_e+1}{2I_g+1} \frac{1}{1+\alpha} \qquad (4)$$

I_g and I_e are the nuclear spins of the ground state and the excited state, respectively, λ is the wave length of the radiation divided by 2π, and α is the internal conversion coefficient. Thus the line shapes for emission and absorption are identical.

A Mössbauer spectrum is obtained as illustrated in Figure 1. The gamma radiation is emitted from a source containing a radioactive parent isotope (e.g. ^{57}Co) which decays to the excited state of the Mössbauer

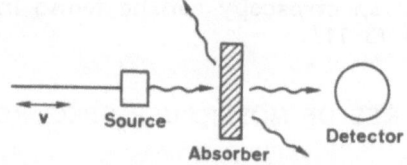

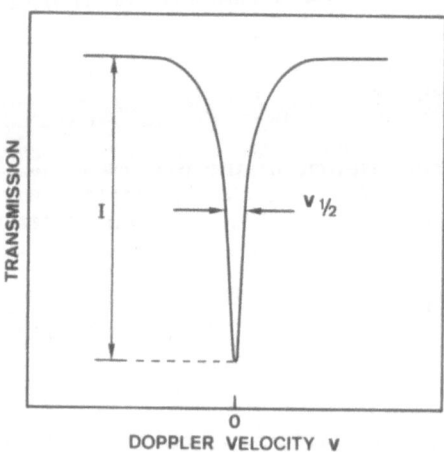

Figure 1.   Schematic illustration of Mössbauer spectroscopy.

isotope (the 14.4 keV state of ^{57}Fe). This is followed by a decay to the ground state by (recoil-free) emission of gamma radiation with an energy E_γ (= 14.4 keV). The decay scheme of ^{57}Co is shown in Figure 2. The transmitted intensity of this radiation through an absorber containing Mössbauer atoms in the ground state (^{57}Fe) is measured as a function of the gamma ray energy, which is varied by moving the source relative to the absorber. This results in a Doppler shift of the gamma ray energy:

$$E(v) = E_\gamma(1+v/c) \tag{5}$$

where v is the velocity of the source, and c is the velocity of light.

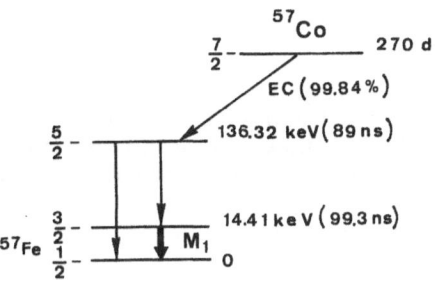

Figure 2. Decay scheme of ^{57}Co.

The Mössbauer spectrum is given by the transmitted intensity versus the velocity of the source. The shape of the absorption line is given by the convolution of the emission and the absorption lines, i.e. a Lorentzian line:

$$I(v) \propto \frac{1}{v^2 + (v_\frac{1}{2}/2)^2} \tag{6}$$

with $v_\frac{1}{2} = 2\ \Gamma_{nat}\ c/E_\gamma$ (= 0.19 mm/s for ^{57}Fe).

3. THE f-FACTOR

When the nucleus of a free atom decays, the energy E_γ will be shared between the gamma quantum and the recoil energy of the atom, E_R. The requirements for conservation of energy and momentum give:

$$E_R = \frac{E_\gamma^2}{2Mc^2} \tag{7}$$

where M is the mass of the atom. For a free ^{57}Fe atom we find that $E_R = 2 \cdot 10^{-3}$ eV, which is much larger than $\Gamma_{nat} = 4.6 \cdot 10^{-9}$ eV. Therefore, resonance absorption is not easily achieved with free atoms.

However, if the atom is bound in a solid, the recoil energy may be taken up by a large number of atoms because E_R is much smaller than the energy required to disrupt the atom to another lattice position. The recoil energy is then reduced by a factor given by the ratio between the mass of the whole crystal and the mass of the Mössbauer atom.

In a rigorous quantum mechanical model it must be taken into account that the gamma emission may result in a change in the occupation numbers of the phonon states. The interesting parameter is the probability, f, for the lattice to be in the same state before and after the decay of the nucleus, i.e. the probability that no phonons are created or annihilated in connection with the decay. This probability is given by:

$$f = \exp\{- k_\gamma \langle x^2 \rangle)\};$$
(8)

where $k_\gamma = E_\gamma / \hbar c$ is the magnitude of the wave vector of the gamma radiation, and $\langle x^2 \rangle$ is the mean value of the squared amplitude of the (thermal) motion of the Mössbauer atom in the direction of emission. $\langle x^2 \rangle$ can be calculated using for instance the Einstein model or the Debye model for the phonon spectrum.

In the Debye model one finds:

$$f = \exp \left\{ - \frac{3E_R}{2k\theta_D} \left[1 + \frac{4T^2}{\theta^2_D} \int_0^{\theta_D/T} \frac{udu}{\exp(u) - 1} \right] \right\};$$
(9)

where $k = 1.38 \cdot 10^{-23}$ J/K is Boltzmann's constant, and θ_D is the Debye temperature. The probability for recoil-free absorption of gamma radiation is also given by Eq. (9).

For $T \gtrsim \frac{1}{2}\theta_D$, the following expression is a good approximation:

$$f \cong \exp \left\{ - \frac{6E_R}{k\theta_D} \frac{T}{\theta_D} \right\}$$
(10)

Because the total area of the Mössbauer absorption spectrum is essentially proportional to f, measurements of the temperature dependence of the area is in good agreement with the expression, (9). However, it must be emphasized that the Debye model is generally a bad approximation of the phonon spectrum. Therefore, the value of θ_D estimated from Mössbauer spectroscopy may differ significantly from those measured by use of other methods such as heat capacity measurements. It should also be noticed that when more than one type of atoms are present in the lattice, the value of θ_D obtained from Mössbauer spectroscopy only reflects the thermal vibrations of the Mössbauer atoms which may differ considerably from the vibrations of other atoms.

4. THE ISOMER SHIFT

The electrostatic interaction between the Mössbauer nucleus and the surrounding electric charges results in an isomer shift and a quadrupole splitting of the nuclear levels.

The electronic charge density at the nucleus $-e|\psi(0)|^2$ gives rise to energy shifts of the nuclear levels which depend on the nuclear radius. As the size of the nucleus is changed during the decay, the effect results in a shift in the nuclear transition energy which is given by:

$$\delta E = \frac{2\pi}{3} Ze^2 |\psi(0)|^2 (\langle r_e^2 \rangle - \langle r_g^2 \rangle) \tag{11}$$

Here, Z is the atomic number of the Mössbauer atom, and $\langle r_e \rangle$ and $\langle r_g \rangle$ are the mean square radii of the nucleus in the excited and the ground states, respectively, defined as:

$$\int \rho_n(\vec{r}) r^2 d\vec{r} = Ze\langle r_n^2 \rangle \; ; \tag{12}$$

where $\rho_n(\vec{r})$ is the charge density of the nucleus ($n = e,g$).

If the Mössbauer atoms in the source and the absorber are present in different chemical surroundings, the Mössbauer line is shifted by an amount:

$$\delta = \frac{2\pi}{3} Ze^2 (|\psi_a(0)|^2 - \psi_s(0)|^2)(\langle r_e^2 \rangle - \langle r_g^2 \rangle) \tag{13}$$

where the indices "a" and "s" stand for "absorber" and "source", respectively. Thus the value of δ (the isomer shift) is proportional to the difference between the electronic charge densities at the source and absorber nuclei. In practice, the isomer shift of an absorber atom is often given relative to a standard material, e.g. metallic iron.

5. QUADRUPOLE INTERACTION

An asymmetric distribution of the charges surrounding the Mössbauer nucleus in combination with a non-spherical distribution of the nuclear charge gives rise to a nuclear quadrupole interaction which partly lifts the $2I+1$ degeneracy of the nuclear states.

The Hamiltonian describing the nuclear quadrupole interaction is given by:

$$\hat{H}_Q = \frac{eQ_n V_{zz}}{4I(2I-1)} [3\hat{I}_z^2 - \hat{I}^2 + \eta(\hat{I}_+^2 + \hat{I}_-^2)/2] \tag{14}$$

where eQ_n (= eQ_e or eQ_g) is the electric quadrupole moment of the nucleus, V_{zz} is the z-component of the electric field gradient (EFG), and

$$\eta = \frac{V_{xx} - V_{yy}}{V_{zz}} \qquad (15)$$

Here, the x,y,z directions are defined as the principal axes of the electric field gradient tensors, and

$$|V_{zz}| \geq |V_{xx}| \geq V_{yy} \qquad (16)$$

When a finite electric field gradient is present, the excited state of ^{57}Fe ($I_e = 3/2$) is split into two levels, whereas the ground state ($I_g = 1/2$) is unaffected by the electric field gradient. Thus, the spectrum in the presence of an electric quadrupole interaction consists of two lines. The quadrupole interaction of ^{57}Fe is illustrated in Figure 3.

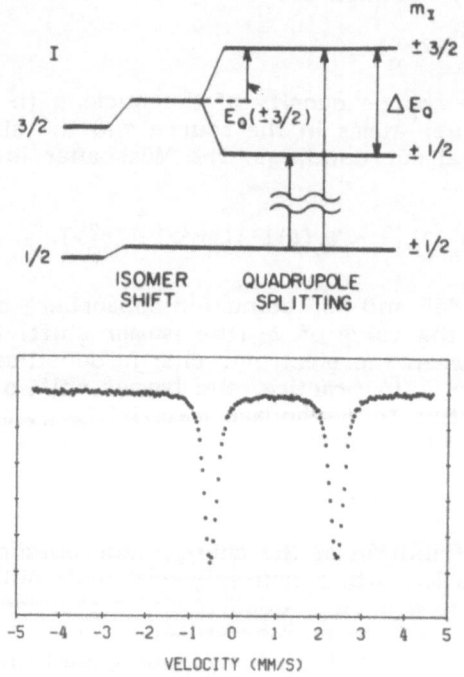

Figure 3. Quadrupole splitting of ^{57}Fe. The figure also illustrates the isomer shift. The Mössbauer spectrum was obtained with an absorber of olivine at 80 K.

6. MAGNETIC HYPERFINE INTERACTION

The Zeeman interaction of the nuclear magnetic dipole moment with a magnetic field B_n at the nucleus completely lifts the degeneracy of the nuclear states. The energies of the nuclear substates are shifted by the amount

$$E_m = - g_n \, \mu_N B_n m \; ; \tag{17}$$

where g_n (= g_e or g_g) is the gyromagnetic factor of the nuclear state, μ_N is the nuclear magneton, and m is an eigenvalue of $\hat{I}_z$. The excited state of ^{57}Fe ($I_e = 3/2$) splits into four states, and the ground state ($I_g = 1/2$) splits into two states. One might then expect eight different transition energies. However, as the gamma transition of ^{57}Fe is a pure magnetic dipole transition, only transitions for which $|\Delta m| \leq 1$ are allowed, and therefore only six transitions are normally observed. The hyperfine interaction as well as a magnetically split Mössbauer spectrum is shown in Figure 4.

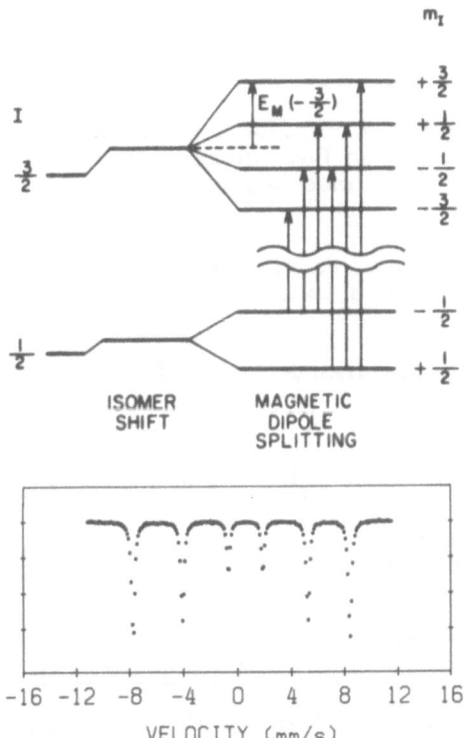

Figure 4. Magnetic hyperfine splitting of ^{57}Fe. The figure also shows a magnetically split spectrum of goethite (α-FeOOH) at 80 K.

When both nuclear quadrupole and Zeeman interactions are present, the Hamiltonian describing the hyperfine interaction is given by:

$$\hat{H}_h = - g_n \mu_N \vec{B}_n \cdot \hat{\vec{I}} + \frac{eQ_n V_{zz}}{4I(2I-1)} [3\hat{I}_z^2 - \hat{I}^2 + \eta(\hat{I}_+^2 + \hat{I}_-^2)/2] \qquad (18)$$

The eigenvalues of $\hat{H}_h$ can generally not be given by simple analytical expressions.

Often, the magnetic field at the nucleus varies in time. If the relaxation time of these fluctuations is comparable to the nuclear Larmor precession time,

$$\tau_L = \omega_L^{-1} = \frac{\hbar}{|g_n \mu_N \vec{B}_n|} \qquad (19)$$

the Mössbauer lines are broadened and shifted in a complicated way. This is discussed in Section 10.

7. POLARIZATION EFFECTS

Consider a Mössbauer experiment in which a single-line source is used, but the absorber gives a magnetically split spectrum. The relative intensities of the absorption lines depend on the angle θ_a between the γ-ray direction and the magnetic hyperfine field. The relative intensities are given in Table 1.

Table 1. Relative line areas with an unpolarized source and a magnetically split absorber spectrum

$m_g^a \rightarrow m_e^a$	θ_a arbitrary	$\theta_a = \frac{\pi}{2}$	$\theta_a = 0$	θ_a random polycrystalline sample
$\pm\frac{1}{2} \rightarrow \pm\frac{3}{2}$	$3(1 + \cos^2\theta_a)$	3	3	3
$\pm\frac{1}{2} \rightarrow \pm\frac{1}{2}$	$4\sin^2\theta_a$	4	0	2
$\pm\frac{1}{2} \rightarrow \mp\frac{1}{2}$	$1+\cos^2\theta_a$	1	1	1

When the magnetic field at the absorber nuclei is zero, but an axially symmetry electric field gradient splits the spectrum into two lines, the relative intensities are given in Table 2.

Here θ_a is the angle between $\vec{k}_\gamma$ and the symmetry axis of the field gradient.

Table 2. Relative line areas with an unpolarized source and a quadrupole split absorber spectrum

$m_g \rightarrow m_e$	Θ_a arbitrary	Θ_a random, polycrystalline sample
$\pm \frac{1}{2} \rightarrow \pm \frac{3}{2}$	$3(1+\cos^2\Theta_a)$	1
$\left. \begin{array}{c} \pm \frac{1}{2} \rightarrow \pm \frac{1}{2} \\[2ex] \pm \frac{1}{2} \rightarrow \mp \frac{1}{2} \end{array} \right\}$	$5-3\cos^2\Theta_a$	1

In the above expressions we have neglected the possible angular
dependence of the f-factor. This can be taken into account by multiplying the relative areas in the tables by the f-factor, $f(\Theta,\phi)$, which
may depend on the emission direction of the radiation relative to the
crystallographic axes.

Even for a polycrystalline sample with a quadrupole splitting, but
without a magnetic hyperfine field, the angular dependence of the f-
factor together with the angular dependence of the emitted intensity
may give rise to an area ratio of the two lines which differs from 1:1.
This is the so-called Goldanskii-Karyagin effect.

8. DIFFUSIONAL BROADENING

Diffusional motion of the Mössbauer atom during the emission or absorption of gamma radiation gives rise to a line broadening. The basic theory for this effect was developed by Singwi and Sjölander in 1960 /12/.
Two cases have to be considered separately, namely diffusion in viscous
liquids (continuous diffusion) and diffusion in solids (jump diffusion).

If the Mössbauer atom diffuses in a liquid, the line width (FWHM)
is increased by:

$$\Delta\Gamma = 2\hbar k_\gamma^2 D \tag{20}$$

where D is the diffusion coefficient.

The diffusional broadening of the radiation emitted from atoms in
a solid is given by:

$$\Delta\Gamma = 2h\{1-\alpha(\vec{k}_\gamma)\}/\tau_0 \tag{21}$$

where $1/\tau_0$ is the average jump frequency, and

$$\alpha(\vec{k}_\gamma) = \int \exp(i\vec{k}_\gamma \cdot \vec{r})P(\vec{r})d\vec{r} \; ; \tag{22}$$

where $P(\vec{r})$ is the probability of finding an atom that originally was at $\vec{r} = 0$ at the position $\vec{r} \neq 0$ after the first jump. The integration is taken over all lattice sites. It can be seen from Eq. (22) that $\Delta\Gamma = 0$ when $\vec{k}_\gamma$ is a reciprocal lattice vector, but generally $\Delta\Gamma \neq 0$. Thus, the line broadning generally depends on the direction of observation.

9. SECOND-ORDER DOPPLER SHIFT

The expression for the Doppler shift (5) is valid only in the non-relativistic limit, but relativistic corrections are negligible for the velocities at which the source is moved in the Mössbauer spectrometer ($v^2/c^2 \approx 10^{-22}$). However, for the thermal motion of the atoms in a solid, the relativistic corrections give rise to a line shift (the second-order Doppler shift) of the magnitude

$$\delta_{sod} = - E_\gamma \frac{\langle v^2 \rangle}{2c^2} \tag{23}$$

where $\langle v^2 \rangle$ is the mean square velocity in the thermal motion of the Mössbauer atom. Since the kinetic energy of the atoms depends on the temperature, the second-order Doppler shift is a function of temperature.

10. TIME-DEPENDENT HYPERFINE INTERACTIONS

The influence of time-dependent hyperfine interactions can be illustrated by considering a paramagnetic ^{57}Fe ion with spin $S = \frac{1}{2}$. When a magnetic field B is applied in the z direction, the ion can be found in eigenstates with $S_z = \frac{1}{2}$ and $S_z = -\frac{1}{2}$, which give rise to the effective fields at the nucleus, $\pm B_n$. For simplicity, we assume that the two eigenstates are equally populated and that the applied field is so small that the nuclear Zeeman splitting due to the applied field is negligible. If the relaxation time τ is long compared to the nuclear lifetime, τ_n, each of the two electronic eigenstates give rise to a normal magnetically split six-line Mössbauer spectrum. These two spectral contributions coincide, because they only differ with respect to the sign of the magnetic hyperfine field. If the relaxation time is comparable to τ_n, the absorption lines become broadened, and when the relaxation time approaches the Larmor precession time, τ_L, of the nuclear magnetic moment in the hyperfine field, the lines are also shifted towards the centroid of the spectrum. When τ is comparable to τ_L, very complex spectra with broad lines can be observed. Finally, for $\tau \ll \tau_L$, the spectrum collapses to one or two lines with line width decreasing with decreasing relaxation time. Figure 5 shows calculated Mössbauer spectra as a function of the relaxation time.

A large number of theoretical papers dealing with the influence of time-dependent hyperfine interactions on the Mössbauer line shape has been published /13-33/. The most general theories are based on stochastic or perturbation models. A formalism in which superoperators

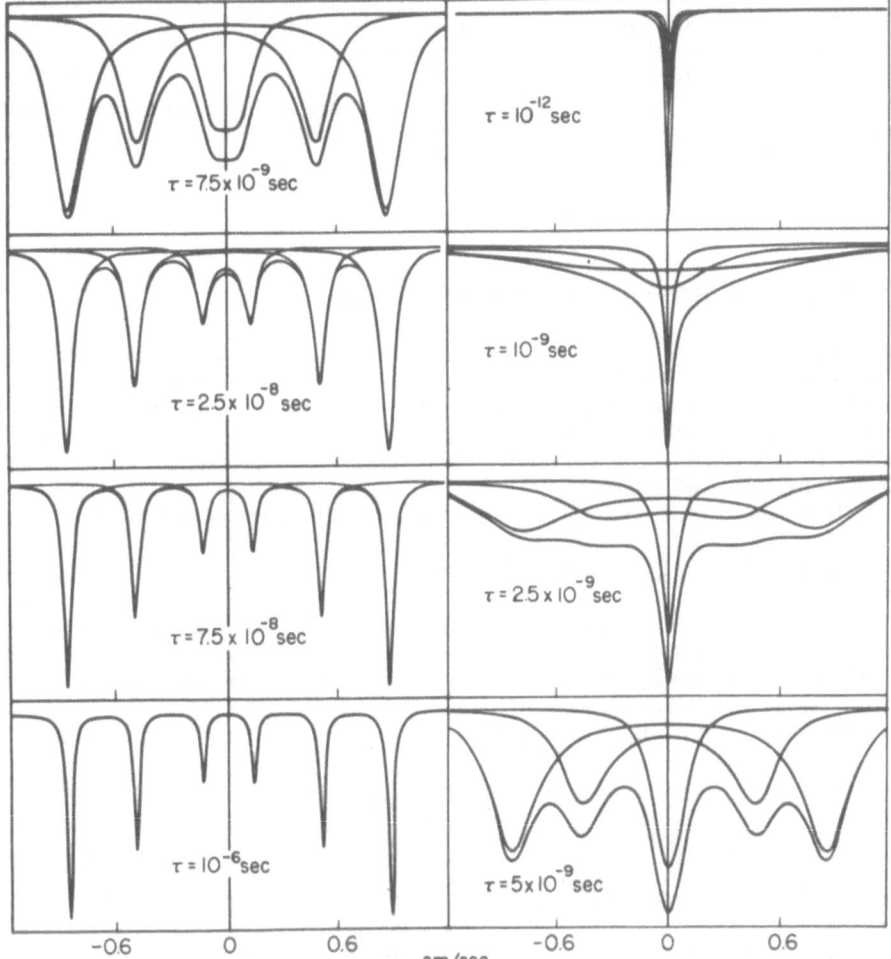

Figure 5.   Calculated ^{57}Fe Mössbauer spectra as a function of the relaxation time.

(Liouville operators) are used has been introduced by Blume /25/. This has facilitated the calculation of complex relaxation spectra, and general expressions for the Mössbauer line shape in the presence of magnetic relaxation have been derived.

In general transitions among ionic states, which may be described as combined nuclear and electronic states must be considered, and the calculation then involves inversion of a matrix of the order $(2I_e+1)$ $(2I_g+1)(2S+1)^2$ which equals 288 in the case of Fe^{3+}. Moreover, the calculation requires a detailed knowledge of the crystal field interaction, the hyperfine interaction, and the relaxation mechanisms. Often, this information is not available, and the number of unknown parameters may be so large that a determination of them based on the Mössbauer relaxation spectra cannot be justified. Therefore, approximate methods which involve a smaller number of parameters in calculation of the spectra are often preferable.

In the following the results of the stochastic theories for the Mössbauer line shape in the presence of relaxation are given, and some special cases and approximate expressions are discussed. The perturbation methods yield expressions that are formally equivalent to the results of the stochastic model /31/.

11. THE STOCHASTIC RELAXATION MODEL

A general stochastic relaxation model has been developed by Clauser and Blume /26/. These authors considered a system described by a Hamiltonian $\hat{H}_0$ containing all the time-independent interactions of the nucleus and the surrounding electrons, and a time-dependent Hamiltonian which gives rise to transitions among the eigenstates of $\hat{H}_0$ at random instants of time.

The results of the calculations is formally expressed by:

$$I(\omega) = \langle \hat{A}^+ (\tfrac{1}{2}\Gamma_{nat} - i\omega - \hat{W} - i\hat{H}_o^{\times})^{-1} \hat{A} \rangle \tag{24}$$

where $\hat{W}$ is a superoperator describing the relaxation among the eigenstates of the system, $\hat{H}_o^{\times}$ is the Liouville operator corresponding to the Hamiltonian $\hat{H}_0$, and $\hat{A}$ is an operator describing the interaction of the system with the electromagnetic field.

Eq. (24) may also be expressed in terms of matrix elements:

$$I(\omega) = \sum_{\substack{\mu,\nu \\ \mu',\nu'}} p(\nu)\langle\nu|\hat{A}^+|\mu\rangle \; (\mu\nu \,| \,(\tfrac{1}{2}\Gamma_{nat} - i\omega - \hat{W} - i\hat{H}_o^{\times})^{-1}|\mu'\nu')\langle\mu'|\hat{A}|\nu'\rangle \tag{25}$$

where $p(\nu)$ is the probability that the system initially is in the state $|\nu\rangle$.

The matrix elements of $(\tfrac{1}{2}\Gamma_{nat} - i\omega - \hat{W} - i\hat{H}_o^{\times})^{-1}$ can be calculated by inverting a matrix of dimension $(2S+1)^2(2I_e+1)(2I_g+1)$.

Thus in the case of $^{57}Fe^{3+}$ the expression (25) involves the inversion of a 288×288 matrix for each point, ω, in the spectrum. Moreover, in many cases one has to take into account interactions which are not considered here. For example, in some cases the magnetic interaction between the Mössbauer ion and the magnetic moments of the ligand nuclei has a significant influence on the Mössbauer spectra. An exact treatment may then cause a substantial increase in the size of the matrix. In practice it may be necessary to use approximate methods. In special cases, the matrix splits up in block matrices, and this simplifies the calculations considerably.

It has, however, been shown by Clauser /27/ that a relaxation spectrum can be expressed by the sum of a number of resonance lines with positions and widths determined by the real and imaginary parts of the eigenvalues of the superoperator $-\hat{H}_o^{\times} + i\hat{W} - \tfrac{1}{2}i\Gamma_{nat}$, and the

amplitudes of the lines are obtained from matrix elements of the eigenvectors of the same superoperator. This greatly facilitates calculation
of theoretical spectra.

12. DIAGONAL HYPERFINE INTERACTIONS

The calculation of theoretical Mössbauer spectra is particularly simple
if the magnetic hyperfine interaction can be described by an effective
magnetic field which fluctuates only along the z-direction and if the
electric field gradient is axially symmetric with its symmetry axis parallel to the same z-direction /22/. The reason is that the hyperfine Hamiltonian in this case is diagonal, i.e. the nuclear eigenstates are not
mixed, and the hyperfine Hamiltonian at a given instant of time commutes with the Hamiltonian at an arbitrary other instant of time. This
implies that the matrix in Eq. (25) splits up in a number of block matrices (with dimension determined by the number of electronic states),
one for each of the nuclear transitions. Under these circumstances
$\hat{H}_h$ is given by:

$$\hat{H}_h = Q'\{3\hat{I}_z^2 - I(I+1)\} - g_n \mu_N \hat{I}_z B_n(t) \tag{26}$$

where

$$Q' = \frac{eQV_{zz}}{4I(2I-1)}$$

and the magnetic field at the nucleus, $B_n(t)$, is a random function of
time, which here is assumed to take discrete values. For an Fe^{3+} ion
in the high spin state (S = 5/2) in a large applied field, B, $B_n(t)$ can
take six different values which are given by:

$$B_n = a_o M + B \tag{27}$$

$$(M = \tfrac{5}{2}, \tfrac{3}{2}, \tfrac{1}{2}, -\tfrac{1}{2}, -\tfrac{3}{2}, -\tfrac{5}{2})$$

The spectral component corresponding to a specific nuclear transition
$m_g \to m_e$ may then be calculated using an expression of the form /18-
23/:

$$I(\omega) = Re \sum_{M,M'} p(M) \, A_{MM'}^{-1}(\omega) \tag{28}$$

Here p(M) is the thermal population of the electronic state $|M\rangle$, and
the matrix elements $A_{MM'}$ are given by:

$$A_{MM'} = [i(\omega_M - \omega) + \tfrac{1}{2}\Gamma_{nat}] \, \delta_{M,M'} - \Omega_{MM'} \tag{29}$$

where $\hbar\omega_M$ is the nuclear transition energy when the ion is in the state M, and $\Omega_{MM'}$ is the transition probability rate for the ionic spin transition $|M\rangle \rightarrow |M'\rangle$, with

$$\Omega_{MM} = - \sum_{M'(\neq M)} \Omega_{MM'} \tag{30}$$

The total spectrum is then obtained by summation of expressions of the type (28) with the proper weight factors (e.g. 3:4:1:1:4:3, in the case of ^{57}Fe, when the hyperfine field is perpendicular to the gamma ray direction).

13. FAST RELAXATION LIMIT

In many cases the relaxation in non-diluted ferric compounds is so fast that the magnetic splitting of the Mössbauer spectrum has collapsed, and the spectra then consist of one or two (non-Lorentzian) lines. This collapse takes place when the relaxation time is of the same order of magnitude as the nuclear Larmor precession time. In the special case when the relaxation is longitudinal and the hyperfine field can assume only the two values $+B_{hf}$ and $-B_{hf}$ with equal probability, the collapse of a given pair of lines takes place at a critical transition probability rate given by /21/:

$$\tau_{cr}^{-1} = |(g_e m_e - g_g m_g)\mu_N B_{hf}|/\hbar \tag{31}$$

The critical relaxation frequency is different for each pair of lines. The inner lines will collapse at a smaller relaxation frequency than the outer lines.

Afanas'ev and Gorobshenko /32/ have developed a general theory for the Mössbauer line shape in the fast relaxation limit. These authors assumed that the relaxation can be described in terms of a fluctuating magnetic field at the nucleus which has different probabilities to point in different directions. They obtained explicit expressions for various cases of relaxation such as longitudinal (only along the z-direction), isotropic transverse (only in the xy plane), anisotropic transverse (only in the x-direction), and isotropic (all directions in space).

14. SLOW RELAXATION LIMIT

When the electronic relaxation time is long compared to the nuclear Larmor precession time, magnetically split spectra are obtained. However, the line width is influenced by the relaxation unless the relaxation times are long compared to the lifetime of the excited nuclear state. In the slow relaxation limit, the width of a Mössbauer line arising from the electronic state $|M\rangle$ is given by /27/:

$$\Gamma = 2\Gamma_{nat} + 2\Omega_{MM} \tag{32}$$

where

$$\Omega_{MM} = - \sum_{M'(\neq M)} \Omega_{MM'} \tag{33}$$

is the probability per unit time for the ion to make a transition from the eigenstate $|M\rangle$ to any other state. Moreover, the line positions are shifted towards the centroid of the spectrum. For longitudinal relaxation between the two states with numerically equal hyperfine fields in opposite directions, the shifts are given by:

$$\alpha = \Omega_{MM}(2\Omega_{MM} + \Gamma_{nat}/2)/2\alpha_0 \tag{34}$$

where α_0 is the Zeeman shift of the Mössbauer line if the hyperfine field is static. The line shifts for more general cases of relaxation have been calculated by Clauser /27/.

REFERENCES

/1/ R.L. Mössbauer, Z. Physik 151, 124 (1958).
/2/ R.L. Mössbauer, Naturwissenschaften 45, 538 (1958).
/3/ H. Frauenfelder: "The Mössbauer Effect" (Benjamin, Inc., New York) 1963.
/4/ G.K. Wertheim: "Mössbauer Effect: Principles and Applications" (Academic Press, New York), 1964.
/5/ H. Wegener: "Der Mössbauereffekt und Seine Anwendung in Physik und Chemie". (Bibliographisches Institut, Mannheim), 1966.
/6/ V.I. Goldanskii and R.H. Herber: "Chemical Applications of Mössbauer Spectroscopy" (Springer Verlag, Berlin), 1975.
/7/ U. Gonser (ed.): "Mössbauer Spectroscopy" (Springer Verlag, Berlin), 1975.
/8/ L. May (ed.): "An Introduction to Mössbauer Spectroscopy" (Adam Hilger, London), 1971.
/9/ R.L. Cohen (ed.): "Applications of Mössbauer Spectroscopy", Vol. I (Academic Press, New York), 1976.
/10/ R.L. Cohen (ed.): "Applications of Mössbauer Spectroscopy", Vol. II (Academic Press, New York), 1980.
/11/ G.J. Long (ed.): "Mössbauer Spectroscopy Applied to Inorganic Chemistry", Vol. 1 (Plenum Press, New York), 1984.
/12/ K.S. Singwi and A. Sjölander, Phys. Rev. 120, 1093 (1960).
/13/ A.M. Afanas'ev and Y. Kagan, Zh. Exper. Teor. Fiz. 45, 1660 (1963). (Sov. Phys. JETP 18, 1139 (1964))
/14/ Y. Kagan and A.M. Afanas'ev, Zh. Exper. Teor. Fiz. 47, 1108 (1964). (Sov. Phys. JETP 20, 743 (1965))
/15/ F. van der Woude and A.J. Dekker, Solid State Commun. 3, 319 (1965).

286

/16/   F. van der Woude and A.J. Dekker, Phys. Stat. Sol. 9, 775
       (1965).
/17/   H. Wegener, Z. Phys. 186, 498 (1965).
/18/   H.H. Wickman, M.P. Klein, and D.A. Shirley, Phys. Rev. 152,
       345 (1966).
/19/   H.H. Wickman: Mössbauer Effect Methodology, Vol. 2, Ed. I.
       Gruverman, Plenum Press, New York, 1966, p. 39.
/20/   H.H. Wickman and G.K. Wertheim in: "Chemical Applications of
       Mössbauer Spectroscopy". Eds. V.I. Goldanskii and R.H. Herber,
       Academic Press, 1968, p. 548.
/21/   M. Blume, Phys. Rev. Lett. 14, 96 (1965).
/22/   M. Blume and J.A. Tjon, Phys. Rev. 165, 446 (1968).
/23/   M. Blume in: "Hyperfine Structure and Nuclear Radiations", Eds.
       E. Matthias and D. Shirley, North-Holland, Amsterdam, 1968,
       p. 911.
/24/   E. Bradford and W. Marshall, Proc. Phys. Soc. 87, 731 (1966).
/25/   M. Blume, Phys. Rev. 174, 351 (1968).
/26/   M.J. Clauser and M. Blume, Phys. Rev. B3, 583 (1971).
/27/   M.J. Clauser, Phys. Rev. B3, 3748 (1971).
/28/   H. Schwegler, Phys. Stat. Sol. 41, 353 (1970).
/29/   H. Schwegler, Fortschritte der Physik 20, 251 (1972).
/30/   L.L. Hirst, J. Phys. Chem. Solids 31, 655 (1970).
/31/   F. Hartman-Boutron, Phys. Rev. B5, 2113 (1974).
/32/   A.M. Afanas'ev and V.D. Gorobschenko, Zh. Eksp. Teor. Fiz.
       66, 1406 (1974). (Sov. Phys. JETP 39, 690 (1974))
/33/   S. Dattagupta, Phys. Rev. B12, 3584 (1975).

CHAPTER 16
Mössbauer Spectral Lineshapes in the Presence of

Electronic State Relaxation

Fernande Grandjean, Institut de Physique B5, Université de Liège,

B-4000 Sart Tilman, Belgium

In Sections 1 and 2, this chapter introduces the reader to the different techniques in computing time dependent Mössbauer spectra. A comparison of the lifetime of the nuclear excited state and the critical time given by the hyperfine splitting with the characteristic time of the relaxation process shows in which time domain Mössbauer spectroscopy can be used to study time dependent phenomena. In practice, for most nuclides, the spectrum is very sensitive to fluctuations in the range of 10^{-6} to 10^{-8} s.

Section 3 develops the technique to compute a spectrum in the presence of a relaxation process described by one or more two-level systems. This technique, which is relatively simple and can be included in a fitting program without increasing dramatically the computational time, may be used in several cases as shown in Section 4.

1. General considerations

Because the basic principles of Mössbauer spectroscopy are covered in another chapter by Mørup (1), this chapter will begin directly with the problem of observing time dependent phenomena by the Mössbauer effect. Time dependent phenomena show up in Mössbauer spectra as peculiar lineshapes, often currently referred to as relaxation profiles, which usually change with temperature or pressure. Hence, it is worthwhile to first describe the Mössbauer spectral lineshape in the absence of relaxation (2). The resonant absorption of a gamma-ray of frequency, ω_0, and wavevector, k, by a nucleus located at position, $\vec{r}$, is easily described. The electromagnetic wave at the nucleus is represented by

$$\Phi(r,t) \propto \exp[i\vec{k} \cdot \vec{r} + (i\omega_0 - \Gamma_n)t] \tag{1}$$

where Γ_n is the natural half-linewidth given by $1/2\tau_n$, where τ_n is the mean lifetime of the nuclear excited state. In this equation, Γ_n is expressed in frequency units, s^{-1}, as is ω_0

Equation [1] indicates that the electromagnetic wave is not monochromatic at the frequency, ω_0, but is spread over a width, $2\Gamma_n$. This equation coresponds to a harmonic oscillator damped by an amount Γ_n. In a Mössbauer experiment, the position of the absorbing nucleus is Doppler shifted by the drive, and consequently,

$$\vec{k} \cdot \vec{r} = -\vec{k} \cdot \vec{v}\, t = -\omega t \tag{2}$$

287

G. J. Long and F. Grandjean (eds.), The Time Domain in Surface and Structural Dynamics, 287–308.

where $\vec{v}$ is the velocity of the drive and ω is the Doppler shift in frequency. A typical value for iron-57 is of the order of $10^7 s^{-1}$. The total Mössbauer absorption is obtained by integrating [1] over all times,

$$\Phi(\omega) \propto \int_0^\infty \exp[- i(\omega - \omega_0)t - (\Gamma_n + \Gamma_I)t]dt \qquad [3]$$

where an additional instrumental linewidth, Γ_I, has been introduced in order to account for all instrumental line broadening effects. The total observed half-linewidth is thus $\Gamma = \Gamma_n + \Gamma_I$. The resultant line shape is Lorentzian and is obtained by taking the real part of equation [3], hence the intensity, $I(\omega)$ is given by

$$I(\omega) = Re \; \Phi(\omega) \propto \frac{\Gamma}{(\omega - \omega_0)^2 + \Gamma^2} \qquad [4]$$

Equation [4] represents a line centered about ω_0 with a half-width at half maximum, Γ. A few typical values of Γ under normal experimental conditions, are given in Table 1.

Table 1. Energy of the transition, lifetime of the nuclear excited state, and full linewidth of some Mössbauer isotopes.

Isotope	Energy (keV)	Lifetime (ns)	Natural Linewidth (mm/s)	Natural Linewidth (MHz)	Typical Experimental (mm/s)
Fe-57	14.4	99.3	0.192	10.0	0.25
Sn-119	23.875	18.3	0.626	54.6	1.2
Sb-121	37.15	3.5	2.1	286.0	2.5
Eu-151	21.532	9.7	1.31	103.0	2.5
Yb-170	84.26	1.57	2.07	637.0	2.7
Ta-181	6.25	6800	0.0065	0.147	-
Np-237	59.54	63	0.073	15.9	0.1

Time dependent effects may modify the lineshape described by equation [4] in two ways. Either, in addition to the change imposed by the Doppler shift of the drive, the position of the nucleus, r, changes, or the frequency, ω_o, of the Mössbauer transition fluctuates. Examples of both cases have been observed. The first case corresponds to situations where the Mössbauer nuclide moves because of dynamic lattice effects, such as small vibrations about the equilibrium lattice position, or because of diffusion within the lattice (3). The second case corresponds to relaxation of the Mössbauer nuclide environment, which induces fluctuations in the hyperfine interactions and thus in the Mössbauer absorption frequency, ω_0.

The three hyperfine interactions, the isomer shift, the quadrupole interaction, and the magnetic hyperfine interaction may fluctuate with time. This chapter will be restricted to the first two interactions, because fluctuations of the hyperfine field are discussed by Mørup (1). The isomer shift and quadrupole interaction may fluctuate because of electron hopping (4,5), valence fluctuation (6,7), high-spin to low-spin relaxation (8,9), or motions of the non-Mössbauer atoms in the lattice (10). The quadrupole interaction may fluctuate because the principal axis of the electric field gradient tensor (11) jumps randomly between x, y and z axes. These are just a few examples and details of these cases will be discussed in section 4.

Of course, another important question is, what is the time scale for relaxation phenomena observed by Mössbauer spectroscopy? When the Mössbauer nuclide is exposed to time dependent effects through the hyperfine interactions, there are several time scales to consider. The first time scale is defined by the Mössbauer nuclide; it is the mean lifetime, τ_n of the nuclear excited state. Table 1 gives typical values for several Mössbauer nuclides. The second time scale is the Larmor precession time, τ_L, which is especially important in the cases of hyperfine field fluctuations (1). The other time scales, τ_R, are related to the fluctuations of the nuclear environment. They are the inverse of the relaxation rates and may be controlled in the laboratory by varying the temperature or pressure. The shape of the spectrum depends on the relative values of the different time scales. As this chapter will not deal with fluctuating magnetic hyperfine fields, it will not consider τ_L and the lineshapes discussed will depend only upon the relative values of τ_n and τ_R.

Some general statements about Mössbauer relaxation spectra can be made without a detailed consideration of their lineshape. If $\tau_n < < \tau_R$, no fluctuation of the nuclear environment occurs during the lifetime of the nuclear excited state. Therefore, the nucleus sees a static environment and a static spectrum is observed as is described by equation [4]. In practice, the spectrum is insensitive to changes in τ_R, when τ_R is greater than a few times τ_n. In other words, taking into account the values of τ_n given in Table 1, small changes in timescale, of the order of $10^{-5} s$, cannot be observed by Mössbauer spectroscopy. In the other extreme, if $\tau_R < < \tau_n$, many fluctuations of the nuclear environment occur during the lifetime of the nuclear excited state. Therefore, the nucleus sees an averaged environment and an averaged Mössbauer spectrum, which shows hyperfine parameters that are the weighted average of the static hyperfine parameters, is observed. For intermediate values of τ_R, peculiar lineshapes are observed. Section 2 explains how these lineshapes can be calculated and in simple cases used to fit experimental spectra. A critical value for τ_R is $h/2\pi\Delta E$, where ΔE is the energy separation between two absorption lines; for this value the spectrum collapses into one line.

2. Lineshapes in the Presence of Relaxation

Equation [4] may be transformed (12) so as to represent the probability of the emission of a photon, P, with wave vector, k, and frequency, ω, by the system which makes a transition from the initial state $|\alpha>$ of energy E_α in s^{-1}, to the final state $|\beta>$ of energy E_β in s^{-1},

$$P_{\alpha\beta}(\vec{k}) = \frac{|<\alpha|H^+|\beta>|^2}{(\omega - E_\beta + E_\alpha)^2 + \Gamma^2} \qquad [5]$$

where H^+ describes the interaction of the solid with the photon, and $|\alpha>$ and $|\beta>$ represent eigenstates of the entire solid. Thus $|<\alpha|H^+|\beta>|^2$ is essentially the Clebsch-Gordan coefficient and may be thought of as the intensity of the transition. Now, this equation will be manipulated to put it into a form suitable for the discussion of time dependent Mössbauer spectra. If one uses the typical relationship for the Laplace transform,

$$[(\omega - E_\beta + E_\alpha)^2 + \Gamma^2]^{-1} =$$

$$\frac{2}{\Gamma} Re \int_0^\infty \exp[i(\omega - E_\beta + E_\alpha)t - \Gamma t] dt$$

one gets

$$P_{\alpha\beta}(\vec{k}) = \frac{2}{\Gamma} \; Re \int_0^\infty \exp(i\omega t - \Gamma t) \qquad [6]$$

$$< \alpha|H^-|\beta > \; < \beta|\hat{U}(t)H^+U(t)|\alpha > \, dt$$

where H^- is the complex conjugate of H^+, $U(t) = \exp(-iHt)$, $\hat{U}$ is the complex conjugate of U, and H is the Hamiltonian for the entire system. The experimentally observed emission probability is obtained by averaging equation [6] over all initial $|\alpha >$ states and summing over all final $|\beta >$ states, yielding

$$P(\vec{k}) = \frac{2}{\Gamma} \int_0^\infty \exp(i\omega t - \Gamma t) < H^- H^+(t) > \, dt \qquad [7]$$

where $H^+(t) = \hat{U}(t)H^+U(t)$ and the brackets denote the average over the occupation of the initial states. Up to this point, no time dependent effect has been introduced and equation [7] may be used to describe a static Mossbauer spectrum.

If the Hamiltonian of the emitting nuclide is a random function of time, another average over the stochastic degrees of freedom in the Hamiltonian must be introduced into equation [7]. Then the lineshape in the presence of a fluctuating Hamiltonian is given by the expression

$$P(\vec{k}) = \frac{2}{\Gamma} \int_0^\infty \exp(i\omega t - \Gamma t)(< H^- H^+(t) >)_{av} \, dt \qquad [8]$$

where $( )_{av}$ denotes the stochastic average. In mathematical terms, equation [8] gives the Mossbauer lineshape as the Laplace transform of a correlation function.

So far, we have essentially followed the work of Blume and Tjon (12) in the treatment of the equations. Expression [8] is complex, but Blume and Tjon (12) discussed in detail its use for a fluctuating magnetic field parallel or perpendicular to the electric field gradient axis. They also discussed its use for a randomly varying electric field gradient (13) in the case of the iron-57 nuclide. The calculations are tedious but, in the end, it is possible to write the lineshape in a rather compact expression, which can be used to fit the spectrum. These examples of relaxation and their treatment by Blume and Tjon (12,13) are now classical and will not be discussed in further detail here. Nevertheless, we should note that the time dependent Hamiltonian, $H(t)$, in several cases commutes with itself at different times (14). This means that it is possible to diagonalize $H(t)$ at some instant in time such that it remains diagonal at a later time. Then, the time variation of $H(t)$ will not induce any transition between two different eigenstates of the system. This theory is called adiabatic. There are circumstances in which this theory is inadequate. For instance, if a fixed magnetic field along the x-axis is superimposed upon a fluctuating field along the z-axis, the Hamiltonian at different times does not commute with itself. The fluctuating term is able to induce transitions between the eigenstates of the time independent part of the Hamiltonian. Thus, the stochastic and quantum mechanical aspects of the problem are no longer separate. This non-adiabatic theory has been developed by Blume (14).

This chapter will not reproduce all the steps of the mathematical treatment of the non-adiabatic theory, but will cover its main features and final result. The problem is to evaluate the stochastic average in equation [8], when the Hamiltonian may randomly fluctuate between the possible operators $V_1, V_2, \ldots V_n$, which need not necessarily commute with each other. Then, the Hamiltonian may be written as

$$H(t) = \sum_{j=1}^{n} f_j(t) V_j,$$ [9]

where V_j is a quantum mechanical operator and $f_j(t)$ is a random function of time. Therefore, the time dependence of H^+ in [8] is given by

$$H^+(t) = \exp[i\int_0^t H(t')dt']H^+ \exp[-i\int_0^t H(t')dt']$$ [10]

Blume (14) has shown that the evaluation of the stochastic average is greatly simplified by using Liouville operators. The definition and properties of the Liouville operators are given in Appendix for convenience. After various algebraic manipulations, the general form of the Laplace transform of equation [8] is

$$F(p) = \frac{1}{2I_1 + 1} \sum_{m_0 m'_0} \sum_{m_1 m'_1} < I_1 m_1 | H^- | I_0 m_0 >$$

$$\sum_a \sum_b p_a < I_0 m_0 I_1 m_1 a | [pE - W - i\sum_{j=1}^{n} V_j^* F_j]^{-1} | I_0 m'_0 I_1 m'_1 b > < I_0 m_0 | H^+ | I_1 m'_1 > ,$$ [11]

where $p = -i\omega + \Gamma$, p_a is the a priori probability of the occurence of the initial value a, W is a stochastic matrix whose off diagonal elements $(a|W|b)$ are equal to the probability per unit time that f(t) makes a transition from a to b and whose diagonal elements are equal to the negative of the sum of the off-diagonal elements in the same row, F_j is a diagonal matrix whose elements are the values of $f_j(t)$, V_j^* is the Liouville operator associated with V_j , I_0 and I_1 are the spin of the ground and excited nuclear states, m_0, m'_0, m_1, m'_1 are the quantum numbers of the ground and excited nuclear states, and E is the unity matrix. Equation [11] may be written in a more compact form by using super operator notation

$$F(p) = Tr\{H^- A H^+\}$$ [11*]

It can be seen that the matrix

$$A = [pE - W - i\sum_{j=1}^{n} V_j^* F_j]^{-1}$$ [12]

is labeled by three indices, two quantum mechanical ones, m_0 and m_1, corresponding to the ground and excited states of the emitting nuclide and one stochastic index, a or b, corresponding to the possible values of $f_j(t)$. Thus, the dimensionality of A is $N = (2I_1 + 1)(2I_0 + 1)$ n, where n is the number of possible values of the Hamiltonian. Thus, for instance, for the 14.4 keV transition of iron-57, the dimensionality of A is 8n, which means the smallest possible dimension of A is 16. The inversion of such a matrix on a computer is easily performed but, in order to compute a line profile, this inversion must be carried out for each data point in the spectrum. Hence, the computing time of a profile may be quite large. Clauser (15) has proposed a reformulation of the line profile which requires only one inversion of the super-matrix [12]. Clauser recognized that the independent variable, ω enters only on the diagonal of A^{-1} . Therefore, the eigenvectors of A^{-1} will be the same whether ω is present or not, and the eigenvalues of A^{-1} with ω removed will differ from those of A^{-1} containing ω by only a constant. Thus, if B and β represent the eigenvectors and eigenvalues of A^{-1} with ω removed , the lineshape is given by

$$I(\omega) = F(p)$$

and

$$F(p) = (H^-B) (\beta - i\omega)^{-1} (B^{-1}H^+) \tag{13}$$

This expression can easily be computed by, first obtaining the eigenvalues and eigenvectors of A^{-1} without ω along its diagonal, and then reintroducing ω in the simple algebraic way indicated in [13]. Thus, it is only necessary to compute the eigenvectors and eigenvalues of A^{-1} once, and the computer time needed to fit relaxation profiles drops drastically. Shenoy and Dunlap (16) reviewed a few applications of this technique to iron-57 in biological systems, ytterbium-170 in gold, and neptunium-237 in NpF_6.

3. The Lineshape in Simple Cases

This section will discuss cases of relaxation between non-magnetic states. The cases of paramagnetic and superparamagnetic relaxation are covered in other chapters (1).

For relaxation between non-magnetic states, it is useful to discuss in detail the lineshape resulting from relaxation within a two-level system, such as is shown in Figure 1. In this case, the environment of the nucleus fluctuates such that the nuclear transition frequency relaxes between two values, ω_1 and ω_2, each occuring with the probability, p_1 and p_2. If $\lambda_{1/2}$ and $\lambda_{2/1}$ are the relaxation rates from 1 to 2 and from 2 to 1 respectively, the lineshape is given by (4,5)

$$I(\omega) \propto \frac{A[\Gamma + 2(\lambda_{2/1}p_1^2 + \lambda_{1/2}p_2^2)] - (\omega - p_1\omega_2 - p_2\omega_1)B}{A^2 + B^2} \tag{14}$$

where

$$A = (\omega - \omega_1)(\omega - \omega_2) - \Gamma(\Gamma + \lambda_{1/2}p_2 + \lambda_{2/1}p_1)$$

and

$$B = (\Gamma + \lambda_{1/2}p_2)(\omega - \omega_2) + (\Gamma + \lambda_{2/1}p_1)(\omega - \omega_1)$$

Furthermore, the relative probabilities, p_1 and p_2 are such that

$$p_1 + p_2 = 1,$$

and

$$\lambda_{1/2}p_1 = \lambda_{2/1}p_2$$

must be fulfilled for the equilibrium populations specified by p_1 and p_2 to be maintained. Equation [14] may be obtained by writing a matrix such as [12]. In the simple case of the two-level system, the result is a 2x2 matrix which can be easily inverted to obtain, after taking the real part of the expression [11], equation [14]. If $p_1 = p_2 = 0.5$, then $\lambda_{1/2} = \lambda_{2/1} = \lambda$, and the lineshape is given by a simpler expression

$$I(\omega) \propto \frac{(\Gamma + \lambda)A - (\omega - \frac{\omega_1 + \omega_2}{2})B}{A^2 + B^2} \tag{15}$$

where $A = (\omega - \omega_1)(\omega - \omega_2) - \Gamma(\Gamma + \lambda)$, and $B = (\Gamma + \frac{\lambda}{2})(2\omega - \omega_1 - \omega_2)$.

In the limit of slow relaxation, λ in equation [15] or $\lambda_{1/2}$ and $\lambda_{2/1}$ in equation [14] go to zero and the resulting lineshape is that of two Lorentzian lines of linewidth, Γ, centered at ω_1 and ω_2, with the relative intensities, p_1 and p_2, respectively. This is quite normal. Indeed, if the relaxation rate is smaller than or close to the natural linewidth, expressed in frequency units, at most one

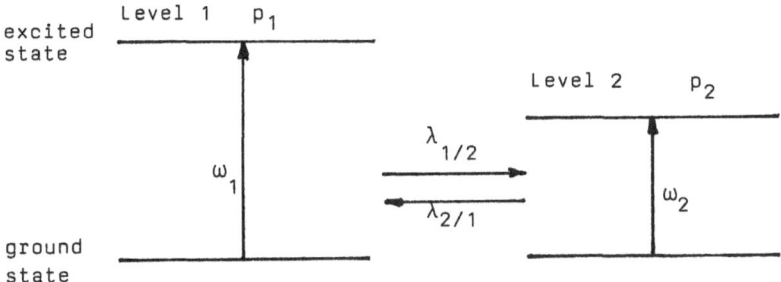

Figure 1. A two-level system representing the relaxation between two frequencies ω_1 and ω_2.

fluctuation will occur in the time window defined by the natural linewidth and thus the system will appear to be essentially static on the Mössbauer time scale.

In the limit of fast relaxation, λ in equation [15] , or $\lambda_{1/2}$ and $\lambda_{2/1}$ in equation [14] , are much larger than the linewidth and the resulting lineshape is a single Lorentzian line of linewidth, Γ, centered on $p_1\omega_1 + p_2\omega_2$. Again, this is quite normal. If the relaxation rate is large, many fluctuations will occur in the time window defined by the natural linewidth and thus, the Mössbauer nuclide will experience an average interaction on the Mössbauer time scale. The limiting behavior for slow or fast relaxation is general and is easily expressed for a two-level system.

Equation [14] can be easily extended to the case of relaxation of a Mössbauer nuclide in an electric field gradient fluctuating between two values, q_1 and q_2. It is usually assumed that there is no mixing of the nuclear spin states as a result of the relaxation and thus each pair of nuclear levels, m_0 and m_1, corresponding to an allowed transition, is a two-level system which can be treated as described by equation [14]. Hence, the observed spectrum is a sum of several expressions such as [14], for the different transitions. Figure 2 shows the situation for an iron-57 nuclide. This technique has been successfully applied by Litterst and Amthauer (4) and by Prietsch et al. (5) to the case of electron hopping between iron ions. Some of their results are discussed in the next section.

A second simple case is the isotropic relaxation of the principal axis of the electric field gradient. The theory of the lineshape has been fully developed by Tjon and Blume (13), who wrote the fluctuating Hamiltonian, used in equation [8], for the iron-57 Mössbauer nuclide. If all directions in space are equivalent, an analytical expression for the lineshape may be written as

$$I(\omega) = Re\ \{\frac{p + 3\lambda}{p^2 + 3\lambda p + 4\alpha^2}\}$$ [16]

where $p = -i\omega + \Gamma$ and $\alpha^2 = Q^2/4$, where Q is the quadrupole splitting. In the limit of slow relaxation, i.e. when the relaxation rate is smaller than the quadrupole splitting, a doublet is observed. In the limit of fast relaxation, i.e. when the relaxation rate is greater than the quadrupole splitting, a singlet is observed. This approach was used to interpret the relaxation spectra of iron(II) ions in the octahedral coordination environment of MgO (11).

In order to interpret the relaxation effects of iron(II) impurities in cubic ZnS, Bonville et al. (18) have extended the theory of Tjon and Blume and have shown how to use the Clauser formalism

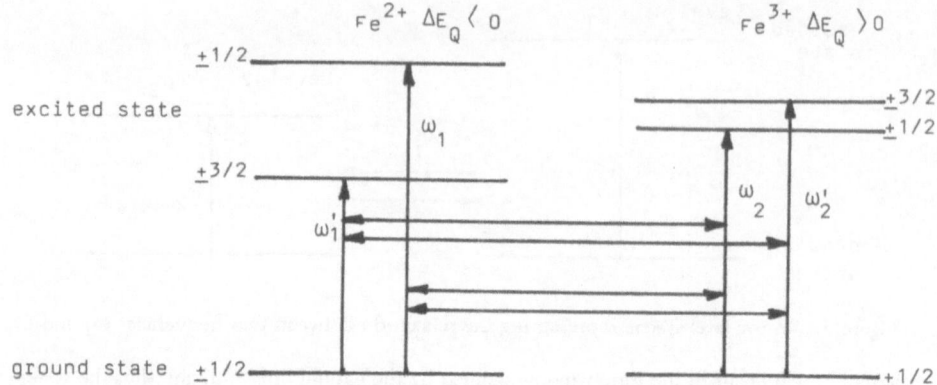

Figure 2. The two-level systems representing the four relaxation processes below 100K in *stage2 − FeCl₃ −graphite*

(15) with Liouville operators. This paper is recommended to the reader because the authors give extensive details of the calculation and the forms of the operators.

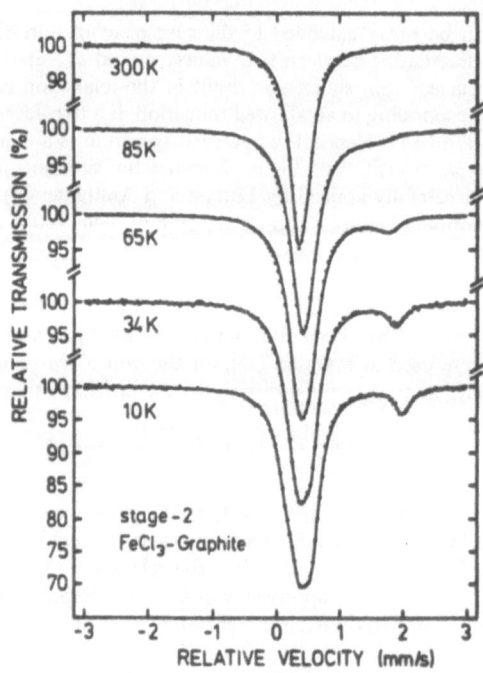

Figure 3. Representative iron-57 Mössbauer spectra of *stage2 − FeCl₃ −graphite* at temperatures from 10 to 300K. The solid lines represent the results of least-squares fits. Reproduced with permission from ref.(5).

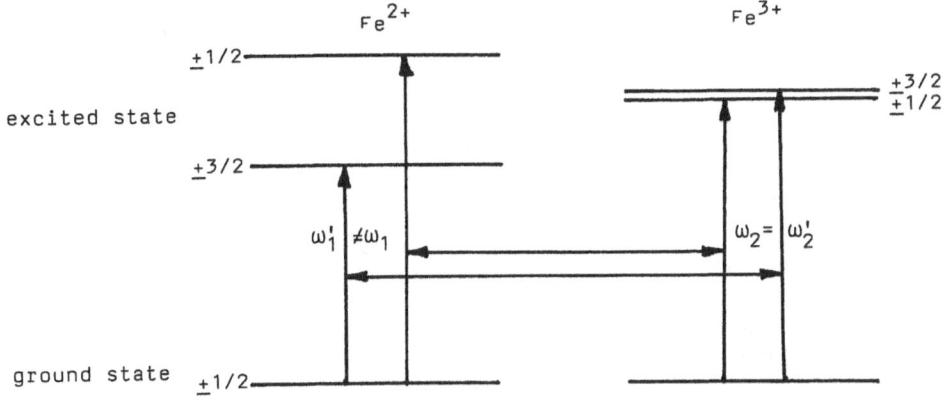

Figure 4. The two-level systems representing the two relaxation processes above 100K in *stage2 − FeCl₃ −graphite* .

4. Examples of Relaxation Between Non-Magnetic States

This section discusses a few typical examples of relaxation processes which do not involve magnetic hyperfine fields. I have tried to select characteristic situations for different Mössbauer nuclei. This is not intended to be a complete review of the literature in this field.

4.1. Electron Hopping Processes in Iron-57 Compounds

The most common Mössbauer isotope is iron-57 and iron may have different valences states. Hence, mixed valent iron compounds are good for studying electron hopping and the consequent relaxation Mössbauer spectra.

Among several recent observations (4,5,19) of electron hopping in different materials, this chapter will discuss in detail the case of intercalated $FeCl_3$ in graphite (5). Figure 3 shows the Mössbauer spectra of *stage2 − FeCl₃ − graphite* obtained from 10 to 300K. At low temperatures, the principal absorption line is due to trivalent ions and the secondary line is due to divalent ions. The latter line is part of a quadrupole doublet, whose other component is hidden under the iron(III) component. As the temperature increases, the iron(II) component broadens, becomes asymmetric, and finally disappears. These features are interpreted as resulting from a thermally activated electron hopping between iron(II) and iron(III) ions. At 10K, the relaxation rate is slow enough to provide a static picture of the iron(II) and iron(III) ions. At temperatures above 100K, the relaxation rate is so high that the different valence states are no longer resolved.

At 10.6K the quadrupole splittings in *stage2 − FeCl₃ −graphite* are respectively 0.18 mm/s for iron(III) and -1.86 mm/s for iron(II). To a first approximation, the iron(III) quadrupole splitting may be neglected. The nuclear quantization axis at the divalent iron site, which is parallel to the c-axis of the unit cell may also be used for the iron(III) site and, as a result, there is no mixing between the nuclear m_I levels. Thus, the spectrum can be interpreted as a sum of two independently relaxing two-level systems as described in Section 3 and illustrated in Figure 4. This simplified model satisfactorily describes the spectra above 100K, in the fast relaxation limit, as shown by the solid line in Figure 3. Below 100K the quadrupole splitting of the iron(III) ion must be taken into

account. For an iron(III) quadrupole splitting which is smaller than the iron(II) quadrupole splitting with different quantization axis, a mixing of the nuclear levels, m_I, occurs and the spectra of
stage2 − FeCl₃ − graphite may be described by four relaxing two-level systems as shown in Figure
2. The fits with these models are good as shown in Figure 3. In order to fit the spectra and to get
reasonable values of the hyperfine parameters, some assumptions were made by Prietsch et al. (5)
For instance, the iron(III) isomer shift was assumed to have the same second-order Doppler shift
as in anhydrous *FeCl₃*. Such assumptions seem to be valid for most fits with a relaxation model.
Usually, the hyperfine parameters are strongly correlated and reliable values can only be obtained
if some additional assumptions are made.

Electron hopping rates in *stage2 − FeCl₃ − graphite* as in many other compounds are assumed
to follow the Arrhenius law

$$\lambda_{2+/3+} \propto \exp(\frac{-E_a}{kT}),$$

where E_a represents the activation energy for the hopping process. Figure 5 shows the logarithmic
plot of the divalent to trivalent relaxation rate, $\lambda_{2+/3+}$, as a function of the inverse of the temperature. The resulting activation energy is 45(20)meV above 70K.

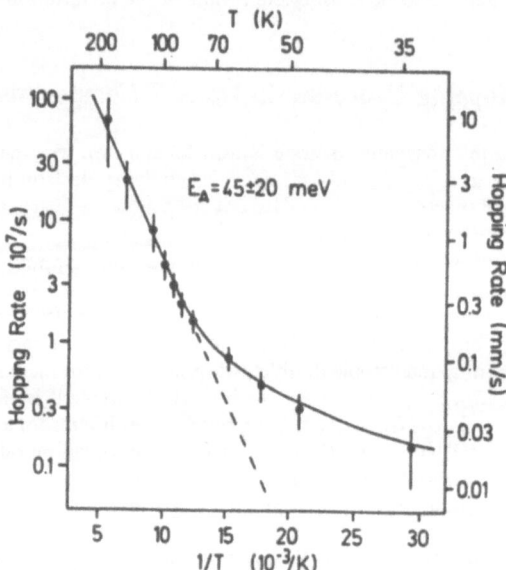

Figure 5. An Arrhenius plot of the electron hopping rate $\lambda_{2+/3+}$. The larger error bars at low or
high temperatures are due to strong correlations between linewidth and relaxation
rate. At low temperatures, tunneling effects contribute to the hopping rate, causing
the observed deviation from a linear relationship. Reproduced with permission
from ref.(5).

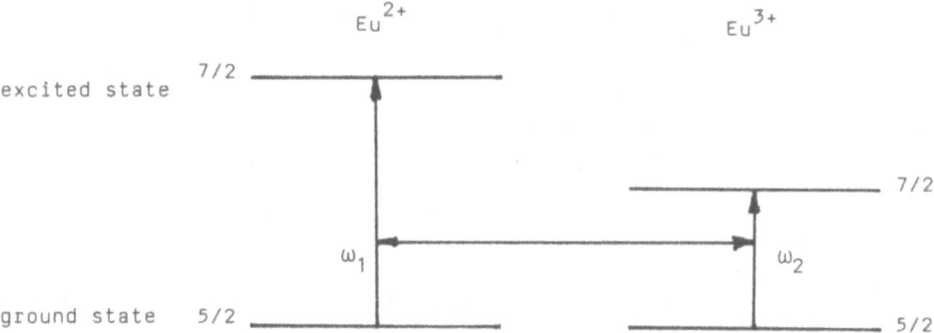

Figure 6. The two-level system representing the valence state relaxation in europium compounds.

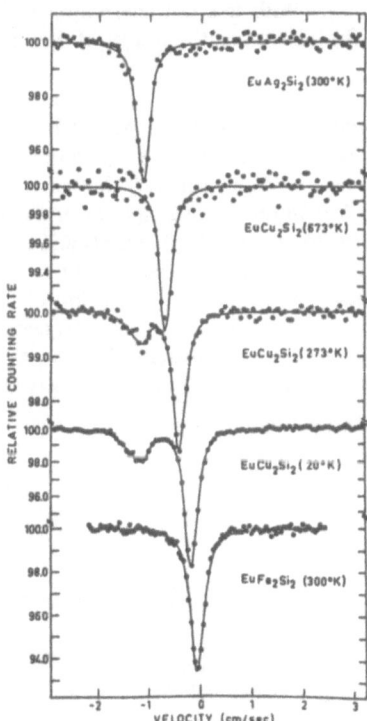

Figure 7. Europium-151 Mössbauer spectra of $EuFe_2Si_2$, $EuCu_2Si_2$, and $EuAg_2Si_2$. Reproduced with permission from ref.(20).

4.2. Electron Hopping Processes in Europium-151 Compounds

Europium-151 Mössbauer spectroscopy is rather easy and europium ions may also have different valence states. They are characterized by quite different isomer shifts. Indeed, europium(II) ions have isomer shifts of about -12 mm/s whereas europium(III) ions have isomer shifts close to zero mm/s. Because the natural linewidth of the resonant absorption of the 21.6 keV line of europium-151 is 1.3 mm/s (Table 1), the two valence states are easily distinguished, by Mössbauer spectroscopy. Consequently, the observation of fluctuating valence in mixed valence europium compounds is quite common (20-22).

If the resonance for europium(III) and europium(II) ions may each be represented by one Lorentzian line, the spectral fitting procedure is again simply a two-level system as shown in Figure 6.

In several mixed valence europium compounds discussed in the literature (20-22), the relaxation times are fast, about $3.5 \times 10^{-11} s$, as compared with the lifetime of europium-151 excited nuclear state, which is $9.7 \times 10^{-9} s$ (Table 1). With this fast relaxation, the spectrum has the shape of a narrow line with an isomer shift which is a weighted average of the europium(II) and europium(III) isomer shifts.

Figure 7 displays spectra of $EuAg_2Si_2$ and $EuFe_2Si_2$ at 300 K and of $EuCu_2Si_2$ at 20, 273 and 673K. The spectra of $EuAg_2Si_2$ show a typical temperature independent isomer shift for europium(II), -11.4(0.1)mm/s. The spectra of $EuFe_2Si_2$ show a typical temperature independent isomer shift for europium(III), 0.7(0.1)mm/s. In the spectra of $EuCu_2Si_2$, two lines are observed at 20 and 273K. The weakest one is found at -12.4(0.2)mm/s and is typical of europium(II). The most intense line is present at all temperatures and its isomer shift is strongly temperature dependent as shown by Figure 8. This behavior is due to the fluctuation of the europium valence state in $EuCu_2Si_2$. At low temperature, nearly all europium ions are trivalent and the isomer shift is equal to -1.8 mm/s, but, at high temperature, the populations of the trivalent and divalent states are approximately equal and an average isomer shift of -7 mm/s is observed.

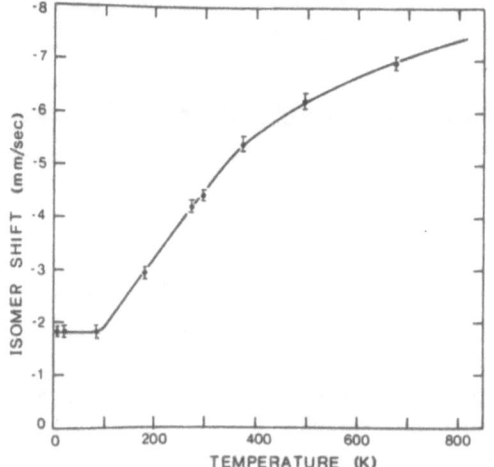

Figure 8. Isomer shift of the intense line in the Mössbauer spectrum of $EuCu_2Si_2$ as a function of temperature. Reproduced with permission from ref.(20).

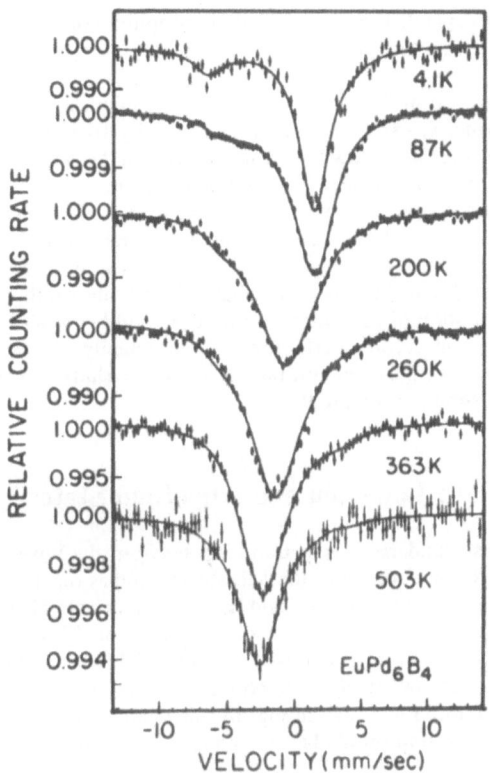

Figure 9. Europium-151 Mössbauer spectra of $EuPd_6B_4$. The solid curves are theoretical spectra with an activation energy of 500K and a level width of 120K. Reproduced with permission from ref. (23).

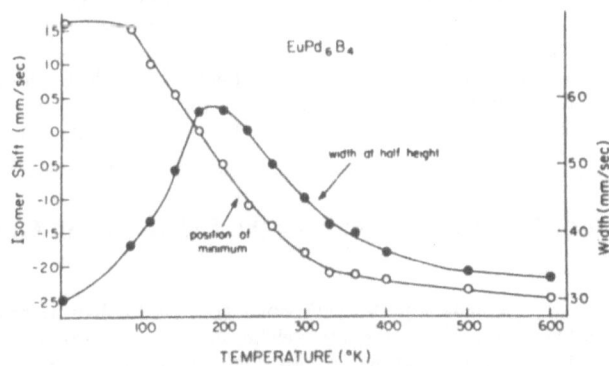

Figure 10. Isomer shift and width of Mössbauer spectrum of $EuPd_6B_4$. Solid curves are given by the theoretical model discussed in ref. (23). Reproduced with permission from ref. (23).

A strongly temperature dependent isomer shift in a europium compound is usually taken as a strong indication of valence state fluctuation.

Apparently, there is no case of fluctuating valence in europium for which a relaxation profile was used to fit the Mössbauer spectra. The line profile used by Bauminger, et al (21) in the case of $Eu_xLa_{1-x}Rh_2$ and by Felner and Nowik (23) in the case of $EuPd_6B_4$ should not be confused with a relaxation profile. As illustrated in Figures 9 and 10, the Mössbauer spectra of $EuPd_6B_4$ show a strongly temperature dependent isomer shift characteristic of valence fluctuation. Furthermore, the spectra around 200K are broadened but this is not relaxation broadening. The spectra are analyzed by assuming fast fluctuations of electrons between a localized 4f level of finite width located above the Fermi level and the conduction band. The finite width of the localized 4f level gives rise to a narrow distribution of the energy of this level, and hence a distribution of the populations of the europium(II) and europium(III) ions. Consequently, there is a distribution of the weighted average of the europium(II) and europium(III) isomer shifts and this distribution broadens the Mössbauer spectra shown in Figure 9.

4.3. Relaxation Between Low-Spin and High-Spin States

Many iron(III) complexes undergo a spin transition between the low-spin (S = 1/2) and high-spin (S = 5/2) states. An extensive review of the Mössbauer studies on these compounds has been written by Gütlich (24). Because the low-spin and high-spin states are characterized by rather different quadrupole splittings, they can easily be resolved in the iron-57 Mössbauer spectra. In spite of this favorable situation, not many papers report spectra which are suitable for fits with a relaxation profile. Indeed, because of low recoil free fractions at room temperature, the spectra are often poorly defined and are therefore not worth fitting. In this section, two cases (8,9) are discussed in which the analysis with a relaxation profile based on a two-level system was successful.

Figure 11 shows the iron-57 Mössbauer spectra (8) of $[Fe(acpa)_2]BPh_4H_2O$ and their fits with a profile equivalent to equation [14]. At low temperature, 189K, the spectrum is typical of the low-spin state with a quadrupole splitting of 2 mm/s. At 286K, the spectrum is broad and asymmetric and results from the relaxation between the low-spin $^2T_{2g}$ state and the high-spin $^6A_{1g}$ state characterized by a small quadrupole splitting. The low-spin to high-spin relaxation rate decreases from $2.9x10^6 s^{-1}$ at 268 K to $7.1x10^4 s^{-1}$ at 189K. These numbers must be compared with the natural linewidth of the iron-57 resonance line, $10^7 s^{-1}$ (Table 1). Another nice example of fits with the same relaxation model is illustrated in Figure 12, which shows the spectra of $^{57}Fe(acpa)_2PF_6$, at temperatures between 110 and 285K. From the fits of these spectra, relaxation rates ranging from $5.26x10^5 s^{-1}$ at 110K to $2.94x10^6 s^{-1}$ at 285K are obtained.

4.4. Relaxation of the Principal Axis of the Electric Field Gradient

Figure 13 shows the Mössbauer spectra (11), obtained with a cobalt-57 source, in AgCl from 80 to 458K. As the temperature increases, the well resolved doublet observed at 80K, broadens around 200K and collapses to a singlet above 250K. The lineshapes are attributed to an isotropic relaxation of the principal axis of the electric field gradient. The iron(II), resulting from the cobalt-57 decay, ocupies a substitutional site bound to a positive ion vacancy which resulted from the removal of a second silver ion whose loss was necessary to conserve charge balance. This vacancy is constrained to occupy one of the three (100) positions which are nearest neighbors to the iron(II) ion. Thus, the principal axis of the electric field gradient may have three directions x, y and

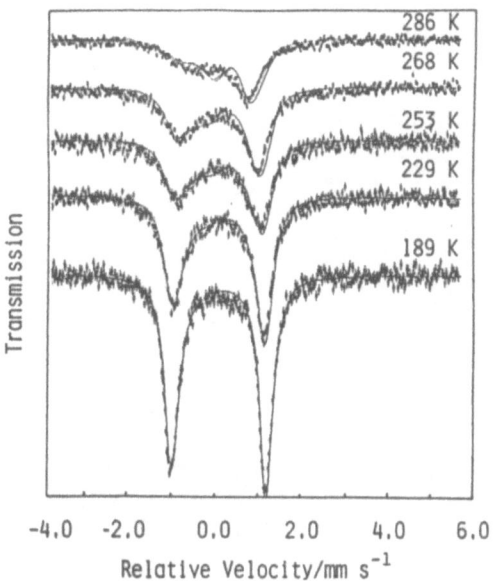

Figure 11. Iron-57 Mössbauer spectra of $[Fe(acpa)_2]BPh_4.H_2O$. The solid curves are the results of fits with a profile equivalent to equation [14]. Reproduced with permission from ref.(8).

z. In a static situation, where no vacancy diffusion occurs, such as at 80K, a quadrupole doublet is observed. With increasing temperature, the vacancy jumps between the equivalent nearest neighbor positions and the principal axis of the electric field gradient relaxes between the three directions. In agreement with the theory of Tjon and Blume (13), as the relaxation rate increases, the spectrum broadens and finally collapses into a singlet.

4.5. Mössbauer Nuclide Motion with Electric Field Gradient Fluctuations

In some compounds (25), a motion of the iron-57 Mössbauer nuclide concomitant with electric field gradient fluctuations is observed. This chapter describes the case of the Chevrel-phase compound, $FeMo_6S_8$. In the high temperature phase of this compound, the small iron cation may occupy two non-equivalent positions. Six are in the inner ring, a hexagon, shown in Figure 14. Six are above and below the hexagon, on the two outer rings, shown also in Figure 14. The low temperature phase of $FeMo_6S_8$ is triclinic and iron occupies only one of the inner ring sites. At a temperature of about 106K, depending upon the stoichiometry, a phase transition occurs to a rhombohedral phase in which the iron ion may occupy one of the outer ring sites. Figure 15 shows the Mössbauer spectra of $FeMo_6S_8$ at 100, 115, and 150K (26) and at 220, 230, and 250K (25). As the temperature increases from 100 to 150K, a second doublet slowly increasing in intensity appears. This is correlated with the increasing population of iron in the outer ring. This iron ion

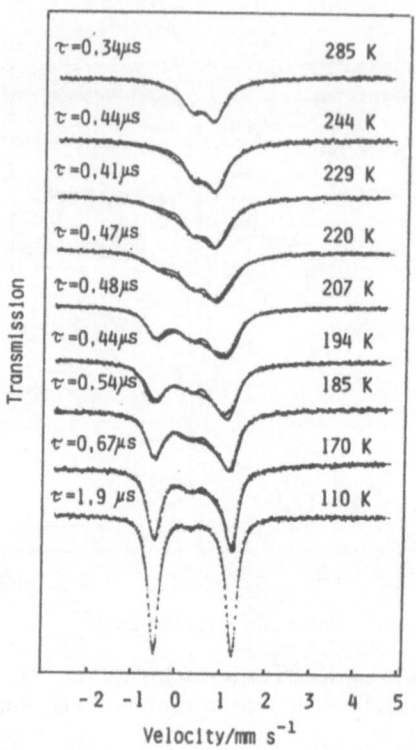

Figure 12. Iron-57 Mössbauer spectra of $^{57}Fe(acpa)_2PF_6$. The solid lines are the results of fits with a profile equivalent to equation [14]. Reproduced with permission from ref. (9).

delocalization takes place on a time scale long compared to the lifetime of the iron-57 excited nuclear state, because no broadening is observed corresponding to relaxation or diffusion up to 200K.

Above 200K, Figure 15 shows that the doublets broaden and become asymmetric. The lineshapes are no longer Lorentzian and the quadrupole splitting drops by about 40% within 50K. In order to explain these features, a relaxation model describing the hopping of the iron atom between the twelve possible positions was proposed. The electric field gradients were assumed to be axial in the inner and outer rings. The main components of the electric field gradient tensor for both rings were kept constant and equal to those measured below 200K. The z direction was kept the same for nearest neighbor inner and outer positions; the canting angle with the hexagonal plane was β. The iron ion jump rate from the inner to the neighboring outer position is γ_2, whereas the jump rate from the outer to the inner position is γ_3 (Figure 14). These jumps lead only to a modulation of the quadrupole splitting. In addition, a jumping process of rate, γ_1, between neighboring inner positions also occurs. This jump results in a change of the z direction of the principal axis of the electric field gradient by 60° , in the azimuthal angle and a change from β to $-\beta$ in the polar angle. A fast motion in the inner ring will reduce the principal component of the electric field gradient from V_{zz} to $[V_{zz}(3\cos^2\beta - 1)]/2$. An analytical expression describing this model can be written and, by fitting the jumping rates γ_1, γ_2 , and γ_3 the distances r_1 and r_2 and the angle α (Figure 14), the solid curves of Figure 15 result Above 200K, the hopping between inner and outer positions in $FeMo_6S_8$ becomes fast ($\gamma_2 \cong \gamma_3 > 10^9s^{-1}$) and their quadrupole splittings are averaged. Simultaneously, a quasielastic Mössbauer absorption line related to the mean time of residence of the iron atom on one site between jumps, appears. The linewidth of this quasielastic absorption depends upon the jump geometry. The motion within the inner ring is much slower with a rate of 2 to 5 $x10^7s^{-1}$ between 220 and 295K. This is reflected in the drop of the quadrupole splitting. The polar angle, β, is estimated equal to 50° . For r_1, the model gives 1.0(1) Å, while from X-ray data, 1.09 Å is observed for $Fe_{1.32}Mo_6S_8$. With r_2 equal to 2.2 Åα is 50° . X-ray data for $Fe_{1.32}Mo_6S_8$ give an r_2 value of 2.3 Å, and an α value of 61° . The asymmetry of

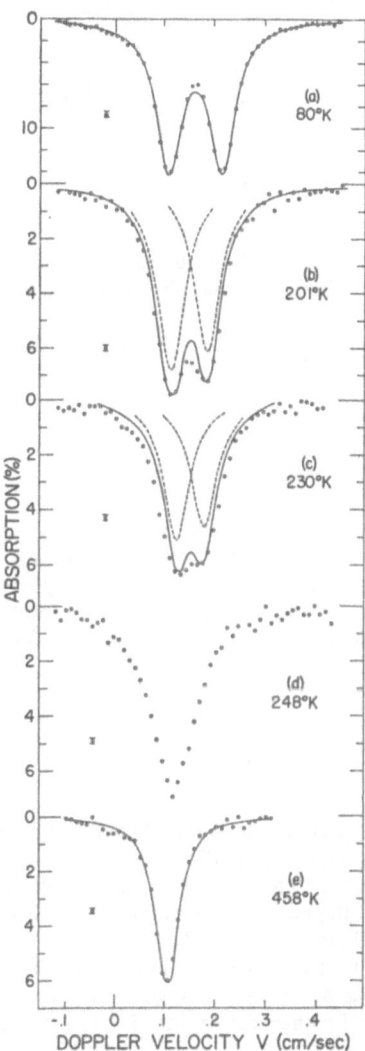

Figure 13. Mössbauer spectra obtained with a source of cobalt-57 in $AgCl$ at different temperatures. The solid lines are the result of fits with Lorentzian lines. Reproduced with permission from ref.(10).

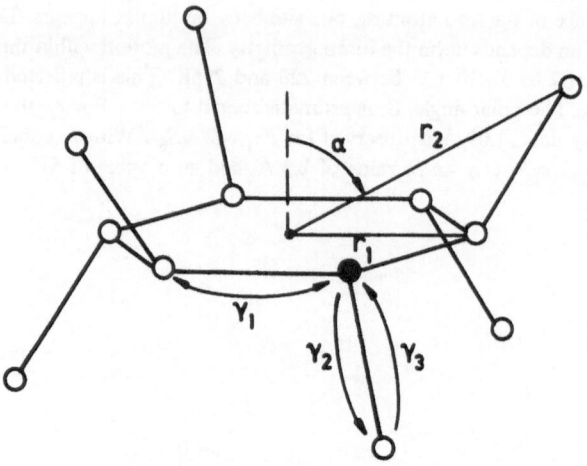

Figure 14. Six inner and six outer sites of the interstitial cluster in $FeMo_6S_8$ populated statistically by one Fe. Reproduced with permission from ref (25).

the doublet is mainly influenced by r_2 and α. In spite of severe assumptions, the model is successful in explaining the spectra over a limited range of temperatures

4.6. An Example with Iodine-129

Mössbauer spectra of iodine-129 impurities, resulting from the decay of implanted tellurium-129m in silicon (27), reveal one component with a highly temperature dependent quadrupole interaction and a strong reduction in the very anisotropic recoil free fraction with increasing temperature. This component was assigned to substitutional iodine atoms in silicon At low temperatures these substitutional iodine atoms are not located exactly at the center of a tetrahedron of silicon atoms because of a trigonal Jahn-Teller distortion. This displacement is responsible for the quadrupole interaction of 452 MHz observed at 4.2K. As the temperature increases, the iodine atom may jump between four equivalent sites around the substitutional site of the undistorted cell. This jump, first, reduces the recoil free fraction as observed and, second, averages to zero the quarupole interaction as observed above 40K

Gorobchenko and Litterst (28) propose a model to describe the line profile due to the jump of the iodine atoms. There is no analytical solution to the problem, but the dimension of the numerical expressions is sufficiently low to allow least-squares fitting of experimental data The computational technique is based on the matrix [12] and the use of Liouville operators. In principle, the dimension of this matrix is 192 for $I_1 = 5/2$, $I_0 = 7/2$, n = 4, for iodine-129 The appropriate reduction of the dimension of this matrix leads to four 12x12 blocks which may be numerically inverted. The computed profiles are shown in Figure 16, for various jump rates given in mm/s, ($1mm/s = 140\ 7x10^6 s^{-1}$). Of course, the results may be extended to several other isotopes with the 5/2-7/2 Mossbauer transition

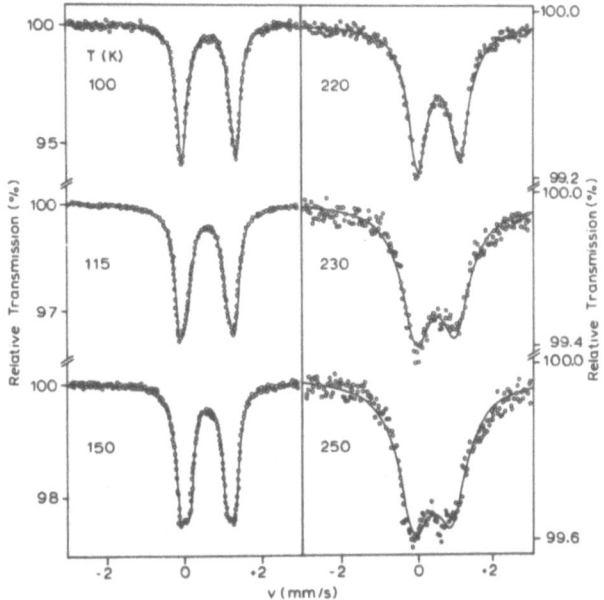

Figure 15. Iron-57 Mössbauer spectra of $FeMo_6S_8$. The solid lines are the results of fits described in the text. Reproduced with permission from ref.(25).

5. *Conclusions*

In this chapter, we have illustrated how by Mossbauer spectroscopy, different relaxation processes, such as electron hopping, diffusion of atoms, high- to low-spin relaxation, motion of the Mossbauer or non-Mossbauer atoms, can be observed. Other examples, such as rotation of ligands may be found in the literature.(29)

Acknowledgments

I am very grateful to Profs. S. Morup and G. J. Long for many fruitful discussions and suggestions and to NATO for a Cooperative Research Grant Number 86/0685.

306

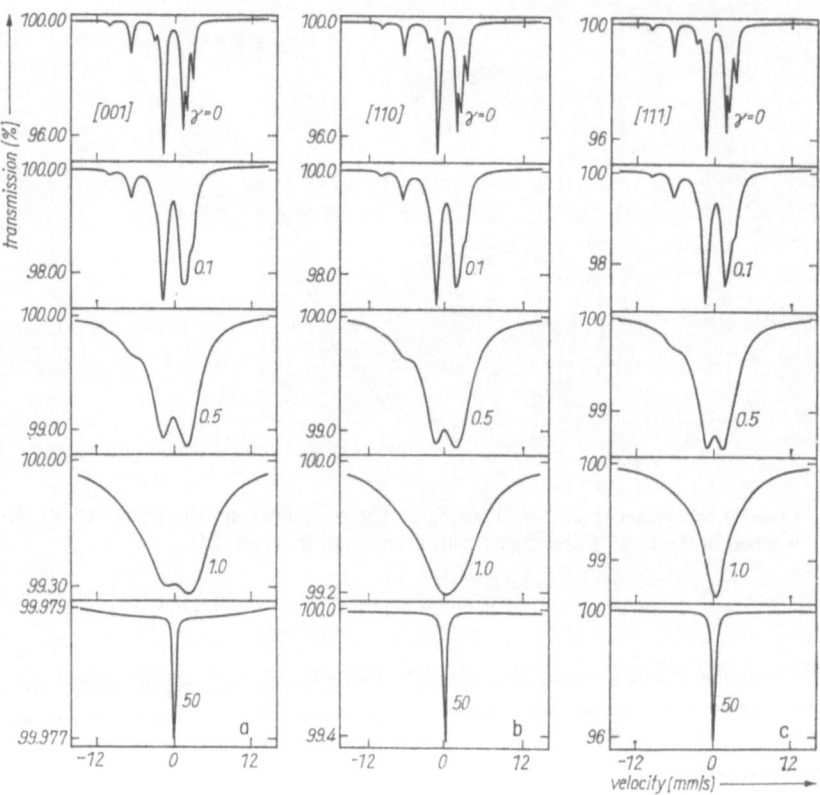

Figure 16. Iodine-129 Mossbauer spectra calculated for various jump rates γ. The gamma-ray direction is along the (a) [001], (b) [110], and (c) [111] axis of *Si*. Reproduced with permission from ref (28)

Appendice A. Liouville operators

With each quantum mechanical operator, A, may be associated the Liouville operator, A^x, which acts on another quantum mechanical operator B, so that $A^x B$ gives the commutator of Λ with B,

$$A^x B = AB - BA = [A,B], \qquad [A1]$$

The operator A^x acts on an ordinary operator B as an ordinary operator acts on a state vector. Therefore, the matrix elements of $A^x B$ may be written as a linear combination of the matrix elements of B

$$< u|A^x B|v > \; = \sum_{u'v'} < uv|A^x|u'v' > \, \cdot \, < u'|B|v' > \qquad [A2]$$

where the elements of A^x are labeled by four indices and those of B by two. The elements of A^x can be expressed in terms of the matrix elements of the operator Λ by using equation $[A1]$

$$< u|(A^x B)|v > \; = \; < u|(AB - BA)|v > \qquad [A3]$$

$$= \sum_{u'} < u|A|u' > \, < u'|B|v > \, - \sum_{v'} < u|B|v' > \, < v'|A|u >$$

Comparing $[A3]$ and $[A2]$ reveals that

$$< uv|A^x|u'v' > \; = \delta_{vv'} < u|A|u' > \, - \delta_{uu'} < v'|A|v > \qquad [A4]$$

The relation $[A4]$ defines the elements of A^x in terms of those of A.

The principal property of the Liouville operator is the relation

$$e^A B e^{-A} = \exp(A^x)B \qquad [A5]$$

The physical significance of the Liouville operator for the Hamiltonian, H^x, is shown by its eigenvalues and eigenoperators. If we have, $H|u > \; = E_u|u >$, $H|v > \; = E_v|v >$, then the transition operators $|u > \, < v|$ are seen to be the eigenoperators of H^x

$$H^x|u > \, < v| = H|u > \, < v - |u > \, < v|H = (E_u - E_v)|u > \, < v|$$

The eigenvalues of the Liouville operator H^x are thus, the differences $E_u - E_v$ of all the energy levels of the Hamiltonian. These differences are physically observable quantities, because they represent the possible spectral lines emitted by the system. It is therefore not surprising that the operator H^x is found useful in describing spectral lineshapes.

References

1. S. Mørup, this volume, p. 271, 309, and 321.

2. S. Dattagupta, "Mössbauer Spectroscopy", Ed. D.P.E. Dickson, F.G. Berry, Cambridge Univ. Press (1986) p. 198.

3. E.R. Bauminger and I. Nowik, "Mössbauer Spectroscopy", Ed. D.P.E. Dickson, F.G. Berry, Cambridge Univ. Press (1986), p.219.

4. F.J. Litterst and G. Amthauer, Phys. Chem. Minerals 10, 250-255 (1984)

5. M. Prietsch, G. Wortmann, G. Kaindl and R. Schogl, Phys. Rev. B 33, 7451-7461 (1986)

6. I. Felner and I. Nowik, Solid St. Comm. 28, 67-70 (1978)

7. R. Nagarajan, S, Patil, L.C. Gupta and R. Vijayaraghavan, J. Magn. Magn. Mat. 54-57, 349-350 (1986)

8. Y. Maeda, N. Tsutsumi and Y. Takashima, Inorg. Chem, 23, 2440-2447 (1984)

9. Y. Maeda, H. Oshio, Y. Takashima, M. Mikuriya and II. Idaka Inorg. Chem. 25, 2958 (1986)

10. D.H. Lindley and P.G. Debrunner, Phys. Rev. 146, 199 (1966)

11. H.R. Leider and D.N. Pipkorn, Bull. Am. Phys. Soc. 11, 49 (1966), Phys. Rev. 165, 494 (1968)

12. M. Blume and J. A. Tjon, Phys. Rev. 165, 446 (1968)

13. J. A. Tjon and M. Blume, Phys. Rev. 165, 456 (1968)

14. M. Blume, Phys. Rev. 174, 351 (1968)

15. M.J. Clauser, Phys. Rev. B 3, 3748-3753 (1971)

16. G.K. Shenoy, B.D. Dunlap, Phys. Rev. B 13, 1353, (1976)

17. F. Gonzalez-Jimenez, P. Imbert, F. Hartmann-Boutron Phys. Rev. B 9, 95 (1974)

18. P. Bonville, C. Garcin, A. Gerard, P. Imbert and G. Jchanno, Phys. Rev. B 23, 4293 (1981)

19. E. De Grave, Sol. St. Comm. 60, 541 (1986)

20. E. R. Bauminger, D. Froindlich, I. Nowik, S. Ofer, I. Felner and I. Mayer, Phys. Rev. Lett. 30, 1053 (1973)

21. E. R. Bauminger, I. Felner, D. Levron, I. Nowik and S. Ofer, Phys. Rev. Lett. 33, 890 (1974)

22. E. R. Bauminger, I. Felner and S. Ofer, J. Magn. Magn. Mat. 7, 317-325 (1978)

23. I. Felner and I. Nowik, Solid St. Comm. 39, 61-63 (1981)

24. P. Gütlich, Chapter 11 in "Mössbauer Spectroscopy Applied to Inorganic Chemistry" vol 1, Ed. G. J. Long, Plenum Press, New York (1984)

25. F. J. Litterst, Hyperfine Inter. 27-28 and 29, 123-134 (1986)

26. J. M. Friedt, C. W. Kimball, A. T. Aldred, B. D. Dunlap, F. Y. Fradin and G. K. Shenoy, Phys. Rev. B 29, 3863 (1984)

27. G.J. Kemerink. J.C. de Wit, H. De Waard, D.O. Boerma and L. Niesen, Phys. Lett. A 82, 255 (1981)

28. V.D. Gorobchenko and F.J. Litterst, phys. stat. sol. (b) 127, 351 (1985)

29. L. Asch, G.K. Shenoy, J.M. Friedt, J.P. Adloff, J. Chem. Phys. 62, 2335 (1975).

CHAPTER 17

MÖSSBAUER SPECTROSCOPY STUDIES OF RELAXATION PHENOMENA IN ULTRAFINE PARTICLES

Steen Mørup
Laboratory of Applied Physics II
Technical University of Denmark
DK-2800 Lyngby, Denmark

1.   INTRODUCTION

The magnetic properties of ultrafine particles are significantly different from those of bulk materials. The most important feature is superparamagnetic relaxation, i.e. spontaneous fluctuations of the magnetization direction. These relaxation phenomena and their influence on the Mössbauer spectra will be discussed in this chapter. Several more comprehensive reviews of Mössbauer studies of magnetic microcrystals have recently been published /1-4/.

2.   MAGNETIC PROPERTIES OF MICROCRYSTALS

The magnetic hyperfine splitting in Mössbauer spectra of magnetic microcrystals may depend on the particle size. Particles with dimensions larger than 50-100 nm essentially exhibit the bulk behavior. However, when the dimensions are below 20-50 nm, the crystallites are single-domain particles. The demagnetizing field will then contribute to the total magnetic field at the nucleus which is given by:

$$\vec{B}_{obs} = \vec{B}_h + \vec{B}_L + \vec{B}_D \tag{1}$$

Here $\vec{B}_h$ is the field due to the electrons surrounding the nucleus,

$$\vec{B}_L = \frac{1}{3} \mu_o \vec{M} \tag{2}$$

is the Lorentz field, and

$$\vec{B}_D = - \mu_o N \vec{M} \tag{3}$$

is the demagnetizing field. Here $\vec{M}$ is the magnetization, and N is the demagnetization coefficient which depends on the particle shape and the direction of the magnetization vector. Thus the magnetic splitting in

309

G. J. Long and F. Grandjean (eds.), The Time Domain in Surface and Structural Dynamics, 309–320.
© 1988 by Kluwer Academic Publishers.

Mössbauer spectra of single-domain particles is different from that of bulk materials. In antiferromagnetic crystals the net magnetization is zero, and therefore $\vec{B}_L = \vec{B}_D = 0$.

3. SUPERPARAMAGNETIC RELAXATION

The magnetic energy of a particle is in general proportional to its volume, and for very small particles the magnetic energy may therefore be comparable to the thermal energy, even well below room temperature. Under these circumstances the magnetization vector is not fixed along one of the easy directions as in large crystals, but fluctuates. The effect of such fluctuations on Mössbauer spectra depends on the frequency compared to the time scale of Mössbauer spectroscopy (which is approximately given by the Larmor precession time τ_L of the nuclear magnetic moment in the magnetic hyperfine field).

In the absence of an applied magnetic field the magnetization vector may fluctuate among the easy directions of magnetization with a relaxation time τ. This process is the so-called superparamagnetic relaxation. Roughly speaking, one observes for $\tau \gg \tau_L$ a magnetically split Mössbauer spectrum (with six lines in the case of ^{57}Fe Mössbauer spectroscopy). For $\tau \ll \tau_L$, a paramagnetic spectrum (with one or two lines) is found. In the intermediate range ($\tau \approx \tau_L$) complex spectra with broadened lines can be observed. In ^{57}Fe Mössbauer spectroscopy, τ_L is of the order of 10^{-8}–10^{-9} s.

For a particle with uniaxial anisotropy energy, the magnetic energy may be expressed by:

$$E(\theta) = KV \sin^2\theta \tag{4}$$

where K is the magnetic anisotropy energy constant, V is the volume, and θ is the angle between the easy direction and the magnetization vector. The superparamagnetic relaxation time is for $KV \lesssim kT$ given by /1,3,5-8/:

$$\tau \cong \frac{M\pi^{\frac{1}{2}}}{K\gamma_o} \left(\frac{KV}{kT}\right)^{-\frac{1}{2}} \exp\left\{\frac{KV}{kT}\right\} \tag{5}$$

where M is the magnetization, γ_o is the gyromagnetic ratio, k is Boltzmann's constant, and T is the temperature.

For microcrystals with other forms of magnetic anisotropy energy (e.g. cubic) the superparamagnetic relaxation is given by a similar expression, and we may in general use the expression

$$\tau = \tau_o \exp\left\{\frac{K_e V}{kT}\right\} \tag{6}$$

where K_e is related to the magnetic anisotropy energy constants.

The superparamagnetic blocking temperature T_B is defined as the temperature below which the superparamagnetic relaxation is slow compared to the time scale of the experimental technique used for the study of the magnetic properties of the small particles, i.e. about 10^{-8}–10^{-9} s for ^{57}Fe Mössbauer spectroscopy. It is seen from Eq. (5) that τ depends critically on the magnitude of the energy barrier, KV, separating the two easy directions. Therefore, Mössbauer studies in the vicinity of T_B give information about this parameter. In practice, a sample of small particles always exhibits a particle size distribution. At a given temperature some particles may therefore be below T_B, whereas others may be above T_B.

A study of the temperature dependence of the spectra thus gives information on the particle size distribution /1,3,9/ or, strictly speaking, on the distribution in KV because K may itself be a function of particle size.

Figure 1 shows spectra of microcrystalline α-Fe$_2$O$_3$ in a tropical red soil. The figure illustrates the typical superparamagnetic behavior of magnetic microcrystals /10/. At 80 K essentially only a six-line component is present, indicating that $T_B > 80$ K for all the particles. At higher temperatures a paramagnetic component appears, and the relative area of this component increases with increasing temperature. This

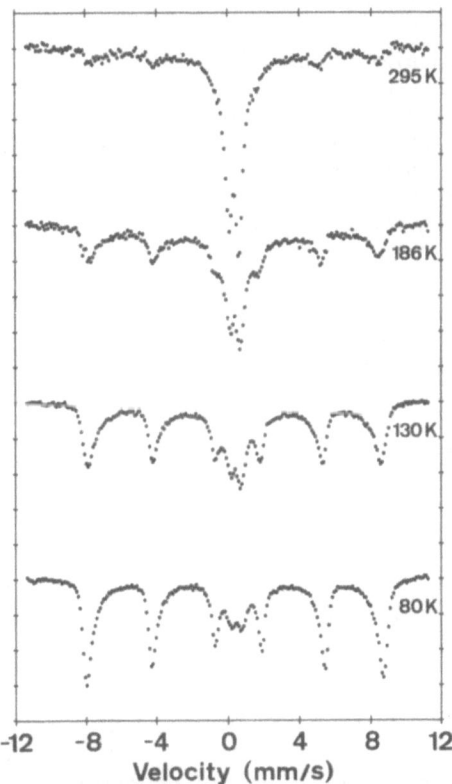

Figure 1.   Mössbauer spectra of microcrystalline α-Fe$_2$O$_3$ in a tropical red soil at various temperatures.

illustrates the distribution in superparamagnetic blocking temperatures which is related to the particle size distribution.

4. MAGNETIC HYPERFINE SPLITTING IN MICROCRYSTALS BELOW THE BLOCKING TEMPERATURE

Below the superparamagnetic blocking temperature the Mössbauer spectra of microcrystals are magnetically split. However, although the probability that the magnetization vector will surmount the energy barrier between the easy directions is negligible, the spectra may still be affected by fluctuations of the magnetization vector.

For a microcrystal with magnetic energy given by Eq. (4), the probability that the magnetization vector forms an angle between θ and $\theta + d\theta$ with the easy direction is given by /1,11/:

$$p(\theta)d\theta = \frac{\exp\{-E(\theta)/kT\}\sin\theta d\theta}{\int_0^{\pi/2} \exp\{-E(\theta)/kT\}\sin\theta d\theta} \tag{7}$$

Thus the magnetization vector may fluctuate in directions close to an easy direction of magnetization, unless the parameter KV is infinitely large compared to the thermal energy. This kind of fluctuations (collective magnetic excitations), which are illustrated in Figure 2, are fast compared to the time scale of Mössbauer spectroscopy ($\approx 10^{-8}-10^{-9}$ s). Therefore, the observed magnetic hyperfine field is the average hyperfine field given by /1,3,11-13/:

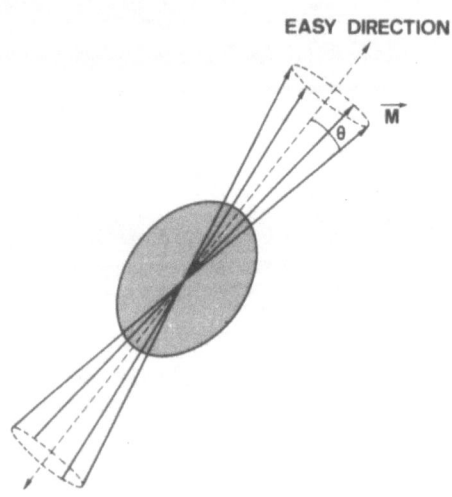

Figure 2. Schematic illustration of collective magnetic excitations in a microcrystal.

$$\vec{B}_{obs} = \vec{B}_o \langle\cos\theta\rangle \tag{8}$$

where $\vec{B}_o$ is the hyperfine field in the absence of fluctuations, and $\langle\cos\theta\rangle$ is given by:

$$\langle\cos\theta\rangle = \int_0^{\pi/2} \cos\theta\, p(\theta)\, d\theta \tag{9}$$

For a particle with magnetic energy given by Eq. (4) one finds when setting $KV/kT = \beta$ /11-13/:

$$\langle\cos\theta\rangle = \frac{1}{2}\beta^{\frac{1}{2}}\frac{\exp(\beta^{-1}) - 1}{\displaystyle\int_0^{\beta^{-\frac{1}{2}}} \exp(x^2)\, dx} \tag{10}$$

For $\beta \ll 1$, one may use the approximation /13/:

$$\langle\cos\theta\rangle \cong 1 - \frac{1}{2}\beta - \frac{1}{2}\beta^2 - \frac{5}{4}\beta^3 - \frac{37}{8}\beta^4 \tag{11}$$

The factor $M\pi^{\frac{1}{2}}/(K\gamma_0)$ in Eq. (5) is typically of the order of 10^{-9}-10^{-11} s. Therefore values of KV/kT below 5 to 10 result in relaxation times shorter than about 10^{-8} s. Consequently, the expression for the reduction in the magnetic hyperfine field due to collective magnetic excitations (Eq. (10)) cannot be applied for values of KV/kT below about 5 to 10 since this results in a superparamagnetic relaxation time which is so short that it significantly affects the spectra /13/.

The temperature dependence of the magnetic hyperfine field of microcrystals is schematically illustrated in Figure 3. The maximum reduction in B_{obs} due to collective magnetic excitations is of the order

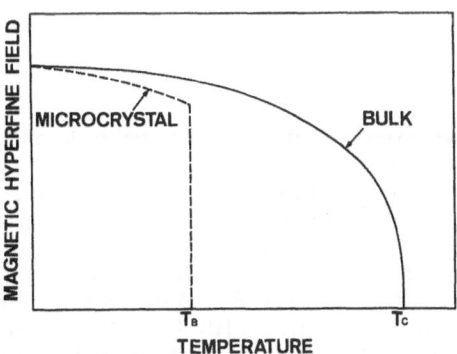

Figure 3. Schematic illustration of the temperature dependence of the magnetic hyperfine splitting in a microcrystal.

of 5-15%. Additional increase in the temperature results in a collapse of the magnetic hyperfine splitting because of fast superparamagnetic relaxation. According to Eqs. (10) and (11), a particle size distribution will result in broadened lines in the six-line component below T_B.

5. APPLIED MAGNETIC FIELDS

In the presence of an applied magnetic field, $\vec{B}$, the magnetic energy of a ferro- or ferrimagnetic particle is given by:

$$E = - \vec{\mu} \cdot \vec{B} + KV\sin^2\theta \tag{12}$$

where $\vec{\mu}$ is the magnetic moment of the particle. For a particle with dimensions of the order of 5-10 nm, μ is typically in the range 10^3-10^5 Bohr magnetons. Therefore, even in moderate external fields, the first term in Eq. (12) is often predominant, and the second term may then be neglected. The average magnetization of the particle is then given by /1,3,13/:

$$\langle M \rangle \cong M \, L \left\{ \frac{\mu B}{kT} \right\} \tag{13}$$

where $L\{\mu B/kT\}$ is the Langevin function.

If the superparamagnetic relaxation time is shorter than about 10^{-9} s, the observed magnetic hyperfine field in the Mössbauer spectrum is given by /1,3,13/:

$$\vec{B}_{obs} \cong \vec{B}_o L \left\{ \frac{\mu B}{kT} \right\} + \vec{B} \tag{14}$$

For $\mu B/kT \gtrsim 4$, we may use the high field expansion of the Langevin function and find:

$$\vec{B}_{obs} \cong \vec{B}_o (1 - \frac{kT}{\mu B}) + \vec{B} \tag{15}$$

Thus a plot of $|\vec{B}_{obs} - \vec{B}|$ as a function of $1/B$ gives a straight line with the slope $B_o kT/\mu$, and therefore $V = \mu/M$ can be determined.

Figure 4 shows spectra of small amorphous particles of $Fe_{1-x}C_x$ at 80 K and 161 K in various applied magnetic fields /14/. At $B = 0$ the alloy particles yield a paramagnetic spectrum. The shoulder at about 2 mms^{-1} is due to an Fe^{2+} impurity. When the magnetic field is applied, a substantial magnetic splitting of the alloy component appears. This shows that the particles are superparamagnetic. Figure 5 shows a plot of $|B_{obs} - B|$ as a function of $1/B$. The experimental points are fitted with straight lines in accordance with Eq. (15). From the slopes an average particle size of 4.2 nm was determined. This method for particle size determination has appeared to be very useful in several studies of ultrafine particles /1,3,10/.

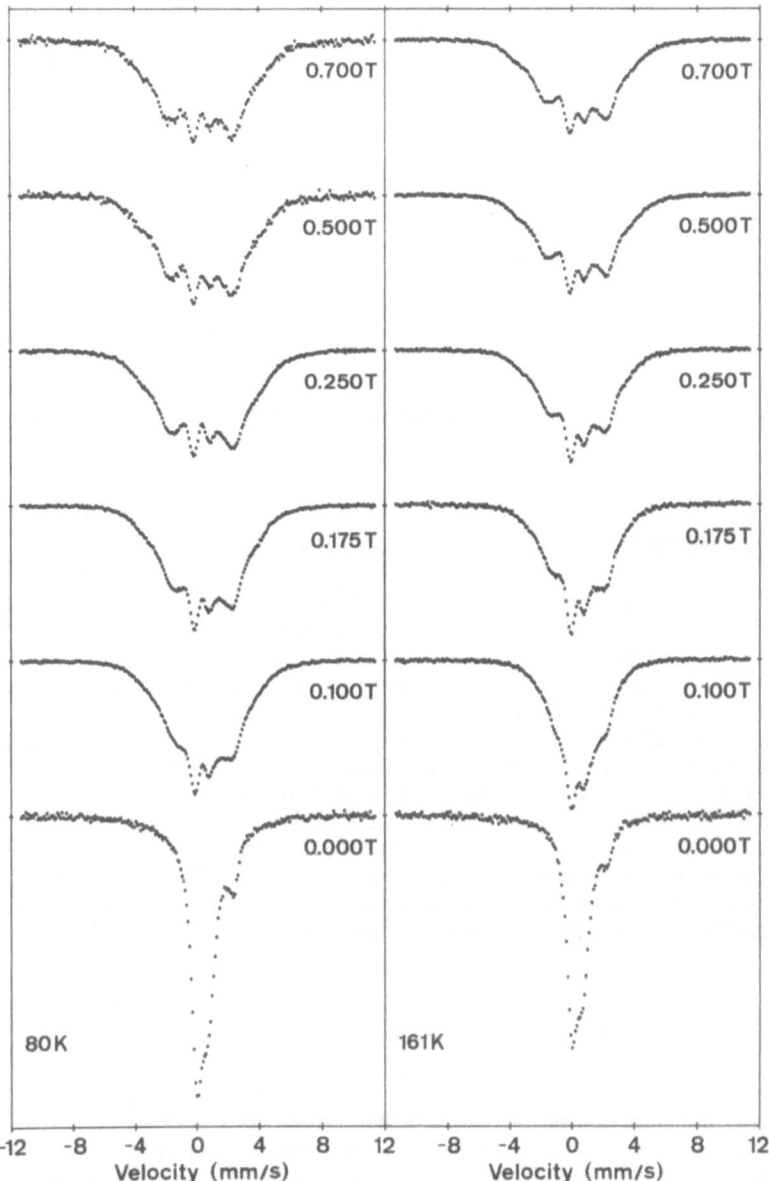

Figure 4. Mössbauer spectra of ultrafine particles of amorphous $Fe_{1-x}C_x$ in various applied magnetic fields at 80 K and 161 K.

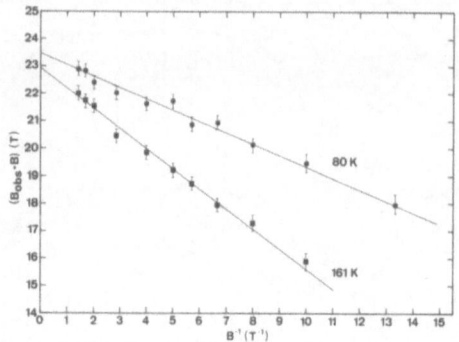

Figure 5.   Induced magnetic hyperfine splitting in the spectra shown
in Figure 4, plotted as a function of the reciprocal applied
magnetic field.

6.   MAGNETIC ANISOTROPY IN MICROCRYSTALS

Since the particle volume V can be determined from the magnetic field
dependence of B_{obs} and the parameter KV can be determined from the
influence of collective magnetic excitations or superparamagnetic relaxation on the Mössbauer spectra, the magnetic anisotropy energy constant
K can be calculated /1,3/.
   Studies of microcrystals of Fe_3O_4 /11,12/, α-Fe, β-Co, and Ni /3/
have shown that the magnetic anisotropy is significantly larger than
the magnetocrystalline anisotropy in macroscopic crystals and that it
changes when different molecules are chemisorbed on the surface of
the particles.   In the studies of Fe_3O_4 microcrystals it was also found
that the anisotropy energy constant increases with decreasing particle
size /12/.   These results suggest that the surface contribution to the
total magnetic anisotropy energy constant is predominant in microcrystals.

7.   MAGNETIC INTERACTIONS AMONG MICROCRYSTALS

In most of the work that has been published on Mössbauer studies of
magnetic microcrystals, the particles have been assumed to be isolated,
i.e. the magnetic interaction among neighboring particles has been neglected.   In practice, however, such magnetic interaction may be quite
important.
   Consider, for example, two α-Fe microcrystals with diameters of
5 nm and separated by 10 nm.   The magnetic interaction energy depends on the directions of the magnetic moments, but is of the order
of $5 \cdot 10^{-21}$ J which is similar to the magnetic anisotropy energy for K ≈
10^5 J m^{-3}, i.e. a typical value of the magnetic anisotropy energy constant for small α-Fe particles /3/.   The interaction energy is of the
same order of magnitude as the thermal energy at room temperature.

Therefore, the magnetic interaction energy has a significant influence on the magnetic properties of the system in this case.

In samples of closely packed magnetic microcrystals, another type of magnetic interaction may also be present. In such samples the exchange interaction between surface atoms belonging to two neighboring crystals may be significant, and this may lead to a large contribution to the magnetic interaction energy.

A model for the magnetic properties of interacting particles has recently been developed /13/. In this model it is assumed that the magnetic anisotropy is negligible compared to the magnetic interaction and that the particles would exhibit fast superparamagnetic relaxation if they were isolated.

The contribution to the magnetic energy of a particle i which arises from the magnetic interaction with the surrounding magnetic particles (j) may be written as:

$$E_i = \sum_j K_m^{ij} \vec{M}_i(T) \cdot \vec{M}_j(T) \tag{16}$$

K_m^{ij} is a magnetic coupling constant for the interaction between the particles i and j with magnetization vectors $\vec{M}_i(T)$ and $\vec{M}_j(T)$, respectively. The magnetic behavior of an assembly of interacting particles can be calculated using a modified Weiss molecular field theory. Thus in Eq. (16), $\vec{M}_j(T)$ is replaced by the average magnetization and is written

$$E_i = - K_m \vec{M}_i(T) \cdot \langle \vec{M}(T) \rangle \tag{17}$$

where

$$K_m = \sum_j K_m^{ij} \tag{18}$$

Below a certain temperature T_p, the magnetic interaction between the particles leads to ordering of the magnetic moments. This is the so called superferromagnetic state.

It is convenient to introduce the reduced average magnetization defined by:

$$b(T) \equiv \frac{\langle M(T) \rangle}{M(T)} \tag{19}$$

The transition temperature above which the system is superparamagnetic is given by:

$$T_p = \frac{K_m M^2(T_p)}{3k} \tag{20}$$

Below T_p, $b(T)$ can be derived from the implicit equation:

$$b(T) = L\left\{\frac{3T_p}{T} \, m^2(T, T_p) \, b(T)\right\} \tag{21}$$

where

$$m(T, T_p) = \frac{M(T)}{M(T_p)} \tag{22}$$

This model has been used to explain the magnetic behavior of samples of microcrystalline goethite (α-FeOOH) /15/. Figure 6 shows some typical Mössbauer spectra of a sample of microcrystalline goethite.

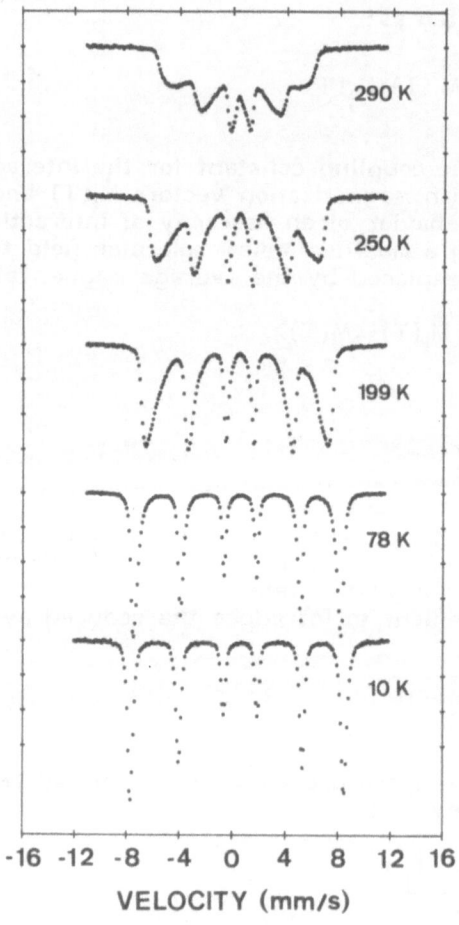

Figure 6. Mössbauer spectra of microcrystalline α-FeOOH at various temperatures.

These spectra are significantly different from those shown in Figure 1. For example, the lines are much more broadened, and there is no paramagnetic component below 290 K, although there is a significant reduction in the hyperfine field. Between 300 K and 320 K the six-line component rapidly collapses to a quadrupole doublet /15/.

The distributions in magnetic hyperfine fields were determined by use of a computer program developed by Wivel and Mørup /16/, and the temperature dependence of the average hyperfine field was determined. It was found that the average hyperfine field at high temperatures was less than 60% of the bulk value. The reduction in B_{obs} in microcrystals due to collective magnetic excitations is not expected to exceed 5-15% (see Section 4). Thus, the result is not in accordance with the theories for collective magnetic excitations and superparamagnetic relaxation in isolated microcrystals discussed in Sections 3 and 4.

Electron micrographs of the α-FeOOH samples showed that the crystallites are in very close contact /15/. Therefore it is likely that the results may be explained by magnetic interactions among the crystallites.

Figure 7 shows a plot of the normalized average hyperfine field, b(T). The solid line represents a fit of the experimental results with the function b(T) calculated from Eq. (21) with T_p = 316 K. The theoretical curve fits the experimental results remarkably well, and thus the model for superferromagnetism seems to describe the magnetic behavior quite well.

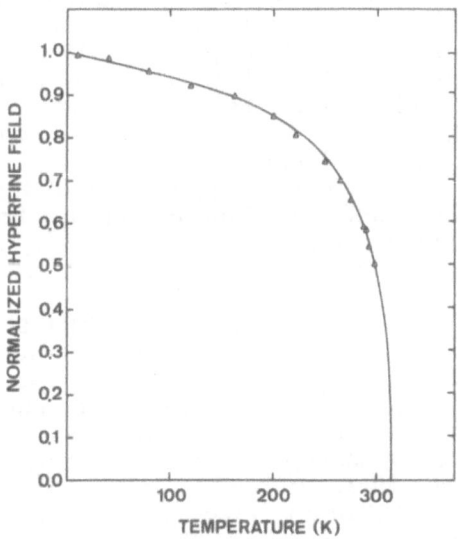

Figure 7. The experimental value of b(T) as a function of temperature for closely packed α-FeOOH microcrystals. The solid line represents fits of the results with the modified Weiss mean field theory discussed in the text.

Further evidence for the importance of the magnetic interaction was obtained in a Mössbauer study of the sample after heating at different temperatures /17/. These studies showed that the heating results in better resolved magnetic hyperfine splittings and larger average values of the magnetic hyperfine field below T_p. This suggests that the magnetic interaction is increased after heating as one would expect when water molecules between the crystallites desorb during heating. Similar results have been obtained by Tronc and Bonnin /18/ in a study of the effect of heating on the properties of ultrafine ferrite particles.

In another study, ultrafine goethite particles were grown in an aqueous solution containing SiO_2 /19/. In this case the Mössbauer spectra indicated the presence of non-interacting or weakly interacting microcrystals. Thus the SiO_2 separate the goethite particles and thus reduce the magnetic coupling among them.

REFERENCES

/1/ S. Mørup, J.A. Dumesic, and H. Topsøe, in: Applications of Mössbauer Spectroscopy, Vol. II, ed. R.L. Cohen (Academic Press, New York, 1980), p. 1.

/2/ J.L. Dormann, Rev. Phys. Appl. 16, 275 (1981).

/3/ S. Mørup, H. Topsøe, and B.S. Clausen, Phys. Scripta 25, 713 (1982).

/4/ A.H. Morrish and K. Haneda, J. Magn. Magn. Mat. 35, 105 (1983).

/5/ W.F. Brown, Jr., J. Appl. Phys. 30, Suppl. 130S (1959).

/6/ W.F. Brown, Jr., Appl. Phys. 34, 1319 (1963).

/7/ W.F. Brown, Jr., Phys. Rev. 130, 1677 (1963).

/8/ A. Aharoni, Phys. Rev. 135, A447 (1964).

/9/ W. Kündig, H. Bömmel, G. Constabaris, and R.H. Lindquist, Phys. Rev. 142, 327 (1966).

/10/ S. Mørup, in: Industrial Applications of the Mössbauer Effect, Ed. G.J. Long and J.G. Stevens (Plenum Press, N.Y., 1986) p.63.

/11/ S. Mørup and H. Topsøe, Appl. Phys. 11, 63 (1976).

/12/ S. Mørup, H. Topsøe, and J. Lipka, J. de Phys. Colloq. 37, C6-287 (1976).

/13/ S. Mørup, J. Magn. Magn. Mat. 37, 39 (1983).

/14/ S. Mørup, B.R. Christensen, J. van Wonterghem, S.W. Charles, and S. Wells, J. Magn. Magn. Mat. 67, 249 (1987).

/15/ S. Mørup, M.B. Madsen, J. Franck, J. Villadsen, and C.J.W. Koch, J. Magn. Magn. Mat. 40, 163 (1983).

/16/ C. Wivel and S. Mørup, J. Phys. E., Sci. Instrum. 14, 605 (1981).

/17/ C.J.W. Koch, M.B. Madsen, S. Mørup, G. Christiansen, L. Gerward, and J. Villadsen, Clays & Clay Minerals 34, 17 (1986).

/18/ E. Tronc and D. Bonnin, J. Physique Lett. 46, L-437 (1985).

/19/ C.J.W. Koch, M.B. Madsen, and S. Mørup, Hyperfine Interactions 28, 549 (1986).

CHAPTER 18

MÖSSBAUER SPECTROSCOPY STUDIES OF PARAMAGNETIC RELAXATION

Steen Mørup
Laboratory of Applied Physics II
Technical University of Denmark
DK-2800 Lyngby
Denmark

1. INTRODUCTION

Before discussing the influence of paramagnetic relaxation effects on Mössbauer spectra, it is appropriate to discuss the ionic eigenstates among which the relaxation takes place. These eigenstates, which in general are mixed nuclear and electronic states, are determined by the electric and magnetic interactions of the ion with the environments. These interactions are usually described by an effective Hamiltonian, the spin Hamiltonian.

2. THE SPIN HAMILTONIAN OF PARAMAGNETIC Fe^{3+} IONS

The spin Hamiltonian, $\hat{H}$, of an Fe^{3+} ion in the high spin state (S = 5/2) is composed of terms due to the crystal field interaction, $\hat{H}_{cf}$, the electronic Zeeman interaction, $\hat{H}_Z$, the magnetic dipole interaction between the Mössbauer ion (i) and the neighboring ions or ligands (j), $\hat{H}_{dd}$, the exchange interaction with neighboring ions, $\hat{H}_{ex}$, the magnetic hyperfine interaction, $\hat{H}_{hf}$, and the nuclear quadrupole interaction, $\hat{H}_Q$ [1,2]:

$$\hat{H} = \hat{H}_{cf} + \hat{H}_Z + \hat{H}_{dd} + \hat{H}_{ex} + \hat{H}_{hf} + \hat{H}_Q \tag{1}$$

where

$$\hat{H}_{cf} = D[\hat{S}_z^2 - \frac{1}{3} S(S+1) + \lambda(\hat{S}_x^2 - \hat{S}_y^2)] + \frac{1}{6} a(\hat{S}_\xi^4 + \hat{S}_\eta^4 + \hat{S}_\zeta^4 - \frac{707}{16}) \tag{2}$$

$$\hat{H}_Z = g\mu_B \, \vec{B} \cdot \hat{\vec{S}} \tag{3}$$

$$\hat{H}_{dd} = \sum_{j\neq i} \hat{H}_{ij}^{dd} = \sum_{j\neq i} g_i g_j \mu_B^2 \, r_{ij}^{-5} [\hat{\vec{S}}_j r_{ij}^2 - 3\vec{r}_{ij}(\hat{\vec{S}}_j \cdot \vec{r}_{ij})] \cdot \hat{\vec{S}}_i \tag{4}$$

G. J. Long and F. Grandjean (eds.), The Time Domain in Surface and Structural Dynamics, 321–333.
© 1988 by Kluwer Academic Publishers.

$$\hat{H}_{ex} = - \hat{\vec{S}}_i \cdot \sum_{j \neq i} 2 J_{ij} \hat{\vec{S}}_j \tag{5}$$

$$\hat{H}_{hf} = A \hat{\vec{S}} \cdot \hat{\vec{I}} \tag{6}$$

$$\hat{H}_Q = \frac{e^2 qQ}{4I(2I-1)} [\hat{I}_z^2 - \frac{1}{3} I(I+1) + \frac{1}{2} \eta (\hat{I}_x^2 - \hat{I}_y^2)] \tag{7}$$

Often the magnetic hyperfine coupling constant A is expressed in terms of the saturation hyperfine field B_0 for the $S_z = 5/2$ state. It is convenient to define the parameters Δ_{cf} and Δ_Z as the total energy splittings due to $\hat{H}_{cf}$ and $\hat{H}_Z$, respectively.

$\hat{H}_{cf}$ gives rise to a splitting of the ionic ground state into three Kramers doublets. The eigenstates of the electronic system depend critically on small perturbations that lift the degeneracy. For $\hat{H}_Z = \hat{H}_{dd} = \hat{H}_{ex} = 0$, the hyperfine interaction alone lifts the degeneracy. The Mössbauer spectrum is then very sensitive to the values of the parameters λ and $\mu = a/D$. The magnetic hyperfine interaction gives rise to a splitting of the order of 10^{-8} eV. Therefore, applied magnetic fields of the order of only 0.001 T result in a similar Zeeman splitting of the ionic states. Such fields therefore have a significant influence on the spectrum and are in fact difficult to avoid. For instance, the dipolar fields from the ligand nuclei can easily be of this order of magnitude, and the dipolar fields due to the neighboring paramagnetic ions may be considerably larger.

If a field of about 0.01 T is applied, the Zeeman interaction is large compared to the hyperfine interaction. Then the electronic eigenstates are essentially determined by $\hat{H}_{cf}$ and $\hat{H}_Z$, and the Mössbauer spectrum consists of three Zeeman components each described by an effective magnetic field. When the electronic Zeeman splitting is comparable to the crystal field splitting, the electronic eigenstates, and therefore also the size and direction of the hyperfine fields, depend on the relative size of Δ_{cf} and Δ_Z. Finally, for $\Delta_Z \gg \Delta_{cf}$, the electronic eigenstates are essentially determined by $\hat{H}_Z$. In this simple case, the Mössbauer spectrum can essentially be described in terms of six Zeeman split spectra with magnetic hyperfine fields in the ratio 5:3:1 arising from the electronic states $S_z = \pm 5/2, \pm 3/2, \pm 1/2$, respectively.

An investigation of the spectrum as a function of the applied field allows a determination of the parameters of the spin Hamiltonian /3,4/.

3. PARAMAGNETIC HYPERFINE SPLIT MÖSSBAUER SPECTRA, AN EXAMPLE

A detailed study of the ferric hexaquo complex $[Fe(H_2O)_6]^{3+}$, in frozen aqueous solutions has been performed by Knudsen /3,4/. The samples were prepared with different glass formers (e.g. glycerol or different nitrates) in order to avoid crystallization during freezing.

Typical spectra obtained at 4.5 K of an absorber containing 0.03 M $Fe(NO_3)_3$, 0.5 M HNO_3, and 5.4 M $LiNO_3$ as a glass forming agent

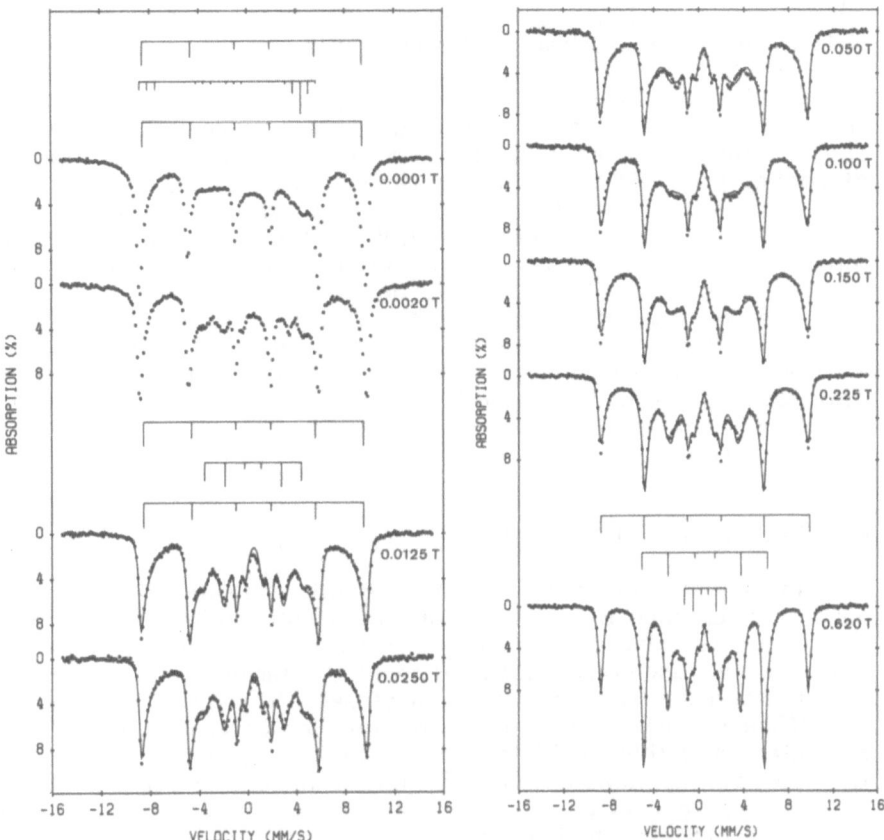

Figure 1. Mössbauer spectra of an absorber of a frozen aqueous solution with 0.03 M Fe(NO₃)₃, 0.5 M HNO₃, 5.4 M LiNO₃ obtained at 4.5 K with various applied transverse magnetic
fields. The full curves show theoretical fits. The bar diagram above the 0.0001 T and the 0.0125 T spectra show an
interpretation in terms of spectra related to the three Kra-
mer doublets. The bar diagram above the 0.620 T spectrum shows an interpretation in terms of three magnetically
split spectra with hyperfine fields in the ratio 5:3:1. The
line intensities are indicated by the lengths of the bars.

are shown in Figure 1. At B = 0.0001 T the spectrum essentially consists of an asymmetric component superimposed to a six-line component.
The spectra are drastically changed when the applied magnetic field is
increased. For B $>$ 0.01 T, the spectra are symmetric, and the spectra
obtained at 0.0125 T and 0.0250 T are essentially identical, but at larger applied fields the spectra again change when the field is increased.

Qualitatively, the results are in accordance with a crystal field
Hamiltonian with the parameters D $\cong$ 0.10 cm^{-1}, λ $\cong$ 0.26 cm^{-1}, and

$a \cong 0.017$ cm^{-1}. The spectrum obtained with B = 0.0001 T can be explained by two magnetically anisotropic Kramers doublets and an isotropic doublet /3,4/. The bar diagrams above the spectrum indicate theoretical line positions. For $A \ll g\mu_B B \ll \Delta_{cf}$, the spectrum is stabilized and is not expected to depend on the external field. The situation is achieved for applied magnetic fields of the order of 0.01-0.03 T. At applied magnetic fields in this range the spectrum consists of the three six-line components indicated by the bar diagram shown above the 0.0125 T spectrum. For B > 0.05 T, the electronic Zeeman interaction becomes comparable to Δ_{cf}, and therefore the spectrum again changes with the applied magnetic field.

The spectrum obtained with B = 0.62 T indicates that Δ_Z can be considered much larger than Δ_{cf}, and this spectrum essentially consists of three six-line components with hyperfine fields in the ratio 5:3:1 (indicated by the bar diagrams).

Similar studies have been made of a number of other Fe^{3+} complexes in frozen solutions /5/.

4.   SPIN-SPIN RELAXATION

In paramagnetic materials the magnetic relaxation frequency is in general determined by contributions from both spin-spin relaxation processes and spin-lattice relaxation processes. The spin-spin relaxation frequency is essentially independent of the temperature whereas the spin-lattice relaxation frequency increases with increasing temperature. At sufficiently low temperatures the spin-lattice relaxation can often be neglected, and this facilitates studies of spin-spin relaxation. Often the spin-lattice relaxation of Fe^{3+} ions is negligible even at 80 K.

Bloembergen et al. /6/ have calculated the probability per unit time for a process which results in a transition of an ion i from the state $|M_a^i\rangle$ to the state $|M_{a+\alpha}^i\rangle$ accompanied by a transition of another ion j from the state $|M_b^j\rangle$ to $|M_{b+\beta}^j\rangle$. Such a so-called cross-relaxation process results in an increase in the crystal field and Zeeman energy of the ion i by an amount E_α^i and a decrease in energy of the ion j by an amount E_β^j. It is assumed that the energy difference $E_\alpha^i - E_\beta^j$ is taken up by the magnetic dipolar energy of the whole spin system /6/. If the line shape functions of the two transitions α and β are Gaussian with second-order moments $(\Delta E_\alpha^i)^2$ and $(\Delta E_\beta^j)^2$, respectively, the expression for the transition probability can be written:

$$\tau_{ss}^{-1} = \frac{(2\pi)^{-\frac{1}{2}}}{\hbar} \frac{|\langle M_{a+\alpha}^i, M_{b+\beta}^j | H_{dd}^{ij} | M_a^i, M_b^j\rangle|^2}{[(\Delta E_\alpha^i)^2 + (\Delta E_\beta^j)^2]^{\frac{1}{2}}}$$

$$\times \exp\left[-\frac{(E_\alpha^i - E_\beta^j)}{2[(\Delta E_\alpha^i)^2 + (\Delta E_\beta^j)^2]}\right] p_j(M_b^j) \tag{8}$$

Here H_{dd}^{ij} is the Hamiltonian describing the magnetic dipolar interaction between the two ions, and $p_j(M_b^j)$ is the probability for the ion j to be in the initial state $|M_b^j\rangle$.

Fast relaxation via the cross-relaxation mechanism may occur when the two neighboring ions i and j have similar energy splittings. Cross-relaxation processes have a much lower probability if the two ions have significantly different energy levels. A schematic illustration of a cross-relaxation process is given in Figure 2.

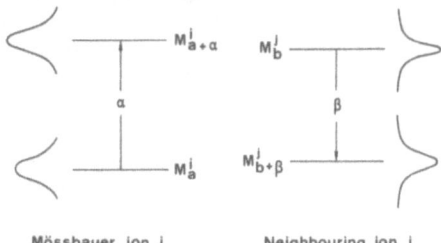

Mössbauer ion i   Neighbouring ion j

Figure 2.   Schematic illustration of a cross-relaxation process by which the ion i performs a transition α, accompanied by a transition β of a neighboring ion, j.

A simple example /7/ illustrating cross-relaxation processes is given in Figure 3 which shows spectra at 80 K of amorphous frozen aqueout solutions with identical concentrations of iron. Spectrum (a) was obtained with an absorber containing iron only in $Fe(H_2O)_6^{3+}$ complexes whereas (b) is a spectrum of an absorber with 50% $Fe(H_2O)_6^{3+}$ complexes and 50% $Fe(H_2O)_5Cl^{2+}$ complexes. The spectrum (b) shows clear satellite lines at about ± 9 mm/s, indicating a relatively long relaxation time, whereas the satellite lines in the spectrum (a) are broader and have smaller intensity, indicating a shorter relaxation time. Thus these two spectra show that the spin-spin relaxation time, τ_{ss}, is relatively short in the sample with only one type of complexes, but τ_{ss} is significantly longer in the sample with two different types of Fe^{3+} complexes. This result is at least qualitatively in accordance with Eq. (8).

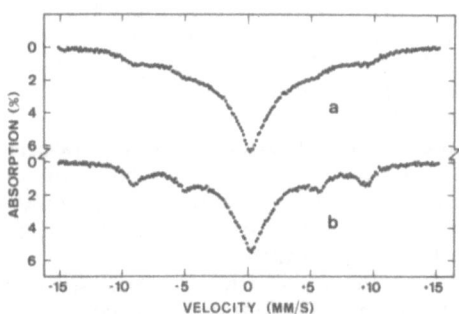

Figure 3.   Mössbauer spectra of frozen aqueous solution absorbers at 80 K. (a) is the spectrum of an absorber containing $Fe(H_2O)_6^{3+}$ complexes. (b) is a spectrum of an absorber containing about 50% $Fe(H_2O)_6^{3+}$ complexes and about 50% $Fe(H_2O)_5Cl^{2+}$ complexes.

326

The energy splittings of Fe^{3+} in amorphous samples at B = 0 may be quite similar for all the ions, but may differ significantly at B > 0, because the energy splitting depends on the angle between the crystal field axes and $\vec{B}$. Therefore, τ_{ss} is expected to increase with increasing magnetic fields. As an example of this effect, Figure 4 shows spectra of an amorphous frozen aqueous solution of $Fe(NO_3)_3$ at B = 0 and B = 1.4 T /8/. The 1.4 T spectrum exhibits clear satellite lines at about ± 9 mm/s, but these lines are not present in the spectrum obtained at B = 0. This difference is a clear indication of an increase in τ_{ss} with increasing applied magnetic field.

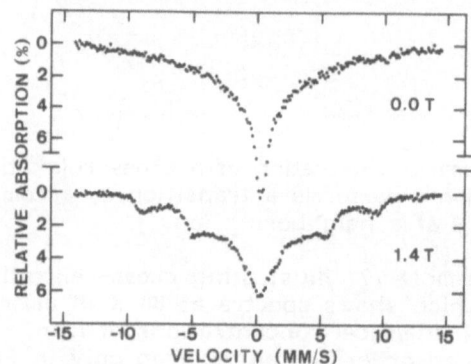

Figure 4. Mössbauer spectra of an amorphous frozen aqueous solution of $Fe(NO_3)_3$ with a concentration of 0.043 moles $Fe(NO_3)_3$ per mole H_2O. The spectra were obtained at 80 K in zero applied field and in a magnetic field of 1.4 T.

Studies of single crystals /9/ of $Fe(NO_3)_3 \cdot 9H_2O$ have illustrated a related magnetic field dependence of τ_{ss}. In this compound the ferric ions are connected in pairs by a screw diad axis which is parallel to the [010] direction, and the directions of the crystal field axes of two such ions may thus be different (although the values of D, λ, and a are identical). Therefore, when a magnetic field is applied, the two ions will in general have different energy splittings unless the field is applied parallel or perpendicular to the [010] direction /9/. Consequently, τ_{ss} should be shorter for these special field directions than for other field directions. Figure 5 shows spectra of single crystals of $Fe(NO_3)_3 \cdot 9H_2O$ at 78 K with an applied field of 1.3 T parallel to the [010] direction and the [11$\bar{2}$] direction, respectively. As expected, the relaxation time is seen to be significantly shorter when the field is parallel to the [010] direction.

The effect of magnetic dilution on the spin-spin relaxation time has also been studied by Mössbauer spectroscopy. Figure 6 shows Mössbauer spectra obtained at 80 K of frozen aqueous solutions of $Fe(NO_3)_3$ in different concentrations. The amount of nitric acid was adjusted to $[Fe(NO_3)_3] + [HNO_3] \approx 7$ M in order to ensure formation of homogeneous glasses. The numbers on the left side of the figure are the approximate relaxation times estimated from the spectra. These measurements

Figure 5.
Mössbauer spectra of single
crystals of $Fe(NO_3)_3 \cdot 9H_2O$
at 80 K with an applied magnetic field of 1.3 T parallel
to the [010] direction (a) and
parallel to the [11$\bar{2}$] direction
(b).

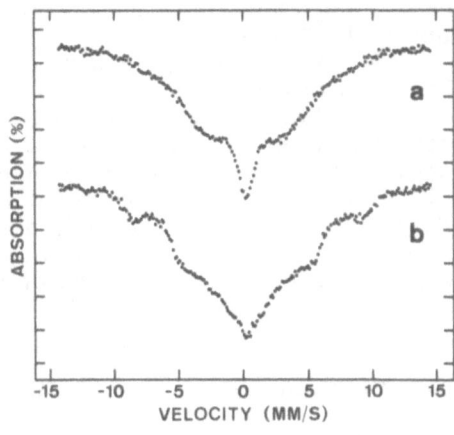

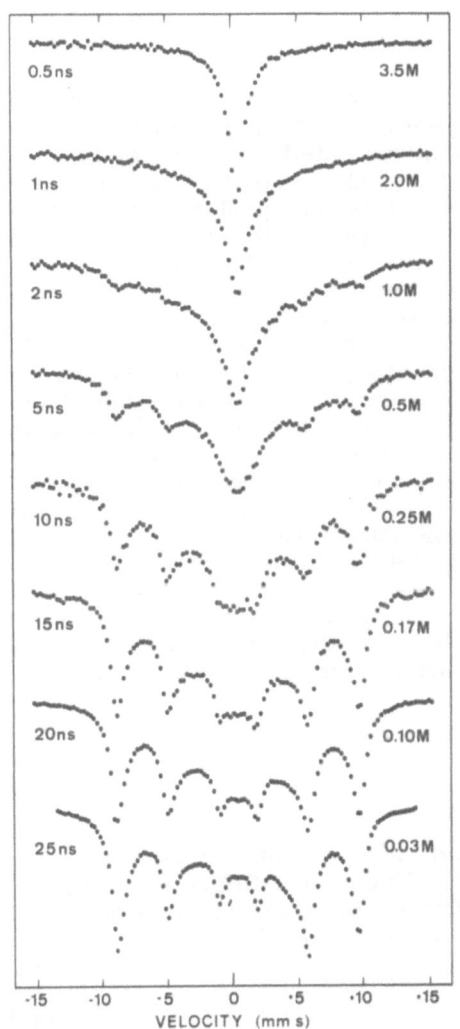

Figure 6.
Mössbauer spectra of some aqueous
$Fe(NO_3)_3$ + HNO_3 glasses. The
iron concentration is indicated.
An estimate of the relaxation time
is given.

clearly illustrate that the spectra are very sensitive to the concentration of Fe^{3+} ions.

An interesting application of the concentration dependence of τ_{ss} has been demonstrated in studies of frozen aqueous solutions /10/. When a solution of a salt is cooled rapidly, it may solidify as a homogeneous glass. However, slower cooling may result in formation of ice crystals. Since the solubility of iron in ice is very low, the iron ions will in the latter case be found in amorphous regions with a relatively high iron concentration, and this results in a short spin-spin relaxation time. It has been shown that this can be used for studies of, for example, the influence of the freezing rate on crystallization and glass formation /10/.

5. SPIN-LATTICE RELAXATION

Spin-lattice relaxation is induced by modulations of the crystal field due to thermal oscillations of the ligands /1,2/. Spin-lattice relaxation of paramagnetic ions is normally investigated by means of EPR. This method is mainly applicable for long relaxation times $\tau \gtrsim 10^{-6}$ s, i.e. at low temperatures. Since ^{57}Fe Mössbauer spectroscopy is sensitive to relaxation effects for relaxation times in the range of 10^{-10} s $< \tau < 10^{-7}$ s, studies of spin-lattice relaxation can be extended to higher temperatures and, as shown in this section, new types of spin-lattice relaxation processes may therefore be observed.

Basically, the spin-lattice processes can be classified as one-phonon and two-phonon processes. In one-phonon (or direct) processes, a spin transition is accompanied by the creation of annihilation of a single phonon, the phonon energy being exchanged with the spin transition energy Δ. For a Debye solid this leads to a relaxation rate $\tau_{s\ell}^{-1} \propto \Delta^2 T$ for $kT \gg \Delta$. If the Zeeman splitting is large compared to the crystal field splitting, $\tau_{s\ell}^{-1} \propto B^2 T$. Direct processes with acoustic phonons are important only at very low temperatures.

At higher temperatures the two-phonon (Raman) processes become the most important ones. In such a process one phonon is annihilated and another phonon is created. The energy difference is taken up in a transition of the electronic spin. In the Debye approximation for the phonon spectrum the corresponding temperature dependence has the form (for $kT \gg \Delta$):

$$\tau_{s\ell}^{-1} \propto \int_0^{\omega_D} \frac{\omega^6 \exp(\hbar\omega/kT) \, d\omega}{\{\exp(\hbar\omega/kT)-1\}^2} \tag{9}$$

which gives $\tau_{s\ell}^{-1} \propto T^7$ for $T \ll 0_D$ and $\tau_s^{-1} \propto T^2$ for $T \gtrsim 0_D$. If $\Delta_Z \gg \Delta_{cf}$, there is no dependence on B. The simple Debye model may hold for acoustic phonons. For optical phonons the Einstein model is more appropriate. This leads to the expression

$$\tau_{s\ell}^{-1} \propto \frac{\exp(-0_E/T)}{\{1-\exp(-0_E/T)\}^2} \tag{10}$$

where Θ_E is the Einstein temperature. For $T \ll \Theta_E$, we find $\tau_{s\ell}^{-1} \propto$ exp$(-\Theta_E/T)$, and for $T > \Theta_E$, we find that $\tau_{s\ell}^{-1} \propto T^2$. For $\Delta_Z \gg \Delta_{cf}$, no dependence on B is expected.

Finally, an Orbach process is defined as a two-phonon relaxation process which occurs via population of an excited electronic state of energy E_0. The temperature dependence is given by

$$\tau_{s\ell}^{-1} \propto \frac{1}{\exp(E_0/kT)-1} \tag{11}$$

The Orbach process is essentially insensitive to applied magnetic fields. This process is not expected to be important for Fe^{3+} because of the lack of appropriate excited electronic states.

If a large magnetic field is applied ($\Delta_Z \gg \Delta_{cf}$), the relative transition probabilities for $\Delta M = \pm 1$ spin-lattice processes are 10:4:0 for the $\pm 5/2 \rightleftarrows \pm 3/2$, the $\pm 3/2 \rightleftarrows \pm 1/2$, and the $\pm 1/2 \rightleftarrows \mp 1/2$ transitions. For $\Delta M = \pm 2$, the relative transition probabilities for the $\pm 5/2 \rightleftarrows \pm 1/2$ and the $\pm 3/2 \rightleftarrows \mp 1/2$ transitions are 5:9 /11/.

Detailed studies of spin-lattice relaxation can be carried out only if $\tau_{ss} \gg \tau_{s\ell}$. Therefore, such studies should be carried out with magnetically dilute samples. As an example /11/, Mössbauer spectra of $NH_4(Fe_{0.02}Al_{0.98})(SO_4)_2 \cdot 12H_2O$ obtained with B = 1.23 T are shown in Figure 7. The temperature dependence of $\tau_{s\ell}^{-1}$ estimated from the spectra could be fitted neither with the Debye model for direct processes nor with the Debye model for Raman processes /11/. A fit based on the Einstein model for Raman processes with $\Theta_E = 450$ K (which corresponds to the frequency of the internal vibrational modes in the $Fe(H_2O)_6^{2+}$ complex) is shown by the broken curve in Figure 8. In the interval 125-250 K the experimental results, however, deviate significantly from the fit. Mössbauer spectra obtained in the presence of various applied fields at 157 K show that $\tau_{s\ell}^{-1}$ decreases with increasing magnetic field (Figure 9), and this result is also in disagreement with the theory for Raman processes. Therefore, a new type of spin-lattice relaxation processes must be present.

$NH_4Al(SO_4)_2 \cdot 12H_2O$ and other alums undergo phase transitions at temperatures $T_c \approx 80$ K /12/. EPR studies of Cr^{3+}-doped alums /13/ have shown that a so-called central mode exists in the phonon spectrum above T_c with a spectral density given by:

$$D(\omega) \propto \frac{\tau_c}{1 + \omega^2\tau_c^2} \tag{12}$$

where the correlation time τ_c can be represented by:

$$\tau_c = \tau_0\exp(T_0/T) \tag{13}$$

EPR studies of $NH_4Al(Fe)(SO_4)_2 \cdot 12H_2O$ /14/ indicate similar critical fluctuations.

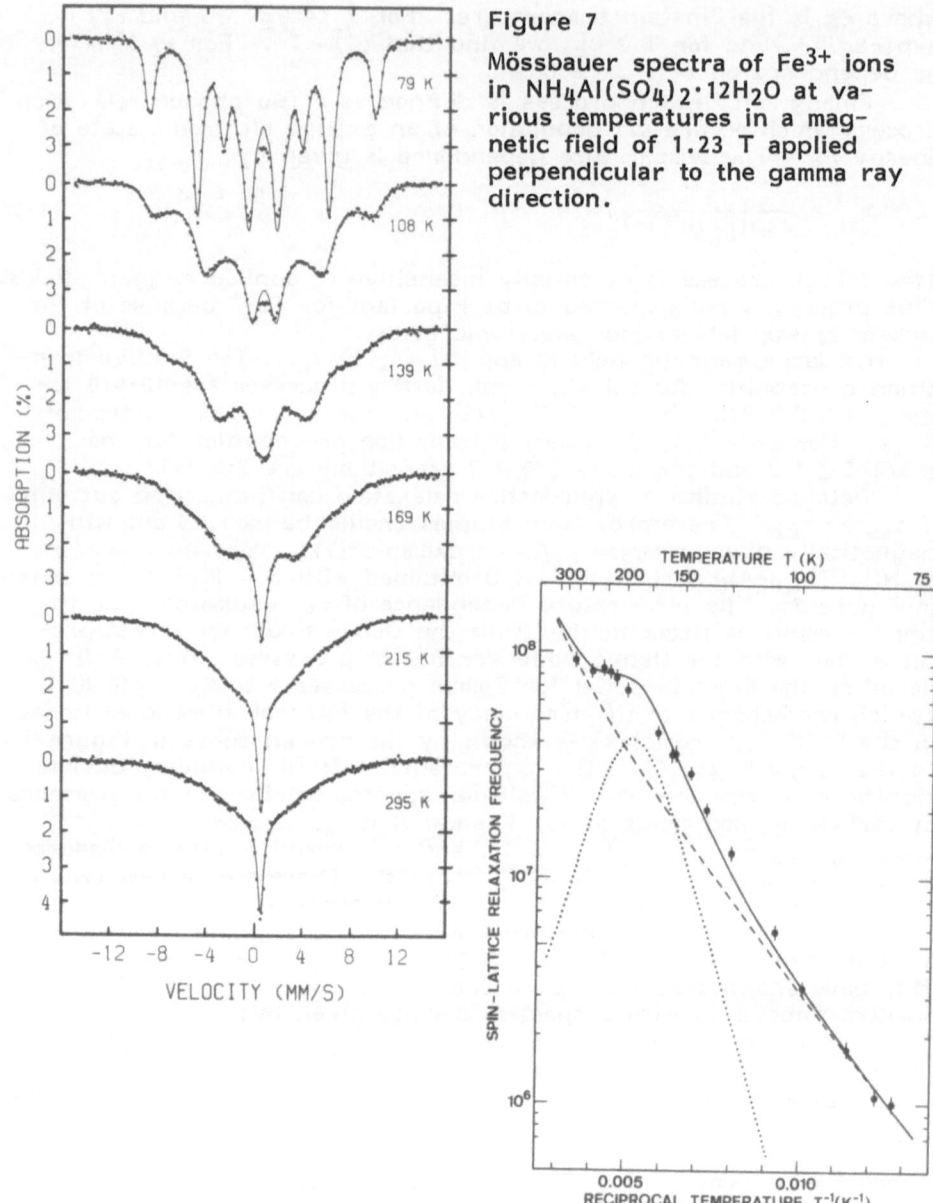

Figure 7

Mössbauer spectra of Fe^{3+} ions in $NH_4Al(SO_4)_2 \cdot 12H_2O$ at various temperatures in a magnetic field of 1.23 T applied perpendicular to the gamma ray direction.

Figure 8. Spin-lattice relaxation frequency of Fe^{3+} ions in $NH_4Al(SO_4)_2 \cdot 12H_2O$ at B = 1.23 T as a function of the reciprocal temperature. The broken curve shows a fit calculated from the Einstein model for Raman processes. The dotted curve corresponds to the contribution from the critical spin-lattice relaxation process. The full curve is the total spin-lattice relaxation rate.

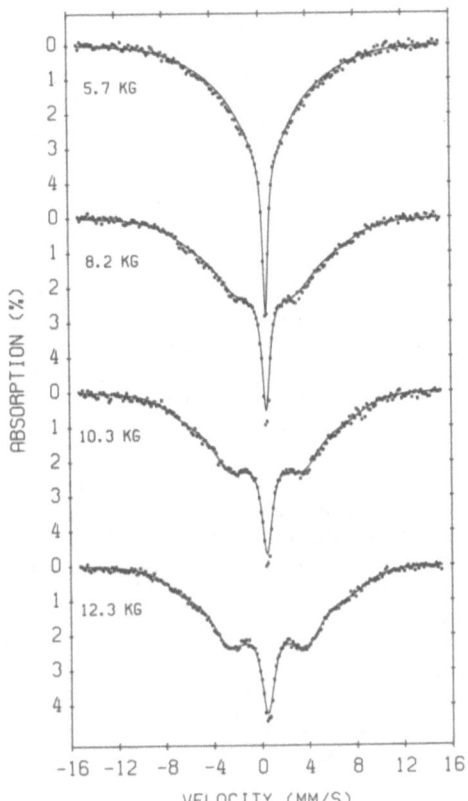

Figure 9.   Mössbauer spectra of Fe^{3+} ions in $NH_4Al(SO_4)_2 \cdot 12H_2O$
obtained with various applied magnetic fields at 157 K.

The probability that the central mode gives rise to direct processes is proportional to the spectral density $D(\omega)$ at the frequency
$\omega = \Delta/\hbar$.   In the present case $\Delta_{cf} \ll \Delta_Z$.   Therefore $\Delta \cong g\mu_B B$.   The
transition rate for a direct process with $\Delta M = 1$ can therefore be ex-
pressed by:

$$\tau_{s\ell}^{-1} \propto \frac{\tau_c}{1 + (\tau_c g\mu_B B/\hbar)^2} \tag{14}$$

A similar expression can be derived for $\Delta M = 2$ processes.
The dotted curve in Figure 8 shows the calculated contribution
from these "critical spin-lattice relaxation processes" to $\tau_{s\ell}^{-1}$ whereas
the full curve shows the total spin-lattice relaxation rate, including
the Raman processes with internal vibrations and the contributions
from the critical spin-lattice relaxation processes.   The addition of
the second contribution improves the fitting considerably.   Moreover,

332

as seen from Eq. (14), the critical spin-lattice relaxation time does, in contrast to the Raman processes, depend on the applied magnetic field. Mössbauer spectra obtained in various applied magnetic fields show in fact a magnetic field dependence of $\tau_{s\ell}$ (Figure 9).

Thus the results can be explained by the presence of two different spin-lattice relaxation processes, namely first-order Raman processes involving an internal vibrational mode of the $Fe(H_2O)_6^{3+}$ complex and critical spin-lattice relaxation processes associated with the phase transition in the alum.

Spin-lattice relaxation has also been studied in a number of other ferric compounds. Suzdalev et al. /15/ studied spin-lattice relaxation in ion-exchange resins. They found that the temperature dependence of $\tau_{s\ell}$ was in better agreement with an Einstein model than with a Debye model. Therefore the internal vibrations seem to be responsible for the spin-lattice relaxation in this system.

In dilute amorphous frozen aqueous solutions of $Fe(NO_3)_3$, $\tau_{s\ell}$ is approximately proportional to T^{-2} between 80 K and the glass transition temperature $T_g \cong 160$ K /16/. This is in accordance with the high-temperature approximation for Raman processes. Thus the lattice modes that are responsible for the relaxation seem to have a characteristic temperature $\Theta \lesssim 160$ K. This result suggests that internal vibrations in the $[Fe(H_2O)_6]^{3+}$ complex, which are characterized by an Einstein temperature of 400-500 K, are unimportant for the relaxation. The results rather suggest that acoustic phonons are responsible for the relaxation. Above T_g, $\tau_{s\ell}$ decreases more rapidly with increasing temperature than below T_g. This is illustrated in Figure 10. Thus a new type of spin-lattice relaxation seems to be present in the supercooled liquid. Recent NMR studies of molecules in solutions /17,18/ have shown that in addition to the translational diffusion, a restricted rotational diffusion may be present. The frequency of these restricted rotations may be of an order of magnitude which can give rise to direct spin-lattice relaxation processes. Therefore, the enhanced spin-lattice relaxation above T_g may be explained by the presence of a new type of spin-lattice processes which are direct processes induced by restricted rotational diffusion of the $Fe(H_2O)_6^{3+}$ complex /16/.

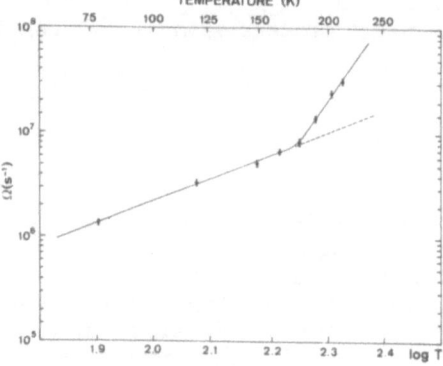

Figure 10.   The spin-lattice relaxation frequency as a function of temperature for a sample of Fe^{3+} in a frozen aqueous solution.

REFERENCES

/1/ G.E. Pake: "Paramagnetic Resonance" (W.A. Benjamin, New York), 1962.
/2/ A. Abragam and B. Bleaney: "Electron Paramagnetic Resonance of Transition Ions" (Clarendon Press, Oxford), 1970.
/3/ J.E. Knudsen: J. Physique Colloq. 37, C6-735 (1976).
/4/ J.E. Knudsen: J. Phys. Chem. Solids 38, 883 (1977).
/5/ S. Mørup and J.E. Knudsen: Acta Chim. Hungarica 121, 147 (1986).
/6/ N. Bloembergen, S. Shapiro, P.S. Pershan, and J.O. Artman, Phys. Rev. 114, 445 (1959).
/7/ S. Mørup and J.E. Knudsen: Chem. Phys. Lett. 40, 292 (1976).
/8/ S. Mørup and N. Thrane: Chem. Phys. Lett. 21, 363 (1973).
/9/ S. Mørup: J. Phys. Chem. Solids 35, 1159 (1974).
/10/ S. Mørup, J.E. Knudsen, M.K. Nielsen, and G. Trumpy: J. Chem. Phys. 65, 536 (1976).
/11/ S.C. Bhargara, J.E. Knudsen, and S. Mørup: J. Phys. C 12, 2879 (1979).
/12/ F. Jona and G. Shirane: "Ferroelectric Crystals" (Pergamon Press, Oxford), 1962.
/13/ R. Chicault and R. Buisson: J. Physique 38, 795 (1977).
/14/ S. Maekawa: J. Phys. Soc. Japan 16, 2337 (1961).
/15/ I.P. Suzdalev, A.M. Afanas'ev, A.S. Plachinda, V.I. Goldanskii, and E.F. Makarov: Zh. Eksp. Fiz. 55, 1752 (1968) (Sov. Phys. JETP 28, 923 (1969)).
/16/ S. Mørup and J.E. Knudsen: J. Phys. C 18, 2943 (1985).
/17/ R.E. London and J. Avitable: J. Am. Chem. Soc. 100, 7159 (1978).
/18/ R.J. Wittebort and A. Szabo: J. Chem. Phys. 69, 1722 (1978).

CHAPTER 19

COMPETITION BETWEEN DELOCALIZATION AND CRISTALLIZATION FOR ELECTRONS
IN THE LOW-TEMPERATURE MODIFICATION OF THE MIXED-VALENCE IRON
OXYPHOSPHATE βFe$_2$PO$_5$

B. Malaman, B. Ech-Chahed and C. Gleitzer
Laboratoire de Chimie du Solide Minéral, UA 158,
Université de Nancy I, B.P. 239
54506 Vandoeuvre-lès-Nancy Cédex
France

1. INTRODUCTION

We have previously prepared αFe$_2$PO$_5$ by the ceramic method, and established the detailed structure. Iron (II) and iron (III) are in face-sharing
octahedra with a rather short iron (II) to iron (III) distance of 2.93Å.
However, no fast electron exchange could be recorded with the Mössbauer
effect even at 600°C, due to a difference in energy between the two
sites [1].

We now have evidenced for a new modification, βFe$_2$PO$_5$, as a result
of soft chemistry techniques. It is metastable and transforms to αFe$_2$PO$_5$
above 600°C. It is isostructural with NiCrPO$_5$, which has been prepared
as a single crystal and shown to have a filled-Lipscombite structure
which is tetragonal with a=5.300Å, c=12.087Å, Z=4, and space group I4$_1$/
amd. In βFe$_2$PO$_5$, the cations are in continuous rows of face-sharing
octahedra connected by the PO$_4$ tetrahedra [2].

The following features, a single crystallographic site for iron,
and a remarkably short iron(II) to iron (III) distance of 2.66Å, are
highly favourable for a fast electron exchange.

2. PROPERTIES

The electrical conductivity is $\sim 10^2$ times lower than in magnetite at
room temperature, the activation energy of 0.32eV, for 303K < T < 625K,
is rather high; below 303K, the phonon spectrum freezes lowering the
activation energy. This implies that 303K $\sim \Theta_D/3$ and $\Theta_D \sim$900K ; indeed
the highest frequency infrared-active longitudinal optical mode,
ν=610cm^{-1}, yields $\Theta_D \sim$875K.

The Zener double exchange is dominated by the antiferromagnetic
coupling between iron ions belonging to different rows. The ferromagnetic direct coupling is weak inside a row. The antiferromagnetic coupling between rows is through an angle, Fe-O-Fe, of $\sim$80°, and hence is
strong. As a consequence T$_N$ is 408K, but, above T$_N$, some short-range order
remains and the inverse susceptibility versus temperature plot does not

335

G. J. Long and F. Grandjean (eds.), The Time Domain in Surface and Structural Dynamics, 335–337.
© 1988 by Kluwer Academic Publishers.

show Curie-Weiss behavior. Below T_N, weak ferromagnetism is apparent and may be due to imperfect order as shown also by the Mössbauer effect (see below), and by the conductivity measurements. There is no break at 408K in $\log\sigma = f(T^{-1})$ although the activation energy should decrease when crossing T_N during heating.

The Mössbauer spectrum at 483K is an asymmetric doublet (Figure 1) with linewidths of 0.35 and 0.59mm/s. This peculiarity may be due to the fast electron exchange, as the isomer shift is 0.61mm/s, and hence ~0.74mm/s at room temperature, a value typical for a mixed-valence $Fe^{2.5+}$. The two components have the same area; however it seems difficult to fit the spectrum with two doublets having the same low-energy peak and adjacent high-energy peaks.

At room temperature the spectrum is complex with a paramagnetic signal remaining near zero velocity. This may indicate a sluggish order-ing for iron (III), the spin relaxation of which is not limited by spin lattice interaction.

At 4.2K, the Mössbauer spectrum shows two sextets for iron (III) (H_1=530, H_2=510kOe) and one for iron (II)(H ~ 400kOe). This is not rela-ted to a lowering of the cell symmetry, according to powder X-ray pat-terns, and may be due to the presence of hydroxyl groups. However, the oxygen 1s signal in XPS has a normal breadth. Neither does the IR spec-trum indicate a significant absorption for hydroxyl groups at ~3400cm^{-1}.

3.   DISCUSSION

Because of the analogy with the octahedral sublattice in magnetite, we have applied the treatment of Ihle, et al.[3] for electron correlations and transfer in Fe_3O_4, and have derived the band diagram of Figure 2, where ν_1 and ν_2 are the effective Coulomb integrals between nearest and next-nearest neighbors. these integrals determine the short and long-range orders respectively. T_t is the transition temperature (~T_v in Fe_3O_4) correlated to ν_2, and J_p^t is the polaronic reduced overlap integral.

The non-reduced overlap integral depends only on the distance, because the total overlap contributions are the same in Fe_3O_4 and Fe_2PO_5, and can be calculated as J=0.24eV. However, the polaronic reduction is lower than in Fe_3O_4 because the electron-phonon coupling is weaker, hence $J_p > 0.035eV$, a value which should be compared with 0.028eV in Fe_3O_4.

More over, because we know the conductivity activation energy, we can calculate $\nu_1 = J_p + E_g$ (at T>T_t)>0.365eV at T>303K and >0.235eV at T<303K valueswhich should be compared with ν_1=0.18eV in Fe_3O_4.

Therefore, both delocalization and localization of electrons are stronger tendencies than in Fe_3O_4. According to the Mössbauer spectra, the transition should lie in between RT and 483K.

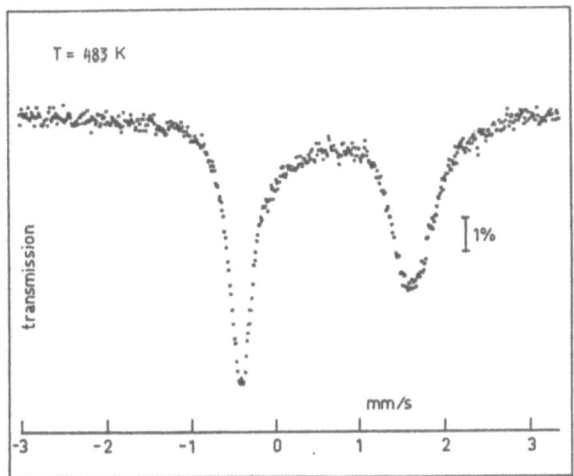

Figure 1.   Fe$_2$PO$_5$ Mössbauer spectrum at 210°C

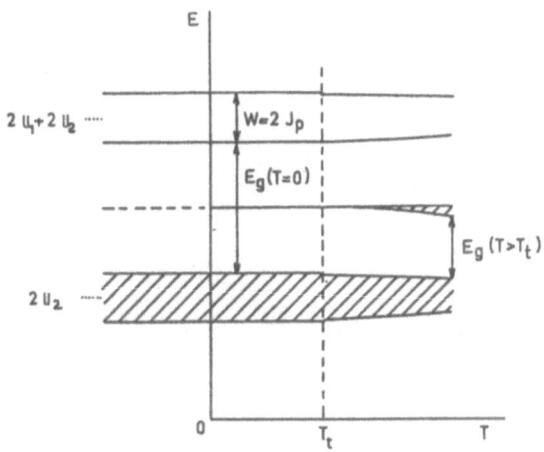

Figure 2.   Band scheme in Fe$_2$PO$_5$

REFERENCES

[1] R. IRALDI, G. LE CAER and C. GLEITZER, Solid State Comm. **40** (1981) 145.
[2] B. ECH-CHAHED, B. MALAMAN and C. GLEITZER, 3rd European Conf. Solid State Chem., Regensburg, 1986.
[3] D. IHLE and B. LORENZ, Phil. Mag. **42B** (1980) 337 ; D. IHLE, Phys. Stat. Sol. **121b** (1984) 217.

CHAPTER 20

ELECTRON TRANSFER IN $CaFe_3O_5$

R. Gérardin, E. Millon, J.F. Brice, O. Evrard
Laboratoire de Chimie du Solide Minéral, Associé au CNRS,
UA 158, Université de Nancy I, B.P.239,
54506 Vandoeuvre-Lès-Nancy Cedex, France
and
G. Le Caer
Laboratoire de Science et Génie des Matériaux Métalliques,
Associé au CNRS, UA 159, Ecole des Mines, 54042 Nancy Cedex,
France.

ABSTRACT. Calcium ferrite, $CaFe_3O_5$, is an oxide in which electron transfer occurs between the iron ions. This intervalence exchange process has been studied by iron-57 Mössbauer spectroscopy and by electrical conductivity measurements. Thermally-activated electron transfer is observed above the Néel temperature. Between 298 and 400K, the process follows the Arrhenius law with an activation energy of about 0.1eV.

1.   INTRODUCTION

Calcium ferrite, $CaFe_3O_5$, is orthorhombic, space group Cmcm, with a=3.021(1)Å, b=10.009(1)Å and c=12.643(1)Å, Z=4 [1]. The structure can be described as a stacking, along the c-axis, of blocks of $CaFe_2O_4$ and FeO respectively. In the former block (Figure 1), the coordination poly-hedron of iron (III) ions can be considered as a distorted oxygen octa-hedron, one Fe-O distance being slightly longer than the remaining five nearest neighbor distances. In the latter block, each iron(II) ion is located at the center of a strongly flattened octahedron. Each iron(II) octahedron shares an edge of length a with two iron(III) octahedra. Because the shortest iron(II)-iron(III) distance is only 3.01Å and because the octahedra are favourably connected, we expect to observe an electron transfer between the iron ions. This intervalence exchange process has been studied by Mössbauer spectroscopy and by electrical conductivity measurements.

2.   MOSSBAUER SPECTROSCOPY

$CaFe_3O_5$ is antiferromagnetic with a Néel temperature, T_N, of 283+2K. Below T_N, the iron-57 hyperfine fields are typical of high-spin confi-gurations for both iron(II) and iron(III), 262 and 504kG at 90K respec-tively [2]. Five spectra have been recorded in the paramagnetic state

339

G. J. Long and F. Grandjean (eds.), The Time Domain in Surface and Structural Dynamics, 339–344.
© *1988 by Kluwer Academic Publishers.*

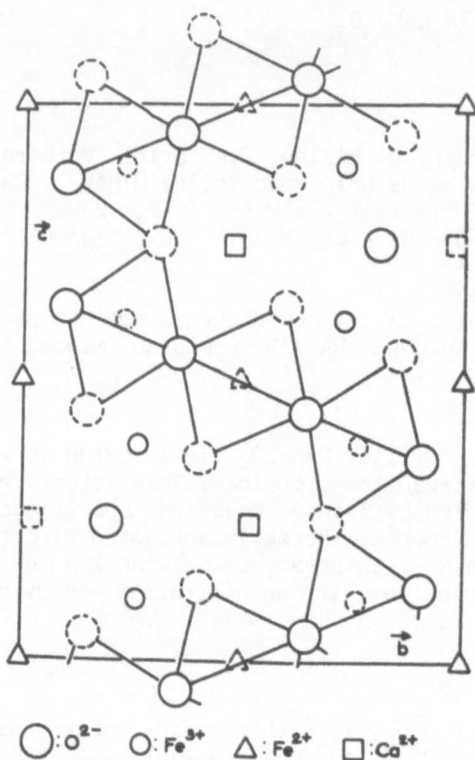

Figure 1. (100) projection of the primitive cell of CaFe$_3$O$_5$. Solid circles are for x=0 atoms while dotted circles are for x=1/2.

between 298 and 630K (Figure 2). The absorption increases in the center of the spectra, at ~0.5mm/s as the temperature increases. All peaks collapse at about 630K at which point only one broad peak is observed, with an isomer shift (with respect to room temperature α-Fe) of 0.32mm/s, a quadrupole splitting of 0.23mm/s and a full linewidth of 0.36mm/s.

The observed temperature evolution of the Mössbauer spectra is reversible and is therefore not due to oxidation of the sample. At 298K, the spectrum, when analyzed with two symmetric doublets, yields quadrupole splittings of 0.31 and 0.78mm/s for iron(III) and iron(II) respectively. The ratio of the area of iron(III) component to the area of iron(II) is 1.65 instead of the expected value of 2, while the ratio is 2.09+0.04 in the magnetic state between 90 and 267K [2]. Moreover, the mean quadrupole splitting, ΔFe^{2+}, calculated from the peak positions in the magnetic state [2] is about 1.10mm/s, except at 227K where it seems to be anomalously small (0.85mm/s, see Figure 3).

The decrease of both the area ratio R and ΔFe^{2+} (Figure 3) can be explained by a fast electron transfer, thermally activated, between an

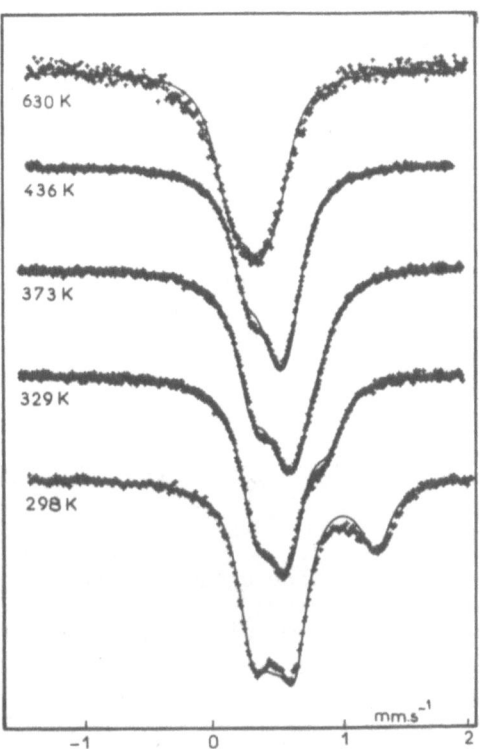

Figure 2. Mössbauer spectra of $CaFe_3O_5$ in the paramagnetic state. The solid lines are calculated with the Wickman model.

iron(II) and aniron(III) neighbor. In order to account for the lineshapes, the Wickman model [3] was used to analyze the spectra using the assumption of an electron jump limited to a trimer of iron(III)-iron(II)-iron(III). This model is similar to the one used by Dziobkowski et al. [4] to explain the Mössbauer spectra of mixed valence iron acetate $[Fe^{II}Fe_2^{III}O(CH_3CO_2)_6L_3]$ where L is H_2O or C_5H_5N. It includes two relaxation times: $\tau(iron(II)<->iron(III))$ and $\tau'(iron(III)<->iron(III))$, four frequencies $v_i(i=1,..,4)$, expressed here in velocity units with v_1 and v_2 being related to iron(III) and v_3 and v_4 to iron(II)(see Figure 4 of [4]). Relaxation involves the frequencies corresponding to the $+1/2 \to +1/2$ transitions v_1 and v_3, and v_1 and v_1 for the iron(II) $\leftarrow - \to$ iron(III) and iron(III) $\leftarrow - \to$ iron(III) processes. For the $+1/2 \to +3/2$ transitions, the processes involve v_2 and v_4, and v_2 and v_2, respectively. Identical transition frequencies for iron(III) $\leftarrow - \to$ iron(III) indicate that there is no actual relaxation between these ions. Dziobkowski et al. [4] arbitrarily set the relaxation time τ' to 200ns, which is not too much larger than the nuclear excited-state lifetime τ_N=99.7ns. They have

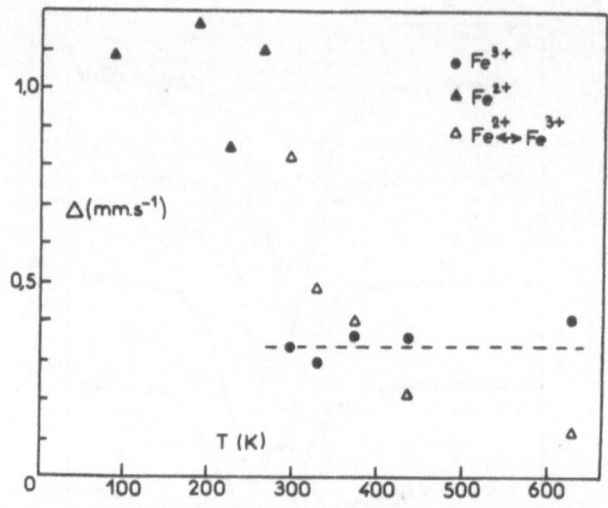

Figure 3. Temperature dependence of the quadrupole splittings Δ of iron (II) and iron (III) ions (only ΔFe^{2+} can be calculated below T_N[2]); iron(II) transforms into mixed valence iron at and above room temperature.

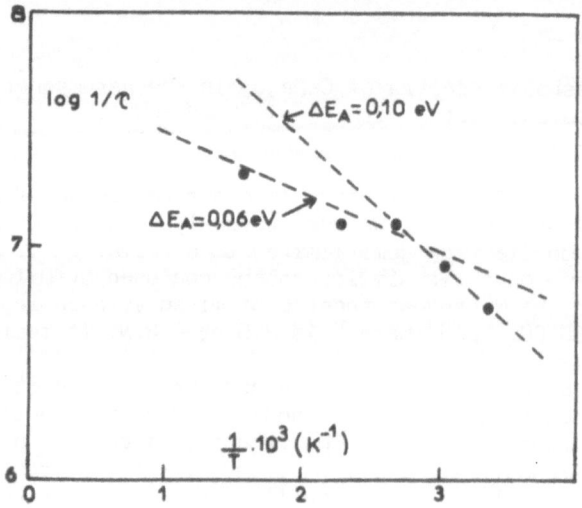

Figure 4. $Log_{10}(1/\tau)$ versus inverse temperature.

observed that arbitrarily longer relaxation times τ' have no obvious effects on the calculated spectra. We also observe no significant changes when τ' varies between 200ns and 2s. Linewidths must also be used in the Wickman model. They have been taken as 0.250 and 0.195mm/s for the 3^+-3^+ and 2^+-3^+ terms respectively. Spectral fits (Figure 2) give τ as 180ns at 298K and 80ns at 400K.

The iron(III) isomer shift decreases linearly with temperature, as explained by the second order Doppler shift. The iron(III) quadrupole splitting remains practically constant about 0.35mm/s while the iron(II) quadrupole splitting decreases very strongly (Figure 3). Figure 4 shows the logarithm of the relaxation frequency $1/\tau$ as a function of the inverse temperature. A linear variation is observed up to 373K with an activation energy, ΔE_A, of 0.10eV. The best least-squares line fit to all experimental points gives an ΔE_A of 0.06eV (Figure 4). The actual activation energy is likely located between these two values. Two explanations can be given for the deviation from a linear variation above 373K. First, the overlap between the various peaks is strong and fits are less sensitive to τ. Second, the exchange model with a trimer must be reconsidered.

3. ELECTRICAL CONDUCTIVITY

$CaFe_3O_5$ powders have been pressed under vacuum at room temperature and sintered for about twelve hours at 1273K. Conductivity measurements have been performed between 200 and 365K. Above 365K, the iron(III) to iron(II) ratio would be modified due to oxidation.

At low temperature (T<235K), the conductivity shows an Arrhenius behavior, $\sigma = A \exp\{-\Delta E_A/kT\}$, with a ΔE_A of 0.30 ± 0.02eV (Figure 5). The conductivity increases rapidly in the temperature range 235 to 333K. The change of slope at 286K corresponds to the magnetic transition temperature.

Although the temperature range between 333 and 365K is narrow, it is nevertheless possible to define a linear part in the observed curve (Figure 5) with an activation energy of 0.09 ± 0.02eV, in very good agreement with the value deduced from the Mössbauer spectra.

4. DISCUSSION AND CONCLUSIONS

$CaFe_3O_5$ is a mixed valence oxide in which electron exchange occurs between different octahedral sites. It may be interesting to follow the evolution of this relaxation process in other ferrites, such as $CaFe_{2+n}O_{4+n}$, and more particularly in $CaFe_4O_6$ and $CaFe_5O_7$, in which the iron(III) octahedra are connected by two or three iron(II) octahedra. A recent Mössbauer study [5], Gérardin et al. to be published, found the existence of a similar electron exchange in $CaFe_5O_7$ above the Néel temperature of 373K, with an activation energy of 0.035 ± 0.004eV, as given by conductivity measurements. A quantitative analysis of the spectra using the Wickman model has not yet been performed as there are three different iron(II) and two different iron(III) sites in the elementary cell.

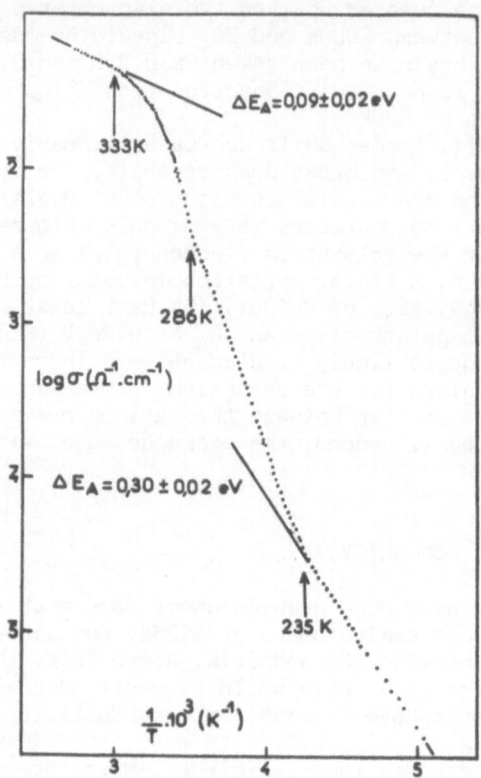

Figure 5. $\text{Log}_{10}\sigma$ versus inverse temperature.

REFERENCES

[1] B. MALAMAN, H. ALEBOUYEH, F. JEANNOT, A. COURTOIS, R. GERARDIN and
     O. EVRARD, Mat. Res. Bull., 1981, 16, 1139.
[2] R. GERARDIN, E. MILLON, J.F. BRICE, O. EVRARD and G. LE CAÉR,
     J. Phys. Chem. Solids, 1985, 46, 1163.
[3] H.H. WICKMAN, Mössbauer Effect Methodology (edited by I.J.
     GRUVERMAN) Plenum Press, New York, 1966, 2, 39.
[4] C.T. DZIOBKOWSKI, J.T. WROBLESKI and D.B. BROWN, Inorg. Chem., 1981,
     20, 679.
[5] A. TAZI, Thesis, 1987, Nancy.

CHAPTER 21

TIME DIFFERENTIAL PERTURBED ANGULAR CORRELATION STUDIES OF DYNAMIC
SYSTEMS

T.Butz
Physik-Department, Technische Universität München
8046 Garching, FRG

Abstract. In this paper a short introduction to time differential perturbed angular correlations (TDPAC) is given with special emphasis to
its application to simple dynamic systems. In paragraph 2 the theory of
TDPAC applied to static systems and in paragraph 3 the TDPAC-spectrometer (including data processing, isotope and sample preparation, and a
comparison of methods) are discussed in order to provide a sound basis
for the assessment of the merits and shortcomings of TDPAC when applied
to dynamic systems. This is discussed in paragraph 4 theoretically, together with some selected examples of recent applications.

1. INTRODUCTION

Time differential perturbed angular correlations (TDPAC) is a hyperfine
spectroscopic technique like Mössbauer spectroscopy (MS), nuclear mag-
netic/quadrupole resonance (NMR/NQR), or μ^+ spin rotation (μ^+SR). Whereas in the former two techniques electromagnetic radiation is passed onto
the sample and its absorption is detected, TDPAC as well as μ^+SR are
based on an entirely different principle. Here, first a probe, i.e. a
radioactive isotope or a polarized μ^+ particle, respectively, is produced by nuclear reactions and then subsequently introduced into the sample. Therefore, the sensitivity of these two techniques is unbeatable.
The probe is so to speak a "wound up" spy which delivers the information
about its fate during its lifetime in the sample via the emission of γ-
quanta or a positron upon deexcitation (radioactive decay) or disitegra-
tion, respectively. Whereas in μ^+SR the arrival of a polarized μ^+ in the
sample sets the beginning (start) of the probe's residence in the sample,
the end (stop) being determined by the disintegration, TDPAC relies on
the emission of two γ-quanta in succession (It is possible to use α,β-
particles or conversion electrons, too). The first signals the start, the
second the stop. In μ^+SR the "exploration time", i.e. the time the μ^+ has
at its disposal to interact with the environment, is the lifetime of the
particle. In TDPAC, this is the lifetime of the intermediate excited nuclear state, i.e. the mean time between the emission of γ_1 and the emission of γ_2. It is clear, the longer this lifetime, the better the reso-

345

G. J. Long and F. Grandjean (eds.), The Time Domain in Surface and Structural Dynamics, 345–368.
© 1988 by Kluwer Academic Publishers.

lution. In μ^+SR, the "signal" is the forward-backward asymmetry in the decay of the polarized μ^+ (parity violation!). In $\gamma-\gamma$-TDPAC, the signal is the anisotropy of the emission probability of γ_2 with respect to the emission direction of the preceeding γ_1, a consequence of angular momentum conservation. Here, due to parity conservation of the electromagnetic interaction we are dealing with even parity only. Since no Boltzmann factors or Lamb-Mössbauer factors or any other temperature dependent factor enters into the signal, both techniques work equally well at any temperature. Both techniques require the measurement of two particles in a "delayed coincidence"; a time histogram of coincidences with successively increasing delay times is recorded. Prompt events show the unperturbed situation, since there was no time to interact at all. Static classical fields will cause the spin of the probe to orient along the quantization axis, the different possible orientations differing in energy, the hyperfine energy. These states are usually called sublevels. Classically, this leads to a spin precession, the precession frequency being proportional to the sublevel energy differences. The proper quantum mechanical description is in terms of a coherent superposition of wavefunctions for the sublevels which leads to "quantum beats", an interference phenomenon. Since these quantum beats are periodic, a periodic modulation of the asymmetry/anisotropy results. It should be emphasized that a superposition of many different fields, which leads to many different precession frequencies, can eventually smear out all oscillations completely (=static inhomogeneous broadening), but always leaves a small fraction of the signal unaffected. This so-called "hardcore" is a consequence of "self-interference", which does not exist classically. In the case of randomly fluctuating fields the coherence is destroyed by random jumps between sublevels and leads to dynamic broadening. For sufficiently long times the asymmetry/anisotropy is lost completely. It is the detailed way of depolarization or loss of anisotropy, respectively, which contains the information on the dynamics of the system. It should be emphasized that in both techniques the hardcore gives an unambiguous discrimination between static inhomogeneous and dynamic broadening: if the hardcore survives for long times, we are definitely in the static regime, otherwise we must have fluctuations in the hyperfine "time window". The lower end of this window, i.e. to shorter fluctuation times, is given by the regime where the probe spin is no longer able to follow the fluctuating field adiabatically but starts to see an average field only (or decouples completely from the environment). The upper end is simply given by the possible time window of observation, i.e. 5 to 10 lifetimes: if the rate of fluctuations is so low that there is a very good chance that the field stays constant during this time of observation, we are simply back in the static regime.

In paragraph 2 an elementary introduction is given to describe how both γ-quanta are correlated via the conservation of angular momentum , and how this angular correlation is perturbed by the interaction of the nuclear moments with static classical fields. We then proceed with a description of the TDPAC-spectrometer, the data processing, the isotope and sample preparation, and a comparison of methods (paragraph 3). In paragraph 4 the theory of time dependent interactions is presented together with some recent studies of dynamic systems.

2. THEORY OF TDPAC FOR STATIC CLASSICAL FIELDS

The theory of TDPAC is fairly well documented in articles by Steffen and Frauenfelder [StefR64] and Frauenfelder and Steffen [FrauH65]. Here, the most important aspects are summarized only.

2.1 Unperturbed angular correlations

The coincidence countrate for two γ-rays emitted into the directions $\vec{k}_1$ and $\vec{k}_2$ can be written in the following form:

$$W(\vec{k}_1,\vec{k}_2) \propto \sum_{m_i,m_f} \sum_{m,m'} \langle m_f \mid \hat{H}_2 \mid m \rangle \langle m \mid \hat{H}_1 \mid m_i \rangle \langle m_f \mid \hat{H}_2 \mid m' \rangle^* \langle m' \mid \hat{H}_1 \mid m_i \rangle^* \tag{1}$$

Here, the matrix element of the operator $\hat{H}_1$, which describes the interaction between the nucleus and the radiation field of γ_1, is taken between (from left to right !) initial states m_i and intermediate states m, followed by a matrix element of H_2 for γ_2 between the same intermediate state m and the final state m_f. In order to obtain an observable probability, the complex conjugated matrix elements, denoted by an asterisk, enter as well in this product. Since none of the sublevels m_i, m, or m_f is observed separately, a sum over all possible sublevels is performed. The angular part of these matrix elements can be factored out using the Wigner-Eckart theorem and we are left with reduced metrix elements, which describe the transition probabilities for the γ-transitions. These will not be considered any further because they enter as proportionality factors only. The angular part contains vector coupling coefficients because the vector describing the initial state spin (orientation and magnitude), the vector describing the γ-emission (direction and multipolarity), and the vector describing the final state spin (orientation and magnitude) have to add up to form a "triangle". This vector coupling can be performed in various different ways. This "triangle" rule is the rule for the conservation of angular momentum. From eq.(1) we arrive at the following expression:

$$W(\vec{k}_1,\vec{k}_2) \propto \sum_{k_1 k_2 N_1 N_2} A_{k_1}(1) A_{k_2}(2) \frac{1}{\sqrt{(2k_1+1)(2k_2+1)}} Y_{k_1}^{N_1^*}(\theta_1,\phi_1) Y_{k_2}^{N_2^*}(\theta_2,\phi_2) \tag{2}$$

with

$$A_k(1) = F_k(L_1 L_1 I_i I) = (-1)^{I_i+I-1}(2L_1+1)\sqrt{(2I+1)(2k+1)} \begin{pmatrix} L_1 & L_1 & k \\ 1 & -1 & 0 \end{pmatrix} \begin{Bmatrix} L_1 & L_1 & k \\ I & I & I_i \end{Bmatrix}$$

$$A_k(2) = F_k(L_2 L_2 I_f I) = (-1)^{I_f+I-1}(2L_2+1)\sqrt{(2I+1)(2k+1)} \begin{pmatrix} L_2 & L_2 & k \\ 1 & -1 & 0 \end{pmatrix} \begin{Bmatrix} L_2 & L_2 & k \\ I & I & I_f \end{Bmatrix}$$

Here, $Y_k^N(\theta_i,\phi_i)$ is a spherical harmonic with θ and ϕ denoting the polar and azimuthal angle of the vector $\vec{k}_i$. The symbols appearing in the definition of the A_k are the vector coupling coefficients, namely the Wigner 3j and 6j symbols. The 3j/6j symbols are tabulated in [RoteM59], the F_k-coefficients are tabulated in [YamaT67] or in [StefR64]. For mixed multipolarities the A_k consist of terms with pure multipolarities and a mixing term. Further details on this topic can be found in [YamaT67]. Note that k on the right hand side of eq.(2) is a summation index. It runs from 0 to k_{max} in steps of 2 (even parity !). The index N runs from $-k$ to $+k$ in steps of 1. (Some useful details are given in the appendix).

The sum over N in eq.(2) can be performed using the addition theorem for spherical harmonics:

$$\sum_N Y_k^{N*}(\theta_1,\phi_1)Y_k^N(\theta_2,\phi_2) = (2k+1)P_k(\cos\theta) \tag{3}$$

$P_k(\cos\theta)$ denotes the Legendre polynomial with θ being the angle between the vectors $\vec{k}_1$ and $\vec{k}_2$:

$$\cos\theta = \cos\theta_1\cos\theta_2 + \sin\theta_1\sin\theta_2\cos(\phi_1-\phi_2) \tag{4}$$

With eq.(3) we finally arrive at the unperturbed angular correlation:

$$W(\vec{k}_1,\vec{k}_2) = W(\theta) = \sum_k A_k(1)A_k(2)P_k(\cos\theta) \tag{5}$$

To lowest order we have:

$$W(\theta) \propto 1 + A_2(1)A_2(2)P_2(\cos\theta) = 1 + A_{22}(\frac{3}{2}\cos^2\theta - 1/2) \tag{6}$$

where A_{22}, the anisotropy, is a shorthand for $A_2(1)A_2(2)$.
A measurement at opposite detectors ($\theta=180°$) would give a factor $1+A_{22}$ more coincidence countrate compared to the isotropic case (e.g. if A_{22} were zero), while a measurement at $\theta=90°$ would give a factor $1-A_{22}/2$ less (if the anisotropy is negative, the words "more/less" have to be interchanged). Since $\theta=0°$ or $180°$ and $\theta=90°$ are the extrema of $P_2(\cos\theta)$ the measurement is usually limited to these angles only. Note that A_{22} depends on nuclear level properties only, i.e. initial state spin, intermediate state spin, final state spin, and the multipolarities of γ_1 and γ_2 (plus eventually mixing ratios). They are constants characteristic for a specific γ-γ-cascade.
Finally, it should be mentioned that we explicitly stated in writing eq(1) that the intermediate state populated after the emission of γ_1 is unchanged at the time of emission of γ_2. This is most easily fulfilled if there was no time to interact, i.e. if the lifetime of the intermediate state is vanishingly short. If this is not the case, we must exclude all perturbations. The coincidence countrate of eq(5) must then be written as a time differential countrate, where the delay time t enters explicitly:

$$W(\theta,t) \propto \exp(\frac{-t}{\tau_N})\sum_k A_{kk}P_k(\cos\theta) \tag{7}$$

with τ_N denoting the lifetime of the intermediate state. The total time integrated countrate is then identical to that of eq(5).
We now turn to the problem when we are unable to exclude all perturbations (this would be the case if we are interested in nuclear level properties), or if we want to study these perturbations on purpose (this would be the case when we apply TDPAC to solid state physics/chemistry, to biology etc. with all nuclear properties already being well known).

2.2 Perturbed angular correlations

In order to have some action of extranuclear perturbing fields on the nucleus we require a finite lifetime of the intermediate state. We then have to write instead of eq(1) a more complicated version in which the intermediate state m_a after γ_1 is changed into a new state m_b before γ_2:

$$W(\vec{k}_1, \vec{k}_2, t) \propto \exp\left(\frac{-t}{\tau_N}\right) \sum_{m_i, m_f} \sum_{m_a, m_b} \sum_{m_{a'}, m_{b'}} \langle m_f \mid \hat{H}_2 \mid m_b \rangle \langle m_b \mid \hat{\Lambda}(t) \mid m_a \rangle \langle m_a \mid \hat{H}_1 \mid m_i \rangle \cdot \tag{8}$$
$$\langle m_f \mid \hat{H}_2 \mid m_{b'} \rangle^* \langle m_{b'} \mid \hat{\Lambda}(t) \mid m_{a'} \rangle^* \langle m_{a'} \mid \hat{H}_1 \mid m_i \rangle^*$$

where $\hat{\Lambda}(t)$ is an unitary operator which describes the evolution of the state m_a during the time interval from its population (=emission of γ_1) until its depopulation (=emission of γ_2). The coincidence countrate of eq(2) now has to be modified to: $W(\vec{k}_1, \dot{\vec{k}}_2, t) \propto$

$$\exp\left(\frac{-t}{\tau_n}\right) \sum_{k_1 k_2 N_1 N_2} A_{k_1}(1) A_{k_2}(2) G_{k_1 k_2}^{N_1 N_2}(t) \frac{1}{\sqrt{(2k_1+1)(2k_2+1)}} Y_{k_1}^{N_1^*}(\theta_1, \phi_1) Y_{k_2}^{N_2}(\theta_2, \phi_2) \tag{9}$$

with the perturbation function

$$G_{k_1 k_2}^{N_1 N_2}(t) = \sum_{m_a m_b} (-1)^{2I + m_a + m_b} \sqrt{(2k_1+1)(2k_2+1)} \langle m_b \mid \hat{\Lambda}(t) \mid m_a \rangle$$
$$\langle m_{b'} \mid \hat{\Lambda}(t) \mid m_{a'} \rangle^* \begin{pmatrix} I & I & k_1 \\ m_{a'} & -m_a & N_1 \end{pmatrix} \begin{pmatrix} I & I & k_2 \\ m_{b'} & -m_b & N_2 \end{pmatrix}$$

This perturbation function contains all information on the perturbing field. The brackets with six quantities are Wigner 3j-symbols which have the following important property: the 3j-symbol vanishes unless the sum of the elements in the bottom row adds up to zero. This condition defines the "primed" quantities.
Let the Hamilton operator $\hat{H}$ describe the type of perturbation and suppose, we have already chosen a representation in which it is diagonal. The matrix elements of $\hat{\Lambda}(t)$ are then simply given by:

$$\langle m_b \mid \hat{\Lambda}(t) \mid m_a \rangle = \exp\left(\frac{-i}{\hbar} E_m t\right) \delta_{mm_a} \delta_{mm_b} \tag{10}$$

where E_m denote the eigenvalues of $\hat{H}$.
The perturbation function then simplifies to: $G_{k_1 k_2}^{NN}(t) = \sum_m (-1)^{2I + m_a + m_b} \cdot$

$$\cdot \sqrt{(2k_1+1)(2k_2+1)} \begin{pmatrix} I & I & k_1 \\ m' & -m & N \end{pmatrix} \begin{pmatrix} I & I & k_2 \\ m' & -m & N \end{pmatrix} \exp\left\{\frac{-i}{\hbar}(E_m - E_{m'})t\right\} \tag{11}$$

An example would be the magnetic dipole interaction with m being the spin projection quantum numbers.
If $\hat{H}$ is not diagonal like in the case of non-axially symmetric electric quadrupole interactions, we require a unitary matrix $\hat{U}$ which diagonalizes the Hamiltonian $\hat{H}$:

$$\hat{U}^{-1} \hat{H} \hat{U} = \hat{D} \tag{12}$$

where $\hat{D}$ is the diagonal matrix with the eigenvalues E_m. Instead of eq(11) we have then:

$$G_{k_1 k_2}^{N_1 N_2}(t) = \sum_{m_a m_b} (-1)^{2I + m_a + m_b} \sqrt{(2k_1+1)(2k_2+1)} \begin{pmatrix} I & I & k_1 \\ m_{a'} & -m_a & N_1 \end{pmatrix} \cdot$$
$$\cdot \begin{pmatrix} I & I & k_2 \\ m_{b'} & -m_b & N_2 \end{pmatrix} \sum_{mm'} \langle m_b \mid m \rangle^* \langle m_a \mid m \rangle \langle m_{b'} \mid m' \rangle \langle m_{a'} \mid m' \rangle^* \exp\left\{\frac{-i}{\hbar}(E_m - E_{m'})t\right\} \tag{13}$$

where $<i \mid j>$ are the matrix elements of $\hat{U}$, i.e. the eigenvectors obtained after solving the eigenvalue equation $(\hat{H} - E_m \hat{E})\hat{\psi} = 0$. The sum over m, m' describes the mixing of the sublevels due to the broken axial symmetry.

In the following we simplify the perturbation functions further restricting ourselves to randomly oriented microcrystals/molecules (=powder) samples, i.e. we average over the Euler angles describing the orientation of the quantization axis in each microcrystal with respect to the laboratory coordinate system (in which $\vec{k}_1$ and $\vec{k}_2$ are defined). This leads to the condition that $k_1=k_2\equiv k$ and $N_1=N_2\equiv N$ (even in the case of a non-axially symmetric interaction) and the "powder" perturbation function is then simply obtained by a sum over certain "single crystal" perturbation functions:

$$G_{kk}(t) = \frac{1}{2k+1} \sum_{N=-k}^{k} G_{kk}^{NN}(t)$$ (14)

Since $G_{kk}(t)$ is independent of N, the addition theorem for spherical harmonics can be applied again and we obtain for the perturbed angular correlation for powder sources:

$$W(\theta,t) \propto \exp\left(\frac{-t}{\tau_n}\right) \sum_{k} A_k(1)A_k(2)G_{kk}(t)P_k(\cos\theta)$$ (15)

For the axially symmetric case the perturbation function of eq(14) can be rewritten. The term with m=m' is easily evaluated as a sum over 3j-products and yields a time-independent contribution. We then have:

$$G_{kk}(t) = \frac{1}{2k+1} + \sum_{m\neq m'} \begin{pmatrix} I & I & k \\ m' & -m & N \end{pmatrix}^2 \cos\frac{(E_m - E_{m'})t}{\hbar}$$ (16)

For half-integer I, the term $1/(2k+1)$ is the only time-independent term, even in the case of quadrupole interactions with degenerate levels ($E_m=E_{-m}$), because the 3j-symbol vanishes for m=-m' for I half-integer. This is not true for quadrupole interactions and integer I. Here, there are additional time-independent terms from terms with m=-m'. When the axial symmetry is broken the time-independent term always increases as a consequence of sublevel mixing. Note, however, that for integer I the $\pm$ degeneracy is lifted (contrary to the case of half-integer I) and hence the additional time-independent terms from m=-m' convert into low-frequency contributions of the type $\cos(E_m-E_{-m})t/\hbar$. Nevertheless, the time-independent term is always at least $1/(2k+1)$, under all circumstances. It is called "hardcore" because it guarantees that the observed anisotropy, on the time average, can never fall below $A_{kk}/(2k+1)$. This will not be true for fluctuating fields, as we shall see in paragraph 4.

To illustrate the perturbation function of eq(16) two examples will be given:

(a) magnetic dipole interaction (e.g. a powder sample of ferromagnetic iron or nickel).

The Hamiltonian is diagonal in the m-representation:

$$H_{m,m'} = \delta_{m,m'}\omega_L\hbar m$$ (17)

with $\omega_L = -\mu B/I\hbar$ denoting the Larmor frequency. B is the internal field acting on the probe in a magnetically ordered sample and μ is the magnetic moment of the intermediate state. The perturbation function then reads:

$$G_{kk}(t) = \frac{1}{2k+1} + \sum_{m\neq m'} \begin{pmatrix} I & I & k \\ m' & -m & N \end{pmatrix}^2 \cos(N\omega_L t)$$ (18)

which can be further simplified using the orthogonality properties of the

3j-symbols to yield:

$$G_{kk}(t) = \frac{1}{2k+1} + \frac{2}{2k+1} \sum_{N=1}^{k} \cos(N\omega_L t) \tag{19}$$

Note that this is true for arbitrary $I \geq 1/2$. The perturbation function for k=2 is displayed in fig.1a and reads:

$$G_{22}(t) = \frac{1}{5} + \frac{2}{5}\cos\omega_L t + \frac{2}{5}\cos 2\omega_L t \tag{20}$$

It consists of the hardcore and equal intensities of the cosine-oscillations with ω_L and $2\omega_L$. (The function $G_{44}(t)$ would have terms up to $4\omega_L$, again with equal intensities).
(b) electric quadrupole interactions (e.g. powder sample with point symmetry at the probe's site lower than cubic).
   The Hamiltonian, in the m-representation, reads:

$$H_{m,m} = \hbar\omega_Q(3m^2 - I(I+1)) \qquad\qquad H_{m,m\pm1} = 0$$

$$H_{m,m\pm2} = \hbar\omega_Q\eta\frac{1}{2}\sqrt{(I\mp m-1)(I\mp m)(I\pm m+1)(I\pm 2)} \tag{21}$$

with $\omega_Q = -eQV_{zz}/4I(2I-1)\hbar$ and $\eta=(V_{xx}-V_{yy})/V_{zz}$.
Here, V_{zz} is the largest component of the electric field gradient tensor (defined in its principle coordinate system, i.e. diagonal), V_{xx} and V_{yy} being the other two components which fulfill $V_{xx}+V_{yy}+V_{zz}=0$ (traceless tensor). eQ denotes the nuclear quadrupole moment of the intermediate state. The asymmetry parameter η is in the range $0 \leq \eta \leq 1$ if $|V_{xx}| \leq |V_{xx}| \leq |V_{zz}|$. It is easily shown that the lowest energy difference $\hbar\omega_o$ is:

$$\hbar\omega_o = \begin{Bmatrix} 3\omega_Q \\ 6\omega_Q \end{Bmatrix} \text{ for } I \begin{Bmatrix} \text{integer} \\ \text{half-integer} \end{Bmatrix}. \tag{22}$$

In the following we shall assume $\eta=0$ for simplicity. Then $\hat{H}$ is diagonal and the eigenvalues are given by the diagonal terms of eq(21). This leads to the following perturbation function:

$$G_{kk}(t) = \frac{1}{2k+1} + \sum_{m\neq m'} \begin{pmatrix} I & I & k \\ m' & -m & N \end{pmatrix}^2 \cos\{3\omega_Q(m^2 - m'^2)t\} \tag{23}$$

This can again be simplified to yield:

$$G_{kk}(t) = \sum_{N=0}^{N_{max}} s_{kN}\cos(N\omega_0 t) \quad \text{with} \quad s_{kN} = \sum_{m,m'} \begin{pmatrix} I & I & k \\ m' & -m & m-m' \end{pmatrix}^2 \tag{24}$$

where the summation sign means that the summation over m and m' should include those terms where m and m' satisfy the condition

$$|m^2 - m'^2| = \begin{Bmatrix} N \\ 2N \end{Bmatrix} \text{ for } I \begin{Bmatrix} \text{integer} \\ \text{half-integer} \end{Bmatrix}. \tag{25}$$

This perturbation function, which now depends on I, is displayed for k=2 and I=5/2 in fig.1b and reads explicitly:

$$G_{22}(t) = \frac{7}{35} + \frac{13}{35}\cos\omega_0 t + \frac{10}{35}\cos 2\omega_0 t + \frac{5}{35}\cos 3\omega_0 t \tag{26}$$

It consists of the hardcore 1/5 and three cosine terms with different intensities and frequencies $\hbar\omega_o = E_{\pm3/2} - E_{\pm1/2}$, $\hbar 2\omega_o = E_{\pm5/2} - E_{\pm3/2}$, and $\hbar 3\omega_o = E_{\pm5/2} - E_{\pm1/2}$. For combined magnetic/electric interactions see [BostL70].

352

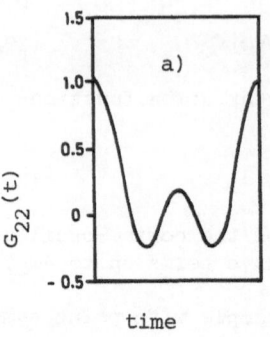
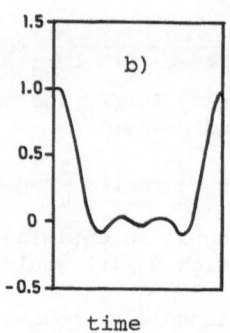

$G_{22}(t)$

time   time

Fig.1
a) $G_{22}(t)$ for magnetic dipole interaction (ran-
dom powder sample, ar-
bitrary I).
b) $G_{22}(t)$ for axially symmetric electric quadru-
pole interaction (ran-
dom powder sample, I=5/2)

A violation of the axial symmetry would lead to modified frequencies ω_i (no longer in the ratio 1:2:3; but always $\omega_3=\omega_1+\omega_2$) and modified intensities.
The effect of field distributions (inhomogeneous broadening) will be discussed next. What is required is a convolution of the distribution function with the perturbation function. For a Gaussian and a Lorentzian
distribution we have the following prefactors in front of each cosine
term:

For $\quad P(\omega) = \sqrt{2\pi\omega_0^2\delta^2} \exp\{\frac{-(\omega-\omega_0)^2}{2\delta^2\omega_0^2}\}$ we get $\sum_N s_{kN} \exp(-\frac{1}{2}\omega_{kN}^2\delta^2 t^2) \cos(\omega_{kN}t)$

for $\quad P(\omega) = (\frac{\delta\omega_0}{2\pi})^{-1}\{\frac{1+(\omega-\omega_0)^2}{(\delta^2\omega_0^2)}\}^{-1}$ we get $\sum_N s_{kN} \exp(-\delta\omega_{kN}t) \cos(\omega_{kN}t)$ (27)

This is easy to remember because the Fourier transfom of a Gaussian is
again a Gaussian, while the Fourier transform of a Lorentzian is an exponential. The width δ is the halfwidth at half maximum (HWHM) for Lorentzians and $\sqrt{2\ln2} \sim 1.177$ times the HWHM for Gaussian distributions.
Note that all cosine-terms are exponentially damped (=line broadening
of the Fourier transform of the prefactors of eq(27)), but the hardcore
($\omega_0=0$) is unaffected. A few selected perturbation functions for I=5/2,
pure quadrupole interaction, are displayed in fig.2 together with their
Fourier transforms.
In actual experiments the finite resolving time of the spectrometer has
to be accounted for in many cases. Here, we require a convolution of the
instrumental time resolution function with the measured coincidence
spectrum, i.e. eq(15). This convolution differs from the previous one
because the $\exp(-t/\tau_N)$ has to be taken into account in addition to the
$G_{kk}(t)$. For a Gaussian time resolution function analytical formulae
were derived by Rogers and Vasquez [RogeJ75]. There are three effects:
(i) the initial form of the pattern is modified due to a "leaking" of
the rise at negative times; (ii) there is a phase shift with respect
to t_0, which leaves the frequencies unaffected; (iii) there is an attenuation of the intensities. The latter effect dominates and can be approximated by (τ_N long compared to τ_{FWHM} = width of time resolution):

$$\cos\omega_0 t \quad \rightarrow \quad \exp(\frac{-1}{16\ln2}\omega_0^2\tau_{FWHM}^2)\cos\omega_0 t \qquad (28)$$

Note, that these factors do not lead to damping, but reduce intensities
only. Again, the hardcore remains unaffected.

3. TDPAC SPECTROMETER, DATA PROCESSING, ISOTOPE AND SAMPLE PREPARATION, AND COMPARISON OF METHODS

3.1 TDPAC spectrometer

Any TDPAC measurement requires the detection of a pair of γ-quanta which are characterized by their energy. Moreover, the time between the emission of γ_1 and γ_2 has to be measured as accurately as possible. Finally, the detection efficiency per detector should be as close as possible to 100%, because the detection probability for a coincidence is the product of two individual probabilities and would rapidly become exceedingly small. In principle the entire solid angle (4π) around the sample should be covered with detectors in order not to waste γ-quanta. A typical compromise between efficiency and cost uses four detectors placed at 90° to each other in a plane. The block diagram of this spectrometer is shown in fig.3. Each detector consists of a scintillation crystal (e.g. NaI(Tl) or BaF_2, 1.75"x2" ∅) with optical coupling to a photomultiplier (RCA8850 or Valvo XP2020, eventually with a quartz window for the UV light of BaF_2). The information from the detectors is processed in two different ways and analyzed independently. In the so-called "slow-circuit" the signals are analyzed with respect to their γ-energies (proportional to the pulse area) and coincidences of suitable pulses from a pair of detectors are formed (here, time resolution plays no role!). In the so-called "fast-circuit" the time difference between two pulses is accurately analyzed (here, energy resolution plays no role). To do this, the time of arrival of a γ in a detector is determined by a "constant-fraction" trigger (CFT) via the rise of the pulse and the logic output pulse is fed into a time-to-amplitude converter (TAC). The analog output signal of the TAC is fed into an analog-digital-converter (ADC). Whenever a valid coincidence for one of

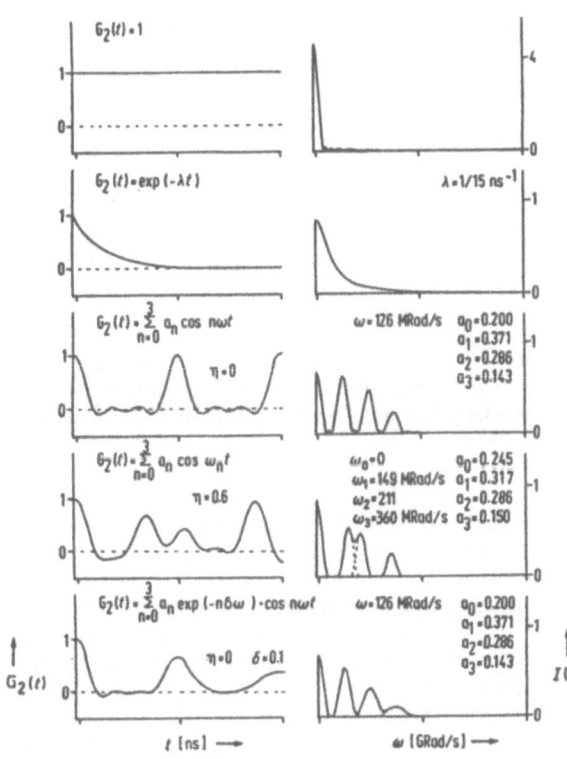

Fig.2 Some examples of perturbation functions for I=5/2, pure quadrupole interaction (left) and their Fourier transforms (right). The topmost pattern is unperturbed.

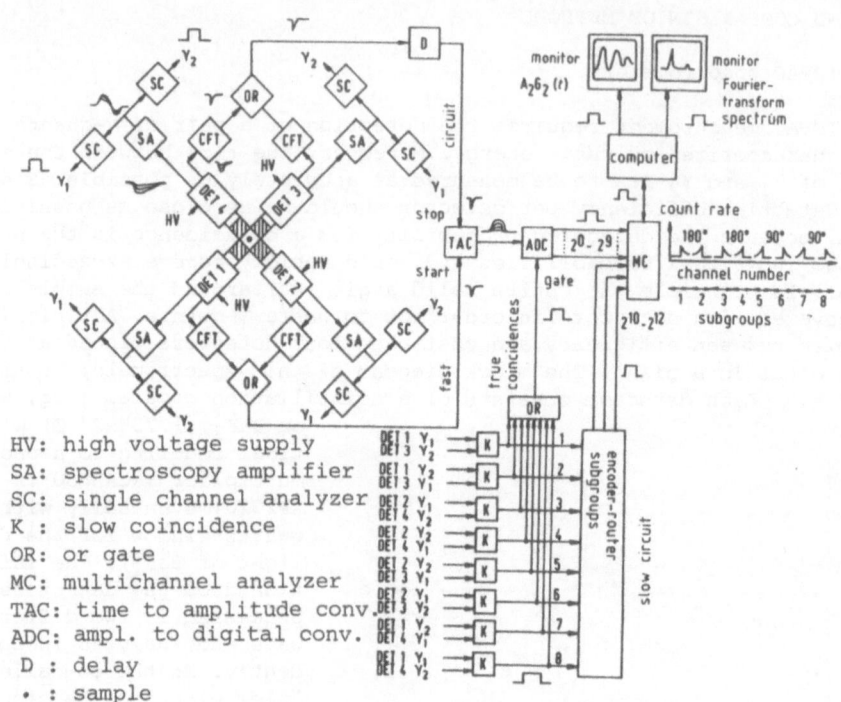

HV: high voltage supply
SA: spectroscopy amplifier
SC: single channel analyzer
K : slow coincidence
OR: or gate
MC: multichannel analyzer
TAC: time to amplitude conv.
ADC: ampl. to digital conv.
 D : delay
 • : sample

Fig.3 Block diagram of a four-detector TDPAC-spectrometer.

the possible detector combinations was found in the "slow-circuit", the memory of a multichannel analyzer is addressed and the content increased by 1. According to the detector combinations the corresponding subgroup of the multichannel analyzer is addressed via a routing unit (encoder). The activity of the sample must be kept sufficiently low ($\sim$10μCi) as to guarantee that no further nuclear decays take place during the pulsed operation of both circuits, including the data storage in the multichannel analyzer (typically 10-20 μs). In this way, eight of the possible 12 coincidence spectra are stored simultaneously. Apart from a constant background, which results from chance coincidences (two nuclei happen to decay at approximately the same time and the corresponding γ-quanta hit two detectors in such a way that γ_1 stems from one nucleus and γ_2 from another nucleus, i.e. they are uncorrelated), one obtains an exponentially decaying histogram of "true" coincidences. The precession signal $A_{22}G_{22}(t)$ is superimposed onto this curve. The exponential histogram decays to the right, if the start detector detects γ_1, and to the left, if it detects γ_2. The zero time is shifted to the center of a multichannel subgroup by a delay line in the stop line. In order to be able to see the "progress" made in data collection, a computer continuously reads out the multichannel analyzer and displays the processed data onto a monitor. The data processing will be described in the next section.

3.2 Data processing

The data reduction starts with the substraction of the chance coincidence background in each subgroup. This background is determined from that
part of the spectrum which contains no "true" coincidences, i.e. the
range of "negative" times. Then all spectra decaying to the left are inverted (i.e. decay to the right), and all eight spectra are shifted to
a common zero time t_o. This t_o of each subgroup is predetermined by measuring prompt coincidences, e.g. using a ^{22}Na ($e^+ \rightarrow 2 \times 511$ keV annihilation
quanta) source. At the same time, the instrumental time resolution τ_{FWHM}
is also determined. Typical values for τ_{FWHM} are 2ns (NaI(Tl)) and 0.5ns
(BaF_2). These procedures (background substraction and shift by linear interpolation) are delicate: a sloppy background correction spoils the
spectrum at late times, where the ratio of "true" coincidences to chance
coincidences becomes low, thus rendering the hardcore determination at
late times problematic; shifting the data by linear interpolation acts
like a low-pass filter, i.e. the sensitivity at high frequencies is reduced. Moreover, the accuracy of the t_o-determination limits the accuracy of the frequency and damping determinations.
We then form the following ratio in order to cancel the detector efficiencies and to cancel the exponential decays (exactly for $\tau_{FWHM}=0$):

$$A_2 G_2 = 2 \frac{\sqrt[4]{W_{13}W_{31}W_{24}W_{42}} - \sqrt[4]{W_{14}W_{41}W_{23}W_{32}}}{\sqrt[4]{W_{13}W_{31}W_{24}W_{42}} + 2\sqrt[4]{W_{14}W_{41}W_{23}W_{32}}} \tag{29}$$

where W_{ij} denotes the "true" coincidence countrate between detectors i
(γ_1) and j(γ_2). This ratio reduces to $A_{22}G_{22}(t)$ for powder samples, provided higher order terms ($A_{44}G_{44}(t)$) can be neglected. This ratio is
displayed on the monitor. If required, a computer fit of the appropriate
theoretical function (e.g. eq(26)) to the data points can be performed
in order to extract various adjustable parameters. In addition to the
time spectrum, the Fourier transformed TDPAC spectrum is shown an a monitor. The power spectrum is often easier to comprehend since it untan-
gles information which is all superimposed in the time spectrum. In order to suppress sidelobes a suitable window function is used, e.g. the
Kaiser-Bessel window function [Kuo F66] with adjustable sidelobe suppression level. Since all oscillatory perturbation functions are even functions (cosines), this phase information should be exploited. A possible
(and simple) way to do this is to transform the time spectrum together
with its mirror image, as described in [ButzT79]. In order to obtain a
fine mesh of Fourier coefficients (and hence a useful lineshape), the
TDPAC data are filled up with an appropriate number of zeros. In this
way all problems with truncation are practically eliminated. The analysis of the Fourier spectra with respect to line-position, linewidth, intensity, and eventually lineshape can all be performed quasi on-line.

3.3 Isotope and sample preparation

The radioactive isotopes are produced either in nuclear reactors (thermal neutron capture), or in cyclotrons and accelerators (particle bom-
bardment). Among the most suitable TDPAC isotopes are: ^{111}In(EC)^{111}Cd,
^{181}Hf(β^-)^{181}Ta, ^{99}Rh(EC,β^+)^{99}Ru and ^{100}Pd(EC)^{100}Rh, ^{111m}Cd, ^{199m}Hg. The

choice of isotope may present problems but should be guided by the following criteria:(i) the use of isomeric probes, like ^{111m}Cd or ^{199m}Hg is strongly recommended for doping or labelling because there is no nuclear transmutation to the daughter isotope, i.e. the chemistry is always the same; however, both isotopes have a halflife of the isomeric state below one hour and consequently require fast and repeated sample preparations; (ii) if the probe is a constituent of the sample under investigation, there are no problems during preparation; however, after the nuclear transmutation there is a foreign atom and all information is obtained from a site which is modified;(iii) if the daughter isotope is a constituent of the sample, it might be difficult to introduce the parent isotope, but the information obtained is from a regular site;(iv) frequently neither the parent nor the daughter isotope is a constituent of the sample; then there might be problems during sample preparation and the information is obtained from a foreign atom site. In all but the isomeric decays electronic- or chemical after-effects can occur whenever the rearrangement of the electronic shell after transmutation cannot be completed sufficiently rapidly, e.g. in semiconductors or insulators.
In many applications of TDPAC a foreign probe atom is introduced on purpose. For example, defects in metals are frequently trapped by these foreign atoms at sufficiently high temperatures and can thus be identified and detected at extremely low concentration levels. Usually, a substitutional incorporation into a solid or a specific binding site in a molecule are desirable. In order to obtain a reasonable TDPAC spectrum about 10^6 to 10^8 coincidences are required. With the typically low efficiencies of the spectrometers this means that 10^{10} to 10^{12} probes are required. Thus very small amounts of samples can be studied. It is even possible to investigate a fraction of a monolayer on a clean single crystal surface. However, working with picogram quantities, i.e. without inactive carrier, is often difficult. Finally, the probes are often produced in the sample via a nuclear reaction, say with a particle beam, or they are implanted with an accelerator. In these cases the radiation damage can cause problems. With longlived probes a subsequent annealing procedure is possible. In-beam measurements with shortlived probes are usually possible only in metals at elevated temperatures where the annealing is fast or in samples with a high rate of self-repair, at least of the near neighborhood of the probe, like e.g. fluorides.

3.4 Comparison of methods

In this section the advantages and disadvantages of TDPAC are compared with the well-known techniques NMR/NQR and MS. The main disadvantage of TDPAC compared to MS is the fact that all radioactive samples have to be prepared by the experimentalist, i.e. one deals with open radioactivities in contrast to commercially available (in many cases) encapsuled sources for MS. Thus a radiochemistry laboratory is needed. However, the required activity levels are low (10-100µCi). The second disadvantage is the fact that TDPAC is a coincidence technique, contrary to MS,with long data collection times. Here, efficient spectrometers are imperative. On the other hand there are enormous advantages compared to MS and NMR/NQR: (1) sample quantities can be extremely small; (2) there is

essentially no limitation to the choice of sample environment: furnaces, cryostats, autoclaves, stirred baths etc. are easily applicable because the γ-energies of the typical TDPAC isotopes are high, contrary to MS; (3) there is no need for applied magnetic fields or RF fields like in NMR/NQR; (4) small interactions are more easily resolved than in MS; (5) contrary to MS and NMR it is practically impossible, for instrumental reasons, to detect only a part of the total hyperfine spectrum (i.e. velocity range or field variation too small), because the time and not the energy spectrum is measured; (6) the signal is constant (=A_{22}) and does not depend on details of the setup, apart from solid angle correction factors (see appendix); this means that all probes are counted with the same sensitivity and hence a quantitative analysis is very simple for multiphase/multisite systems; (7) contrary to MS and NMR there is practically no way to influence the linewidth/damping by instrumental means, i.e. small site inhomogeneities can be reliably detected; (8) reorientational motion is normally unambiguously discriminated from static inhomogeneous broadening via the hardcore behaviour.
For all these reasons TDPAC is very well suited for in situ applications in Chemistry, Materials Science, and Biophysical Chemistry.

4. THEORY OF TDPAC FOR FLUCTUATING CLASSICAL FIELDS WITH EXAMPLES OF RECENT STUDIES OF DYNAMIC SYSTEMS

As in paragraph 2, we restrict the discussion to classical fields, i.e. we treat the nuclear spin system quantum-mechanically only. We consider stochastic processes and regard the extranuclear system as a "thermal bath" which acts on the nucleus in the form of fields jumping at random as a function of time between a finite or infinite number of possible stochastic states:

$$\hat{H}(t) = \sum_j \hat{V}_j f_j(t) \tag{30}$$

with the set of random functions $\{f_j\}$ being defined in such a way, that at a given moment τ only one of them, say $f_a(\tau)=1$, while all other $f_j(\tau)$ are zero. This defines jump processes; we do not discuss pulse processes where the interaction takes place at random instances only. Throughout the following discussion we assume stationary Markovian processes. Note that the establishment of thermal equilibrium with the bath requires a great number of jumps. This means that the limiting cases of very slow relaxation, which will be presented below, where a few jumps per precession period take place only, are not easily justifiable. A plausible argument for the validity at low jump rates is that the ensemble average over a great number of possible sequences (TDPAC is a counting technique) is effectively equivalent to the time average over a large number of jumps. For further details see [BlumM71, WinkH76, BoscF77, DattS81].
In section 4.1 we present the generalization of eq(9) for discrete jump processes. In section 4.2 the special case of continuous isotropic fluctuations will be discussed. Finally, in section 4.3 a few selected examples of recent applications of TDPAC to dynamic systems will be given.

4.1 Discrete jump processes

In eq(10) the time evolution operator $\hat{\Lambda}(t)$ was obtained by solving the Schrödinger equation:

$$\frac{d}{dt}\hat{\Lambda}(t) = \frac{-i}{\hbar}E\hat{\Lambda}(t) \tag{31}$$

which, for constant E, leads formally to

$$\hat{\Lambda}(t) = \exp\left(\frac{-i}{\hbar}Et\right) \tag{32}$$

or, more precisely, to eq(10) expressing the matrix elements of $\hat{\Lambda}(t)$ in the representation where the Hamilton H is diagonal. Thus there is merely a phase factor in the time evolution (=quantum beats). We have also seen that the energy eigenvalues always enter in pairs as differences (this is a consequence of the product of matrix elements with their complex conjugates). Now, for stochastic fluctuations, we mix substates stochastically, thus destroying the phase coherence. It is convenient to introduce a superoperator, the so-called Liouville operator, which, first, deals with energy differences and, secondly, which spans the space for all the different Hamilton operators for the different stochastic states, i.e. the different $\hat{V}_j$. Its definition is as follows. Adjoined to the Hamiltonian we have the Liouville operator $\hat{H}^{\times}$ in such a way, that its action on any ordinary operator $\hat{A}$ is equivalent to the commutator:

$$\hat{H}^{\times}\hat{A} = [\hat{H}, \hat{A}] = \hat{H}\hat{A} - \hat{A}\hat{H} \tag{33}$$

For a superoperator we now require 2 times 2 indices to describe its matrix elements. For a specific stochastic state we have:

$$\langle m_b m_{b'} \mid \hat{V}_j^{\times} \mid m_a m_{a'} \rangle = \delta_{m_{a'}, m_{b'}} \langle m_b \mid \hat{V}_j \mid m_a \rangle - \delta_{m_a m_b} \langle m_{a'} \mid \hat{V}_j \mid m_{b'} \rangle \tag{34}$$

An example would be the case of I=1/2, magnetic interaction, with $\hat{V}$ being diagonal with the elements $\hbar\omega_L m$. Hence, adjoined to

$$\hat{V} = \hbar\omega_L \begin{pmatrix} 1/2 & 0 \\ 0 & -1/2 \end{pmatrix} \text{ we have } \hat{V}^{\times} = \hbar\omega_L \begin{pmatrix} 0 & 0 & 0 & 0 \\ 0 & 1 & 0 & 0 \\ 0 & 0 & -1 & 0 \\ 0 & 0 & 0 & 0 \end{pmatrix} \begin{array}{cc} m_b = 1/2 & m_b' = 1/2 \\ -1/2 & 1/2 \\ 1/2 & -1/2 \\ -1/2 & -1/2 \end{array} \tag{35}$$

$$\begin{array}{l} m_a = 1/2 - 1/2 \quad 1/2 - 1/2 \\ m_b = 1/2 \quad 1/2 - 1/2 - 1/2 \end{array}$$

$\hat{H}$ finally consists of a block matrix, each block describing a particular stochastic state j. We now require 3 times 3 indices. The matrix elements of H are given by:

$$\langle n' m_2 m_{2'} \mid \hat{H}^{\times} \mid n m_1 m_{1'} \rangle = \delta_{n,n'} \langle m_2 m_{2'} \mid \hat{V}^{\times} \mid m_1 m_{1'} \rangle \tag{36}$$

Evidently, the situation is particularly simple whenever all $\hat{V}$ are diagonal or commute: $[\hat{V}_i, \hat{V}_j] = 0$. The mixing between different substates is described by the relaxation matrix $\hat{R}$, which is diagonal in the nuclear variables. For a situation, where the system starts from the clastic state n and turns over to the stochastic state n' the matrix elements of $\hat{R}$ are given by (assuming detailed balance):

$$\langle n' m_b m_{b'} \mid \hat{R} \mid n m_a m_{a'} \rangle = \delta_{m_a m_b} \delta_{m_{a'} m_{b'}} w_{n \to n'}$$

$$\langle n m_b m_{b'} \mid \hat{R} \mid n m_a m_{a'} \rangle = -\delta_{m_a m_b} \delta_{m_{a'} m_{b'}} \sum_{n'} w_{n \to n'} \tag{37}$$

with $w_{n \to n'}$ denoting the transition probability. An example would be a two state system:

$$\hat{R} = \begin{pmatrix} -\gamma & \gamma \\ \gamma & -\gamma \end{pmatrix} \quad \text{with} \quad \gamma = w_{n \to n'} \tag{38}$$

The simple $I=1/2$ up/down model with jumps between B_z and $-B_z$ would then be described by the following superoperator:

$$-\frac{i}{\hbar}\hat{H}^\times + \hat{R} = \begin{pmatrix} -\gamma & 0 & 0 & 0 & \gamma & 0 & 0 & 0 \\ 0 & -i\omega_L - \gamma & 0 & 0 & 0 & \gamma & 0 & 0 \\ 0 & 0 & i\omega_L - \gamma & 0 & 0 & 0 & \gamma & 0 \\ 0 & 0 & 0 & -\gamma & 0 & 0 & 0 & \gamma \\ \gamma & 0 & 0 & 0 & -\gamma & 0 & 0 & 0 \\ 0 & \gamma & 0 & 0 & 0 & i\omega_L - \gamma & 0 & 0 \\ 0 & 0 & \gamma & 0 & 0 & 0 & -i\omega_L - \gamma & 0 \\ 0 & 0 & 0 & \gamma & 0 & 0 & 0 & -\gamma \end{pmatrix} \begin{matrix} \\ \\ \text{up} \\ \\ \\ \\ \text{down} \\ \\ \end{matrix} \tag{39}$$

$$\text{up} \qquad\qquad\qquad \text{down}$$

The eigenvalues are (doubly degenerate): $\lambda_1=0$, $\lambda_2=-2\gamma$, $\lambda_3=-\gamma+i\sqrt{\omega_L^2-\gamma^2}$, and $\lambda_4=-\gamma-i\sqrt{\omega_L^2-\gamma^2}$. These eigenvalues appear in the exponent. Hence, for $\gamma=0$ we have the normal hardcore plus the the $\cos\omega_L t$ oscillations, whereas for finite $\gamma<\omega$ we get damped oscillations with a renormalized (reduced) frequency. At $\gamma=\omega$ we have the aperiodic limit where the oscillations disappear, and for $\gamma\gg\omega$ we have exponential damping only. For $\gamma\to\infty$, we recover the unperturbed situation ($\lambda_4\to 0, \lambda_2, \lambda_3\to -\infty$).
The generalization of eq(31) then leads to the Schrödinger- von Neumann equation:

$$\frac{d}{dt}\hat{\Omega}(t) = \left(\frac{-i}{\hbar}\hat{H}^\times + \hat{R}\right)\hat{\Omega}(t) \tag{40}$$

with the formal solution:

$$\hat{\Omega}(t) = \exp\left(\frac{-i}{\hbar}\hat{H}^\times + \hat{R}\right)t \tag{41}$$

The generalization of the perturbation function of eq(9) now reads:

$$G_{k_1 k_2}^{N_1 N_2}(t) = \sum_{m_a m_b} (-1)^{2I+m_a+m_b} \sqrt{(2k_1+1)(2k_2+1)} \begin{pmatrix} I & I & k_1 \\ m_{a'} & -m_a & N_1 \end{pmatrix}$$

$$\begin{pmatrix} I & I & k_2 \\ m_{b'} & -m_b & N_2 \end{pmatrix} \sum_{nn'} P_n \langle n' m_b m_{b'} \mid \exp\left(\frac{-i}{\hbar}\hat{H}^\times + \hat{R}\right)t \mid n m_a m_{a'}\rangle \tag{42}$$

Here, P_n denotes the a priori probability of finding the environment at $t=0$ in the stochastic state n. For an equal probability we would have $P_n = 1/$number of states $\mathcal{N}$. Since we do not know in what stochastic state the system was at $t=0$ we have to sum over all possible initial and final states n and n'.
The task to perform now is the determination of the eigenvalues and eigenvectors of $\hat{\Omega}(t)$. In general, the eigenvalues of this so-called Blume-matrix are complex, but appear in conjugate pairs. Moreover, their

real part is always less or equal to zero. If we denote the eigenvectors by $\vec{M}_q$ corresponding to the eigenvalue $\lambda_q = -\gamma_q + i\omega_q$, the eigenvalue belonging to $\vec{M}_q^*$ is $\lambda_q = -\gamma_q - i\omega_q$. This finally leads to the following compact definition of the perturbation function [WinkH73]:

$$G_{k_1 k_2}^{N_1 N_2}(t) = \sum_{m_a m_b} (-1)^{2I+m_a+m_b} \sqrt{(2k_1+1)(2k_2+1)} \begin{pmatrix} I & I & k_1 \\ m_{a'} & -m_a & N_1 \end{pmatrix}$$
$$\begin{pmatrix} I & I & k_2 \\ m_{b'} & -m_b & N_2 \end{pmatrix} \sum_{nn'} P_n \langle n'm_b m_{b'} \mid \vec{M}_q \rangle \langle \vec{M}_q' \mid nm_a m_{a'} \rangle \exp\left\{(-\gamma_q + i\omega_q)t\right\}$$

(43)

Here, M_q and M_q' are the "ket"- and "bra"-vectors of the unitary matrix built up by the eigenvectors. Compared to eq(13) we have increased the size of the unitary matrix to $(2I+1)(2I+1)\mathcal{N}$ instead of $(2I+1)$, which can pose some problems for high spins and many states. The little trick with the superoperator formalism, however, condensed a pair of matrix elements into a single one. The powder perturbation function is again obtained by eq(14). We note, that ultimately we have a sum over exponentially damped cos-oscillations. The fast relaxation case $(\gamma \to \infty)$ is particularly simple: it can be shown (see [WinkH73]) that in this limit we have

$$G_{kk}(t) = \exp(-\gamma_k t)$$

(44)

with
$$\gamma_k = \frac{1}{3}\tau_c \omega_L^2 k(k+1)$$

for magnetic interactions and with

$$\gamma_k = \frac{3}{5}\tau_c \omega_Q^2 k(k+1)[4I(I+1) - k(k+1) - 1]$$

for quadrupole interactions. Here, the correlation time $\tau_c = 1/\mathcal{N}\lambda$. This is the so-called Abragam-Pound limit [AbraA53]. Note that $\gamma_4/\gamma_2 = 10/3$ for the magnetic case and $\gamma_4/\gamma_2 = \frac{10}{3}(4I(I+1)-21)/(4I(I+1)-7)$ for quadrupole interactions. This helps to distinguish between the two types of interaction in the fast relaxation limit.

We now give a particular example for the case of three stochastic states with an axially symmetric electric field gradient whose largest component jumps between the x,y,and z-axis, as discussed by [WinkH73] for I=5/2. We have:

$$\begin{aligned} V_1 &= \hbar\omega_Q\{3I_x^2 - I(I+1)\} \\ V_2 &= \hbar\omega_Q\{3I_y^2 - I(I+1)\} \quad \text{and} \quad \hat{R} = \begin{pmatrix} -2\gamma & \gamma & \gamma \\ \gamma & -2\gamma & \gamma \\ \gamma & \gamma & -2\gamma \end{pmatrix} \\ V_3 &= \hbar\omega_Q\{3I_z^2 - I(I+1)\} \end{aligned}$$

(45)

The results obtained by expanding the exponential operator $\hat{\Omega}(t)$ into a series up to 50 terms (rather than to diagonalize the 6·6·3=108 times 108 matrix) is shown in fig.4a. For $\lambda=0$ we have the static case (cf.fig 1b), for $\lambda \sim \omega_q$ (i.e. $\lambda \sim 1$ in units of t^{-1}) the oscillating behaviour is lost (note the small shift to lower frequencies !). This is the case when the frequencies ω_q in eq(43) become imaginary. Finally, the pattern approaches more and more a single exponential decay, i.e. the Abragam-Pound limit. For exceedingly fast relaxation rates the nucleus is unable to follow the fluctuations adiabatically and hence remains essentially unperturbed, i.e. $G_{kk}(t) \to 1$, irrespective of the instantaneous interaction and even in cases when the average interaction does not vanish. In other words, the difference in interaction strength between stochastic

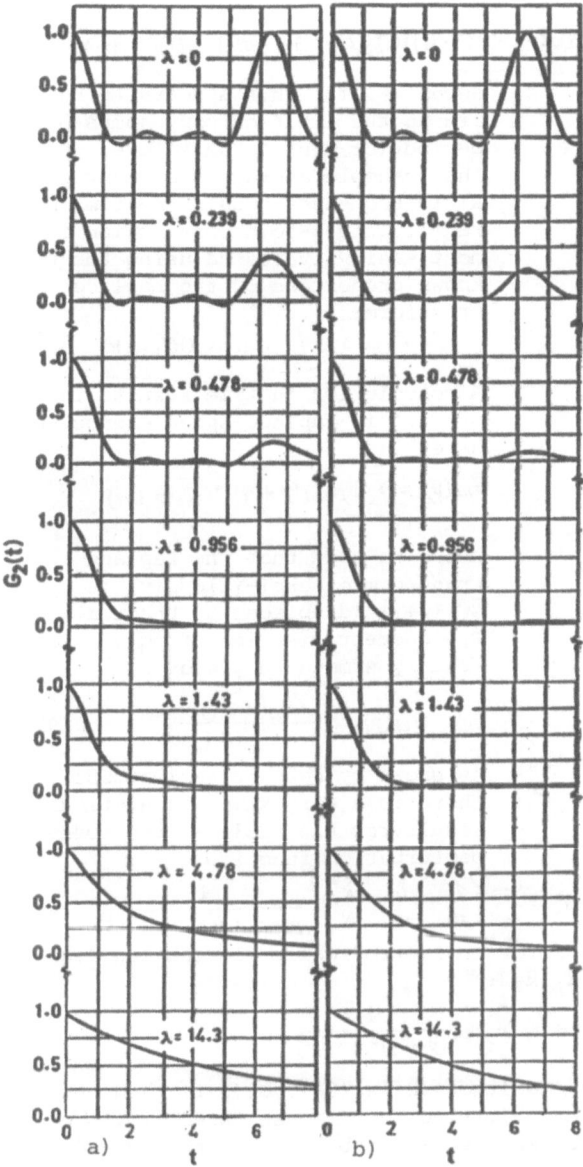

states is averaged out first to a possibly non-zero mean value (=motional averaging), but ultimately the nuclear system decouples from the bath.

A second example is the case of a four state model with the electric field gradient maintaining its orientation but jumping between different asymmetry parameters $\eta = \pm 0.2$ and ± 0.6 [WinkH73]. Here, for computational purposes the range of η is extended from $-1 \leq \eta \leq 1$. The results are shown in fig.5. Note, that relatively few jumps are required to average η to zero. The most spectacular effect is the conservation of the hard-core. This is simply a consequence of the lack of any reorientational motion. If the reorientational motion is spatially restricted, say to within a cone, we have a partial loss of the hard-core. We would have to perform an average over angles in complete analogy to the solid angle correction factors (see appendix). It thus appears difficult to distinguish between static, in-homogeneous broadening and dynamic broadening <u>without</u> reorientational motion.

4.2 Continuous isotropic processes

For isotropic processes it is possible, according to Blume [BlumM71] to write the perturbation function for s jumps/collisions as an s-fold convolution of the static perturbation function (denoted by $G^0_{KK}(t)$):

Fig.4
(a) $G_{22}(t)$ for the discrete x,y,z model for axially symmetric electric field gradients, $I=5/2$, $\omega_0 = 6\omega_Q = 2\pi$.
   (λ and ω in units of t^{-1}) [WinkH73].
(b) Same as (a), but for the isotropic continuous model [DattS81].

$$G_{kk}(t) = \sum_{s=0}^{\infty} \gamma^s \int_0^t dt_s \cdots \int_0^{t_2} dt_1 \exp\{-\gamma(t-t_s)\} G_{kk}^0(t-t_s) \cdots \exp(-\gamma t_1) G_{kk}^0(t_1) \quad (46)$$

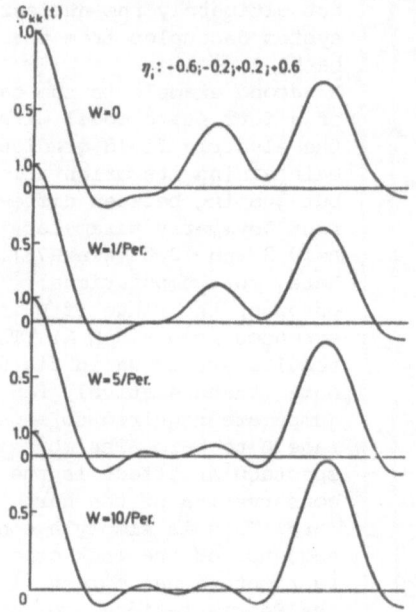

$\eta_i: -0.6; -0.2; +0.2; +0.6$

$G_{kk}(t)$

W=0

W=1/Per.

W=5/Per.

W=10/Per.

Fig.5 $G_{22}(t)$ for I=5/2, electric field gradient fluctuating between $\eta=\pm0.2$ and ±0.6 without reorientation [WinkH73].

Here, $\gamma=1/\tau_c$. This appears to be valid in the limit of many jumps per period (or, as stated before, when the ensemble average over many possible sequences equals the time average). This s-fold convolution is most easily evaluated using the well-known properties of the Laplace-transform:

$$\tilde{G}_{kk}(p) = \int_0^{\infty} \exp(-pt) G_{kk}(t) dt \quad (47)$$

A convolution in time results in a product in Laplace-space. Hence we have:

$$\tilde{G}_{kk}(p) = \sum_{s=0}^{\infty} \{\gamma G_{kk}^0(p+\gamma)\}^s \tilde{G}_{kk}^0(p+\gamma) \quad (48)$$

where we used that the Laplace transform of $\exp(-\gamma t) G_{kk}^0(t)$ is $\tilde{G}_{kk}^0(p+\gamma)$ (displacement theorem). The geometric series of eq(48) is readily summed to yield:

$$\tilde{G}_{kk}(p) = \frac{\tilde{G}_{kk}^0(p+\gamma)}{1-\gamma G_{kk}^0(p+\gamma)} \quad (49)$$

As shown by Dattagupta [DattS81], there is a very simple analytical result for the Laplace transformed perturbation function:

$$\tilde{G}_{kk}^0(p+\gamma) = \sum_{n_1 n_0} \begin{pmatrix} I_1 & I_0 & k \\ n_1 & -n_0 & N \end{pmatrix}^2 \frac{1}{p+\gamma+i(V_{n_1}-V_{n_0})} \quad (50)$$

A specific example is the case for an isotropically fluctuating electric field gradient for I=5/2, k=2:

$$\tilde{G}_{22}^0(p+\gamma) = (p+\gamma) \sum_{n=0}^{3} s_{2n} \frac{1}{(p+\gamma)^2 + n^2\omega_0^2} \quad (51)$$

The coefficients s_{2n} are the amplitudes of eq(26). $\tilde{G}_2(p)$ is then obtained by substituting eq(51) into eq(49). Here, we discuss three limiting cases. The Laplace transforms required are tabulated in the appendix.

(a) $\gamma=0$: In this case

$$\tilde{G}_{22}(p) = \tilde{G}_{22}^0 = p \sum_{n=0}^{3} s_{2n} \frac{1}{p^2 + n^2\omega_0^2} \quad (52)$$

from which we recover the static perturbation function:

$$G_{22}(t) = \sum_{n=0}^{3} s_{2n} \cos(n\omega_0 t) \quad (53)$$

(b) $\gamma \to 0$: Here,

$$\tilde{G}_{22}^0(p+\gamma) \approx (p+\gamma) \sum_{n=0}^{3} \frac{s_{2n}}{p^2 + 2p\gamma + n^2\omega_0^2}$$

$$\approx (p+\gamma)\sum_{n=0}^{3}\frac{s_{2n}}{p^2+n^2\omega_0^2}(1-\frac{2p\gamma}{p^2+n^2\omega_0^2}) \tag{54}$$

$$\approx p\sum_{n=0}^{3}\frac{s_{2n}}{p^2+n^2\omega_0^2}+\gamma\sum_{n=0}^{3}\frac{s_{2n}(n^2\omega_0^2-\gamma^2)}{(n^2\omega_0^2+\gamma^2)^2}$$

(neglecting terms like $s_{2n}\cdot s_{2n'}$, which is not very well justified). From this and $\tilde{G}_{22}(p)\approx\tilde{G}_{22}^0(p+\gamma)$ we finally get:

$$G_{22}(t) = (1-\gamma t)\sum_{n=0}^{3}s_{2n}\cos(n\omega_0 t)\approx \exp(-\gamma t)\sum_{n=0}^{3}s_{2n}\cos(n\omega_0 t) \tag{55}$$

This is an approximate result for the slow relaxation limit: all s_{2n}, including the hardcore, are multiplied with the same exponential damping factor.

(c) $\gamma\to\infty$: Here we have:

$$\tilde{G}_{22}^0(p+\gamma)\ \longrightarrow\ \frac{1}{p+\gamma}\sum_{n=0}^{3}s_{2n}(1-\frac{n^2\omega_0^2}{(p+\gamma)^2})=\frac{1}{p+\gamma}[1-\frac{\omega_0^2}{(p+\gamma)^2}\sum_{n=0}^{3}s_{2n}n^2]$$

$$=\frac{1}{p+\gamma}[1-\frac{14}{5}\frac{\omega_0^2}{(p+\gamma)^2}]\qquad\text{(with }\sum_n s_{2n}=1\text{)} \tag{56}$$

Expanding now the expression for $\tilde{G}_{22}(p)$ to first order in the numerator and to third order in the denominator we get:

$$\tilde{G}_{22}(p)\approx\frac{\frac{1}{p+\gamma}}{1-\frac{\gamma}{p+\gamma}(1-\frac{14}{5}\frac{\omega_0^2}{(p+\gamma)^2})}=\frac{1}{p+\gamma\frac{14}{5}\frac{\omega_0^2}{(p+\gamma)^2}}\approx\frac{1}{p+\gamma\frac{14}{5}\frac{\omega_0^2}{\gamma^2}}=\frac{1}{p+\frac{14}{5}\frac{\omega_0^2}{\gamma}} \tag{57}$$

This ultimately leads to the Abragam-Pound limit for I=5/2, k=2 (cf. eq(44)):

$$G_{22}(t) = \exp(-\frac{14}{5}\frac{\omega_0^2}{\gamma}t) \tag{58}$$

As pointed out by Bosch and Spehl [BoscF77], this result is unphysical at $t\to 0$. For a finite interaction strength $G_{22}(t)$ should always start with a vanishing slope at t=0. For intermediate γ, the inverse Laplace transform has to be computed numerically. The results of this calculation are shown in fig.4b [DattS81], and are very similar to those of the discrete x,y,z-model (fig.4a).

4.3 Examples of recent TDPAC studies of dynamic systems

The first example is the case of a fluctuating magnetic field in moderately dilute ^{181}Hf in nickel. The measured spectra at various temperatures below the Curie temperature of nickel are shown in fig.6 [GerdE76]. The authors tried to fit their data with several models: (i) 6 stochastic states $\pm x,\pm y,\pm z$; (ii) 2 stochastic states up/down; (iii) two stochastic states on/off. Only the latter one was able to reproduce the measured data. It was concluded that at the concentration of 10^{-3}Hf

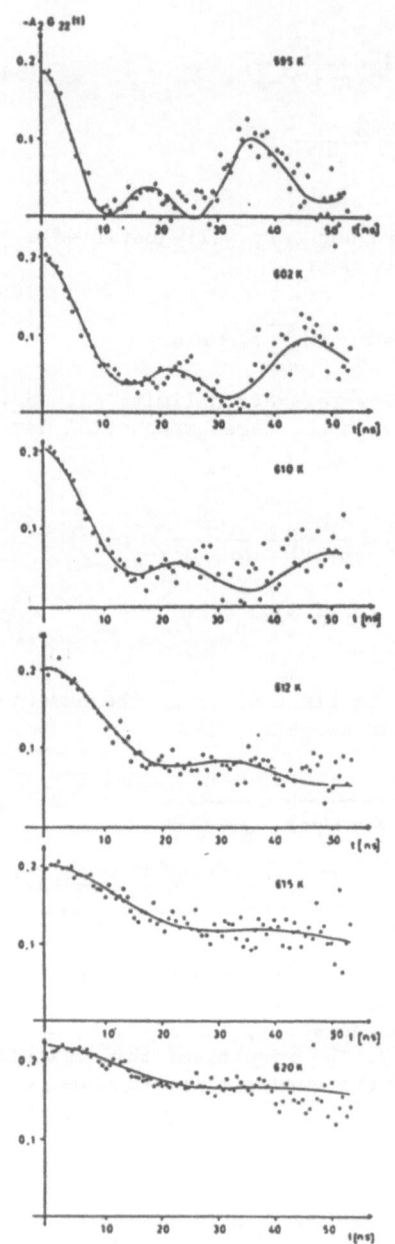

Fig.6 TDPAC spectra for 10^{-3}Hf
in nickel close to the
Curie temperature [GerdE76]

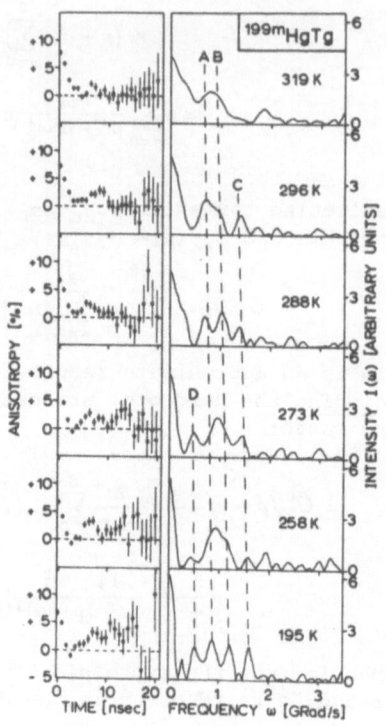

Fig.7 TDPAC spectra (left) and
their Fourier transforms
(right) for ^{199m}Hg labelled
immobilized trypsinogen at
various temperatures.
[ButzT83]

in nickel there are paramagnetic domains around the impurity atoms which
are fluctuating between the paramagnetic and ferromagnetic state. Thus
TDPAC appears to be very useful to
study details of magnetic fluctuations.

The second example is the case of a
fluctuating electric field gradient
at a ^{199m}Hg label, introduced into
the activation domain between CYS191
and CYS220 in bovine trypsinogen
[ButzT82,83]. The enzymatic activity
of the digestive enzyme trypsin is
believed to be blocked in the proenzyme trypsinogen by the flexibility

of the activation domain. Single crystal studies of Hg-labelled trypsinogen showed that the flexibility is not lost by the labelling process. On the other hand, when treated with pancreatic trypsin inhibitor
(plus the dipeptide ILE-VAL as effector), the flexibility is lost and
the activation domain becomes rigid [WaltJ82]. The TDPAC spectra and
their Fourier transforms for precipitated, ^{199m}Hg-labelled trypsinogen
are shown in fig.7 for various temperatures. At low temperatures, there
is a series of discrete frequencies due to different Hg-conformational
substates. The pattern is completely static. Upon warming up, we progressively enter the slow relaxation regime and gradually loose the discrete structures. Since the molecules as a whole were immobilized, we
observe intramolecular reorientational motion with a wide range of possible electric field gradient orientations. The reorientational correlation time, which is about 10 ns at 300K with an activation energy of
8 kcal/mole, is most readily extracted from the broadening of the zero
frequency peak, i.e. the Fourier transform of the hardcore (cf.eq(55)).
A control experiment with the rigidified activation domain yielded
static spectra up to the highest possible temperatures. Thus TDPAC is
very well suited to study intramolecular motion in biomolecules or
other macromolecules.

4.4 Summary

TDPAC is a very powerful tool for studying dynamic systems, especially
because it allows to discriminate between inhomogeneous static and dynamic broadening via the hardcore. This is particularly true for systems
with reorientations of the magnetic field/electric field gradient. Fluctuating fields without a reorientation of the quantization axis are difficult, if not impossible, to distinguish from static inhomogeneous
broadening. If the full relaxation range from motional averaging down
to the quasi-static regime is observable, a whole wealth of detailed
information about the dynamics of the system is obtainable.

5. ACKNOWLEDGEMENT

The continuous interest and support of Prof.G.M.Kalvius is gratefully
acknowledged. It is a pleasure to thank Dr.F.J.Litterst for various
stimulating discussions concerning relaxation theory. S.Saibene performed the miracle to get all these nice formulae out of the computer.
I am grateful to W.Tröger for his help in preparing the figures.

APPENDIX

I)   Some useful 3j-symbols [EdmoA57]

$$\begin{pmatrix} J & J & 2 \\ M & -M & 0 \end{pmatrix} \qquad (-1)^{J-M} \frac{2[3M^2 - J(J+1)]}{[(2J+3)(2J+2)(2J+1)(2J)(2J-1)]^{\frac{1}{2}}}$$

3j-symbols (continued)

$$\begin{pmatrix} J & J & 2 \\ M & -M-1 & 1 \end{pmatrix} \qquad (-1)^{J-M}(1+2M)\left[\frac{6(J+M+1)(J-M)}{(2J+3)(2J+2)(2J+1)(2J)(2J-1)}\right]^{\frac{1}{2}}$$

$$\begin{pmatrix} J & J & 2 \\ M & -M-2 & 2 \end{pmatrix} \qquad (-1)^{J-M}\left[\frac{6(J-M-1)(J-M)(J+M+1)(J+M+2)}{(2J+3)(2J+2)(2J+1)(2J)(2J-1)}\right]^{\frac{1}{2}}$$

II) Some useful $F_2(L,L',I,5/2)$ coefficients [StefR64]

L	1	1	2	2	3	3	4
L'	1	2	2	3	3	4	4
I							
1/2	0.0000	0.0000	-0.5345	-0.3780	-0.8018	0.0000	0.0000
3/2	0.3742	-0.9487	-0.1909	-0.5873	-0.4410	-0.3788	-0.8113
5/2	-0.4276	-0.5071	0.1909	-0.4980	0.0267	-0.4374	-0.3245
7/2	0.1336	0.6944	0.3245	-0.0714	0.4009	-0.2901	0.1711
9/2	0.0000	0.0000	-0.1909	0.5297	0.4009	0.0391	0.4897
11/2	0.0000	0.0000	0.0000	0.0000	-0.3341	0.4173	0.3924
13/2	0.0000	0.0000	0.0000	0.0000	0.0000	0.0000	-0.4130

For mixed transitions we have:

$$A_{kk}=(F_k(L,L,I_1,I_2) + 2\delta F_k(L,L',I_1,I_2) + \delta^2 F_k(L',L',I_1,I_2))/(1+\delta^2) \quad (A1)$$

where δ is the mixing ratio [YamaT67].
Eq(A1) is valid for the second transition in a γ-γ-cascade, i.e. γ_2.
For the first transition, i.e. γ_1, the interference term in eq(A1)
should have an additional factor $(-1)^{L+L'}$, which normally is a minus
sign!

III) Solid angle correction factors [YateM64]

The finite size of the detectors smears out the angular correlation
partly. This effect can be accounted for by multiplying each $A_k(i)$ by
an appropriate solid angle correction factor Q_k. Because of geometrical
effects and angular dependent absorption coefficients for the full energy peak of a spectrum, these Q_k are strongly energy dependent. We have:

$$Q_k=J_k/J_0 \text{ with } J_k = \int P_k(\cos\theta)\mu(\theta)\sin\theta d\theta \qquad (A2)$$

where $\mu(\theta)$ denotes the angle dependent absorption coefficient and the
integration is carried out from 0 to the cone-angle θ_{max}.
For $\mu(\theta)=1$ (say for very low γ-energies) we simply have e.g.:

$$Q_2=\cos\theta_{max}\cos^2(\theta_{max}/2) \qquad (A3)$$

IV) Some useful Laplace-transforms

$$1 \leftrightarrow 1/p \quad \Big| \quad e^{-\gamma t} \leftrightarrow 1/(p+\gamma) \quad \Big| \quad \cos\omega t \leftrightarrow p/(p^2+\omega^2) \quad \Big| \quad -t\cos\omega t \leftrightarrow (\omega^2-p^2)/(\omega^2+p^2)^2$$

REFERENCES

[BlumM71] M.Blume, Nucl.Phys.A167(1971)81
[BoscF77] F.Bosch and H.Spehl, Z.Physik A280(1977)329
[BostL70] L.Boström, E.Karlsson, and S.Zetterlund, Physica Scripta 2
           (1970)65
[ButzT79] T.Butz, A.Vasquez, and A.Lerf, J.Phys.C12(1979)4509
[ButzT82] T.Butz, A.Lerf, and R.Huber, Phys.Rev.Lett.48(1982)890
[ButzT83] T.Butz, A.Lerf, and R.Huber, Hyperf.Interact.15/16(1983)869
[DattS81] S.Dattagupta, Hyperf.Interact.11(1981)77
[EdmoA57] A.R.Edmonds,"Angular Momentum in Quantum Mechanics", Princeton
           University Press, Princeton, NJ 1957
[FrauH65] H.Frauenfelder and R.M.Steffen in:"Alpha-, Beta-, and Gamma-
           Ray Spectroscopy", K.Siegbahn ed., North-Holland, Amsterdam
           1965
[GerdE76] E.Gerdau, H.Winkler, B.Giese, W.Gebert, ahd J.Brausfurth, Hy-
           perf.Interact.1(1976)469
[Kuo F66] F.F.Kuo and J.F.Kaiser,"System Analysis by Digital Computer"
           John Wiley, NY 1966, pp232
[RogeJ75] J.D.Rogers and A.Vasquez, Nucl.Instr.& Meth.130(1975)539
[RoteM59] M.Rotenberg, R.Bivins, N.Metropolis, and J.K.Wooten,Jr.,
           "the 3-j and 6-j symbols", Technology Press MIT, Cambridge,
           Mass. 1959
[StefR64] R.M.Steffen and H.Frauenfelder in:"Perturbed Angular Correla-
           tions", E.Karlsson, E.Matthias, and K.Siegbahn eds., North-
           Holland, Amsterdam 1964, p.3-89
[WaltJ82] J.Walter, W.Steigemann, T.P.Singh, H.-D.Bartunik, W.Bode, and
           R.Huber, Acta Cryst.B38(1982)1462
[WinkH73] H.Winkler and E.Gerdau, Z.Physik 262(1973)363
[WinkH76] H.Winkler, Z.Physik A276(1976)225
[YamaT67] T.Yamazaki, Nuclear Data, Section A, Vol.3, nr.1, Academic
           Press, NY 1967,p1-23
[YateM64] M.J.L.Yates in"Perturbed Angular Correlations", E.Karlsson,
           E.Matthias, and K.Siegbahn eds., North-Holland, Amsterdam
           1964, p.453

Further suggested references:

TDPAC Reviews:

H.H.Rinneberg, At.Energy Rev. 17(1979)477
M.Forker and R.J.Vianden, Magn.Reson.Rev.7(1983)275
A.Lerf and T.Butz, Angew.Chem.Int.Ed.Engl.26(1987)110

Relaxation theory:

M.Blume, Phys.Rev.174(1968)351
H.Gabriel, Phys.Rev.181(1969)506
C.Scherer, Nucl.Phys.A157(1970)81
A.G.Marshall and C.F.Meares, J.Chem.Phys.56(1972)1226

F.Bosch and H.Spehl, Z.Physik <u>268</u>(1974)145

Relaxation: TDPAC experiments:

H.F.Wagner, M.Popp, and M.Forker, Z.Physik <u>248</u>(1971)195; refs. therein

E.Gerdau, J.Birke, H.Winkler, J.Braunsfurth, M.Forker, and G.Netz, Z.Physik <u>263</u>(1973)5

A.Baudry, P.Boyer, J.D.Fabris, and P.Vuillet, Hyp.Interact.<u>10</u>(1981)1057 and refs. therein

P.W.Martin, S.El-Kateb, U.Kuhnlein, J.Chem.Phys.<u>76</u>(1982)3819 and refs. therein

CHAPTER 22

MUON SPIN RELAXATION: A NOVEL MAGNETIC RESONANCE TECHNIQUE TO STUDY DYNAMIC AND STATIC MAGNETISM

Carel Boekema
San Jose State University
Department of Physics
One Washington Square
San Jose, CA 95192-0106
USA

ABSTRACT.   The novel muon-spin-relaxation technique is discussed in some detail with emphasis on the application in Solid State Physics, and especially with regard to studies in dynamic and static magnetism. A comparison with other magnetic resonance techniques is given.   A future outlook on muon spin research in magnetic dynamics is presented.

1.   INTRODUCTION

The muon-spin-relaxation (μSR) technique is a relatively new experimental tool in solid-state investigations.   Since about 1974, μSR has experienced considerable growth and has been successfully applied to problems of interest in solid-state research, most notably in the area of dynamic magnetism.   The fact that a muon can be implanted in any kind of material forms the basic key to success of the μSR method.

The status of the μSR technique is given in several review articles[1-6] and the proceedings of the first four International Topical Meetings on Muon Spin Rotation: μSR1,[7] Rohrschach (CH) 1978; μSR2,[8] Vancouver (Canada) 1980; Yamada-μSR83,[9] Shimoda (JPN) 1983; μSR86,[10] Uppsala (SW).

In μSR, the positive muon is used as a local magnetic spin probe in the interstitial regions of the sample under investigation.   When a spin-polarized muon beam is incident on a target-sample, the muons are slowed to thermal velocities within 10^{-9} - 10^{-11} sec.   In this fast process, the initial muon spin polarization is maintained.[3]   The fate of the muon depends upon the host and the temperature.   At low enough temperatures, the muon may stop at an interstitial site as has been observed in several solid materials.   Experimental evidence[7] and theoretical calculations[11] have shown that the stopping site is at least 1000 Å away from the damage produced by the thermalization process.   Thus, the immediate environment sampled by the stopped muon is that of the undamaged host.

369

G. J. Long and F. Grandjean (eds.), The Time Domain in Surface and Structural Dynamics, 369–375.
© 1988 by Kluwer Academic Publishers.

In the next two sections, we will describe the measurement of muon spin polarization, giving information about the static and dynamic magnetic fields. Basically, two experimental arrangements are available with the applied field either parallel (longitudinal) or perpendicular (transverse) to the initial muon spin polarization. In the last section, we will also give a brief review and examine the future outlook of μSR studies in solid state physics, especially magnetism.

2. μSR METHODS

As a result of the parity-violating pion decay ($\pi^+ \rightarrow \mu^+ + \nu_\mu$), a spin-polarized muon beam can be obtained. The pions are created by accelerators when a high-energy proton beam strikes a production target. In the rest frame of the pion, the muon is fully spin polarized parallel to its momentum direction. Some relevant physical muon properties are given in Table I. After implantation of the muon in a target, the muon decays via the parity-violating weak interaction ($\mu^+ \rightarrow e^+ + \nu_e + \bar{\nu}_\mu$) emitting the decay positron preferentially in the direction of the muon spin. Precession of the muon spin is caused by the local magnetic environment of the muon due to internal fields of the host or an externally applied field. These directional changes are observed by measuring the positron counting rate at time t after muon implantation, which is given by

$$\frac{dN_e}{dt} = N_0 \exp(-t/\tau_\mu)[1 + A_S\, G(t)\, \cos\theta]\, d\Omega$$

where θ is the angle between the initial muon spin direction and the positron emission direction, τ_μ is the mean muon lifetime (2.2 μsec), A_S is the initial (energy-averaged) asymmetry. $G(t)$ is the depolarization (or relaxation) function which yields information about variations in the local effective field.

TABLE I. Some Physical Properties of the Muon Relevant to μSR

Mass	106 MeV
	207 m_e
	0.113 m_p
Spin	½
Lifetime	2.197 μs
Magnetic moment	4.49 x 10^{-26} J/T
	3.183 μ_p
Gyromagnetic ratio	13.5537 MHz/kOe

The transverse field arrangement for μSR measurements has been commonly used. In Fig. 1, a μSR-spectrometer set-up for this mode of operation is depicted. In transverse-field μSR, a magnetic field is applied perpendicular to the initial muon spin direction, which causes the muon spins to precess at their Larmor spin precession frequency f. The μSR signal, observed in a time histogram of muon decay events, is superimposed on the muon decay. The time-distribution function can be written as follows:

$$N(t) = N_0 \exp (-t/\tau_\mu)[1 + A_S G_\perp(t) \cos (2\pi ft + \phi)] + BG$$

where ϕ is the phase angle between the initial muon spin direction and the positron-counter telescope. The term BG represents background events. The Larmor frequency f is determined by the magnetic field (B) at the muon site, and is given by $f(MH_z) = 13.55 B(kOe)$. The function $G_\perp(t)$ describes the transverse muon relaxation.

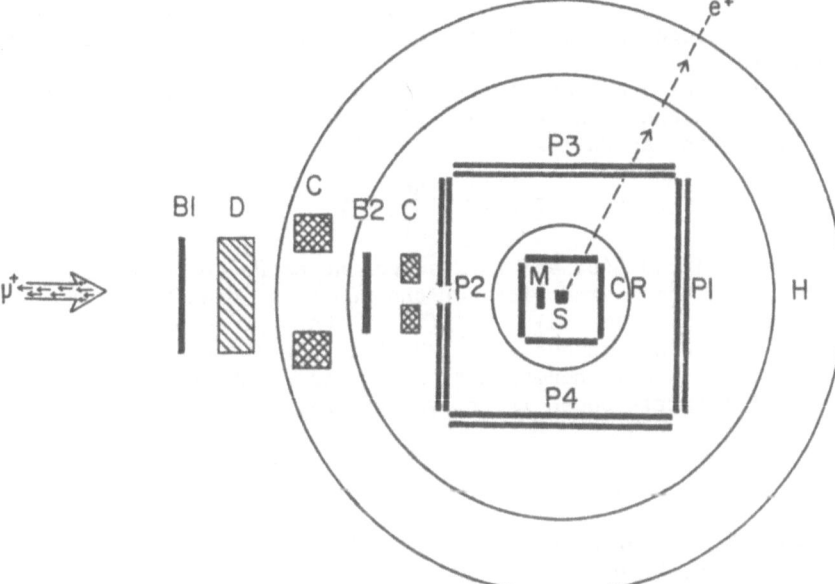

Figure 1
Schematic of a typical μSR spectrometer in transverse-field arrangement. The degrader (D) reduces the initial muon (μ^+) beam momentum such that a maximum amount of muons stop in the sample (S). The event $B1.B2.M.\overline{P1}.\overline{P2}.\overline{P3}.\overline{P4}$ defines a muon stop in the sample. Positron (e^+, from the muon decay) events are registered in the P1 (front), P2 (back), P3 (upper) and P4 (lower) telescopes. The collimators are denoted by C. H represents the Helmholtz coil pair producing the transverse field. The same spectrometer can be used in the longitudinal field arrangement by changing the field direction and recording data from the front and back telescopes only. The inner counters of the muon and positron telescopes are located inside the vacuum jacket of the cryostat (CR) ensuring high-efficiency collection of muon-decay events occurring in the sample under study.

In the longitudinal-field arrangement, the positron counters are fixed at $\phi = 0$ and π radians with respect to the initial muon-spin direction. Typically, the measurements are carried out in zero applied field although an external field can be applied parallel to the initial muon-spin direction if desired. The two time histograms are described by

$$N_{\phi=0}(t) = N_0 \exp(-t/\tau_\mu)[1 + A_S G_\parallel(t)] + BG$$

$$N_{\phi=\pi}(t) = N_\pi \exp(-t/\tau_\mu)[1 - A_S G_\parallel(t)] + BG$$

The longitudinal muon-spin-relaxation function ($G_\parallel(t)$) can be obtained in a relatively simple way from these histograms. $G_\parallel(t)$ contains useful information about effective dynamic as well as static magnetic fields. For example, one can extract correlation times not accessible by other magnetic resonance techniques (see next section).

Using standard μSR techniques, time histograms are recorded for the upstream and downstream counter telescopes. For the transverse field arrangement, upper and lower counter telescopes are also used (see Fig. 1).

For the zero and longitudinal field measurements, the μSR signals in the front and back time histograms determine the polarization as follows:

$$P_{exp}(t) = \frac{F(t) - \alpha B(t)}{a_b F(t) + a_f \alpha B(t)}$$

where $F(t)$ and $B(t)$ are the background corrected number of events and a_f and a_b the asymmetries for the front and back histogram respectively. The value $P_{exp}(t)$ is the experimental determination of $G_\parallel(t)$. The parameters $\alpha = N_0/N_\pi$ the front/back ratio, a_b and a_f are determined in normalization runs in which short-lived μSR signals from the sample under investigation are recorded.

3. PROBING MAGNETISM

As a magnetic spin probe method, the μSR technique used in transverse, zero or longitudinal mode is complementary to other spin probes exploited in well-established magnetic resonance techniques like MES and NMR. In Table II, a comparison between μSR, MES, NMR and ESR is made. Also, note that with respect to spin fluctuation and correlation times, the μSR technique is complementary to other techniques, as seen in Table III.

Information regarding magnetic dynamics with characteristic times shorter than 10^{-4} sec comes from experiments using these microscopic probe techniques. Due to the limitations and differences (see Tables II and III) in technique, the experimental results obtained from different methods will most likely differ in spin dynamics studies; especially when an applied field is used. However, results from several experimental tools used should converge to a clearer picture in magnetism studies.

TABLE II

A comparison of μSR with other magnetic resonance techniques

	μSR	MES	NMR, ESR
Spin-probe site	interstitial	substitutional	substitional
Site determination	easy/not easy	easy	easy
Applied field	zero/non-zero	zero/non-zero	non-zero
Quadrupole interactions	no	yes	sometimes
Spin probe - spin probe interactions	no	no	yes
Separation static and dynamic effects	yes	no	yes
Diffusion effects	sometimes	no	sometimes

TABLE III

Characteristic measurement times for various magnetic probes and the typical spin-correlation (local field fluctuation) times scales for different experimental techniques.

Technique	Measurement Time (sec.)	Correlation Time Scales
μSR	10^{-5}–10^{-6}	10^{-4}–10^{-12}
MES	10^{-8}–10^{-9}	10^{-8}–10^{-11}
NMR	10^{-6}–10^{-8}	10^{-7}–10^{-12}
ESR	10^{-9}–10^{-11}	10^{-9}–10^{-12}
Neutron Scattering	10^{-9}–10^{-14}	10^{-9}–10^{-14}
AC Susceptibility	10^{-1}–10^{-5}	10^{-1}–10^{-5}

Since the first observation of muon spin precession in Ni and Fe, much µSR work has been performed to study magnetically ordered materials. Perhaps one of the most successful studies has been the understanding of the origin of the muon hyperfine fields at interstitial sites in elemental ferromagnets[4,8]. The major characteristics that distinguish µSR in these studies are the following:

a) The muon samples a local magnetic field at an interstitial site yielding information which is rather difficult to obtain by other resonance techniques.

b) µSR signals remain in most cases observable through magnetic phase transitions, due to the very short dead time inherent to the technique.

c) The muon perturbs (not damages!) the host in its local environment; this may complicate the problem. For the magnetic oxides, this complication has been resolved[12]. Further, this also provides a sensitive test for theories of impurities in magnetic materials.

Zero-field muon precession frequencies have been observed in magnetic oxides such as α-Fe_2O_3. In these oxide materials, the existence of a muon-oxygen bond (analogous to hydrogen bonding) has been inferred from the data[12,13,14].

µSR investigations of spin glasses near the glass temperature at which spins become frozen in random orientations have revealed details of the spin-freezing process. Further experimental studies and theoretical analysis should be most helpful in addressing the question of whether the spin-glass transition is a true cooperative phase transition.

In spin glasses, dynamic effects are also observed below the glass temperature. Magnetic moments produce a local field at the muon site. Time-dependent fluctuations of these moments are reflected in the time-dependence of the local magnetic field.

These spin fluctuations can be observed through the muon depolarization (relaxation) rate. The method used is quite similar to nuclear spin-lattice relaxation measurements in NMR such that much of the relevant theory can be taken from NMR studies.

µSR studies in spin glasses have concentrated on understanding the nature of spin fluctuations and freezing processes below as well as in the neighborhood of the glass temperature. For more detail, the reader is referred to my third lecture on "Muon Spin Research on Spin Glasses." Here we note that there is a good overlap between the "correlation-time window" available to µSR (see Table III) and the region important for spin-glass phenomena near and below the glass temperature which is inaccessible to NMR.

The theoretical and experimental situation is not so advanced in the study of magnetic dynamics as in the studies in static magnetism. However, the intrinsic interest in dynamic phenomena is high, and much progress in the interpretation and understanding of dynamic magnetic effects is being made.

In conclusion, μSR appears to be a most useful magnetic probe to study spin-fluctuation phenomena in several exciting systems of current interest[10]. For the near future, the following list of possible items is given:

1. Studies of the new high T_c Superconducting Oxides;
2. The effect of disorder on critical dynamics, near percolation limits;
3. Spin fluctuations in systems without phase transitions;
4. Spin fluctuations in itinerant magnetic systems;
5. Studies of nonlinear excitations (solitons);
6. Continued studies on
   a. Heavy-Fermion Superconductors and
   b. Spin-Glasses.

4. REFERENCES

1. A. Schenck 1976 p. 159 Nuclear and Particle Physics at Intermediate Energies (ed. J.B. Warren, Plenum Press, New York).
2. A. Seeger, 1978 p. 349, Hydrogen in Metals I (ed. G. Alefeld and J. Volkl, Springer Verlag).
3. J.H. Brewer and K.M. Crowe, 1978 Ann. Rev. Nucl. and Part. Sci. 28 (1978) 239.
4. A.B. Denison et al., Helvetia Physica Acta 52 (1979) 469.
5. 'Muons and Pions in Materials Research,' eds. J. Chappert and R.I. Grynszpan, North Holland, Amsterdam (1984).
6. 'Muon Source for Solid-State Research,' Subcommittee on Muon Sources for Solid-State Research, Solid State Sciences Committee, National Research Council, National Academy Press, Washington D.C. (1984).
7. 'Muon Spin Rotation,' Hyperfine Interactions 6 (1979).
8. 'Muon Spin Rotation II,' Hyperfine Interactions 8 (1981).
9. 'μSR '83 - Yamada Conference,' Hyperfine Interactions 17-19 (1984).
10. 'μSR '86,' Hyperfine Interactions 31-32 (1986).
11. D.K. Brice, Phys. Lett. 66A (1978) 53.
12. C. Boekema, Proc. Yamada - μSR '83 Conference (April 1983 Shimoda, JPN), invited contribution, Hyperfine Interactions 17-19 (1984) 305.
13. C. Boekema et al., J. Mag. Mag. Mat. 36 (1983) 111.
14. C. Boekema et al., Phys. Rev. B31 (1985) 1233.

CHAPTER 23

MAGNETIC DYNAMICS IN OXIDES, AS OBSERVED BY MUON SPIN RELAXATION

Carel Boekema
San Jose State University
Department of Physics
One Washington Square
San Jose, CA 95192-0106
USA

ABSTRACT

Recent and earlier magnetic-oxide investigations by means of the
muon-spin-relaxation (μSR) technique are reviewed. The dynamics of
the processes in the observed phase transitions, as probed by the
muon, as well as of the behavior (probing, bonding and motion) of the
muon "spy" will be discussed. The main emphasis will be on the μSR
studies on α-Fe_2O_3 (hematite), Fe_3O_4 (magnetite) and V_2O_3
(karelionite). Electron-phonon interactions appear to be major
ingredients in the physical origins of the magnetic and
metal-to-insulator phase transitions seen in karelionite and
magnetite.

1. INTRODUCTION AND MUON BEHAVIOR IN OXIDES

As discussed in my first lecture on the technique of muon spin
relaxation (μSR), this experimental method is rather new; yet μSR
shows much progress. Therefore this lecture focuses on recent
historical background as well as the most current interests in
μSR-oxide studies. The μSR-oxide investigations started around 1974
and were undertaken to study the muon chemical properties in oxides.
At present, oxide spin glasses and the high-T_c superconducting oxides
are being studied.

In this section, muon probing, bonding and motion in magnetic
oxides will be discussed. How the muon, functioning as our spy,
behaves in oxides has been learned from a variety of studies,
including μSR work on hematite (α-Fe_2O_3), karelionite (V_2O_3) and
magnetite (Fe_3O_4). The magnetic dynamics in physical processes,
present in the host, will be highlighted in the following section.

Since the first use of muons as probes in solids, the μSR
technique had to deal with a major problem, namely the unknown muon
localization and diffusion in interstitial regions. In probing these
regions, it is most advantageous to also know the chemical effect the
muon has on its environment.

After thermalization of the implanted muon in a solid, a
muonium-like state (analogous to a hydrogen atom) is sometimes

377

G. J. Long and F. Grandjean (eds.), The Time Domain in Surface and Structural Dynamics, 377–389.
© *1988 by Kluwer Academic Publishers.*

formed, depending strongly on the conduction-electron density.[1,2] Muonium is not found in metals, because the positive muon charge is screened by the conduction electrons, such that no electron is bound to the muon. In the early stages of μSR, several investigators assumed muonium formation in insulators, even though the experimental results indicated that it only occurs in a few cases, such as in the oxide-material SiO_2 (quartz).[3] However, it has become clear[4] that immediately after muon implantation "hot" Mu may be formed and that chemical bonding determines the fate and state of the muon in insulators.

When discussing the fate of a muon implanted in a particular host, one has to consider that the muon site may be determined by the formation of chemical muon bonds. In the insulating transition-metal oxides, muon-oxygen bonds do occur.[5] Since the first measurements in α-Fe_2O_3, much progress has been made in understanding muon behavior (state, site, bonding and motion) in the magnetic oxides. μSR studies on corundum-structured magnetic oxides,[6] magnetite,[7] the rare-earth orthoferrites[8] and other magnetic oxides[9] have shown that at temperatures below 500 K the muon forms a muon-oxygen diatomic complex which is analogous to a hydrogen bond.

An outstanding example of μSR research in magnetic oxides is the study of α-Fe_2O_3. The temperature dependence of the μSR signals observed in zero field is illustrated in Fig. 1. Above 120 K, the effective field follows the magnetization curve, including the Morin transition, a well-known spin reorientation for α-Fe_2O_3. Below 120 K, three separate signals are seen, suggesting strongly that the muons are localized at different local energy minima. This effect was the first and most direct evidence indicating formation in solids of metastable muon states whose lifetimes are longer than the muon decay time.[2,10]

The muon-oxygen bond formation as revealed in magnetic oxides can be best described as a "muoxyl bridge" with a bond length of 1.0 Å. Potential energy calculations have resulted in two kinds of probable stopping sites: the so-called Rodriguez (R) and Bates (B) sites shown in Fig. 2a and 2b. Similar so-called Holzschuh and Hofmann sites have been found in the rare-earth orthoferrites.[11] The exact positions of the energy minima depend critically on lattice relaxation and polarization effects.[6] Zero-point muon motion must be taken into consideration as the relative distances between the minima are of the order of 1 Å.

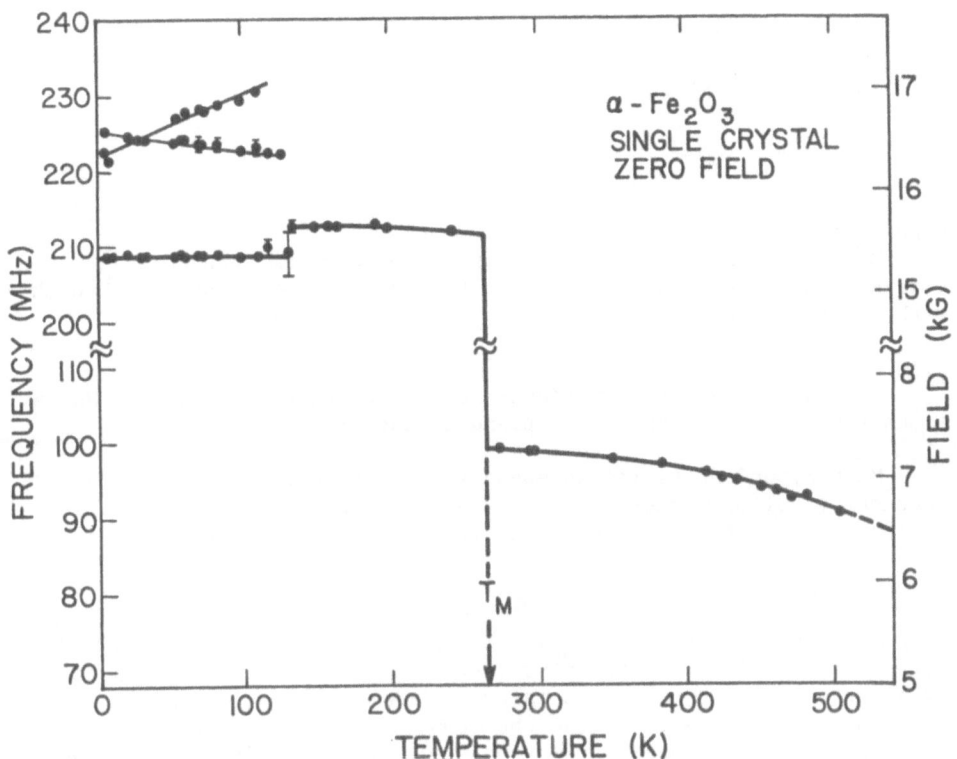

Figure 1
The temperature dependence of the frequencies of the μSR signals in zero applied field for α-Fe$_2$O$_3$. The lines serve as a guide to the eye. Above 500 K, the μSR signal disappears due to the onset of global muon diffusion. Above 700 K, a μSR signal with zero frequency reappears.[6]

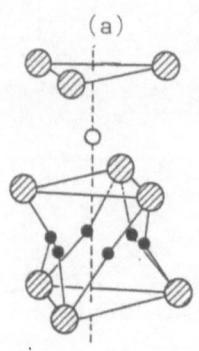 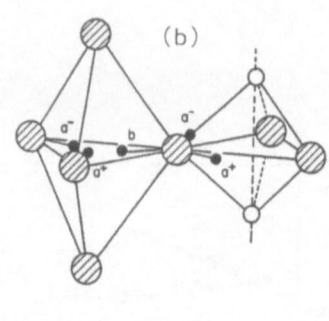

Figure 2a
    Muon stopping site candidates for the corundum structure:
Rodriguez sites are located on the vertices of a distorted octahedron
with an oxygen octahedron.

Figure 2b
    Muon stopping site candidates for the corundum structure: Bates
sites are located in the basal plane around an oxygen.

    Muon hyperfine fields measured in various magnetic oxides
frequently follow magnetization temperature trends. In zero applied
field the local field at the muon site (B_μ) is given by[5]

$$\vec{B}_\mu = \vec{B}_{int} = \vec{B}_{dip} + \vec{B}_{hpf}$$

where the internal field (B_{int}) has two origins: dipolar (B_{dip}) and
hyperfine (B_{hpf}). The hyperfine (or Fermi-contact) field is caused
by a non-zero electron-spin density at the muon site. When a
covalent bond occurs involving the muon, this field (B_{hpf}) is
primarily a supertransfer hyperfine field (B_{sthf}).[5,12] The dipolar
term has the following contributions[5]

$$\vec{B}_{dip} = \vec{B}_{dip,res} + \vec{B}_{lor} + \vec{B}_{dem} \ ,$$

where B_{lor} is the Lorentz field, B_{dem} the demagnetization and
$B_{dip,res}$ is the residual dipolar field of the magnetic moments within
the Lorentz field. These residual dipolar field contributions are
readily calculated as shown[6-8,11] for the rare-earth orthoferrites,
hematite, and magnetite.
    Muon-hyperfine-field calculations have been most helpful in
locating the site and other properties of muon states in oxides. The
idea that the muon behaves similarly in differently structured
(magnetic) oxides has also been strongly enhanced by a major advance
in understanding muon motion in oxides. In the temperature region
above room temperature a muon hopping model, using an
Abragam-formalism,[13] was successfully applied to interpret
muon-spin-relaxation data in α-Fe_2O_3.

In the following section, magnetite (Fe_3O_4) will be discussed serving as an example where the physical properties of both the spy and host have been studied. In the third section μSR on karelionite (V_2O_3) will be addressed. In section four, muon motion effects as observed in several magnetic oxides will be briefly discussed. In the final section, future aspects of, and conclusive remarks on, μSR studies of dynamics in oxide materials showing a variety of phase transitions will be presented.

2. MAGNETITE (Fe_3O_4)

Fe_3O_4 is a familiar ferrimagnetic oxide ($T_{FN} = 858$ K) that undergoes a metal-to-insulator transition at the well-known Verwey[14] temperature (T_V) near 123 K. At RT and above, magnetite can be studied to investigate muon bonding and motion effects; below RT, the poorly-understood Verwey phase transition and its relative phenomena (precursors) can be investigated.

Magnetite can be written as $(Fe^{3+})[Fe_2^{3+}e^{-1}]O_4$, which is not only a chemical but also a structural description; the square brackets refer to the B sublattice and the parentheses to the A sublattice in the inverse spinel structure. The extra 3d electron shown in the square brackets is responsible for the electric conduction above T_V. At RT, the iron magnetic moments are ordered along the <111> direction, due to a weak magnetic anisotropy. Below T_V, the easy axis of magnetization is parallel to the <100> direction.

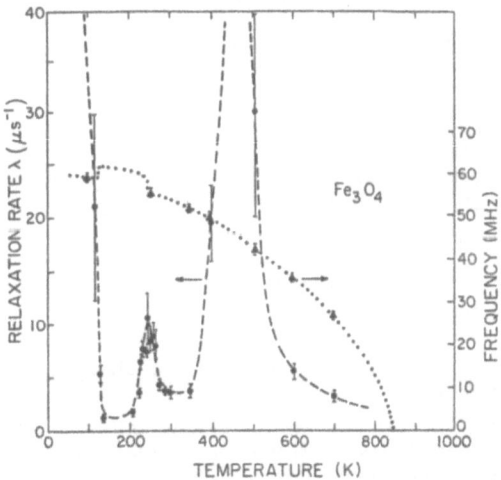

Fig. 3.
Temperature dependences of the μSR frequency and relaxation rate in magnetite.

   The temperature dependence of the observed µSR frequency and
relaxation rate for magnetite is depicted in Fig. 3. As can be
observed the general trend follows the magnetization curve. A
discontinuity occurs at T_V, as expected, and a reversal of this
internal field change occurs just below 250 K (for a more detailed
picture, see Fig. 4). Also the muon relaxation rate shows
departures: near T_V a drastic increase with decreasing temperature; a
maximum (not a divergence) near 250 K; and finally, a divergence and
disappearance of the signal near 360 K.

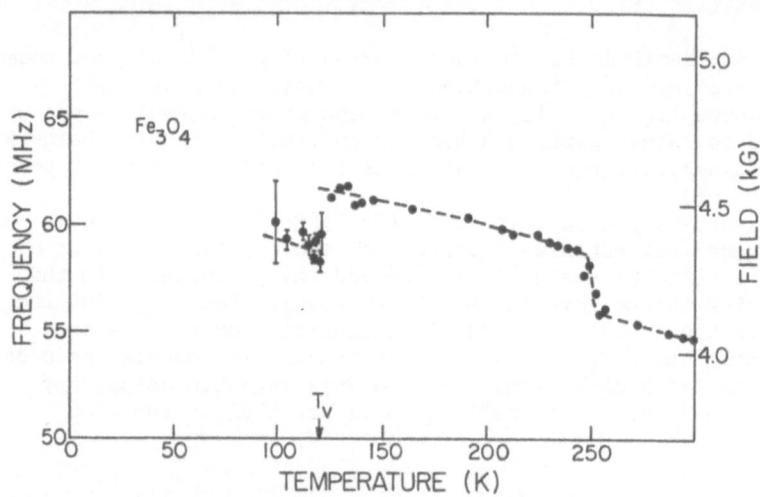

Fig. 4.
    µSR frequencies observed in magnetite in the
Verwey-transition temperature region.

   The latter behavior is a manifestation of muon motion effects.
At temperatures well above 500 K, the muon diffuses quite rapidly
throughout the entire lattice and is "motionally narrowed" by virtue
of the fact that it samples many sites during its lifetime. At lower
temperatures, this sampling will be less and the relaxation rate will
increase. Around 350 K, a transition takes place from global to
local muon diffusion. This local diffusion will cause the decrease
in the rate as observed.
   Of course, of more interest is the anomaly near 250 K (about
twice T_V). Careful experiments have shown that this anomaly is not a
muon state effect but a solid state effect. In fact, the anomaly
occurs at the highest temperature seen for anomalies or "precursors"
above the Verwey transition. There are known dynamic processes in
magnetite in this temperature regime which provide insight into the
observations made. The current conduction model for magnetite above
T_V is phonon-assisted electron hopping along iron cation chains in
the B sublattice.[15] Up to about 250 K, correlated atomic-group
motion (molecular polarons) have been observed indirectly through
neutron scattering experiments.[16] These polarons affect the
electron-hopping time and therefore change the magnetic environment
for the muon. Further, electron-phonon interactions are most likely

key factors in establishing Verwey order below T_V. The Verwey transition itself represents a phase change in which charge ordering occurs with a simultaneous small lattice distortion. This condensation of electron order with small atomic displacements appears to be a result of the temperature dependence of the electron-phonon interaction. As has been suggested by Mott and others,[15,17] above T_V magnetite may be considered as a Wigner glass, in which the conduction electrons randomly occupy one-half the iron B sites. Due to the random fields produced by this disorder, each conduction electron (or small polaron) then occupies an Anderson-localized state.

The anomaly near 250K may be due to cross-relaxation produced when the muon Larmor frequency equals the temperature-dependent hopping rate for the conduction electrons. This interpretation of the µSR data predicts an electron-hopping time (τ_e) of 2.8 nsec ($\omega_L \tau_e \cong 1$; $\omega_L \cong 352$ MHz), which is in good agreement with the estimate of 3 nsec from Mössbauer measurements. If this cross-relaxation explanation is correct, a shift of the anomaly to higher temperatures would occur in an external field (then ω_L is larger).

Temperature sweeps at 2 and 3 kOe with fields parallel to the <111> axis were performed below RT. In figure 5, the results for the temperature dependence of the muon frequency and relaxation rate are depicted. As can be seen, a field-dependent shift in the anomaly temperature can be observed. The external-field data suggest two peaks in the relaxation rate: one near 250K in both sweeps and the second at the "cross-relaxation" temperature.

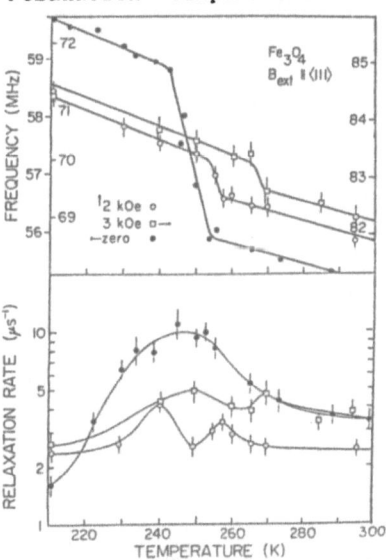

Fig. 5.
Temperature dependencies of the µSR frequency (a) and the relaxation rate (b) near the anomaly seen in magnetite in zero (·), 2 kOe (o) and 3 kOe (□) field applied in the <111> direction.

The molecular polarons, as deduced from diffuse neutron scattering, cease to exist above 250K. This result is consistent with the µSR data, assuming that below 250K molecular polarons provide short-range order. In a simple description, the change in muon frequency at the anomaly is associated with a change in the way the muon probes the local fields due to the neighboring iron moments. The valence and magnetic moment fluctuations of the iron ions is much slower than the muon frequency due to the short-range order, but becomes much faster at higher temperatures.

Within the model introduced above, one might expect to observe two frequencies with temperature-dependent intensities: one as observed above the anomaly where the local fields are influenced by independent small polarons, and the second as observed below where these fields are influenced by short-range order. In contrast, a dependence of the magnitude of the jump in muon frequency on applied field (parallel to the <111> direction), indicated new measurements were necessary with the magnetic field applied in different directions.

Figure 6 shows the µSR frequency as a function of temperature in the presence of an external field with its direction along the <110>. Two fields were used, at 720 and 2500 Oe respectively. The frequency scales are shifted by a constant amount to account for the difference in applied field. As the temperature is dropped below 250K, the µSR line splits, indicating the onset of two local fields. The upper line exists for zero and larger magnetic fields; we may say the lower line is field induced because its frequency has not been observed below 500 Oe, which about 2.5 times the demagnetization field (B_{dem}) of the sample involved.

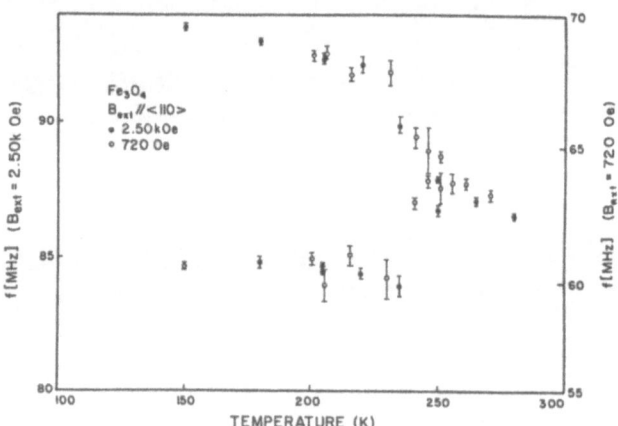

Fig. 6.
Temperature dependences of the µSR frequency observed in magnetite in 720 and 2500 Oe fields, applied along the <110> direction.

The transition from one muon magnetic environment to two magnetically different sites should be considered in light of dynamic effects. All these features are also observed through the muon relaxation rate. As an external field is applied along the <110> direction, the structure of the curve of muon relaxation rate vs temperature becomes complicated near 250K: the peak appears to split in two, as also was observed in the cross-relaxation measurements (see above).

The results of the longitudinal measurements are uninspiring, but confirm the interpretation. The plot of ln (λ) vs $10^3/T$ where λ is the muon relaxation rate in a 3.5kOe longitudinal field ($\vec{B}_{ext}$ // <111>) shows essentially a straight line with no apparent break near 250K.

These observations can be explained within standard relaxation theory[18] if one considers that the average field the muon spin sees is parallel to the induced magnetization ($\vec{M}$) direction in the presence of the externally applied field. In all the cases investigated, the local field direction (B_z) probed by the muon is parallel to $\vec{B}_{ext}$ for $B_{ext} > B_{dem}$. Cross relaxation is a spin-lattice phenomenon which can be expressed through

$$\lambda \sim (\langle B_x^2 \rangle + \langle B_y^2 \rangle) \frac{\tau_c}{1 + \omega_0^2 \tau_c^2}$$

where ω_0 is the muon frequency and τ_c the characteristic correlation time of the local magnetic field. For the transverse case, the initial muon-spin polarization is perpendicular to B_z and local fluctuations in $\vec{M}$ producing B_x and B_y allow sufficient relaxation. In the longitudinal case, B_x and B_y are so small that the spin-lattice effect cannot be observed.

Summarizing this section, the μSR results in magnetite strongly support the idea of phonon-assisted electron hopping above the Verwey transition. A narrow polaron-band scheme can be used to describe the conduction-electron behavior. In that case as proposed by Mott, a Wigner glass forms in which electrons near the Fermi energy are in Anderson-localized states induced by other localized electrons. Molecular polaron formation appears to be a major ingredient in the interpretation of the observed anomaly near 250K. In this interpretation, a model of an extra phase transition is used involving the onset of short-range order about 125K above the well-known Verwey transition. Short-range order or correlated atomic motion is indicated by the observed two inequivalent magnetic sites for the muon below 250K when an external field is applied along the <110> axis. Work is in progress to determine the axes and spatial symmetry of the proposed atomic correlations. As one may conclude from the above, the μSR technique has been most helpful in addressing an old problem: the Mott-Wigner glass phase in magnetite above the Verwey temperature has been clearly established.

3. KARELIONITE (V_2O_3)

The transition metal oxide vanadium sesquioxide (V_2O_3) exhibits most intriguing temperature-dependent electric and magnetic behavior.[19] Stoichiometric V_2O_3 is an insulating antiferromagnet below about 155K, while above this transition temperature (T_{mi}) this oxide behaves as a paramagnetic high-resistivity metal. Concurrent with the first-order metal-insulator transition is a slight crystal distortion with a change of the c/a ratio in the corundum structure.

he low-temperature magnetic structure, in which the V^{3+} ($3d^2$) ions have an unusual low magnetic moment of 1.2 μ_B, has sheets of parallel-aligned spins within the hexagonal layers, the spins making an angle of +/- 71° with the c-axis. V_2O_3 is conducting above T_{mi} through an itinerant band scheme with the band edge being sensitive to change in parameters such as temperature, pressure and doping.[19] As with magnetite (see above), electron-phonon interactions play an important role in determining the electric properties of karelionite.

As discussed, the magnetic field at a localized muon site in a magnetic oxide is a sum of the externally applied field, the dipole contribution from magnetic ions, and a covalent contribution of supertransfer fields due to transfer of unpaired spin density from the metal ions to the muon through the intermediate oxygen. In V_2O_3, the V^{3+} ion has its two 3d electrons in the t_{2g} states; therefore π bonding between the cation d orbitals and the oxygen p orbitals will dominate the spin transfer. As observed for Cr_2O_3 (Cr^{3+}, $3d^3$), these effects can be neglected; this is in contrast with α-Fe_2O_3 (see Fig. 1) where covalent effects are the cause of the observed temperature dependence of the splitting below 120K. Thus for V_2O_3, the observed fields are expected to be mainly of dipolar nature.

The μSR frequency (or the local field) vs temperature for a single crystal V_2O_3 sample in zero applied field follows the known magnetization curve; furthermore, the MI and magnetic phase transition is clearly observed at T_{mi}.[20] As expected, a single signal with zero frequency was seen in the paramagnetic phase above T_{mi}. Below T_{mi}, two signals were observed; one signal continues to have a zero frequency, while the other has a value of 15 MHz when extrapolated to zero Kelvin. The 15 MHz signal is stronger in intensity by a ratio of 3:1. The magnitude of the local field is about 1.1 kOe and external-field data show that the internal field directions (in zero applied field) are at 7°, 66°, 116° and 174° with respect to the c axis.

Based upon the above discussed knowledge (the magnitudes, directions and mainly dipolar character of observed fields), a muon site search has been performed to reproduce the experimental results. The muon sites which satisfy these search conditions are shown in Fig. 7. These sites are in the same regions as those located in α-Fe_2O_3; however, only a subset of the Bates and Rodriguez sites are found for V_2O_3. Each oxygen in V_2O_3 has less than the five sites (3B and 2R) expected from symmetry and potential considerations.

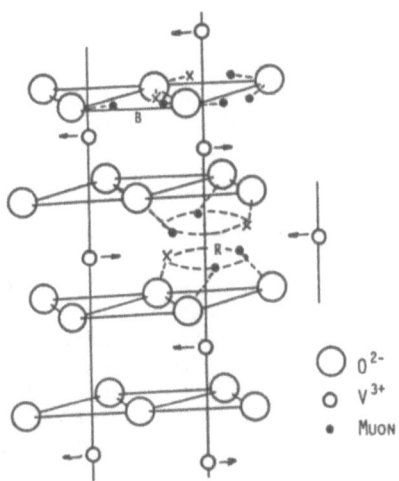

Fig. 7.
    Muon sites are shown for the Bates (B) and Rodriguez (R) regions
in the corundum structure of V_2O_3. Sites "missing" for the
monoclinic phase are marked x. The arrows indicate the displacements
of the V^{3+} ions during the MI phase transition.

    The missing fraction provides a signature of the structural
changes at the phase transition. Goodenough suggested,[21] confirmed
by indications of X-ray studies,[22] that V-V pairing is the
responsible mechanism for the phase transition in karelionite. This
pairing should occur across the shared octahedral edge with an
increased V-V on the edge together with a simultaneous
increase of the V-V distance across the octahedral face; the V-V pair
must then rotate toward the octahedral void. The arrows on the V
ions in Fig. 7 represent these motions. In both the Rodriguez and
Bates regions, the electric potential is raised for sites lying
closer to the V^{3+} ions and is lowered for the remaining sites,
removing the equivalence expected for the corundum structure. Thus
this μSR-oxide study provides an indirect proof of the driving
mechanism of the phase transition in karelionite.
    The changes in μSR hyperfine parameters observed across the MI
transition for V_2O_3 can be compared with those seen across the Verwey
MI transition for magnetite. A large change in μSR frequency is
observed in V_2O_3, while only a 6% change occurs for Fe_3O_4. This is
basically due to the antiferromagnetic ordering in V_2O_3 from a
high-temperature paramagnetic state, while Fe_3O_4 remains magnetically
ordered across the Verwey transition. Fe_3O_4 reveals a drastic change
in relaxation rate leading to the disappearance of the μSR signal
below about 100K. In V_2O_3 the changes in muon relaxation rate above
and below T_{mi} are not that dramatic and are relatively small. The
contrast here is due to the differences in the 3d electronic states
responsible for the magnetic fields. In Fe_3O_4, these electrons are
in localized polaron states, while in V_2O_3 these electrons are in
extended itinerant band states above T_{mi}.

In conclusion, the μSR results on V_2O_3 show that V-V pairings forced by the electron-phonon interaction are the main physical cause of the metal-insulator transition. Because of the interstitial position of the muon probe, the proof could be more direct than existing indications; a result which is promising for the other μSR-oxide work in progress.

4. MUON MOTION IN OXIDES

If one compares the muon relaxation rate as a function of temperature for magnetically and structurally different oxides (α-Fe_2O_3, Fe_3O_4, and $YFeO_3$) for which no magnetic transitions occur between RT and 600K, a sharp increase in relaxation rate starts above 360K (see also Fig. 3 for Fe_3O_4). These data suggest that the muon is behaving similarly in various magnetic oxides owing to the overall presence of the oxygen ions (and thus not because of impurities, as was proposed in the early μSR days).

A consistent explanation is that the muons are localized well below RT in μ-O bonding states; as temperature increases ($>$ 100K) local diffusion sets in; above 500K global diffusion through the whole lattice takes place. The activation energies of this thermally activated global diffusion process are of the order of 400 meV, determined from temperature dependencies in relaxation data. Activation energies of the local diffusion occurring below 360K are of the order of 100 meV.

Such activation energies are characteristic for hydrogen- (muon) oxygen bonding. This picture of muon dynamics as described above and in the first section, is substantiated further by the combined μSR results and their interpretation for the rare-earth orthoferrites[11] and for a well-known insulating diamagnetic oxide corundum (α-Al_2O_3).[4] Further comparing α-Fe_2O_3, Fe_3O_4 and V_2O_3, we conclude that μ-O bonding and muon motion takes place in insulating as well as metallic and semimetallic oxides. Thus overall, muon behavior, including motion, appears to be similar in various oxides.

5. CONCLUDING REMARKS

From μSR studies on well-known magnetic oxides, some of them showing phase transitions, new and relevant aspects in magnetic dynamics have been learned, in addition to understanding the general muon behavior in oxides. Muon dynamics in oxides is quite interesting and potential energy calculations are being undertaken to better understand in detail the global and local muon diffusion characteristics. This may be useful knowledge for hydrogen behavior (storage!) in ceramic oxides.

In the case of magnetite, a definite proof has been provided by μSR that above T_V the conduction mechanism could be described by phonon-assisted electron hopping. Electron-phonon interactions play an important role in the onset of the polaronic phase transition just below 250K. For karelionite, μSR provided a more direct proof for the physical origin of the metal-insulator transition.

Although in karelionite and magnetite different energy-band features are present, both μSR studies revealed the overall importance of the electron-phonon interaction in the physical origins of their phase transitions. Thus, application of μSR to the investigation of high-T_c superconducting oxides appears to be full of promise.

6. REFERENCES

1. T.L. Estle, Hyperfine Interactions **8** (1981) 365.
2. A.M. Stoneham, Helvetia Physica Acta **56** (1983) 449.
3. A. Schenck 1976 p. 159 Nuclear and Particle Physics at Intermediate Energies (ed. J.B. Warren, Plenum Press, New York).
4. C. Boekema et al., Hyperfine Interactions **32** (1986) 667.
5. C. Boekema, Hyperfine Interactions **17-19** (1984) 305. A.B. Denison, J. Appl. Phys. **55** (1984) 2278.
6. C. Boekema et al., J. Mag. Mag. Mat. **36** (1983) 111.
7. C. Boekema et al., J. Mag. Mag. Mat. **31-34** (1983) 709; Hyperfine Interactions **15-16** (1983) 529; Phys. Rev. **B31** (1985) 1233; Phys. Rev. **B33** (1986) 210.
8. E. Holzschuh et al., Phys. Rev. **B27** (1983) 5294; C. Boekema et al., Phys. Rev. **B30** (1984) 6766.
9. Y.J. Uemura et al., Hyperfine Interactions **17-19** (1984) 339.
10. A. Browne and A.M. Stoneham, J. Phys. **C15** (1982) 2709.
11. T.K. Lin et al., Hyperfine Interactions **31** (1986) 475.
12. G.A. Sawatzky and F. van der Woude, J. Phys. **C6** (1974) 47.
13. K.J. Rüegg et al., Hyperfine Interactions **8** (1981) 574.
14. E.J.W. Verwey and P. Haayman, Physica **8** (1941) 979.
15. "Proc. Int. Mtg. on Magnetite and Other Materials Showing a Verwey Transition", Phil Mag. **B42** (1980).
16. S.M. Shapiro et al., J. Phys. Soc. Jpn. **39** (1975) 839.
17. N.F. Mott, Festkörper Probleme **19** (1979) 331.
18. A. Abragam, "Principles of Nuclear Magnetism" Oxford Press (1961).
19. J.M. Honig, J. Solid State Chem. **45** (1982) 1.
20. A.B. Denison et al., J. Appl. Phys. **57** (1985) 3743.
21. J.B. Goodenough, "Magnetism and the Chemical Bond", John Wiley & Sons, New York (1963) p. 262.
22. P.D. Dernier and M. Marezio, Phys. Rev. **B2** (1970) 3771.

CHAPTER 24

MAGNETIC SUSCEPTIBILITY STUDIES OF TIME DEPENDENT PHENOMENA:
APPLICATION TO THE MAGNETIC RELAXATION PROCESSES IN FINE PARTICLES AND
IN SPIN-GLASSES

D.FIORANI
I.T.S.E.-C.N.R.
Area della Ricerca di Roma-P.B.10
00016 Monterotondo Stazione(Rome)-P.B.10
Italy

ABSTRACT. The application of magnetic susceptibility measurements to
the study of time-dependent phenomena is reviewed. In particular, the
measurements of the A.C. susceptibility, $\chi(a.c.)(\omega)=(dM/dH)(H \rightarrow 0)$
represents a powerful technique for dynamic studies. It gives a
measure of the susceptibility at a given observation time $t_m(=1/\omega)$, ω
being the angular frequency of the external oscillating magnetic
field. The attention is focused on the study of magnetic relaxation in
fine particles and spin-glass systems. Their main dynamic properties,
are reviewed: the time dependence of the susceptibility, the time
decay of the remanent magnetization and the frequency dependence of
the blocking and freezing temperature.

1. INTRODUCTION

Susceptibility measurements are among the principal measurements
carried out on magnetic systems to study their static and dynamic
properties.

If a material containing magnetic ions is placed in a magnetic
field H, it acquires an induced magnetization M which is proportional
to H:

$$(1.1) \qquad \vec{M} = \chi \vec{H}$$

The proportionality constant χ is the susceptibility, which is a
quantitative measure of the response of a material to an applied

391

G J. Long and F. Grandjean (eds.), The Time Domain in Surface and Structural Dynamics, 391–428
© 1988 by Kluwer Academic Publishers.

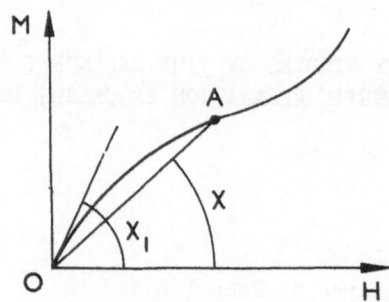

Fig.1 Definition of the susceptibility $\chi = M/H$ and of the initial susceptibility $\chi(i) = (dM/dH)(H\rightarrow0)$

magnetic field. $\chi = M/H$ is usually called static susceptibility (d.c.) as it is measured in a static external magnetic field. For magnetically isotropic samples, where the H and M vectors are collinear, the susceptibility is a scalar quantity. For anisotropic materials χ is a symmetric tensor. χ is dimensionless, but is expressed as emu/cm^3.

In some cases, when M varies linearly with the field (e.g. for paramagnets if H/T<<1), χ is field independent. In general χ is a function of H: $\chi =f(H)$(Fig.1). Therefore the ordinary susceptibility M/H may be different from the initial susceptibility defined as:

(1.2) $\chi_o(d.c.)=(M/H)(H\rightarrow0)$ static initial susceptibility

(1.3) $\chi(a.c.)=(dM/dH)(H\rightarrow0)$ dynamic initial susceptibility

$\chi(a.c.)$, called the dynamic or differential susceptibility, is measured by means of applying a very small oscillating magnetic field H(t)=hcos ω t which can also be written H(t)=Re[hexp(i ω t)] ($\omega =2\pi\nu$ is the angular frequency of the a.c. field). The magnetic field induces a magnetization varying in the time M(t)=m(ω)exp(iωt). The corresponding susceptibility is a complex quantity:

(1.4) $\chi(a.c.)= \chi'(\omega) - i\chi''(\omega)$

The real part χ' is called the dispersion, and χ'' the absorption.
In the presence of magnetic relaxation the system can not be able to follow the changes of the external field immediately. In this case $\chi(a.c.)(\omega)$ is different from the initial static susceptibility (d.c). The frequency dependence of $\chi(a.c)$ will depend on the value of the frequency ω of the oscillating field with respect to the

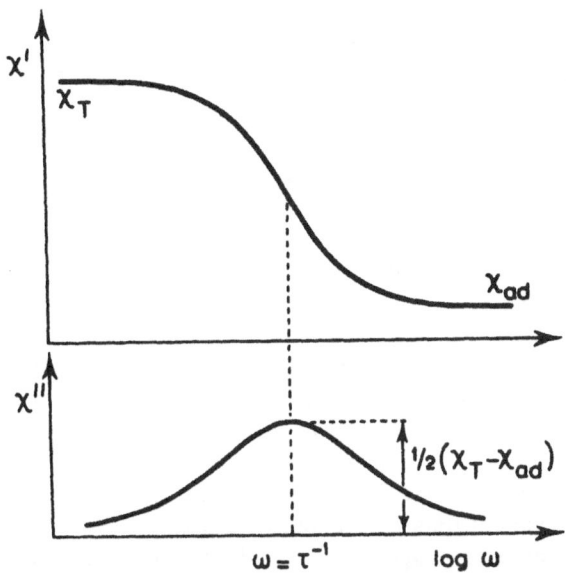

Fig.2 Frequency dependence of χ' and χ'' according to the Eq.(1.5) (from ref.2).

relaxation time τ (i.e., generally speaking, the time necessary for a magnetic system, constituted by an assembly of spins, to move from one configuration of minimum energy to another).

When a magnetic field is applied,the system of magnetic ions will tend to reach an equilibrium distribution over the different energy levels depending on the value of the magnetic field.

If $\tau \ll 1/\omega$ the magnetic system follows instantaneously the changes of the magnetic field and the observed magnetic properties will be the result of a time average. In the low-frequency limit the system reaches equilibrium conditions and χ' corresponds to the low field static value $\chi_o(dc)$ and it is called isothermal susceptibility $\chi(T)$.This means that the spins are in thermal contact with their surroundings.

If $\tau \gg 1/\omega$ the magnetic system is not able to follow the changes of the magnetic field and the observed magnetic properties will correspond to those of an energy configuration far from the equilibrium. The high frequency limit of χ' is called the adiabatic susceptibility $\chi(ad)$, which means that the assembly of spins is uncoupled from its surroundings.

The two above defined susceptibilities are related as /1/:

$$(1.5) \quad \chi(\omega) = \frac{\chi(T) - \chi(ad)}{1 + i\omega t} + \chi(ad)$$

where:

$$(1.5a) \quad \chi'(\omega) = \frac{\chi(T) - \chi(ad)}{1 + \omega^2 \tau^2} + \chi(ad)$$

$$(1.5b) \quad \chi''(\omega) = \omega\tau\frac{(\chi(T) - \chi(ad))}{1 + \omega^2 \tau^2}$$

The frequency dependence of χ' and χ'' is reported schematically in Fig.2 /2/. $\chi' = \chi(T)$ if $\tau \ll 1/\omega$ and $\chi' = \chi(ad)$ if $\tau \gg 1/\omega$. χ'' tends to zero at these limits and shows a maximum at $\omega \simeq \tau^{-1}$, whose height is $1/2(\chi(T) - \chi(ad))$.

The measurements of the initial susceptibility have the important advantage, with respect the measurements at high fields, of reducing the perturbation on the magnetic system due to the external field itself. In general this allows much clearer evidence of the magnetic phase transitions to be obtained.

The most common apparatus used to measure the high-field susceptibility are the magnetic balances (Gouy and Faraday), based on the force method /3/ and the vibrating sample magnetometer /4/, where fields of the order of magnitude of several kOe are used.

For low field susceptibility measurements the most common apparatus are the SQUID magnetometer /5/ for the static susceptibility and the Mutual Inductance Bridge for the dynamic susceptibility /6/, where fields of the order of magnitude of the Oersteds or tenths of an Oersted are used.

The A.C. susceptibility measurements are particularly useful in studying magnetic relaxation phenomena, as it is possible, by using the same technique, and just varying the frequency of the oscillating field, to change the experimental time window over a large range (usually $10s < (tm=1/\omega) < 10^{-4}s$). This is a great advantage with respect to other techniques as it makes possible a direct comparison of the results obtained at different frequencies. Caution is necessary instead in comparing results obtained by different techniques, each having its characteristic time constant, as the observed properties cannot be exactly the same (e.g. difference between a sum of local properties and an average of them). In examining the magnetic susceptibility studies of time dependent phenomena we shall restrict ourselves to the analysis of the dynamic properties of fine particles and spin-glasses.

2. MAGNETIC RELAXATION PROCESSES IN SMALL PARTICLES

Magnetically ordered materials below a critical size consist of a single domain, because there would not be an energy gain by splitting it into smaller domains /7,8/. The critical size depends on the type of magnetic material. For most of them it is of the order of a few hundred Å. The exchange interactions within a single domain particle may lead, depending on their sign, to a ferromagnetic, ferrimagnetic or antiferromagnetic ordering.

The easy magnetization directions are separated by anisotropy energy barriers arising from different contributions: magnetocrystalline, shape, stress, surface, interaction, exchange. For fine particles the energy barrier is low (and decreases with decrease in the particle size) and it may be comparable to thermal energy kT. In this case the magnetic moment of a particle, considered as a whole, may relax, with a given relaxation time τ , between the directions of easy magnetization, the relative orientation between the spins remaining unchanged. The observed magnetic behaviour will depend on the characteristic measuring time tm of the experimental method used. If $\tau \ll$ tm the observed magnetic properties will result in a time average of the magnetization fluctuations. Hence a material constituted by an assembly of independent particles will behave as a paramagnet, although one with a very large magnetic moment. This is why the term "superparamagnetism" has been used to describe the apparently disordered magnetic behaviour of fine particles /8/.

On the contrary if $\tau \gg$ tm the magnetization of a particle will be "seen", during the experiment, fixed along one easy direction, as for the magnetically ordered materials.

The temperature at which $\tau =$ tm is defined as the blocking temperature Tb and depends on the time scale of the experiment: Tb(tm). Below Tb the magnetic behaviour is characterized by irreversibility, remanence and a time dependence of the susceptibility.

Susceptibility and magnetization measurements are very fruitful in studying relaxation phenomena in fine particle systems. Such measurements allow accurate determination of Tb to be made. In particular A.C. susceptibility measurements at different frequencies allow the measuring time dependence of Tb to be studied with the advantage of using the same technique in a large range of time-scales.

A single domain behaviour is observed in different kinds of

materials, such as ferrofluids (colloidal suspensions of small magnetic particles), magnetic rocks (containing iron oxide-rich minerals in finely divided form embedded in a non magnetic matrix), catalysts (where magnetic particles are precipitated in a non magnetic matrix), magnetic recording media, permanent magnets, etc.. By studying the relaxation properties it is possible to obtain informations which are useful for applications in different fields: e.g. catalysis (the particle volume distribution), archeology (dating of ceramics), paleomagnetism, atmospheric pollution, etc..

2.1 The superparamagnetic state

Consider an assembly of identical, mutually independent, single domain ferromagnetic particles, each having a magnetic moment $Mp=MsV$(Ms is the saturation magnetization and V is a volume of a particle). In the absence of a magnetic field the resulting magnetization is zero. Let's consider for simplicity the case of uniaxial particles with their axes aligned parallel to H. This does not represent the situation in real systems, where a distribution of orientations of anisotropy axes with respect to the field direction is present. The total energy per particle is given by:

$$(2.1) \quad E(\vartheta) = KaVsin^2\vartheta - HMsVcos\vartheta$$

where Ka is the anisotropy energy constant and ϑ is the angle between the magnetization direction and the easy direction of the magnetization.

The equilibrium value of the magnetization parallel to the field direction is given by:

$$(2.2) \quad M = \frac{Ms \int_0^\pi exp[-E(\vartheta)/KT] \cos\vartheta \sin\vartheta d\vartheta}{exp[-E(\vartheta)/KT]\sin\vartheta d\vartheta}$$

when it is averaged over a time longer than the relaxation time of the magnetization ($tm \gg \tau$).

When the anisotropic energy is negligible with respect to that due to the magnetic field and such that $KaV/kT \ll 1$, the magnetization is described by the Langevin function:

$$(2.3) \quad <M> = MsL(MsHV/kT)$$

where $L(MsHV(kT) = coth(MsHV/kT) - kT/(MsHV)$

In the limit of low and high field the following approximation is possible:

$$(2.4) \quad <M> = M_s^2 HV/3kT \qquad \text{for } M_sHV/kT \ll 1$$

$$(2.5) \quad <M> = M_s(1-kT/M_sHV) \qquad \text{for } M_sHV/kT \gg 1$$

The equations (2.3)-(2.5) are formally identical to the corresponding ones for classical paramagnetism. The difference arises from the fact that M_sV is the moment of the particle, m_p, which may contain up to several thousands of atoms.

If KaV is not small compared to kT, the equation (2.3) is not appropriate and another expression should be used /9/.
For large values of KaV/kT a good approximation is given by the equation:

$$(2.6) \qquad <M> \simeq M_sVtanh(M_sVH/kT) \qquad \text{for } KaV \gg kT$$

which for $M_sHV/kT \ll 1$ becomes:

$$(2.7) \qquad <M> = M_s^2 HV^2/kT \qquad \text{for } M_sHV/kT \ll 1$$

On the other hand for a statistical distribution of the anisotropy axes or for a cubic symmetry the approximation for low and high fields, given by the equations (2.4) and (2.5) remain valid independently of the ratio KaV/kT /10/.

In the real systems the particle volume is far from uniform and there is a distribution of anisotropy axes. The magnetization is a unique function of M_sHV/kT. Therefore if the effective particle volume is temperature independent and once the correction for the temperature variation of M_s has been made, magnetization curves measured at different temperatures superimpose (Fig.3), as long as H/T is sufficently small that $tm \gg \tau$. No hysteresis is observed in the M vs H curves. The initial susceptibility follows a Curie law.

The superposition of the M vs H/T curves, and the observation of a Curie-Weiss law are considered evidences for the presence of superparamagnetism. Both are valid independently of a distribution $f(V)$ of the particle volume.

Deviations from the superposition of the M vs H/T plots may have different origins. The most important are:
- The Curie temperature and the M_s vs T variation for a ferromagnetic particle may be different from those of the bulk material, at least

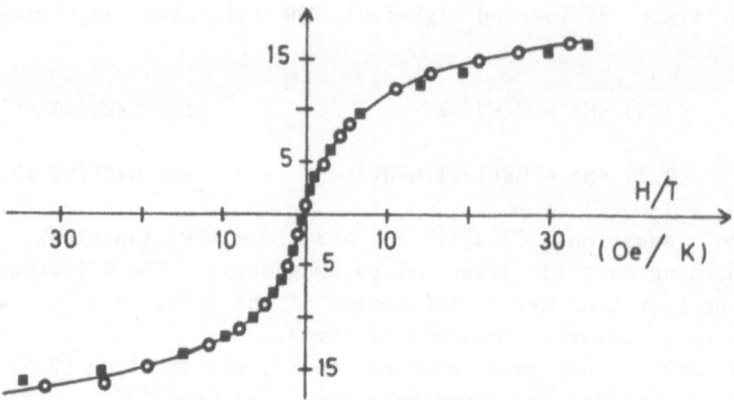

Fig.3 Magnetization curves of iron particles (average diameter:45Å) in mercury: o: 77K; □: 200K (from ref.10)

below a given dimension.
- Interactions between particles may be present. These produce an internal field and therefore in such a case the initial susceptibility is expected to follow a Curie-Weiss law $\chi = C/(T-\vartheta)$, where the Weiss constant gives a measure of the internal field.

2.2. Temperature dependence of the susceptibility

Consider an array of isolated ferromagnetic particles. It is assumed that the Curie temperature of each particle is high compared to the temperature range examined, so that the saturation magnetization Ms can be considered independent of temperature.

The relaxation time is given by the Arrhenius law:

$$(2.8) \qquad \tau = \tau_0 \exp(Eb/kT)$$

where τ_0 is a time constant (of the order of 10^{-9} for ferromagnetic particles). Its formulation was given by Brown /11/. Eb is the anisotropy barrier height, given by Eb=KaV for uniaxial anisotropy. The temperature dependence of τ_0 and Ka are considered negligible in the measured temperature interval.

It is assumed that there is a distribution in the volume of the particles. In presence of an alternating magnetic field with angular frequency ω the susceptibility of the system is given by /12/:

$$(2.9) \qquad \chi(T, \omega) = \int_0^\infty \chi_v(T, \omega) f(V) dV$$

where $f(v)$ is the volume distribution function and $\chi_v f(V)dV$ is the contribution to the total susceptibility of particles with volumes in the range dV about V.

The complex susceptibility is:

$$(2.10) \qquad \chi = \frac{\chi(T) + i \omega \tau \chi(ad)}{(1 + i \omega \tau)}$$

whose the real part is given, for particles with volume V, by:

$$(2.11) \qquad \chi_v(T, \omega) = \frac{(\chi(T) + \omega^2 \tau^2 \chi(ad))}{1 + \omega^2 \tau^2}$$

$\chi(T)$ is the superparamagnetic susceptibility corresponding to thermal equilibrium. It is given by a Curie law:

$$(2.12) \qquad \chi(T) = M_s^2 V/3kT$$

$\chi(ad)$ is the susceptibility of a collection of single domain particles in the blocked state. It is temperature independent:

$$(2.13) \qquad \chi(ad) = \frac{<\sin^2 \varphi > M_s^2}{2Ka}$$

where φ is the angle between the applied field and the easy direction of the magnetization. The average is over all the particles.

The general form of $\chi(T, \omega)$ is obtained by substituting Eq.(2.10) and (2.8) in Eq. (2.9).

In the high temperature limit kT>>KaV and then $\omega^2 \tau^2 << 1$. The susceptibility becomes:

$$(2.14) \qquad \chi(T, \omega) = M_s^2 \bar{V}/3kT$$

where $\bar{V} = \int_0^\infty V f(V) dV$ is the average volume. In the low temperature limit kT<<KaV and only particles whose volume V<<$\bar{V}$ contribute to the temperature-dependent part of the susceptibility.

The susceptibility given by Eq. (2.9) exhibits a maximum. The temperature Tb at which the maximum occurs depends on the form of f(V) (FIG. 4) /13/. Gittleman et al /12/ have calculated Tb for different distribution functions (e.g. for a Poisson distribution function Tb = 1.8KaV/k|ln($\omega \tau_0$)|.

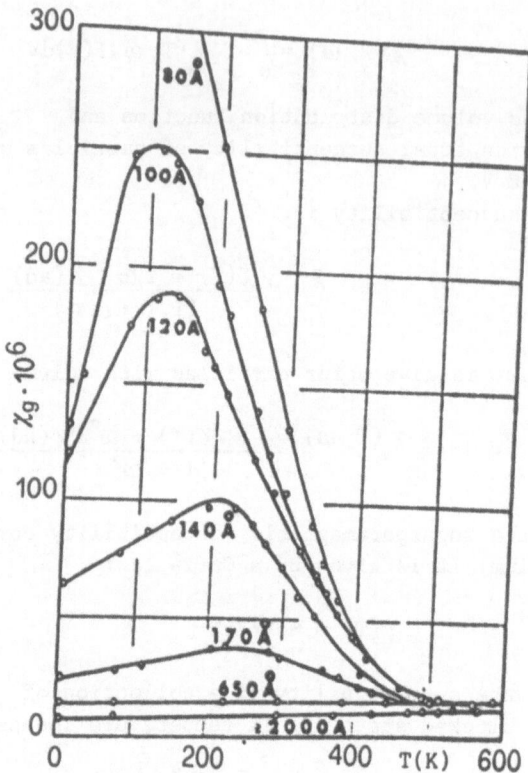

Fig.4 Temperature dependence of the susceptibility of NiO particles (
from ref.13)

Below Tb irreversibility effects are observed. While the low-
field susceptibility measured after cooling down in zero field
(ZFC) shows a maximum at Tb, the susceptibility measured after
cooling down in an applied field χ(FC) splits from χ(ZFC)
approximately at Tb and continues to increase with decreasing
temperature (Fig.5)/14/. The difference between χ(FC) and χ(ZFC) is
due to the fact that the cooling down through Tb in an applied
magnetic field leads to a blocking of the particles magnetizations in
easy directions close to that of the field.
On increasing the magnetic field the height of the
susceptibility maximum decreases and the maximum shifts towards lower
temperatures (Fig.6) /14/. Above a given value of the magnetic field
(H=2Ka/Ms for uniaxal anisotropy) there is only one energy minimum
and relaxation is no longer present. Taking into the account the

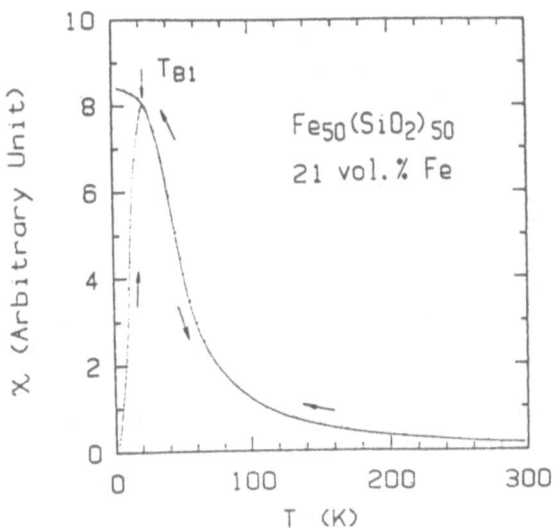

Fig.5 χ(ZFC) and χ(FC) vs T (H = 50 Oe) for iron particles (average diameter:25Å) dispersed in silica matrix (from ref.14).

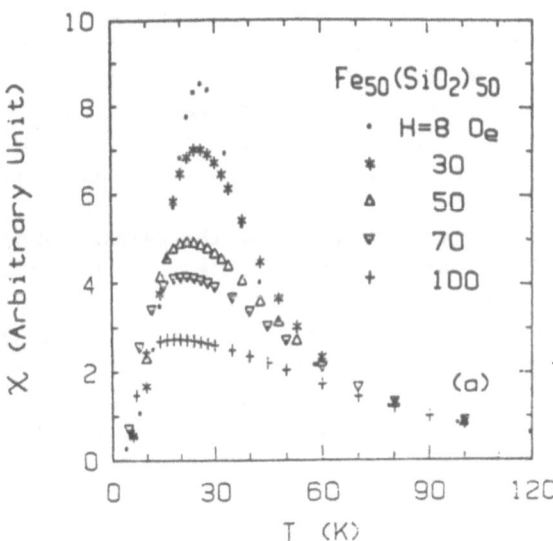

Fig.6 Temperature dependence of the susceptibility at different magnetic fields for iron particles dispersed in silica (from ref.14)

effect of the magnetic field on the superparamagnetic relaxation time /11/, characteristic lines in the H,T plane have been derived by numerical calculation for a constant τ by Wenger and Mydosh /15/ assuming that the anisotropy axes are parallel to the field direction. The dependence $t_h \simeq h^{2/3}$ was found for a wide range of field values, where t_h is the reduced blocking temperature ($t_h = $ [T(0)- T(H)]/T(0) and h the reduced field (h $= \mu$H/2E), where μ is the magnetic moment of the particle and E is the energy barrier for H=0). The same kind of H,T relationship has been predicted for spin-glasses by de Almeida and Thouless /16/. In the superparamagnetic model a crossover to a h^2 dependence is predicted as h $\rightarrow$ 0. Recently, calculations have been reported for the case of a random distribution of anisotropy axes, which represents the real situation in fine particle systems /17/. For low fields the same dependence $t_h \simeq h^2$ was found, while for high fields h $\geq$ 0.04, $t_h \sim$h(2-h). In the range 0.05<h<0.7 a coincidence with a $h^{2/3}$ dependence was found. The introduction of an interaction field has allowed calculations of the H,T lines to be extended to the case of interacting particles /17/.

2.3. Dependence of the blocking temperature on the measuring time .

In the previous section we have pointed out that Tb depends on the characteristic time scale (tm) of the experiment. For an assembly of independent particles the dependence of Tb on the measuring time is described by the Arrhenius law: $\tau = \tau_o$exp(Ea/kT). Linear log(tm) vs 1/T plots are expected, from which τ_0 and Ea can be determined. If the mean particle volume is known the effective anisotropy constant Ka can be obtained.

The Arrhenius law can not describe the relaxation process in the presence of interactions between particles. In such case unphysical values of τ_0 are obtained, as in the case of interacting iron particles dispersed in alumina matrix ($\tau_o \simeq 10^{-18}$s) /18/ (Fig. 7).

Shtrikman and Wohlfarth /19/ calculated the effect of the interactions on the frequency dependence of Tb. This leads to a modification of the Arrhenius law to the Fulcher law, where T is replaced by T-To:

$$(2.15) \qquad \tau = \tau_o \exp[Ea/k(T-To)]$$

The temperature To is a measure of the inter-particle interactions. This law was initially proposed by Tholence /20/ for

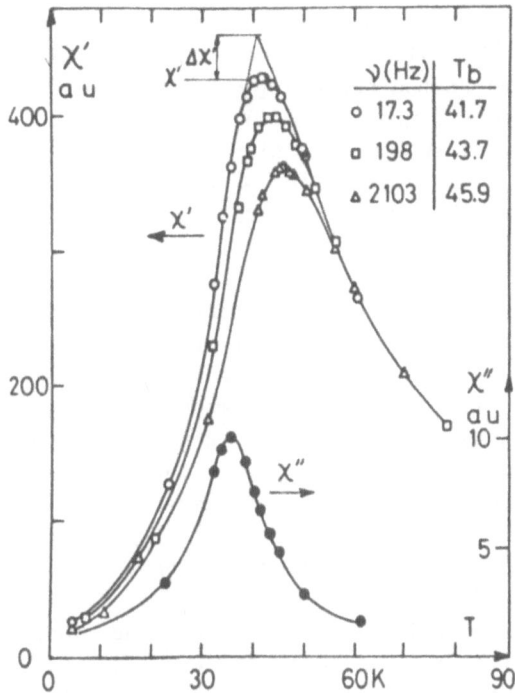

Fig.7 Temperature dependence of the A.C. susceptibility for iron particles (average diameter:50Å) dispersed in alumina. χ "(ν = 17Hz) (from ref.18)

describing the relaxation process in spin-glasses and was interpreted by Shtrikman and Wolfarth in terms of interactions between magnetic clusters.

Two cases are distinguished, the weak and the strong coupling regime. In order to distinguish between them the temperature To is compared to T_k = KaV/k such that To <<T_k for weak coupling and To >>T_k for strong coupling. The particles are assumed to be identical, with volume V, uniaxial anisotropy constant Ka and saturation magnetization Ms. They are taken to interact ferromagnetically.

For the weak coupling regime the concept of an interaction field has been used. It has been proposed that the statistical mean value of the interaction field Hi=Hitanh(HiMV/kT) $\simeq$ HiMsV/kT be used. Hence the contribution of the interaction energy to the energy barrier is Ei =VHiMs (To is proportional to the square of the interaction field : To =E_i^2/KaT).

For the strong coupling regime Shtrikman and Wohlfarth have used the concept of an effective, temperature dependent, volume of particles (related to the correlation function), which increases with decreasing temperature as $V(eff) = V_0[(T-T_0)/T_0]^{-1}$.

In both limits an approximation to a Fulcher law is obtained, whose validity is restricted to temperatures much higher than To, the temperature at which the relaxation time is predicted to diverge, signaling a transition.

The Fulcher law gives account of the experimental frequency dependence of Tb in a system of interacting iron particles dispersed in an alumina matrix /18,21/. However the dynamic properties of such a system are better explained in terms of a model implying a transition at OK /21/ (Fig.8). This model, assuming a distribution in volumes and in the relaxation times, includes the contribution to the energy barrier coming from the magnetic interactions which has been calculated applying the Boltzman statistics.

2.4. Relaxation of the remanent magnetization

Consider an assembly of independent and identical monodomain ferromagnetic particles in which the easy magnetization axes are randomly distributed. In the superparamagnetic state, for $\tau \ll$ tm the magnetization follows the variation of the applied magnetic field in a reversible and practically instantaneous manner.

Conversely if τ), tm the magnetization is time dependent and increases with the time as:

$$(2.16) \qquad M(t) = M[1-\exp(-t/\tau)]$$

Below Tb the magnetization relaxes slowly and then, when the field is withdrawn, a component in the direction of the field is left(isothermal remanent magnetization: IRM) which decreases with the time as

$$(2.17) \qquad IRM = M(o)\exp(-t/\tau)$$

Let us assume that the field is cycled (from H=0 to H=Hm to H=-Hm). If $\tau \gg$ tm a symmetrical hysteresis cycle is observed with IRM=Ms/2. This is due to the angular distribution of the easy axes relative to that of the field (the factor 1/2 is the average value of $\cos \vartheta$ in one half space.) (FIG. 9) /22/.

If the magnetic field is applied during a cooling down from T > Tb to T < Tb and then is withdrawn, a remanent magnetization is left

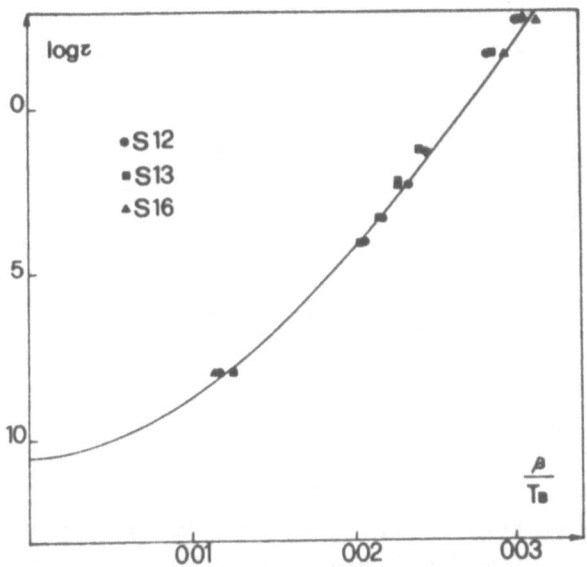

Fig.8 Variation of the blocking temperature Tb in the plot log τ vs
1/Tb   for different compositions of iron-alumina granular materials:
S12(50% Fe); S13(55% Fe); S16(70% Fe).The line represents the best fit
of the model to the experimental results. (from ref.21)

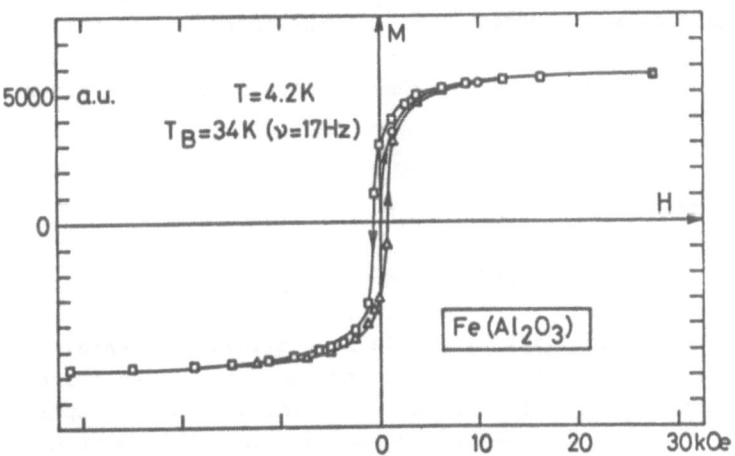

Fig.9 Hysteresis cycle at T = 4.2K after zero field cooling for iron
particles dispersed in alumina(from ref.22).

(thermoremanent magnetization: TRM). This is due to the fact that below Tb the relaxation time to move from one easy direction to another is very high and therefore the magnetic moment of each particle maintains the orientation that it had at the temperature Tb, i.e. along an easy direction close to that of the field.

The time decay of the thermoremanent magnetization is very slow. TRM can be very stable with time. This is the reason why clays and lavas preserve for an indefinite time a remanent magnetization parallel to the direction of the terrestrial magnetic field that acted on them during the cooling. This fact has been used to follow the variation in the time of the terrestrial magnetic field. The remanent magnetization decreases with temperature vanishing at Tb.

Studies on the relaxation of TRM in a magnetic fluid of frozen fine cobalt particles have shown that a logarithmic decay is followed for small measurement times /23,24/. Deviations from the logarithmic law were instead observed for large measurement times, where a decay as $1/t$ was observed /24/. This behaviour was explained by a model based on the summation of relaxation times over a very narrow range of particles sizes distribution /25/ (Fig. 10). The time dependence coefficient $A = - d(TRM)/d\ln t$ was found to vary with the applied field H, reaching the maximum value for the coercitive field Hc, in agreement with the proposed model /26/.

A decay of IRM and TRMs (saturated thermoremanent magnetization) following a power law t^{-a} has been found in a system of interacting iron particles dispersed in an alumina matrix /27/ (Fig. 11). The coefficient was found to increase with the magnetic field. The temperature dependence of TRMs was found to be exponential (TRMs $\exp(-bT)$).

3. MAGNETIC RELAXATION PROCESSES IN SPIN-GLASS

In the field of random magnetism spin-glasses have been reveiving an immense interest since the observation, in 1971, of a sharp peak in the a.c. susceptibility of dilute AuFe alloys /28/ (Fig. 12).

This observation has raised the question of an associated phase transition, that challenged theoreticians and experimentalists. However no long range order was detected by neutron diffraction experiments and no anomaly was observed in the specific heat at the temperature of the susceptibility maximum, the so called freezing temperature Tf /29/. A hyperfine splitting was observed, roughly at the

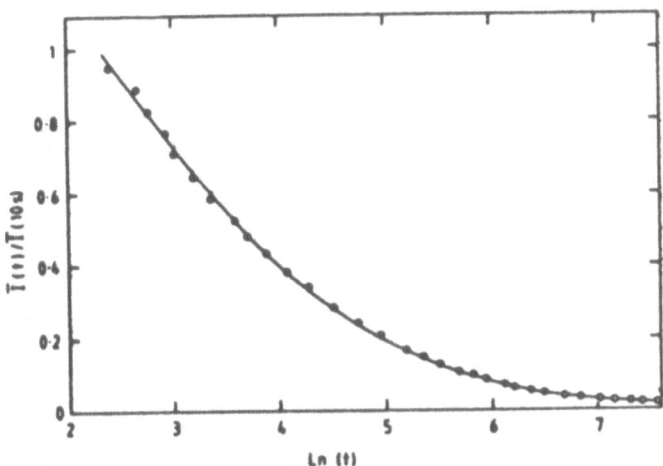

Fig.10 Time decay of IRM of frozen cobalt fine particles (average diameter:40Å).The solid line represents the best fit theoretical curve (from ref.25)

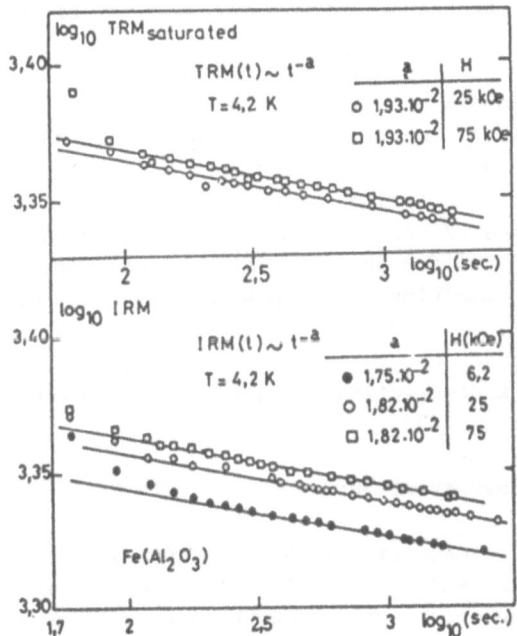

Fig.11 Time decay of TRMs and IRM of iron particles (average diameter:50Å) dispersed in alumina (from ref.27)

same temperature as the susceptibility peak,in Mossbauer spectra /30/ whose analysis suggested a random alignment of frozen spins at low temperature.

Two models were originally proposed to describe the passage from the paramagnetic state to the spin-glass state:

- a superparamagnetic model, inspired by the Néel's theory of small particles /31/, which describes the passage to the spin-glass state as a progressive blocking of independents moments, produced in an extended temperature range /32/.

- a phase transition model, developed within the mean field theory /33/ which describes this passage as a true thermodynamic transition, occurring at a well defined temperature accompanied by critical phenomena.

The controversy about the existence or not of a spin-glass transition is partly due to the large variety of magnetic behaviours observed experimentally. This is due to the fact that the properties considered characteristic of spin-glasses (i.e. the peak of the low-field susceptibility at Tf, the frequency dependence of Tf, the irreversibility and the magnetic relaxation below Tf...) are exhibited by a wide variety of compounds: diluted metallic alloys, concentrated semiconductors and insulators, crystalline as well as amorphous materials.

At present an increasing number of experimental evidences, static and dynamic, of critical behaviour, support the idea of a new type of phase transition in spin-glasses. However, while these evidences are very clear for dilute metallic alloys (where the validity of the description in terms of phase transition is reinforced by the existence of universality classes among the measured critical exponents) /34/ and for some insulating systems /35,36/,the occurrence of a phase transition has to be ruled out for some insulating compounds where, although the macroscopic characteristics of the S.G. state are observed, no critical behaviour has been found /37/. For the last materials a description in terms of a spin-cluster model seems to be more appropriate /38/.

3.1 The spin glass state

Spin-glasses are magnetic systems characterized by a random freezing of the moments below a given freezing temperature Tf, without long range order. The above concept may be expressed as follows /33/:

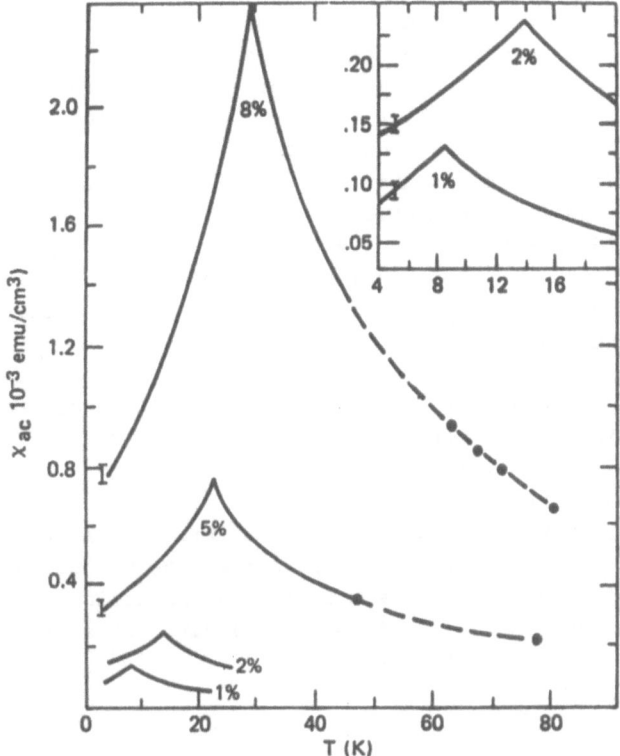

Fig.12   Temperature dependence of the A.C. susceptibility in AuFe
alloys for various Fe concentrations (from ref.28)

$$(3.1) \quad [\langle \vec{S}_i \ \vec{S}_j \rangle] \to 0 \quad \text{for } i \neq j \quad \text{as } r(ij) \text{ tends to infinite}$$

$$(3.2) \quad q = [\langle \vec{S}(t=0)\vec{S}(t \to \infty)\rangle] \quad \text{for } T<T_f ; \quad q=0 \text{ for } T>T_f$$

where $\langle \ \rangle$ represents the thermal average and $[\ ]$ the configurational
average over all the spins. The first equation indicates that although
short range magnetic order may be present, there is no correlation
between the directions of the moments at long distances. The second
equation, which is a measure of the correlation between the
orientations of one moment over long time intervals, indicates that
each spin has a certain orientation below T_f (random static freezing).
Edwards and Anderson /33/ proposed this quantity as a convenient order
parameter for spin-glasses(q). Although the macroscopic moment is zero

in a paramagnet as well as in a spin-glass, i.e. the magnetization cannot serve as an order parameter (differently than is the case of a ferromagnet), the non-vanishing value of q differentiates between the two states: the spin-glass has a "memory" of local spin orientations,whereas the paramagnet not.

Two ingredients are essential for stabilizing a spin-glass state: the randomness and the frustration.

The disorder may be topological, as in glassy and amorphous materials, or may be substitutional in type, as in the case of a random distribution of magnetic ions in magnetically diluted crystalline materials.

The frustration, i.e. the existence of a conflict between magnetic interactions /37/ has a different origin in dilute metallic alloys and in insulating materials.

In dilute metallic alloys, i.e. typically 1 at %, the frustration is related to the oscillatory character of the long range RKKY (Ruderman, Kittel, Kasuya, Yosida) interaction /38/. A magnetic moment diluted in a metallic matrix polarizes the conduction electrons and this polarization is then felt by another magnetic impurity. The interaction between two spins Si and Sj, located at a distance r(ij), is described for large distances by:

$$(3.3) \qquad J(ij) = \frac{J_0 \cos(2K_f r + \varphi)}{(K_f r)^3}$$

where J_0 and φ_0 are constant and K_f is the Fermi wave number of the host metal.

The sign and the size of the interaction is very sensitive to the inter-moments distance r(ij) (which is higher than the lattice distances), oscillating between positive and negative values. Since the magnetic impurities are considered to be randomly positioned, these oscillations introduce a conflict among competing positive and negative alignments.

In insulating materials, where the interactions are short-ranged and therefore the spin-glass behaviour is observed at high concentrations of magnetic ions (above the percolation threshold) the frustration may be of two types:
- it may be due to the competition between interactions of opposite sign, as in Eu(0.4)Sr(0.6)S /39/, where nearest-neighbour interactions are ferromagnetic and next-nearest neighbour interactions are antiferromagnetic.

- it may have a topological origin in systems with antiferromagnetic interactions only, as in some amorphous materials (e.g. cobalt-aluminosilicate /40/ or in some particular lattices /41/ (e.g. f.c.c., as in $Cd(x)Mn(1-x)Te$ /42/; octahedral sublattice of a spinel, as in $ZnCr(1.6)Ga(0.4)O(4)$. /43/.

In general the conflict between interactions, independently of its origin and mechanism, and their random distribution, avoids any conventional order to be established. In this situation it is not possible to find a spin configuration satisfying all the interactions at the same time. Many degenerate states can exist, where some bonds remain necessarily unsatisfied.

Because of the coexistence of randomness and frustration, a large number of almost degenerate thermodynamic states exist in spin-glasses, having the same macroscopic properties but with different microscopic configurations. They are separated by high free energy barriers. These different states can be characterized at a microscopic level by the value of the magnetization at each point $m(i,s)$, where the index i runs over the points of the lattice and s labels the different states. The low temperature spin-glass state may be considered to be made by a large number of metastable states and a small fraction of asymptotically stable states. This multivalley structure is responsible for the existence of a broad spectrum of long relaxation times in spin-glasses. The experimental features of spin-glasses indeed reveal the existence of very long times of approach to equilibrium, which are necessary to overcome the energy barriers encountered in moving from one minimum to another one. If the hopping time is much longer than the time scale of the experiment, the system will appear locked in the vicinity of one state, corresponding to a local free energy minimum. As the time elapses, the system relaxes between progressively deeper minima lying closer to equilibrium. Recent results show however that under normal experimental conditions the system fails to reach thermodynamic equilibrium.

3.2 Spin-glass dynamics

Spin-glass dynamics has received since the beginning a continuous and growing interest among both experimentalists and theoreticians. Measurements of magnetic relaxation phenomena in spin-glasses are sensitive to the "window" of the experimental measuring time. The spin-glass state is characterized by the existence of a large distribution of relaxation times. When the relaxation times increase their distribution widens.The attention has been mainly focused on the following aspects: a) the time dependence of the susceptibility; b) the time decay of the remanent magnetization; c) the frequency dependence of the freezing temperature. In the following sections we shall review some recent experiments on relaxation in spin-glasses.

Time dependence of the susceptibility.   The most significant physical property of spin-glasses is the characteristic behaviour of the low field magnetic susceptibility (Fig.13). This provides the clearest determination of the freezing temperature Tf and it is of foundamental importance in investigating the nature of this freezing.

The susceptibility behaviour depends on the magneto-thermal history: the low-field susceptibility (M/H) measured after cooling down the sample in zero field (χ ZFC) shows a sharp maximum at Tf, while, when it is measured after cooling down in the applied magnetic field (χ FC) does not exhibit a large anomaly at Tf (it remains constant below Tf or shows a small maximum at Tf (Fig.13)/45/.

The freezing temperature marks the passage from a state, where the magnetization is time independent and reversible (the paramagnetic state), to a state where the magnetization is time dependent and irreversible.

Fig. 14 shows that for T>Tf the application of a magnetic field instantaneously determines the presence of a magnetization, which remains constant as long as the field is applied, and vanishes instantaneously when the field is withdrawn (this is related to the existence of only one minimum of energy for H≠0 and H=0 respectively). For T<Tf the magnetization (ZFC) increases with time, M(ZFC)(T,H,t), towards its equilibrium value and when the field is withdrawn a remanent magnetization is left which decreases with time, IRM(T,H,t). The zero field cooling procedure leads to non-equilibrium conditions corresponding to local metastable states. The slow time evolution of the magnetization is due to the high energy barriers separating the different energy minima.

On the other hand the field-cooling procedure leads to quasi - equilibrium conditions, corresponding to asymptotically stable states (the equilibrium susceptibility is experimentally inaccessible below Tf). The magnetization M(FC) remains almost stable as long as the applied field is kept constant (a part from a weak relaxation, corresponding to a decrease of ~1%) and it is almost completely reversible. The remanence, TRM(T,H,t), is responsible for the splitting between χ (FC) and χ (ZFC) below Tf.

The susceptibility at a given observation time can be measured by

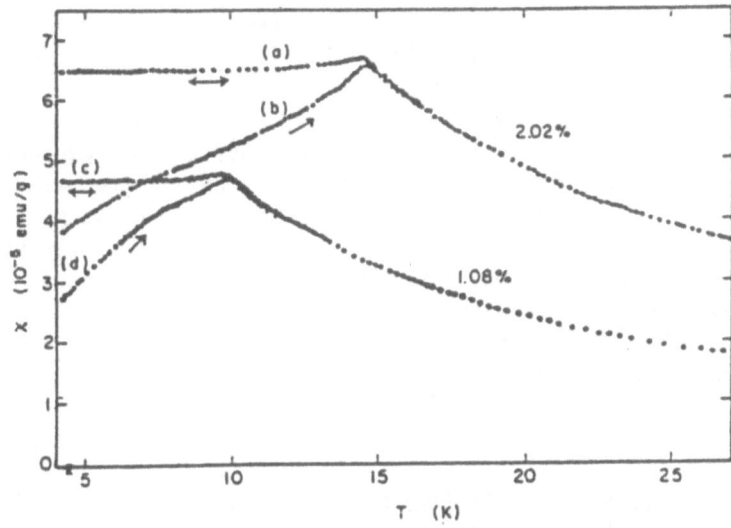

Fig.13 The field cooled and the reversible susceptibility, lim(H→0) [M(FC)(T,H) - TRM(T,H,t)]/H of CuMn (from ref.45).

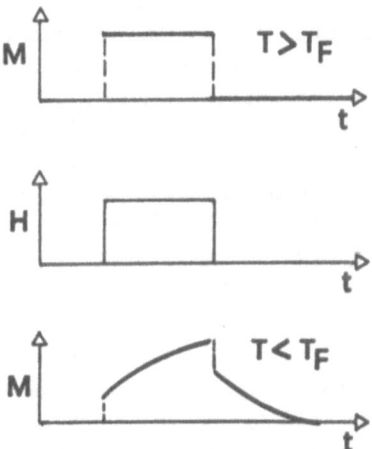

Fig.14 Magnetization as function of the time at T>Tf and T<Tf.

both D.C.(ZFC) experiments, by following the change of the magnetization with time, and A.C. experiments, varying the angular frequency of the external oscillating field. Fig. 15 /46/ shows that the in phase component χ' has a sharp maximum at Tf, which shifts towards higher temperatures with increasing frequency. The magnitude of χ' decreases with increasing frequency. The frequency dependence of χ' disappears just above the maximum, where all the curves measured at different frequencies merge into a single curve with Curie-like behaviour. χ'' shows a maximum with an inflection point in correspondence of the maximum of χ'. The magnitude of χ'' increases with increasing frequency.

It has been shown that D.C.(ZFC) and A.C. susceptibility measurements are related through /47/:

$$(3.4) \qquad \chi'(\omega) \simeq M(t)/H \qquad\qquad t=1/\omega$$

$$(3.5) \qquad \chi''(\omega) \simeq \pi(\delta M/\delta \ln t)/2H \qquad t=1/\omega$$

Thus combining the results of A.C. and D.C. experiments is possible to cover many decades in time of the S.G. dynamics.

Relaxation phenomena in spin-glasses present some degree of resemblance with relaxation in ordinary glasses and glassy polymers /48/. Aging is a general property of the glassy state. In spin-glass it has been found that the relaxation of the zero field magnetization /49/ and the time decay of the thermoremanent magnetization /50/ depend on the time the sample is kept below Tf prior to any field variation (denoted as the waiting time tw).

Fig.16a shows that χ(ZFC) relaxes upwards with time and that for the same observation time decreases with increasing tw . Fig.16b shows the time evolution of the relaxation rate $S(t)=(1/H) \delta M/\delta \ln t$ for different waiting times /46/. A maximum in S(t) is observed for $t \simeq tw$ for all waiting times. It is evident that the aging process has a dominating influence on the spin-glass relaxation in this time window. Only at observation times much smaller than tw the experimental data are representative of spin-glass relaxation at thermodynamic equilibrium.

In a.c. susceptibility measurements at constant temperature and constant frequency, aging is detected as a time dependence of $\chi'(\omega)$ and $\chi''(\omega)$ /51/.

It has been found that χ(FC) depends on the rate of cooling through Tf. With decreasing the rate of cooling the system approaches a state closer to the equilibrium one. In metallic spin-glasses M(FC) is smallest for the slowest cooling rate /52/, whereas the opposite

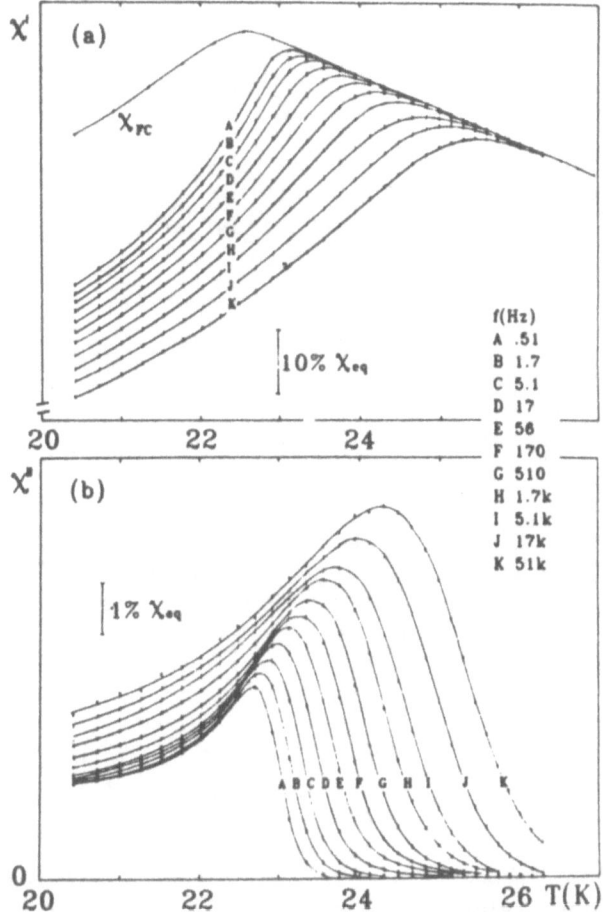

Fig.15 A.C. susceptibility χ'(Fig.a) and χ''(Fig.b) at different
frequencies and χ(FC)(Fig.a) of [Fe(0.15)Ni(0.85)](0.75)P(16)B(6)Al(3)

was observed in insulating spin-glasses, such as cobalt-
aluminosilicate /53/(Fig.17). This difference is related to the type
of magnetic interactions responsible for the freezing.

Time decay of the remanent magnetization. The study of the time decay
of the thermoremanent magnetization TRM has received since the
beginning a great deal of attention from both a theoretical and
experimental point of view. Reported results were found to obey
different types of laws.

416

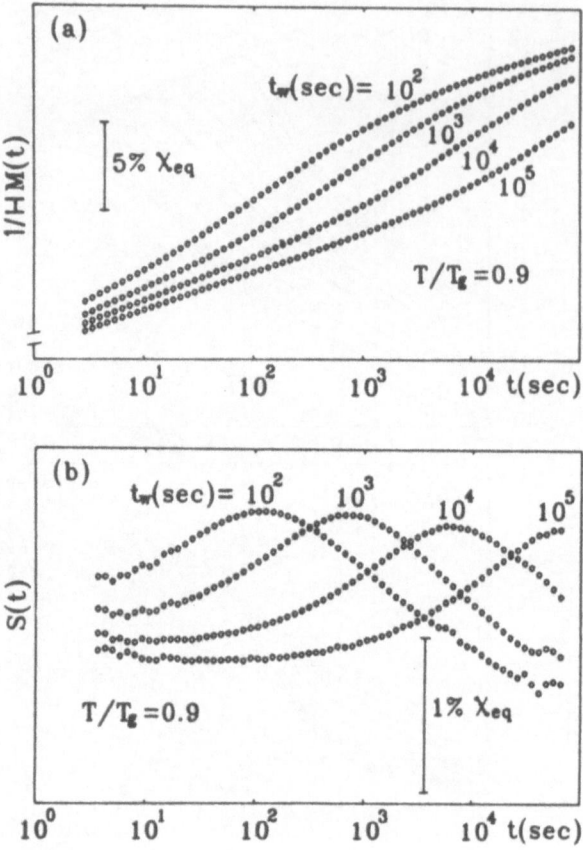

Fig.16 Time dependence of $\chi(ZFC) = M(t)/H$ (H=0.1G) (Fig.a) and the corresponding relaxation rate $S(t) = (1/H) \delta M/ \delta \ln t$ (fig.b) at different waiting times (from ref.46).

The first experiments on the relaxation of TRM, performed on AuFe alloys, showed that the decay was logarithmic /54/:

(3.6) $TRM(t) = TRM - S\ln t$

where S is a constant.

Experiments performed on CuMn alloys showed that the time and temperature dependence of TRMs (the saturated thermoremanent magnetization)are associated in a unique relation /55/:

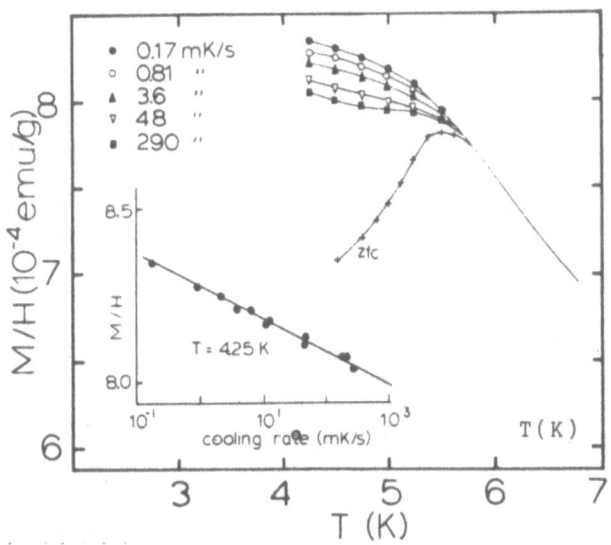

Fig.17 χ(FC)(=M/H) as a function of various cooling rate and χ(ZFC) (+) of a 14.3at% cobalt-aluminosilicate glass (from ref.53).

$$(3.7) \qquad TRM_s = TRM_0 \exp \left[- (T/To) \ln(t/\tau_0) \right]$$

where T is a constant and τ_0 is the characteristic relaxation time. These results have been described by a phenomenological model which assumes a distribution of asymmetrical double well potentials and a thermal activation process to overcome the energy barriers.

Measurements on Eu(0.4)Sr(0.6)S /56/ showed that the time decay of TRM follows a power law:

$$(3.8) \qquad TRM(t) \sim t^{-a}$$

where the exponent "a" depends on both temperature and field. This result was consistent with previous Montecarlo silmulations /57/.

In the last few years the experiments have shown that the decay of TRM is affected by aging, depending on the time the sample has been kept below Tf before the field is switched off.

Experiments on both metallic and insulating spin-glass /49,58/ have shown that the time decay of the thermoremanent magnetization is characterized by a stretched exponential form:

$$(3.9) \qquad TRM = TRM \exp[-(t/\tau_p)]^{1-n}$$

at least for relatively small applied field and times in the range of $5-10^3$ s. It has been found that the characteristic relaxation rate

$1/\tau_p$ decreases exponentially with tw: $1/\tau_p = \omega_0 \exp(-tw/to)$ (Fig.18) /59/ and that it varies exponentially with the inverse of the reduced temperature Tf/T: $1/\tau_p = A\exp(-\alpha Tf/T)$ over nearly the entire temperature range of measurements. It increases more rapidly as the freezing temperature is approached from below /50/(Fig.19). The waiting time dependence of $1/\tau_p$ indicates that the spin-glass system is relaxing between states with ever increasing time constants.

Nordlab et al /60/ have found that the total relaxation of the magnetization may be described assuming a pure logarithmic decay superimposed by a stretched exponential form that accounts for the influence of the aging process:

$$(3.10) \quad RM = SH\ln(t) + M_1 + Ma(T,tw) \exp(-t/\tau_p)$$

where S is the relaxation rate at dynamic equilibrium, M_1 is the intercept at log (t)=0 for the logarithmic decay and Ma the magnitude of the superimposed stretched exponential. Recent experiments performed on a very long time scale /61/ showed that the time decay for TRM has the form, at very small fields, of the product of a stretched exponential law by a power law:

$$(3.11) \quad TRM = TRM(0)(t/\tau_p)^{-\alpha}\exp[-(t/\tau_p)^{1-n}]$$

A time scaling of the aging effect was found by using convenient reduced effective units.(Fig.20)

The study of the field effect on the relaxation of TRM has shown that the increase of the applied field increases the characteristic relaxation rate $1/\tau_p$(Fig.21) of the system and hence accelerates the time decay of TRM. Recent measurements performed on AgMn(2.6%)Sb(0.46%) /58,61/ and CdCr(1.7)In(0.3)S(4) /62/ at different cooling fields and temperatures showed a linear increase of $\log(1/\tau_p)$ with $1/(1-n)$. This is in agreement with recent calculations of the dynamics of the infinite range model of Ising spin-glasses /63/.

The frequency dependence of the freezing temperature. In the last few years a great interest has been devoted to the study of the frequency dependence of Tf. This study has been performed mainly by A.C. susceptibility measurements at different frequencies (typically in the range 1Hz-100kHz). In few cases the "window" of the experimental measuring time has been extended by using other techniques (e.g.neutron diffractions, Faraday rotation, Mossbauer spectroscopy...).

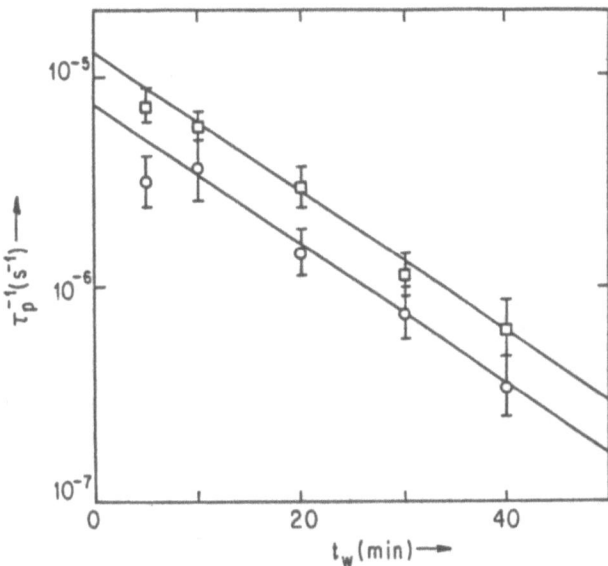

Fig.18   The waiting time dependence of the characteristic   relaxation
rate $(1/\tau_p)$ of AgMn(2.6at%) at T=4.35K.□ powder sample; o sample in
foils.The drawn lines are   the   best fits   of   the exponential form,
$1/\tau_p = \omega_o \exp(-tw/to)$ to the experimental results (from ref.59).

        Tf   is   found to increase with increasing the frequency but   this
effect   can   be   very different from one   system   to   another.   It   was
initially   suggested   that   a frequency sensitivity (f.s.)   of   Tf   be
defined   by the ratio $\triangle Tf/(Tf\triangle \log \nu)$ (the relative variation of   Tf
per decade of time in a given frequency range, namely 10Hz and 100Hz).
In general a low,   concentration independent,   frequency sensitivity is
found in dilute metallic alloys (e.g.   f.s.=0.005 in CuMn /64/,   while
higher,   concentration   dependent   (e.g.   in Eu(x)Sr(1-x)S   /20/),
frequency sensitivity, varying greatly from system to system, is found
in insulating materials (e.g.   f.s.=0.05 in ZnCr(1.6)Ga(0.4)O(4) /43/.
A comparison between the frequency sensitivities has been suggested as
a   criterion   for distinguishing spin-glasses "with transition"   from
superparamagnetic-like   systems (where a high frequency dependence   of
Tf is expected), but this criterion does not allow a clear distinction
to be made between these two behaviours for systems with intermediate
frequency dependence.

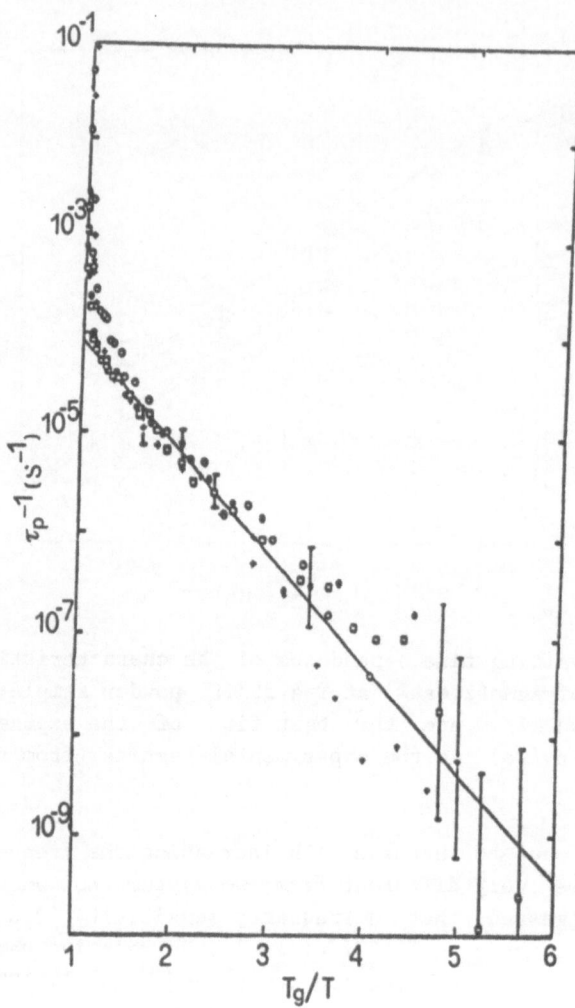

Fig.19   Temperature   dependence   of   the   characteristic
relaxation rate $1/\tau_p$ plotted as $\log(1/\tau_p)$ vs   Tf/T .The line
represents the fit of the exponential form $1/\tau_p = A\exp(-\alpha Tf/T)$ to
the experimental data for   Tf/T>1.5. (o AgMn(2.6at%)+Sb(0.4at%) ;
•Ag:Mn(2.6at%); □ Ag:Mn(4.1at%); + CuMn(4.0at%)

Phenomenological laws. Untill recently the frequency dependence of Tf
has been mostly understood in a phenomenological approach which does
not account for a phase transition.
        A simple picture of thermal activation over constant energy

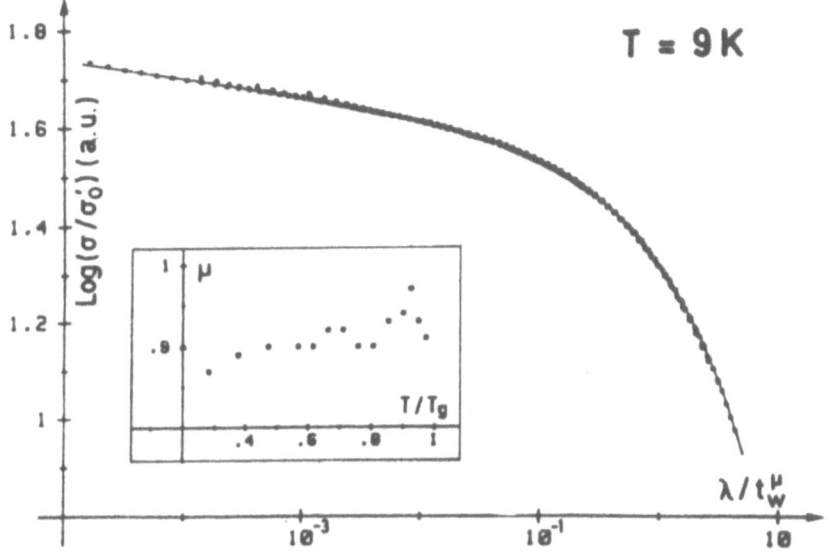

Fig.20 Ag:Mn(2.6%). Time decay of TRM.Master curve where the curves obtained at different tw values (10^3 s<tw<10^5 s) are merged (from ref.61).

barriers for a collection of independent clusters was found to be inadequate: the typical relaxation time increases faster than predicted by an Arrhenius law as T is lowered and the resulting frequency dependence of Tf is also weaker than the Arrhenius behaviour. In the cases where a linearity is observed in the typical $1/Tf$ (ν) vs $\ln(1/\nu)$ plots the deduced characteristic relaxation time was unphysical (e.g. $\tau_0 = 10^{-37}$s for CuMn /65/. This fact suggested that the spin-glass state is not a result of a simple blocking process of independent clusters, but that a cooperative mechanism is responsible for the freezing.

In 1980 Tholence /20/ observed that the use of the empirical Fulcher law allowed to describe the frequency dependence of Tf with physical value for τ_0 and Ea in many spin-glasses, metallic as well as insulating. A good scaling with the concentration was found in the case of dilute metallic alloys. The introduction of the characteristic temperature To introduces a pathology which might signal a transition. In fact the Fulcher law assumes the existence of energy barriers which diverge at To. In the real systems instead it is generally observed that the Fulcher variation of Tf(ν) is not exact down to To.

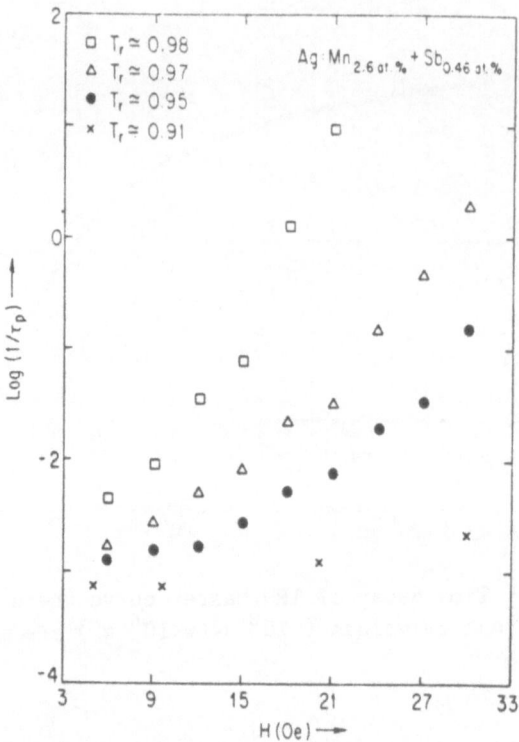

Fig.21 The field dependence of the characteristic relaxation rate
(1/τ_p) of AgMn(2.6at%)+Sb(0.46at%) at different temperatures below
Tf(Tr=Tf/T) (from ref.59).

Therefore To is ruled out as a good candidate for the transition
temperature Tc. At very low frequencies Tf is found to be frequency
independent /56,66/ and this fact was interpreted as providing
evidence of cooperative freezing.

Dynamic scaling laws. In the last few years there have been attempts
to describe the dynamic properties of spin-glasses within the
framework of critical phenomena.
    Two alternative pictures for the slowing down of the relaxation
time have been proposed:
    1) Model of transition at T=Tc
Near a transition at finite temperature the relaxation time exhibits

the standard critical slowing down /67/:

$$(3.12) \qquad \tau/\tau_0 \sim \xi^z \sim \varepsilon^{-z\nu}$$

where z is the dynamic exponent which describes the slowing down of the relaxation, ν is the critical exponent which describes the growth of the correlation length and $\varepsilon = (T-Tc)/T$. Recent computer simulations give $z\nu = 7.2\pm1$ /68/. The field dependence of is given by:

$$(3.13) \qquad \tau/\tau_0 \sim \varepsilon^{-z\nu} f(h\varepsilon^{-\Delta})$$

where $h = \mu H/KT$ and $\Delta = (\beta + \gamma)/2$ /69/. β and γ are the conventional exponent for the non linear susceptibility.

2) Model of transition at T=0

Near a transition at T=0 the relaxation time is predicted to diverge as /70/:

$$(3.14) \qquad \ln(\tau/\tau_0) \sim T^{-z\nu}$$

and similarly:

$$(3.15) \qquad \ln(\tau/\tau_0) \sim T^{-z\nu} f(HT^{-\Delta})$$

where $\Delta = 1 + (\beta+\gamma)/2$. In the Ising model $z\nu = 2$ and $z\nu = 4$ in two and three dimensions respectively.

The experiments devoted to verify these models are performed by studying the dependence of the observed freezing temperature on the A.C. excitation frequency and on the magnitude of an additional static field. Actually in spin-glass there is not a single relaxation time, but a broad distribution of them. It is assumed that all the relaxation times of the distribution will follow the equations (3.12) and (3.14) with their own prefactor $\tau' = a'\tau$. Therefore in order to apply the above reported dynamic scaling relations it is necessary to choose an appropriate criterion of analysis which defines an experimental freezing temperature allowing the A.C. frequency to be related to a characteristic relaxation time of the system. The phenomenological definition of Tf, that is the temperature corresponding to a maximum of χ(a.c.) does not satisfy this requirement. Some authors have chosen as characteristic relaxation time the average time $\bar{\tau} = \int_0^\infty \tau P(\tau)d\tau'$, while others the edge maximum

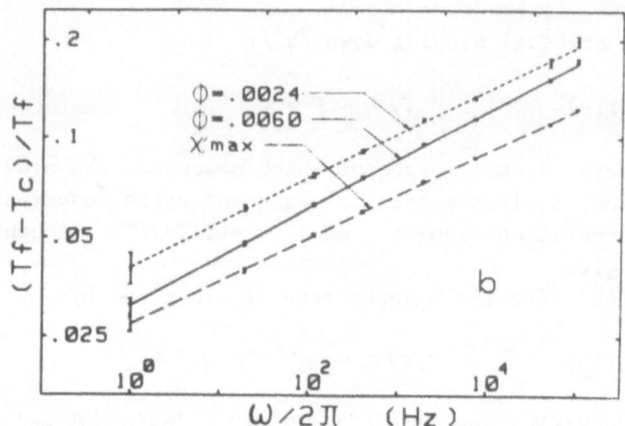

Fig.22   CdCr(1.7)In(0.3)S(4).   Fits of Tf(ω) to the dynamic scaling
relation $\omega/\omega_o = [(Tf-Tc)/Tf]^{-z\nu}$ for three definitions of Tf(ω) : Tf =
T(χ''/χ' = φ = 0.0024; Tf = T(φ = 0.006) and T = T(χ' maximum)
(from ref.35).

of the distribution P(τ).

The first ones /35,36/ have taken Tf(ω) as the temperature at
which $\chi''(\omega)/\chi'(\omega)$ is equal to a very low constant value. This is
related to the fact that the susceptibility can be expressed:

(3.16) $\chi' = \chi_o \int P(\tau)/(1+\omega^2\tau^2) \, d\tau$

(3.17) $\chi'' = \chi_o \int P(\tau) \, \omega\tau/(1+\omega^2\tau^2) d\tau$

and that in the limit $\omega^2\tau^2 \ll 1$ the ratio $\chi''/\chi' \simeq \omega\tau$.(Fig.22)

The second ones have taken Tf(ω) as the temperature at which the
difference   between   the χ(a.c.) and χ(d.c.) (field cooled),
considered as the equilibrium susceptibility χ(eq) becomes zero
($\Delta\chi = [\chi(eq) - \chi(ac)] / \chi(eq)$). It is assumed that the normalized quantity
is only a function of τ max(H,T) in the limit $\Delta\chi \to 0$. For $\Delta\chi$ constant
and close to zero,a set of coupled values ω_i,Ti has been obtained,
which is assumed to fulfill the condition τ max(Ti) $\simeq$ 1/ω_i (Fig.23).
In computer simulations τ is taken as the average   relaxation
time./68/

For some systems the data can be fitted by both laws,(Tc = 0 and
Tc $\neq$ 0) but an increasing number of recent experiments support the

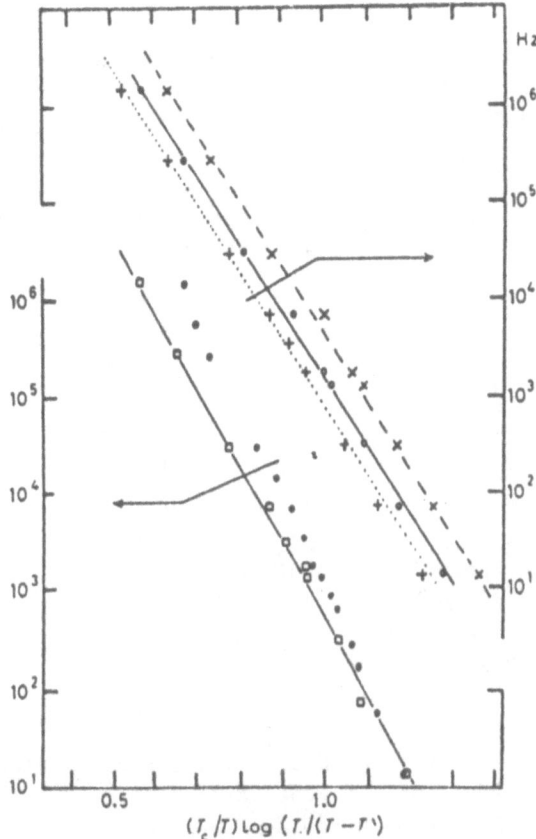

Fig.23 (MnF2)(0.65)(BaF2)(0.15)(NaPO3)(0.20). vs (Tc/T)log[T/(T-Tc)].
The data are taken for different $\Delta\chi=(\chi(eq)-\chi)/\chi(eq)$ = constant
values(x $\Delta\chi$= 2%; o $\Delta\chi$= 1%; + $\Delta\chi$= 0.5%;•$\Delta\chi$ = T(max) and $\Delta\chi$= T(1-Tc/T)$^\beta$
($\square$), β= 0.85 (from ref.72).

idea of a transition at finite temperature. In fact for systems where
the occurrence of a phase transition was clearly established by static
measurements, the power law give a better account of the experimental
results with $z\nu$ exponent close to the calculated one, with a
reasonable value for τ_o and critical temperatures corresponding to
the static values (e.g. $z\nu \simeq 8$ in Eu(0.4)Sr(0.6)S /36/; $z\nu \simeq 7$ in
CdCr(1.7)In(0.3)S4/35/; $z\nu$ =9-10 in Cd(0.45)Mn(0.55)Te /71/; $z\nu$ =8.5
in manganese fluorophosphate/72/

Moreover the model of transition at finite temperature allowed a good scaling of the field-temperature Tf(H,ω) lines with values of (β+γ) (e.g. 3.5 in CdCr(1.7)In(0.3)S(4)/35/;4 in Eu(0.4)Sr(0.6)S/36/ in agreement with those deduced from static measurements.

On the other hand the generalized Arrhenius law (T=0 transition) is not appropriate to describe the experimental results, giving in some cases a non-unique value of the $z\nu$ exponent(very high and unphysical for dilute metallic alloys /47/) and less satisfactory values of τ_0 .

REFERENCES

1) H.B.Casimir and F.K.du Pré,Physica 5,507(1987)
2) R.L.Carlin and A.J.van Duyneveldt,"Magnetic properties of transition metal compounds", Springer Verlag(1977)
3) C.J.O'Connor,Progr.Inorg.Chem.29,203(1982)
4) D.Feldan and R.P.Hunt,Z.Instr.72,259(1964)
5) R.P.Giffard, R.A.Webb, J.C.Wheatley, J.Low Temp. Phys.6,553(1972)
6 S.C.Whitmore, S.R.Ryan and T.M.Sanders Jr,Rev.Sci.Instrum. 49,11(1978)
7) L.Néel,C.R.Hebd.Sean.Acad.Sci.228,664(1949); Review papers: J.L.Dormann,Rev.Phys.Appl.16,275(1981); S.Morup,J.A.Dumesic and H.Topsoe,"Applications of Mossbauer spectroscopy",R.L.Cohen Ed.,Vol.2,Academic Press,p.1(1980)
8) C.P.Bean,J.Appl.Phys.26,1381(1985)
9) C.P.Bean and J.D.Livingstone,J.Appl.Phys.30,120(1959)
10)C.P.Bean and I.S.Jacobs,J.Appl.Phys.27,1448(1956)
11)W.F.Brown,Jr.,Phys.Rev.130,1677(1963)
12)J.I.Gittleman, B.Abeles and S.Bozowski, Phys.Rev.B 9,389(1974)
13)J.T.Richardson and W.O.Milligan,Phys.Rev.B 102,1289(1956)
14)G.Xiao,S.H.Liou,A.Levy,J.N.Taylor and C.L.Chien,to be published
15)L.E.Wenger and J.A.Mydosh,Phys.Rev.B 29,4156(1984)
16)J.R.L.de Almeida and D.J.Thouless,J.Phys.A 11,983(1978)
17)J.L.Dormann, D.Fiorani and M.El Yamani, Phys. Lett.A 120,951987)
18)J.L.Dormann, D.Fiorani, J.L.Tholence and C.Sella,J.Magn. Magn. Mat.35,117(1983)
19)S.Shtrikman and E.P.Wolfarth,Phys.Lett.85A,467(1981)
20)J.L.Tholence,Solid State Commun.35,113(1980)
21)J.L.Dormann,L.Bessais and D.Fiorani,to be published
22)D.Fiorani,J.L.Tholence and J.L.Dormann, J.Magn.Magn.

Mat.31-34,947(1983)

23)K.O'Grady,R.W.Chantrell,J.Popplewell and S.W.Charles,
IEEE Trans.Magn.172943(1981)

24)A.T.Cayless,S.R.Hoon and B.K.Tanner, J. Magn. Magn.
Mat.30,303(1983)

25)R.W.Chantrell, S.R.Hoon and B.K.Tanner, J. Magn. Magn.
Mat.30,303(1983)

26)K.O'Grady,R.W.Chantrell,J.Popplewell and S.W.Charles,IEEE
Trans.Mag.172943(1981)

27)D.Fiorani,Doctoral Thesis(Grenoble)1984

28)V.Cannella,J.Mydosh and J.I. Budnick,J. Appl.
Phys.42,1689(1971)

29)Recent reviews on spin-glasses: K.Binder and
A.P.Young,Rev.Mod.Phys.56,801(1987);C.Y.Huang,J.Magn.Magn.Ma
t.53,1(1985)

30)R.J.Borg,Phys.Rev.B 1,349(1970)

31)L.Nèel,Ann.Geophys.5,99(1949)

32)J.L.Tholence and R.Tournier,J.Phys.35,C4-229(1974)

33)S.F.Edwards and P.W.Anderson,J.Phys.F 5,965(1975)

34)N.de Courtenay,H.Bouchiat,H.Hardespint and A.Fert,to be published

35)E.Vincent and J. Hammann, J. Phys. C, in press; E.Vincent,
J.Hammann and M.Alba, Solid State Commun.58,57(1986)

36)N.Bontemps, J.Rajchenbach, R.V.Chamberlin and R.Orbach,
J.Magn.Magn.Mat.54-57,1(1986)

37)J.Hammann, D.Fiorani, M.El Yamani and J.L.Dormann,
 J.Phys.C.19,6635(1986)

38)J.L.Dormann and M.Nogues,3th Int.Conf. on Physics of Magnetic
Materials(Szcyrk-Bila,Poland),1986

39)H.Maletta and H.Felsch,Phys.Rev.B 20,1245(1979)

40)L.E.Wenger and J.A.Mydosh,Phys.Rev.Lett.49,238(1982)

41)K. de Seze,J.Phys.C 10,2353(1977); J.Villain,Z.Phys.B 33,31(1979)

42)R.R.Galazka, S.Nagata and P.H. Keesom, Phys. Rev.B 22,3344(1980)

43)D.Fiorani, S.Viticoli, J.L.Dormann, J.L.Tholence and
A.P.Murani,Phys.Rev.B 39,2776(1984)

44)G.Parisi,Phys.Lett.A 73,203(1979); Phys.Rev.
Lett.43,1574(1979)

45) S. Nagata, P.H.Keesom and H.N.Harrison,Phys.Rev.B 19,1633(1979)

46)P.Svedlindh, P.Grondberg, P.Nordlab, L.Lundgren and
H.S.Chen,Phys.Rev.B 35,268(1987)

47)L.Lundgren, P.Svedlindh and O.Beckman, Phys.Rev.B 26,3990(1982)

48)K.L.Ngai,A.K.Rajagopal and C.Y.Huang, J. Appl.
Phys.55,1714(1984)

49)L.P.Lundgren,L.Nordlab,P.Swedlindh and O.Beckman,
J.Appl.Phys.57,3371(1985);Phys.Rev.Lett.51,911(1983)

50)R.V.Chamberlin,G.Morzukewich and R.Orbach, Phys.
Rev.Lett.55,111(1985)

51)L.P.Lundgren,P.Swedlindh and O.Beckman,J. Magn. Magn.
Mat.31-34,1348(1983)

52)H.Bouchiat and D.Mailly,J.Appl.Phys.57,3453(1985)

53)T.Wang,H.V.Bohm and L.E.Wenger,J.Magn.Magn.Mat.54-57,89(1986)

54)C.N.Guy,J.Phys.F 8,1306)(1978)

55)J.J.Prègean,J.Physique 39,C6-907(1978) J.J.Prèjean and
J.Souletie,J.Physique 41,1335(1980)

56)J.Ferrè,J.Rajchenbach and H.Maletta, J.Appl.
Phys.52,1697(1981)

57)K.Binder and K.Schroder,Phys.Rev.B 14,2142(1976)

58)R.Hoogerbeets,Wei-Li Luo,R.Orbach and D.Fiorani,
Phys.Rev.B 33,6531(1986)

59)R.Hoogerbeets,Wei-Li Luo and R.Orbach,to be published

60)P.Nordlab,P.Swedlindh,L.Lundgren and L.Sandlund,
Phys. Rev.B 33,645(1986)

61)M.Ocio,M.Alba and J.Hammann,J.Phys.Lett.46,L-1101(1985)

62)M.Alba,E.Vincent,J.Hammann and M.Ocio,Conference on
Magn.Magn.Mat.,Baltimore,1986

63)C. De Dominicis,H.Horland and F.Lainee,J.Phys.Lett.46,L-463(1985)

64)C.A.M.Mulder,A.J. van Duyneveldt and J.A.Mydosh,Phys.Rev.B
23,1384(1981)

65)J.L.Tholence,J.Appl.Phys.50,7310(1979)

66)M.Ayadi,P.Nordlab,J.Ferré,A.Mauger and R.Triboulet, J.Magn.
Magn. Mat.54-57,1223(1986)

67)C.Holenberg and B.I.Helperin,Rev.Mod.Phys.49,435(1977)

68)A.T.Ogielski,Phys.Rev.B 32,7384(1985)

69)H.Sompolinski,Phil.Mag.851,543(1985)

70)K.Binder and A.P.Young,Phys.Rev.B 29,2864(1984)

71)M.Saint Paul,J.L.Tholence and W.Giriat,Solid State
Commun.60,621(1986)

72)P.Beauvillain,J.P.Renard,M.Matecki and J.J.Préjean,
Europhys. Lett.2,23(1986)

CHAPTER 25

MUON SPIN RESEARCH ON SPIN GLASSES

Carel Boekema
San Jose State University
Department of Physics
One Washington Square
San Jose, CA 95192-0106
USA

ABSTRACT

Major results of muon-spin-relaxation (μSR) studies on canonical and
oxide spin glasses are discussed. The metallic spin glasses AgMn,
AuFe, CuMn, and CuFe and the oxide spin glass Fe_2TiO_5 are highlighted.
The μSR work shows clearly that below the spin-glass temperature (T_g)
both slow and rapid spin fluctuations occur. Single-crystal μSR
investigations indicate that the freezing process in spin glasses is
spatially inhomogeneous, whereas polycrystalline-sample μSR results
point toward a dominantly homogeneous freezing process. Currently,
the model descriptions of the dynamics in spin glasses observed above
T_g are not unanimous; anisotropy appears to be the necessary
ingredient to attempt to solve this problem in spin dynamics.

1. INTRODUCTION

In recent years, great interest has existed in the properties of
solids showing random magnetism. Magnetic solid-state studies have
revealed a new basic type of magnetism, namely the spin-glass state.
Various unusual characteristics indicate a "glassy" state, in which
random freezing of the spins occurs. A major unanswered question in
the spin-glass work is how the spin system enters into the "glass"
state. μSR continues to provide answers to this and other questions
on dynamic magnetism, due to the unique "time window" in
investigating the dynamical and static critical phenomena.

In the next section, μSR in metallic spin glasses will be
addressed, followed by a section on μSR in oxide spin glasses,
especially Fe_2TiO_5 (pseudobrookite) which may be the ideal spin glass
for theorists.[1] This lecture will be open-ended, because work is
still in progress, as mentioned in the concluding section of my first
lecture on the μSR technique.

G. J. Long and F. Grandjean (eds.), The Time Domain in Surface and Structural Dynamics, 429–438.
© *1988 by Kluwer Academic Publishers.*

2. μSR IN METALLIC SPIN GLASSES

In this section, recent results of μSR studies on metallic spin glasses are reviewed in some detail. These μSR studies on spin dynamics have contributed significantly to the present knowledge and will continue to provide important experimental data on these so-called canonical spin glasses.

A qualitative description of the metallic spin glass is straightforward. At high temperatures, the spins (magnetic moments of the metal ions) are in essence independent and fluctuate very rapidly. As the temperature is lowered, the effect of the random RKKY and dipolar spin-spin interactions becomes increasingly important, especially near and below the spin-glass temperature (T_g). At temperatures well below T_g, the spin-glass system reveals a ground state configuration, for which the spin directions are frozen but random.

μSR appears to be an ideal experimental tool for spin-glass studies.[2,3,4] The zero-field technique is advantageous since even low fields, which are needed for ESR and NMR, influence the spin-glass state. Furthermore, μSR is able to investigate and separate both spin-lattice and spin-spin relaxation processes. To obtain information about magnetic field distributions and spin fluctuations from μSR data, a model must be formulated for a muon stopped in a metallic spin glass. The μ^+ is known[3] to come to rest at an octahedral site in Cu well below room temperature. Similar localization is expected in other fcc noble metals like Au and has been observed in Ag.[5] The muons randomly occupy these interstitial sites, and the local magnetic field will therefore vary in magnitude, direction and time dependence as a function of site.

To describe these magnetic fields, the distribution at each site and the overall distribution have to be modelled.[6,7] For the case when the local field is static, the zero-field μSR relaxation function can be given by

$$G_{\parallel}(t) = \frac{1}{3} + \frac{2}{3}(1 - \ell t)\exp(-\ell t)$$

The 1/3-component represents muons stopped at a site where the local field is parallel to the initial muon spin direction. The damping of the 2/3-component is due to the random perpendicular field components, ℓ is the Lorentzian width of the field distribution. Introduction of fluctuations in the local field leads to modulation of the 1/3-component. Assuming one correlation time τ, we can write

$$G_{\parallel}(t) = \frac{1}{3}\exp\left(-\frac{2t}{3\tau}\right) + \frac{2}{3}(1 - \ell t)\exp(-\ell t) \text{ for } \ell\tau \ll 1$$

$$G_{\parallel}(T) = \frac{1}{3}\exp\left(-\frac{2t}{3\tau}\right) \text{ for } \ell\tau \gg 1$$

Here we would like to make the following notes: (1) Any dynamic process slower than about 10 times the muon decay time will look static in the µSR measurements. (2) In the limit of rapid fluctuations, instead of an exponential decay of the 1/3-component, an exponential square root decay has been found (exp - $(\lambda t)^{1/2}$ with $\lambda = 4\ell^2\tau$). (3) When the magnetic moments are not dilute in the spin-glass systems, gaussian-like damping of the form

$$\frac{1}{3} + \frac{2}{3} \left(1 - (\ell t)^2 \right) \exp \left\{ \frac{-(\ell\tau)^2}{2} \right\}$$

provides a better description. (4) Note in particular that from measurements with no externally applied field, information can be extracted about homogeneous (τ, 1/3-component) and inhomogeneous (ℓ, 2/3-component) broadening processes. This feature is important for spin glasses, because the spin-glass dynamics and statics are sensitive to an applied field.

µSR data on AgMn (1.6%), AuFe (1.4%) and CuMn (1.1%) show that near T_g the correlation time increases drastically with decreasing temperature, about 4 orders of magnitude for a 20% change in temperature. The temperature variation of the Lorentzian width (ℓ) for AgMn[4] was found to be decreasing with nearing T_g from below, whereas an increase is expected based upon the single-correlation time model. This result implied a spectrum of fluctuation modes. Consequently, a more detailed model describing slow fluctuations and rapid (limited amplitude) fluctuations in the spin-glass system yields a much better description of the data for AgMn and CuMn.[8]

Recent µSR measurements performed below T_g in AgMn spin glasses have revealed that the spin correlations decay algebraically with time as t^{-g} (g = 0.5).[9] This form is in contrast with the usually assumed exponential decay, but in complete agreement with theoretical calculations for a disordered spin-glass system.[9] The value g = $\frac{1}{2}$ is in agreement with the prediction of a mean-field dynamic theory. In this model, both slow and rapid spin fluctuations are taken into account below the spin-glass transition.

It appears that the spin-glass dynamics are quite different below and above T_g.[4] The field dependence of the transverse relaxation ($\lambda_\parallel$) rate above T_g must be better understood. For temperatures just approaching T_g from above, $\lambda_\parallel$ increases faster than the bulk magnetization. This suggests a nonlinear local response of the spin system to the applied field and indicates a homogeneous cooperative phenomenon as opposed to models of spin clustering. At temperatures above $3T_g$, the spin dynamics have changed to nearly complete reorientation.

We have discussed some results of µSR experiments in canonical spin glasses performed by the LAMPF collaboration. To obtain a more complete picture and for a review from a somewhat different perspective we refer to the µSR research of the TRIUMPH-KEK collaboration. In their Phys. Rev. paper[10] it is concluded that (1) the combined results from zero-field µSR, neutron-spin-echo, AC-susceptibility and MES measurements in AuFe and CuFe have shown rapid slowing down of the impurity spins above T_g and the appearance of long time persisting static spin polarization below T_g; (2) the combined results also indicate that strong inhomogeneous spin freezing can be ruled out for metallic spin glasses.

From the above, it is clear that the results of the µSR studies of the LAMPF and TRIUMPF-KEK groups indicate that homogeneous cooperative spin freezing is dominant over spin-clustering effects. On the other hand, the "cluster" option for spin-glasses is still being recognized[1,11] and its concept has created recent theoretical interests.[12]

The µSR-Braunschweig-SIN group provided from their pure and Au-doped CuMn investigations[13] strong evidence for a spatially inhomogeneous spin-freezing process. This inhomogeneous model introduces a number of sites where the Mn magnetic moments produce only a fast fluctuating local field. From the remaining sites (the so-called clusters) both dynamic and static fields are contributed. In these µSR studies single crystals were used, which improves the quality of the µSR signal.

Summarizing their results: (1) The homogeneous spin freezing cannot account for the relaxation data in the temperature range $0.8 < TT_g^{-1} < 1.5$. (2) Both spin-freezing descriptions can be used well for $TT_g^{-1} < 0.8$ and $TT_g^{-1} > 1.5$. (3) The fraction fc of spin clusters is temperature dependent: for $TT_g^{-1} = 1$, fc is near to 1, and close to $TT_g^{-1} = 1.5$, fc is below 0.1; further, fc is approximately linear with TT_g^{-1}. (4) Doping with a small amount of nonmagnetic Au changes the spin-glass features well below T_g but does not change the freezing process above T_g; this substantiates their interpretation of their inhomogeneous spin-freezing model.

In conclusion, µSR studies on metallic spin glasses have brought new and relevant information, for instance that both slow and rapid spin fluctuations occur below the spin-glass transition. In combination with other experimental methods µSR, especially in zero applied field, has provided detailed information on basic internal properties of the canonical spin glasses.

3. EARLY µSR STUDIES IN OXIDE SPIN GLASSES

Previous µSR measurements in oxide spin glasses have been restricted to a few very limited cases. Two alumino-silicate glasses, $(MnO)_{40}(Al_2O_3)_{20}(SiO_2)_{40}$[14] and $(CoO)_{40}(Al_2O_3)_{10}(SiO_2)_{50}$,[15] have been studied along with one concentration within the Fe-Ti sesquioxide system, $(FeTiO_3)_{88}(Fe_2O_3)_{12}$.[16]

The alumino silicates were one of the first insulating oxides showing such spin-glass features as a cusp in the AC susceptibility. At that time, one expected to observe muonium signals in insulators as was reported for Al_2O_3 and SiO_2. However, a full μ^+ signal was observed, which can now be understood in terms of chemical muon bonding (see second lecture). Although the alumino silicates are oxide spin glasses, structural glassy effects also appear in the experimental results. It is often difficult to separate structural and magnetic effects.

The µSR work in the 88% ilmenite - 12% hematite solid solution was done using a polycrystalline sample. Neutron-diffraction and static magnetic susceptibility measurements on a single crystal of the mixed oxide showed highly anisotropic behavior, and evidence of clustering. A cusp in $\chi_\parallel$ was evident, but only a broad peak at higher temperatures was seen in $\chi_\perp$.[17] One can assume that μSR data taken in a single crystal sample should show similar anisotropy, but that polycrystalline averaging would smear out these effects so that no such detail was observed.

The μSR data show that the fluctuations of the Co and Mn moments in the alumino silicates and of the Fe moments in the Fe-Ti oxide gradually slow down between $5T_g$ and T_g. This is on contrast to the sharp spin-glass transition observed for the canonical spin glasses like AgMn. A comparison of the scaled spin-correlation behavior ($^{10}\log (\tau_c T_g)$ vs TT_g^{-1}) for these three oxide spin glasses with a power law (homogeneous spin freezing) and a Vogel-Fulcher law (inhomogeneous process) suggest that the inhomogeneous spin-freezing is a reasonable description. However, these results should not be taken as typical of what will be obtained for properly prepared and well-characterized single crystals due to the complicating factors just discussed.

4. SPIN DYNAMICS IN PSEUDOBROOKITE OBSERVED THROUGH μSR

The pseudobrookite system $Fe_{2-x}Ti_{1+x}O_5$ appears to be an excellent vehicle to pursue answers to questions in spin-glass studies. Current issues include: the relative importance of long-range versus short-range interactions in determining spin-glass features (in metallic spin glasses, the RKKY interactions are long range and dominant); the controversy in describing spin glasses by means of an Ising or Heisenberg Hamiltonian (anisotropy!); the nature of the spin-freezing process, in which random and competing magnetic interactions are present, causing so-called "spin frustration". As discussed below in some detail, μSR measurements on $Fe_{1.75}Ti_{1.25}O_5$ seem to provide some answers and steps leading in the right direction.

$Fe_{2-x}Ti_{1+x}O_5$ are insulating or semi-conducting oxides which show remarkable anisotropic magnetic properties[18,19]. The magnetic susceptibility is strikingly anisotropic at temperatures well below RT. Only the c-axis susceptibility exhibits a cusp at the spin-glass temperature (T_g is about 50K for x = 0), while the perpendicular components show paramagnetic behavior till about 10K; indications of transverse spin ordering at about 6K have been observed in low-field susceptibility measurements.[19]

Neutron-diffraction experiments[20] have not revealed any magnetic long-range order. Together with Mössbauer studies[21], the combined results indicate that the Fe^{3+} and Ti^{4+} ions are randomly distributed at specific sites in the orthorhombic structure. For x > 0, when there is a surplus of Ti^{4+} ions, Fe^{2+} ions are also present. Oriented single-crystal Mössbauer spectra show that below T_g all the Fe-moments are antiparallel to the c-axis.[21] ESR results from experiments performed in polycrystalline Fe_2TiO_5, suggest a true thermodynamic transition into the spin-glass state.[22]

According to Tholence et al.[23] the frequency dependence of T_g, as determined by χ- and neutron-diffraction measurements on Fe_2TiO_5, can be described by a Vogel-Fulcher law, or by a power law, characterizing the spin-glass transition. Fitting an Arrhenius law to the data yields unrealistic non-physical values of for instance the activation energy, as also found for other spin-glasses.

$$\tau = \tau_0 \cdot \exp \left\{ \frac{E_a}{k(T - T_g)} \right\}$$

in which E_a is the averaged activation energy and τ_0 the characteristic correlation time, is a semiphenomenological approach. The basic idea behind this law describing inhomogeneous spin freezing consists of building a magnetic superlattice of blocked moments, when T_g is approached from higher temperatures. The Vogel-Fulcher law stems from structural glass research and has been suggested for other glassy materials, including spin glasses.

Another way to explain critical slowing down in spin dynamics is the standard power law,[2-5] and is given by

$$\tau = \tau_0 \cdot \left\{ \frac{T}{(T - T_g)} \right\}^\varepsilon .$$

Using the scaling variable $\dfrac{T_g}{T - T_g}$, or the nonlinear scaling variable $\dfrac{T - T_g}{T}$, one can show that both laws are truncated asymptotic forms of each other. The corrections are small for $T > 2T_g$.

For Fe_2TiO_5, Tholence et al.[23] found $\tau_0 = 2.5 \times 10^{-12}$ sec; $E_a/k = 170K$ and $T_g = 43.5K$ (Vogel-Fulcher (VF)); and $\tau_0 = 10^{-13}$ sec and $\varepsilon = 9.5$ (power law). This may be compared with $\tau_0 = 10^{-13}$ sec and $E_a/k = 80K$ (VF) calculated[24] from the μSR data of $(FeTiO_3)_{0.88}$-$(Fe_2O_3)_{0.12}$. For the metallic spin-glass CuMn (5%), μSR results[2-4] revealed $\varepsilon \simeq 2.9$ and $\tau_0 \simeq 0.8 \times 10^{-12}$ sec (power law) which can be compared with $\varepsilon = 5$ and $\tau_0 = 10^{-12}$ sec obtained from other experimental data on CuMn (5%). Assuming $\tau_0 = 10^{-13}$ sec, Tholence[23] determined from neutron and susceptibility data that in a Vogel-Fulcher description E_a/k is in the range from 40 to 80K for RKKY spin-glasses, and varies from 5 to 25K for insulating spin glasses. Clearly, the analytic expressions in describing the spin dynamics of the spin-freezing process above T_g are not unanimous.

The single crystals of $Fe_{1.75}Ti_{1.25}O_5$ used in our study were prepared by the floating-zone technique. The single crystals were checked by means of X-ray diffraction, Auger analysis and resistivity measurements. Zero-field experiments in a standard longitudinal arrangement were performed as a function of temperature. μSR studies on well-known oxides like α-Fe_2O_3, Fe_3O_4 and $RFeO_3$ (see second lecture) have shown that at temperatures around and below RT the muon forms a muon-oxygen diatomic complex, $(\mu O)^-$. This bond formation can be best described as a "muoxyl" bridge with a bondlength of 1A. In the $Fe_{2-x}Ti_{1+x}O_5$-system, the muon is expected to bond, probe and behave very similarly as in the above-mentioned magnetic oxides.

The temperature dependences of the asymmetry (A_s) and muon-spin-relaxation rate (λ), obtained for single crystalline $Fe_{1.75}Ti_{1.25}O_5$ in zero field, are depicted in Figs. 1a and 1b. As can be seen from the A_s-reduction to 1/3, and the divergence in λ, T_g can readily be determined and is 44.0(±0.5)K, about 6K lower than found earlier for Fe_2TiO_5.[23] This reduction in T_g can be explained by the higher concentration of Ti^{4+} ions, which behave like magnetic vacancies, and the extra presence of the Fe^{2+} ions.

A distinct exponential λ-behavior is observed above T_g (see Fig. 1b), while below, a square-root exponential decay is seen. The latter indicates that below T_g fast fluctuation modes exist in the "frozen" state. This is consistent with what has been found for metallic spin glasses. Near 8K, a maximum in λ is observed, which can be due to the recently proposed transverse spin ordering at these temperatures (see above), as deduced from χ measurements. To explain the origin of this transverse effect, the spin Hamiltonian needs at least an anisotropy term; in addition to its random nature, this term must have a short-range character.

Even near T_g, λ is rather low ($<1\mu s^{-1}$), probably due to the well-defined ($\mu O)^-$ state in the single crystal. For comparison, Uemura et al. found λ (at 1.3 T_g) $\approx$ 2.5 μs^{-1} in the polycrystalline Fe-Ti sesquioxide sample, whereas we observed λ (at 1.3 T_g) = 0.22 μs^{-1} for $Fe_{1.75}Ti_{1.25}O_5$. Our very low λ values allow a comparison between spin-freezing models as discussed above in some detail.

Using $\lambda \sim \tau_c$,[2-5] our results above 50K are consistent with both the VF law and the power law. Below 50K $\approx$ 1.1 T_g, the errors are too large for a useful comparison. In the equation

$$\ln \lambda = \lambda_0 + \frac{E_a}{k} \cdot (T - T_g)^{-1},$$

E_a/k is estimated to be 20(±5)K for the temperature region 50K < T < 100K, consistent with the reported temperature range for insulating spin glasses, (see above). It is, however, a factor of 8 lower than reported for Fe_2TiO_5.[23]

Note that for $Fe_{1.75}Ti_{1.25}O_5$ the drastic change in λ occurs between T_g and $2T_g$ (see Fig. 1b); this is sharper than reported for the ilmenite-hematite sample (T_g < T < $3T_g$),[16,17] but less than the sharpness seen for the canonical spin glasses (T_g < T < 1.4T_g).[2-5] This may have to do with differences in ranges of the competing magnetic interactions.

These results suggested that careful and high-statistical measurements between 1.1 T_g and T_g should be performed, because the VF and power laws are quite different in this range, such that a differentiation could be made. Surprisingly, in these experiments non-exponential muon spin relaxation was observed. Tentatively, the relaxation data are well described by a fit of two exponentials, indicating inhomogeneous spin freezing. Preliminary analysis shows

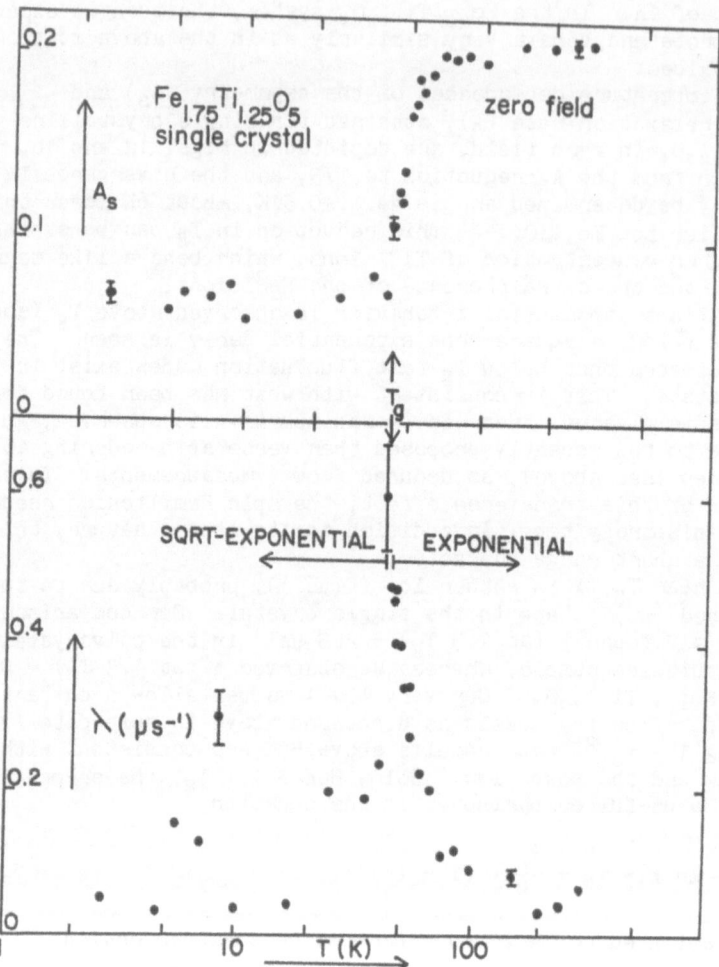

Fig. 1.
    Temperature dependences of the asymmetry (A_S; upper part) and
the muon-spin-relaxation rate (λ; lower part b) of the zero field μSR
signal observed for $Fe_{1.75}Ti_{1.25}O_5$.
    a) At the spin-glass temperature (T_g) of about 44K, A_S drops
sharply to the 1/3 value due to fast increasing internal fields.
    b) At T_g and near 8K, drastic λ-increases are observed,
characterizing the spin-glass transition and the transverse spin
ordering, respectively.

that at T_g half of the atomic sites are in spin clusters; above 1.33 T_g ($\approx$ 60K) no cluster fraction can be observed; the cluster fraction depends linearly on the temperature. (Compare with μSR results on Au-doped and pure CuMn). Because a two-exponential fit is simplistic, work is in progress for a better detailed data-analysis.

Finally, single-crystal studies have again indicated that inhomogeneous spin freezing is dominant just above T_g; it may well be possible that the use of polycrystalline samples will wash or smear out the observability of the spin clusters, which were not seen in earlier μSR studies. The observation of spin clustering in both metallic and insulating single-crystal spin glasses may be described by Griffiths' phase[25] (work along these lines is in progress). Griffiths' phase describes theoretically the occurrence of small spin clusters, which freeze before the bulk spin system freezes.

As one may see, this lecture has been written in the style of a progress report, presenting two contrasting views on spin freezing, as will happen in contemporary physics. You as young witnesses may act as jury, realizing also that your judgment will not be final (neither will mine). Time (and our "spies") will tell, and isn't time the most important aspect in dynamics?

6. ACKNOWLEDGMENTS

These three lectures on μSR-oxide studies could not have been written without the collaboration of, and discussions within, the μSR groups at LAMPF and SIN. A member of both groups is Art Denison from the University of Wyoming; I owe him much. Vic Brabers from the Technical University of Eindhoven has provided most of the quality crystals used in our research; through the courtesy of George Honig from Purdue University and Joe Remeika from AT&T Bell Laboratories, other important single crystals were obtained.

I wish to thank Walter Hofmann, Walter Kündig, Peter Meier, Kurt Rüegg and Bruce Patterson from the University of Zürich; special thanks are due to WH and KR, as their bricks were essential in building our muoxyl bridges. The Los Alamos group (Wayne Cooke, Bob Heffner, Dick Hudson and Mario Schillaci) supported and encouraged our μSR-oxide studies. From the LAMPF-μSR users, I acknowledge gratefully Stan Dodds, Tom Estle, Grant Gist, Doug MacLaughlin and Jean Oostens. From Texas Tech: the contributions of Roger Lichti and our graduate students Kwai-chow Chan and Tsai-ku Lin have been crucial in the successful progress of the oxide work featuring the muon spy. The condensed-matter group at San Jose State is gratefully acknowledged for their discussions and valued support.

The research at Los Alamos is supported by the U.S. Department of Energy. Research by this author at Texas Tech and San Jose State is supported by the Robert A. Welch Foundation and the San Jose State Foundation.

438

7. REFERENCES

1. J.A. Mydosh, Hpf. Int. 31 (1986) 347.
2. D.E. MacLaughlin, Hpf. Int. 8 (1981) 749.
3. Y.J. Uemura and T. Yamazaki, J. Mag. Mag. Mat. 31-34 (1983)
1359; Physica 109 & 110B (1982) 1915; Y.J. Uemura, Hpf. Int. 8 (1981)
739.
4. R.H. Heffner et al., J. Appl. Phys. 53 (1982) 2174.
5. R.H. Heffner, Hpf. Int. 8 (1981) 655.
6. R. Kubo, Hpf. Int. 8 (1981) 731.
7. M. Leon, Hpf. Int. 8 (1981) 781.
8. R.H. Heffner et al., J. Mag. Mag. Mat. 31-34 (1983) 1363.
9. D.E. MacLaughlin et al., Phys. Rev. Lett. 51 (1983) 927.
10. Y.J. Uemura et al., Phys. Rev. B31 (1985) 546.
11. J.A. Mydosh, 1983, p. 38, Lecture Notes in Physics 192
(Springer-Verlag, New York).
12. J.A. Hertz, Phys. Rev. Lett. 51 (1983) 1880. S. Nambu, Progr.
Theor. Phys. 74 (1985) 446.
13. H. Pinkvos et al., Hpf. Int. 31 (1986) 363.
14. L.H. Biemann et al., Hpf. Int. 4 (1978) 861.
15. Y.J. Uemura et al., Hpf. Int. 8 (1981) 757.
16. Y.J. Uemura et al., J. Mag. Mag. Mat. 31-34 (1983) 1379.
17. Y. Ishikawa et al., J. Mag. Mag. Mat. 31-34 (1983) 1379.
18. U. Atzmony et al., Phys. Rev. Lett. 43 (1979) 782.
19. Y. Yeshurun et al., Phys. Rev. B31 (1985) 3191; J. Mag. Mag.
Mat. 54-57 (1986) 203.
20. U. Atzmony et al., J. Mag. Mag. Mat. 15-18 (1980) 115.
21. E. Gurewitz and U. Atzmony, Phys. Rev. B22 (1980) 6093
22. N.M.L. Köche et al., Solid State Comm. 52 (1984) 781.
23. J.L. Tholence et al., J. Mag. Mag. Mat. 54-57 (1986) 203.
24. C. Boekema et al., "LAMPF Progress Report E854" (Nov. 1983).
25. R.L. Lichti, private communication; R.B. Griffiths, Phys. Rev.
Lett. 23 (1969) 17.

CHAPTER 26

ADSORPTION STUDIES USING LOW-ENERGY ELECTRON DIFFRACTION

Renee D. Diehl
Department of Physics
University of Liverpool
Liverpool L69 3BX
England

ABSTRACT. Low-energy electron diffraction(LEED) has been the most
widely-used experimental technique for the characterization of surface
structures for the past 25 years. More recently, it is being used as a
means of studying surface dynamics. The kinetics of ordering and of
domain growth are two dynamical processes which have been studied using
LEED. This chapter provides an introduction to the types of structures
which are studied using LEED, a description of the experimental aspects
of LEED, and examples of some LEED surface studies.

1. STRUCTURES ON SURFACES

1.1 Experimental Techniques Used to Study Surfaces

There are a variety of experimental techniques used to study surfaces,
most of them abbreviated by acronyms. Most of them use either photons,
electrons, ions, or neutral particles to probe the surface and then
analyze either the same particles or different particles which are
produced at the surface, as illustrated in Figure 1. A sample of surface
techniques is given in Table I. In addition to these techniques, there
are others which do not use a probe particle, but analyze the surface
itself. Two examples of these are FEM, field emission microscopy, and
STM, scanning tunneling microscopy, which use high electric fields to
extract electrons from a surface which acts as a field emitter(FEM), or
to extract electrons from a field emitter to a relatively flat
surface(STM).
   Surface techniques can be divided up roughly into techniques to
study various properties of surfaces:
a. geometrical structure(LEED, neutron diffraction, EXAFS, ISS)
b. electronic structure(AES, UPS, IPES)
c. elemental analysis(AES, SIMS)
d. chemical states(XPS,HREELS)
e. vibrational properties(HREELS, INS, RAIRS)
Many techniques can be used for more than one aspect of surface study,
and in most surface studies, more than one technique is used to obtain

G J. Long and F. Grandjean (eds.), The Time Domain in Surface and Structural Dynamics, 439–465.
© 1988 by Kluwer Academic Publishers.

complementary information about the surfaces.

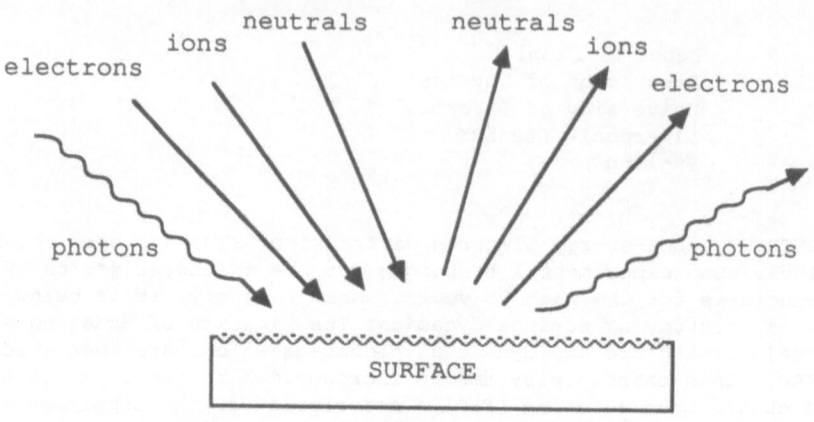

Figure 1. Particles incident on and emerging from a surface.

TABLE I. Some experimental techniques used in surface science.

ACRONYM	TECHNIQUE	PARTICLES IN	OUT
LEED	Low-Energy Electron Diffraction	electrons	electrons
ISS	Ion Scattering Spectroscopy	ions	ions
UPS	Ultra-violet Photoemission Spectroscopy	photons(UV)	electrons
AES	Auger Electron Spectroscopy	electrons or xrays	electrons
HREELS	High-resolution Electron Energy Loss Spectroscopy	electrons	electrons
IPES	Inverse Photoemission Spectroscopy	electrons	photons(UV)
INS	Inelastic Neutron Scattering	neutrons	neutrons
EXAFS	Extended Xray Absorption Fine Structure	photons(xray)	xrays
LEAD	Low-Energy Atom Diffraction	atoms	atoms
RAIRS	Reflection Absorption Infrared Spectroscopy	photons(IR)	photons(IR)
SIMS	Secondary Ion Mass Spectrometry	ions	ions

1.2. The Need For Ultra-High Vacuum

The development of surface science is relatively recent, even though the foundations for many of the experimental techniques were established in the 1920's and 1930's. One of the reasons for this is that to study surfaces, it is usually necessary to have ultra-high vacuum(UHV), and large-scale ultra-high vacuum systems were not easily available until the 1960's. One of the major applications of surface techniques, the study, preparation, and manufacture of microcircuitry, contributed to the development of UHV and of surface analytical techniques in the 1960's and 1970's. The need for ultra-high vacuum arises from the need to keep the surface clean while it is being studied. The need for long mean-free paths of the incident and outgoing particles is also a consideration, but generally speaking, it does not require ultra-high vacuum. The maximum pressures at which experiments can be carried out is illustrated by the empirical relation for the number of molecules at room temperature hitting a surface per unit area per unit time, given by

$$N \approx 10^{20} \text{ molecules cm}^{-2}\text{sec}^{-1} \times P(\text{torr}),$$

where 1 torr = 1.3 mbar. A normal surface density of atoms is $\approx 10^{14}$ molecules cm^{-2}; so therefore there are $10^6 \times P$ molecules hitting each surface atom each second. For example, at $P = 10^{-6}$ torr, there is one molecule hitting each surface atom per second! If each molecule which hits the surface sticks to it, then in one second the surface will be completely covered. In an experiment, if a maximum of 5% impurities on the surface can be tolerated and the experiment takes 1 hour, then the required $P \approx 10^{-11}$ torr. In practice, room temperature sticking probabilites are closer to 0.5 than to 1, but this changes the result by only a factor of 2.

A pressure of 10^{-11} torr is a typical base pressure for an ultra-high vacuum system used for surface studies. From the preceding calculation, it is apparent that surface experiments must be carried out over times as short as possible to minimize surface contamination. The time of an experiment depends on the details of the surface being studied and on the technique, but in experiments on single crystals, one to two hours is a typical time for an experiment, after which a clean surface must be prepared again.

1.3. Surfaces

Surface experiments can be carried out on single crystals, poly-crystalline, or even amorphous materials, but most structural experiments are carried out on either single crystal or polycrystalline materials. And in many experiments on polycrystalline materials such as polycrystalline graphite, the interest is mainly with the crystalline parts of the material and not the boundaries between crystallites. Surface experiments are carried out on clean surfaces and surfaces on which one or more substance have been adsorbed. Many of the studies on clean surfaces are concerned with how the physical and electronic structures at the surface differ from those of the bulk material. The region at the surface in which some properties differ from the bulk crystal can extend into the crystal by many layers of atoms. This region often has a different, "reconstructed" structure relative to the truncated bulk structure as a result of the interactions being different at the surface. In adsorption experiments, adsorbate molecules or atoms are adsorbed onto a substrate surface by evaporation, condensation, or simply by exposure to a gas. Here we will mainly be concerned with the structures of adsorbates on simple single-crystal substrates.

In two dimensions, simple substrates have square, hexagonal, or rectangular symmetry. On atomically rough surfaces, the third dimension becomes important in the surface structure, and more complicated symmetries are possible. Even on a flat surface, however, there is a potential modulation across the surface due to the surface atoms that causes the molecules which adsorb on the surface to move preferentially to particular "sites" on the surface, for instance on top of an atom, on a bridge between two atoms, or in a hollow between atoms, as indicated in Figure 2.

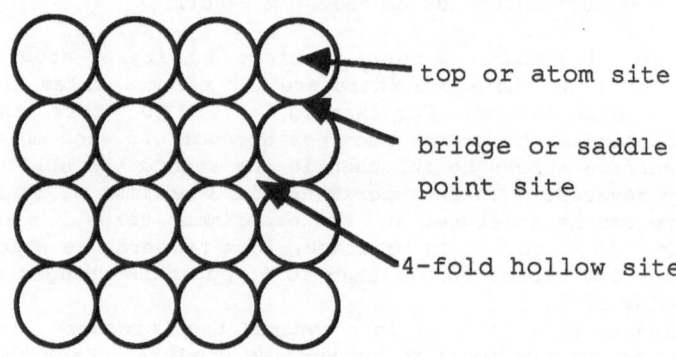

top or atom site

bridge or saddle point site

4-fold hollow site

Figure 2. A substrate surface showing various adsorption sites.

1.4. Structures in Adsorption Systems

When atoms or molecules adsorb on the surface, if the temperature is low enough and there is enough mobility of the adsorbed atoms to reach an equilibrium site, the atoms will form an ordered structure on the surface. There are many types of ordered adsorbate structures. Some examples of these are commensurate structures, incommensurate structures, and structures in which the molecular axes are orientationally ordered, as shown in Figure 3. Incommensurate structures may be uniform or have density modulations such as domain walls. While commensurate structures, by their definition, must line up with some symmetry direction of the substrate, incommensurate overlayers are often rotationally oriented in a different direction.

Adsorption systems can be roughly divided into two groups, physisorption and chemisorption, based on the strength of the interaction between the adsorbate atoms and the substrate atoms. This adsorption potential energy is generally about 0.1 to 1 eV/atom for physisorbed systems and about 1 to 10 eV/atom for chemisorbed systems although there is actually a continuum of adsorption energies and there is no strict division between physisorption and chemisorption. The type of ordered structure that occurs for any particular substrate-adsorbate combination at any particular temperature and coverage depends on the details of the interactions between the adsorbed atoms or molecules and the substrate and of the interactions between adsorbate atoms or molecules with each other. This interaction can be attractive or repulsive and have multipole terms. There can also be interactions between the adsorbate molecules via the substrate, i.e. an

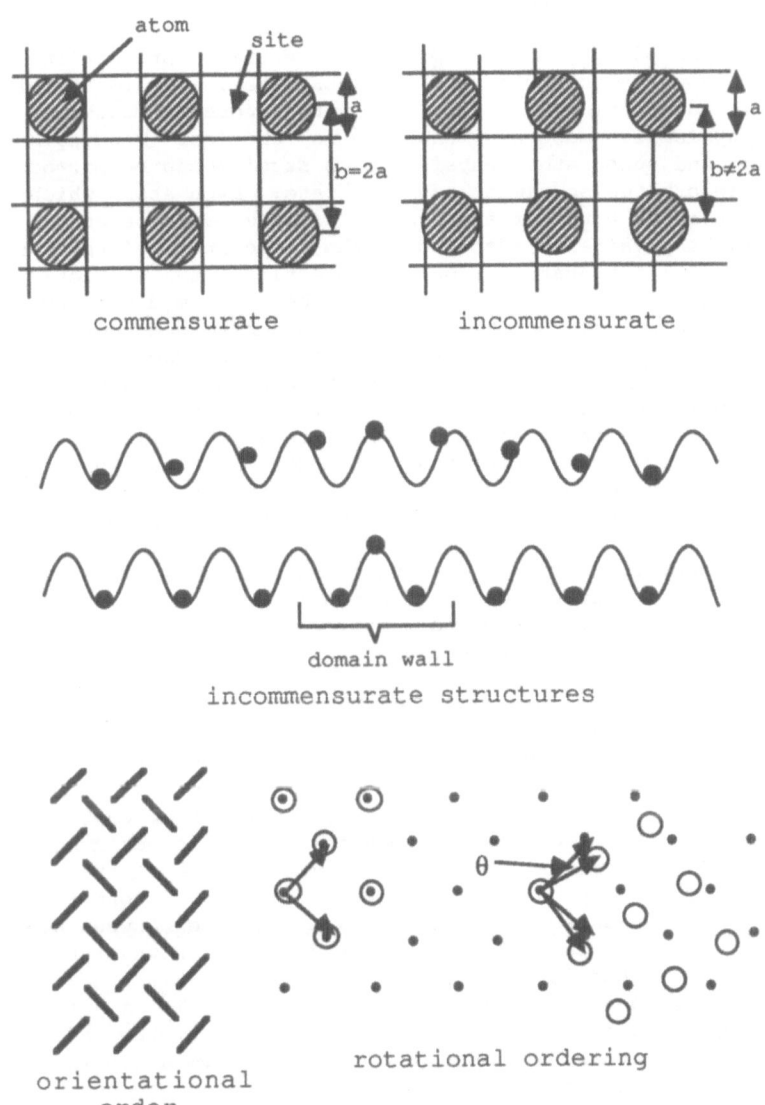

Figure 3. Examples of ordered surface structures.

interaction that arises from the adsorbate molecule at one place causing the electronic structure of the substrate to change in the vicinity of that adsorbate molecule, thus affecting the behavior of other admolecules. These interactions are anisotropic and oscillatory and their range can extend well beyond the nearest neighbors[1].

1.5. Physisorption Systems

In physisorption systems, the primary attractive interaction between the adsorbed atoms or molecules and the substrate is the relatively weak van der Waals interaction[2]. This is why the atoms or molecules are said to be "physically" adsorbed; there is no "chemical" bonding between the adsorbate and substrate. Physisorption studies can be performed with various adsorbates on almost any substrate. Even atoms which will chemically react with certain substrates at some temperatures can sometimes be physically adsorbed if the temperature of the substrate is kept below the temperature required to activate the chemical reaction.

Although physisorption studies can be performed on many types of substrates, most physisorption studies that have been performed so far have used graphite as a substrate[3]. Although graphite provides an extremely good substrate for physisorption studies since it is relatively inert and has very large perfect areas, its greatest strength is that it can be prepared in different forms which makes it compatible with a large number of experimental techniques. Thus, one physisorption system can be studied by as many as ten or twenty techniques! This makes physisorption systems some of the best-studied systems in surface science.

For adsorption of rare gas atoms and simple small molecules on graphite, the substrate-adsorbate potential energy for one adsorbate molecule can be fairly accurately represented by assuming pairwise interactions with a 6-12 potential:

$$U(r) = 4 \varepsilon [(\sigma/r)^{12} - (\sigma/r)^{6}]$$

between each carbon atom in the substrate and the adsorbed molecule and summing over every atom in the substrate[2]. There is a small potential modulation across the surface which means that while certain sites on the surface are favored adsorption sites, there is not much to hinder mobility of atoms across the surface. The surface of the basal plane of graphite has a hexagonal symmetry and consists of a honeycomb array of carbon atoms, as shown in Figure 4. The distance between nearest neighbor carbon atoms is 1.42Å and the distance between the centers of the carbon hexagons is 2.46Å.

For rare gases and simple molecules such as N_2 on graphite, the adsorption energies for the center of the hexagon, the saddle point between carbon atoms, and on top of a carbon atom are different. There is a slight preference for the atoms or molecules to move to the centers of the hexagons but the difference between the potential energy of an adatom at the center of a hexagon and an adatom on top of a carbon atom is small compared to the potential energy of adsorption. Table II lists some potential energies for some atoms and molecules on graphite.

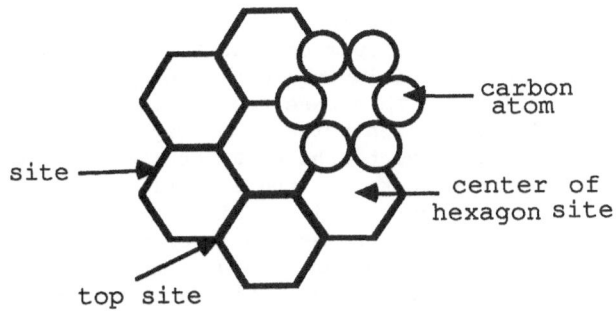

Figure 4. Graphite Substrate and Adsorption Sites.

Table II Potential Energies for some Adsorbates on Graphite[4].

Adsorption site	Adsorption Energies (eV)						
	He	Ne	Ar	Kr	Xe	N₂	CO
center of hexagon	0.0217	0.0443	0.0983	0.1184	0.1585	0.0945	0.0950
saddle point	0.0203	0.0418	0.0955	0.1151	0.1557	0.0919	0.0924
top of carbon atom	0.0200	0.0415	0.0951	0.1149	0.1553	0.0917	0.0921

Like the substrate-adsorbate interaction, the main interactions between rare gas atoms adsorbed on the graphite surface are the attractive van der Waals potential and the repulsive Pauli interaction at short distances. So when a few atoms adsorb on the surface, they are attractive to each other and form 2D islands on the surface, as long as the temperature is high enough so that the mobility of the atoms is not severely hindered. The temperature at which the atoms become immobile over experimental times is related to the size of the lateral potential modulation across the surface - in this case, the size of the potential barrier that a molecule has to overcome to move to the next center-of-hexagon site(see Table II). The lateral modulation energy also affects the way the atoms order on the surface. An example of this behavior is the system of krypton adsorbed on graphite.

1.6. Krypton/Graphite Commensurate-Incommensurate Transition

When less than one monolayer of krypton orders on graphite, it forms a structure which is defined by the substrate periodicity, shown in Figure 5[5]. If the substrate potential modulation were smaller, it is likely that the separation of the Kr atoms would be about 6% smaller, which is close to its spacing in 3D solid Kr. If the surface is filled such that

one out of three hexagons is occupied by a Kr atom(as in Figure 5) and
then as more Kr is added to the surface, the Kr lattice is observed to
compress until it reaches a spacing close to the spacing of bulk Kr[6].
The competing interactions in this case are the Pauli repulsion of the
Kr atoms, the lateral interaction across the graphite surface, and the
Kr-graphite attraction relative to the Kr-Kr attraction. If the
Kr-graphite attraction were weaker or the lateral potential modulation
due to the graphite were larger, it is possible that the first layer of
Kr would remain commensurate as the second layer formed on top.
However, LEED and xray diffraction experiments carried out on the
Kr/graphite system showed that over a particular temperature range, the
Kr compresses via a power law dependence of the chemical potential
change until the density approaches that of a close-packed plane of Kr
atoms in the 3D solid[7]. The second layer forms on top of this
compressed layer.

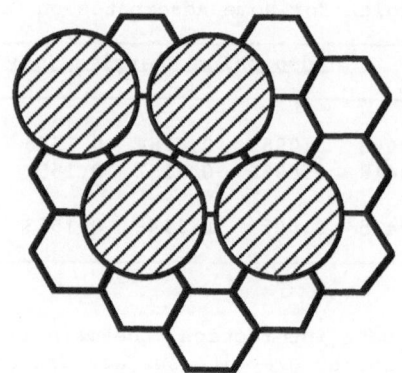

Figure 5. Commensurate Kr on Graphite

1.7. Chemisorption Systems

The majority of recent studies of the statistical mechanics of adsorbed
layers have been concerned with physisorption rather than chemisorbed
species. The reason for this is that the interactions in chemisorption
systems are not as well understood as those in physisorption systems,
mainly because they are stronger, longer range interactions in which the
substrate plays a more active role than in the case of physisorption[8].
These "chemical" interactions are responsible for the catalytic
properties of surfaces. In many cases, there is also an
adsorbate-induced "reconstruction" of the substrate surface due to its
altered electronic structure[9]. Because chemisorption involves some
sort of chemical bond between the adsorbed atoms or molecules and the
substrate, the structures which occur depend on the details of the
chemistry of the adsorbate and the substrate. For instance, when CO is

adsorbed on platinum, it adsorbs preferentially in the "top" sites, above a Pt atoms[10]; whereas on Pd, CO adsorbs preferentially in the bridge sites, between Pd atoms[11]. Chemisorption atom-surface interactions are often grouped into ionic, polar-metallic, or metallic bonding groups, depending on whether the shifted electronic level of the adatom is above, at the same level, or below the Fermi level, respectively[12]. But beyond the details of the chemistry of the interactions, some systems are observed to display behavior which is consistent with the same types of models which might describe 2D physisorption systems. An example of one such system is O/Ni{111}.

1.8. Order-Disorder Transition of (2x2) O/Ni{111}

On the Ni{111} surface, the oxygen atoms occupy the three-fold hollows as shown in Figure 6[13]. These three-fold hollows include both fcc and hcp sites, and presumably there is some small difference in the binding energy between them. The lattice of adsorption sites has hexagonal symmetry if this difference in binding energy is zero, and triangular symmetry if it is not zero. The order-disorder transition of this structure was studied using LEED, and the critical exponents(see section 2.7) associated with the disordering were extracted from the data and are shown in Table III[14].

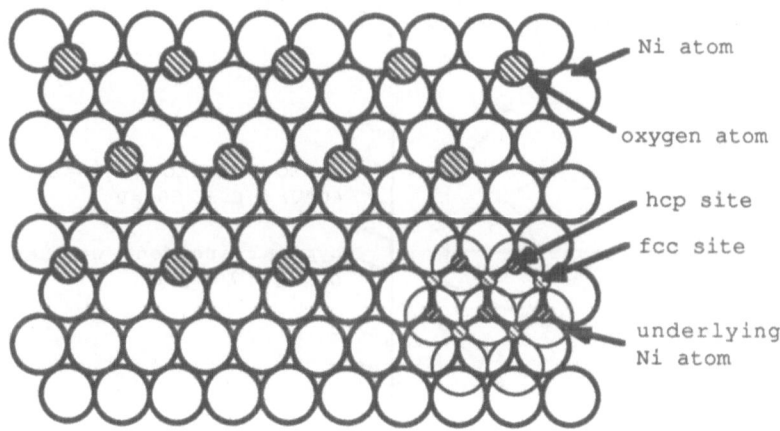

Figure 6. p(2x2) Phase of O/Ni{111}[13]

If the lattice of adsorption sites has triangular symmetry, which was expected to be the case, the order-disorder transition is predicted to be in the universality class of the four-state Potts model[8]. From the table, it can be seen that the measured exponents(see section 2.7) are in fact closer to the Ising model exponents. Two possible explanations for this result have been advanced. One is that if the energy difference between the sites is very small, the system would behave as

if all sites were equivalent(fcc and hcp) until very close to the
transition. In this case, the model describing the system would be the
Heisenberg model with corner-cubic anisotropy and might display an Ising
transition[15]. The other explanation is that defects on the surface
may be causing a rounding of the experimental data which has not been
taken into account in the theory[8].

TABLE III   Exponents for (2x2) O/Ni{111}[8,14]

exponent	experiment	four-state Potts	Ising
β(long-range order)	0.14±0.02	1/12	1/8
γ(short-range order)	1.9±0.2	7/6	7/4
ν(susceptibility)	0.94±0.10	2/3	1

2. LOW-ENERGY ELECTRON DIFFRACTION

2.1 Introduction and Historical Notes

Low-energy electron diffraction(LEED) studies involve directing an
energy-selected beam of electrons in the energy range 40-500eV onto a
well-defined surface and observing the elastically backscattered
electrons, as shown in Figure 7.

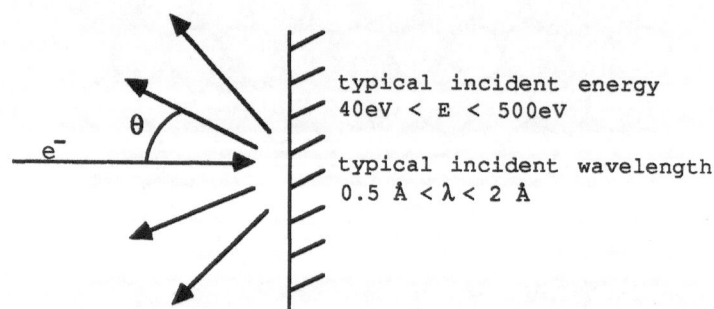

typical incident energy
40eV < E < 500eV

typical incident wavelength
0.5 Å < λ < 2 Å

Figure 7. Electrons incident on and diffracted from a surface.

The big breakthrough which led to the understanding of LEED occurred in
1925 when C. Davisson, who was studying the scattering properties of
low-energy electrons from metals, had an accident in his lab[16]. His
oxygen bottle broke, causing the polycrystalline Ni sample which he had
been heating to form a very thick oxide layer. In order to clean this
sample, it was heated to a very high temperature which caused it to
crystallize into relatively large crystals. In his further studies,
Davisson noted that the backscattered electrons were no longer producing

smoothly varying intensities as a function of scattering angle, but very sharp peaks instead. Until these experiments, the wave nature of electrons was not understood, and the backscattered peaks were hypothesized to arise from the "shells" of electrons around each atom of the solid. There were some other ideas that the peaks were actually caused by the electrons producing x-ray emission which then produced a diffraction pattern. What was not universally accepted at the time(and certainly not demonstrated) was that electrons themselves could diffract and display wave characteristics. Further study of the electron backscattering patterns by Davisson and L. H. Germer led them to discover(and publish in 1927) that if they assumed that electrons were producing the backscattered peaks and that the crystalline structure of the solid(not electron levels) was producing the diffraction grating, then the de Broglie hypothesis, $\lambda = h/p = \sqrt{150/E}$ (which was only published in 1925) was consistent with the results[16].

2.2. Electrons Scattered by Solids

For incident electrons having energies between about 40eV and 500eV, electrons scattered and emitted from a solid have an angular distribution which consists of a number of sharp peaks on a smoothly varying background, while the energy spectrum mostly consists of a broad low-energy(<50eV) peak and a relatively sharp elastic peak, as shown in Figure 8.

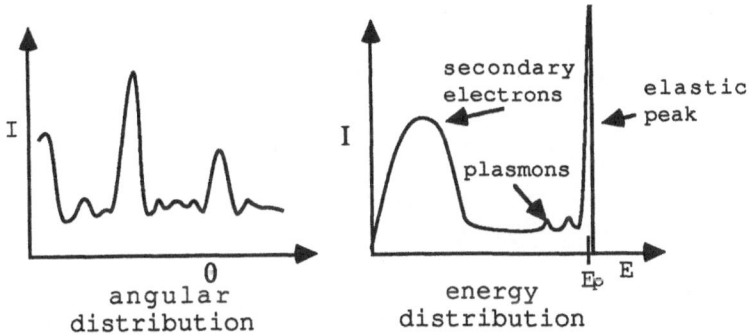

Figure 8. Angular and Energy Distribution of Electrons Scattered from a Solid

Between these peaks, there are also small features due to characteristic energy losses by primary electrons caused by plasmon, interband, and atomic excitations. The "elastic" peak corresponds to electrons which either have scattered elastically or by interactions with phonons($\Delta E < 0.1 eV$). The reason that LEED is surface sensitive is because the probabilities for inelastic scattering are high. Typically for a primary beam energy of about 100eV, only 1% of the incident electrons are scattered into the "elastic" peak. The mean free path of the elastically scattered electrons is only about two to three atomic

layers. The broad low-energy peak of the energy spectrum of scattered
electrons arises from secondary electrons scattering from primary
electrons and with other secondaries in a "cascade-type" mechanism. In
a typical LEED experiment, the collected electrons have energies greater
than about 80% of the primary beam energy.

2.3. Instrumentation for LEED

The basic elements of a LEED diffractometer are an electron source, a
sample goniometer, and a detector for the scattered electrons, as shown
in Figure 9. The typical guns used for LEED have a very simple design.
They consist of a thermionic cathode, an extraction electrode, focusing
electrodes, and electrostatic deflection electrodes. Attempts to
improve electrostatic guns include methods of decreasing the beam
diameter and the beam divergence and minimizing the energy spread of the
beam[17].

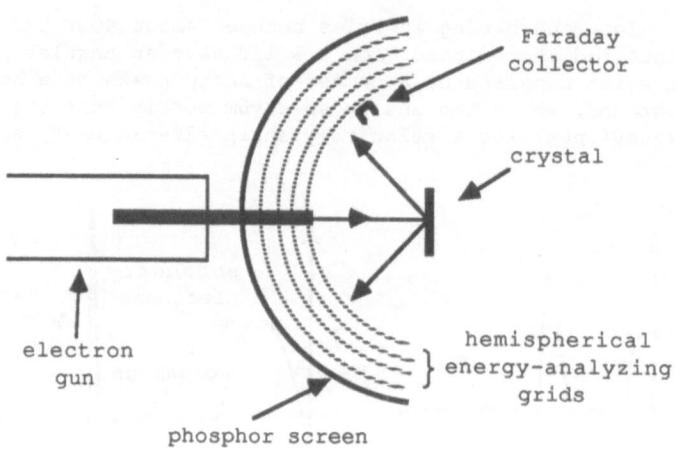

Figure 9. A typical LEED optics.

Signal detection in LEED requires the measurement of
energy-resolved and angle-resolved current at a fixed energy between 0
eV and 1000 eV. Two types of detectors are commonly used, a Faraday cup
and a set of hemispherical grids with a fluorescent screen. The major
advantage of the grid-screen system is that it provides a visual display
of essentially all diffracted beams, making a rapid determination of the
surface unit mesh possible. The major disadvantage is in signal
processing, in which case the image will need to be reconverted to
digital or analogue output. This can be done with light photometers,
video cameras, or by using a microdensitometer to anaylze photographic
film. In some experiments it is desirable to minimize the electron
current to the sample crystal. This is the case in studies of weakly

adsorbed or highly mobile adsorbates where the electron beam might desorb the adsorbate or cause it to diffuse away from the beam. In these cases, the addition of a channel electron multiplier array (CEMA) to the detector improves the sensitivity by about 10^4 for a single plate or 10^6 for a chevron array. Increasing the sensitivity of the detector is also necessary in experiments on kinetic processes, where data must be collected quickly. The sensitivity can be further increased by using a position-sensitive pulse detector in place of the fluorescent screen[17]. The best combination of sensitivity and resolution which is currently available is the use of a Faraday cup detector with a channel electron multiplier[17], but this has the drawback of detecting only one point in momentum-transfer space at a time.

2.4. Periodicities in 2D

The periodicities on surfaces are characterized by translation operators $T=ms_1+ns_2$, shown in Figure 10, and the point operations of rotation and reflection. A surface structure would be completely defined if the positions were known for all the atoms in the surface region between the actual surface and the plane where the bulk structure is established. There are five possible Bravais lattices in 2D and they are summarized in Table IV. A crystal structure is formed by associating every lattice point of a Bravais lattice with a unit assembly or basis or atoms.

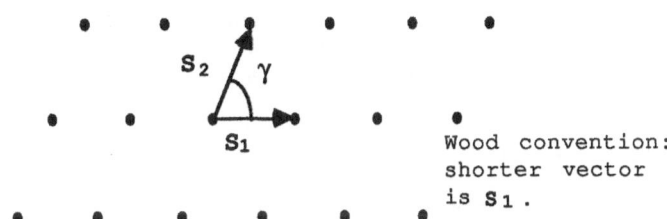

Wood convention:
shorter vector
is S_1.

Figure 10. Surface periodicity[18].

2.5. LEED Patterns

For typical incident beams in LEED, one incident electron does not overlap appreciably with another, and consequently a LEED pattern results from essentially isolated electron-lattice interactions. The speed of the electrons is of the order of 10^7m sec^{-1}, which is much faster than most thermally-induced surface motions. The effect of thermal motions on the LEED pattern is therefore just a reduction in intensity of the diffracted beams.

TABLE IV. Bravais Lattices in 2D[19]

Lattice	Unit Cell	Axes
1. oblique	parallelogram	$s_1 \neq s_2, \gamma \neq 90°$
2. primitive or	rectangle	$s_1 \neq s_2, \gamma = 90°$
3. centered rectangular		
4. square	square	$s_1 = s_2, \gamma = 90°$
5. hexagonal	60° rhombus	$s_1 = s_2, \gamma = 120°$

Since the electrons are in effect infinitely far removed from the crystal both before and after the diffraction, the effect of the interaction is shown by the change in the wave vector of the electron for eigenstates:

$$\Psi_k(\mathbf{r}) = \exp(i\ \mathbf{k} \cdot \mathbf{r})$$

of Schrödinger's equation:

$$H\ \Psi_k = E(k)\ \Psi_k$$

in the absence of a crystal. Given the wave vector **k** for the incident electrons, the wave vectors **k**′ for the diffracted electrons are determined by the conservation of energy and momentum parallel to the surface:

$$E(\mathbf{k}') = E(\mathbf{k}) \qquad \text{and} \qquad \mathbf{k}'_{||} = \mathbf{k}_{||} + \mathbf{g}(hk)$$

where $\mathbf{g}(hk) = h\mathbf{s}_1{}^* + k\mathbf{s}_2{}^*$ represents a vector of the reciprocal lattice net. The conservation of $\mathbf{k}_{||}$ equation is, in fact, equivalent to the Bragg equation:

$$n\lambda = d_{hk}(\sin\phi' - \sin\phi)$$

where ϕ and ϕ' are the incident and scattered angles, respectively, measured relative to the surface normal and d_{hk} is the spacing of a set of planes (hk) in the crystal. The hk in these equations refer to the Miller indices of the reciprocal lattice of the crystal structure. The set of all wave vectors **g** that yield plane waves with the periodicity of a given Bravais lattice is known as its reciprocal lattice. In 2D, the reciprocal net corresponding to the surface net defined by $\mathbf{s}_1$ and $\mathbf{s}_2$ is defined by

$$\mathbf{s}_1{}^* = 2\pi(\mathbf{s}_2 \times \mathbf{z})/(\mathbf{s}_1 \cdot \mathbf{s}_2 \times \mathbf{z}); \quad \mathbf{s}_2{}^* = 2\pi(\mathbf{s}_1 \times \mathbf{z})/(\mathbf{s}_2 \cdot \mathbf{s}_1 \times \mathbf{z})$$

where **z** is the unit vector perpendicular to the surface. The relationship between vectors in the reciprocal lattice and planes of points in the direct lattice is as follows: for any family of lattice planes separated by a distance d, there are reciprocal lattice vectors perpendicular to the planes, the shortest of which has a length of $2\pi/d$. Conversely, for any reciprocal lattice vector g, there is a family of lattice planes normal to g and separated by a distance d, where $2\pi/d$ is the length of the shortest reciprocal lattice vector parallel to g.

The Miller indices can be used for labeling diffraction beams in LEED, as shown in Figure 11. Consider a rectangular surface unit mesh, such as for an fcc{110} plane. Then by choosing the real space lattice vectors to be $s_1 = ax$ and $s_2 = \sqrt{2}ay$, we calculate the reciprocal vectors $s_1^* = (2\pi/a)x$ and $s_2^* = (\sqrt{2}\pi/a)y$. These vectors and their linear combinations will be the allowed diffraction beams in the LEED experiments, and the beams are specified by (hk) where

$$g = h\, s_1^* + k\, s_2^*.$$

In the case of reconstructed surfaces or adsorbed structures, diffraction beams are indexed with respect to the reciprocal vectors of the substrate s_1^* and s_2^* as in

$$\begin{bmatrix} a_1^* \\ a_2^* \end{bmatrix} = \begin{bmatrix} P_{11} & P_{12} \\ P_{21} & P_{22} \end{bmatrix} \cdot \begin{bmatrix} s_1^* \\ s_2^* \end{bmatrix}$$

or

$$a^* = P\, s^*.$$

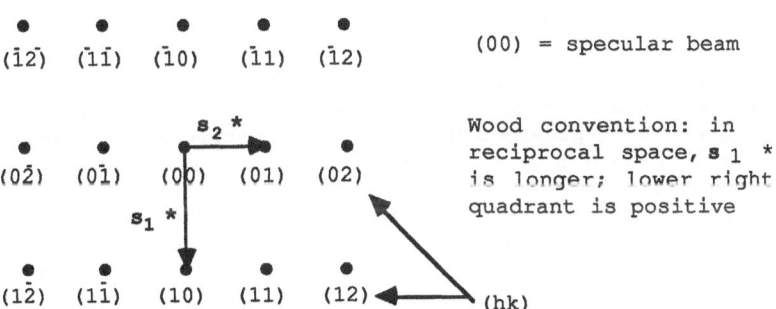

Figure 11. Reciprocal Lattice of a rectangular-symmetry surface[18].

For example, the LEED pattern from commensurate sodium on Ni{100} has beams in the positions (1/2,0), (0,1/2), etc., with respect to clean Ni{100}[20]. The P matrix for this overlayer relative to the substrate, is

$$\begin{bmatrix} 1/2 & 0 \\ 0 & 1/2 \end{bmatrix}.$$

When the angle between s_1^* and s_2^* equals the angle between a_1^* and a_2^*, which it does in this case, then a shorthand notation is used:

$$(s_1^*/a_1^* \quad x \quad s_2^*/a_2^*) \, R\delta$$

where only the magnitudes of the vectors are used and δ is the angle of rotation between s and a. In the preceding case of Na/Ni{100}, this gives an overlayer having a 2x2 periodicity.

2.6   Intensities of LEED beams

The intensities of LEED beams are of interest because they contain information about atomic positions within the unit cell of the surface structure. Intensities of LEED beams may be measured as functions of the electron energy E or the angle of incidence. Experimentally, it is usually easiest to measure the intensities as a function of E with the angle of incidence kept constant. In the resulting $I(k,E)$ curves there are certain symmetry properties which reflect the symmetry of the surface being studied. The fact that $I(k,E)$ must be identical for symmetry-equivalent beams provides a test for the validity of the experimental data.

<u>2.6.1. Kinematic Theory of LEED.</u> The Born approximation provides a simple model for problems in which scattering cross-sections are small(e.g. x-ray diffraction)[21]. An analysis utilizing the Born approximation is termed a kinematical analysis, and such an analysis for one reciprocal lattice beam in LEED shows only peaks in the I(E) curves corresponding to Bragg conditions expressed by $k'=k+g(hkl)$, or

$$k_{||}' = k_{||} + g(hk0) \qquad \text{and} \qquad k_\perp' = k_\perp + g(001)$$

where $g(001)$ represents a reciprocal vector perpendicular to the surface. A convenient graphical representation of the conditions for Bragg peaks is provided by the Ewald construction in reciprocal space, shown in Figure 12. In a purely 2D diffraction experiment, the reciprocal lattice would consist of uniform rods perpendicular to the surface and the intensity of the LEED beams would not be modulated with respect to the incident energy. The kinematic peaks in the I(E) curves arise from the modulation of the reciprocal lattice rods that occurs when the third dimensional periodicity is introduced, i.e. from the interference of the diffracted amplitudes from each layer of the substrate. Other peaks in experimental I(E) curves, excluding the ones associated with inelastic phenomena at very low energies, arise from multiple scattering. Since elastic cross-sections in LEED can be large, there is a reasonably high probability that diffracted electrons may be able to act as incident beams of other diffraction processes before escaping from the surface. This implies a coupling between diffracted beams. As the energy of the primary beam increases, both the elastic and the inelastic cross-sections decrease, and in general, a high-energy region of the I(E) spectrum appears kinematic to a reasonable approximation. The onset of this region is typically around 200eV, although it varies depending on the surface[22].

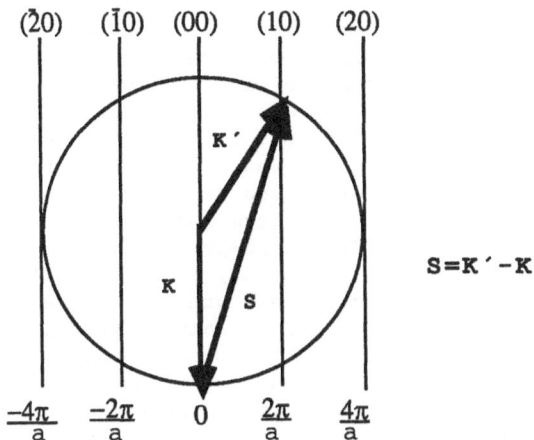

Figure 12. Ewald construction in 2D for a perfect 2D lattice and for incident electrons at normal incidence. The lines represent reciprocal lattice rods perpendicular to the surface.

2.6.2. Dynamical Theories of LEED.

In x-ray crystallography for bulk structure analysis, the interaction of the incident x-rays with the scatterers of the solid is weak so that a single scattering or kinematical theory may be used. In LEED, however, the scattering cross-section for incident electrons by the surface ion cores is large(as much as $1Å^2$, or 10^{10} times stronger than in x-ray scattering) so that multiple scattering events are important[22]. As a result, a rather complex calculation is required to determine the diffracted intensities to be expected from theory. It is necessary to follow an indirect procedure of guessing a structure, computing the anticipated intensities and comparing them with experiment. This procedure is not very systematic in that it relies on one of the guesses being the correct structure, and a good fit is not a guarantee that the solution is unique. This problem is minimized by making the data set for the experiment and theory as large as possible[22].

Quite a number of computational schemes have been developed for LEED, but they all contain the same basic components. They are (a) a realistic description of the individual ion core scattering properties, (b) several orders of multiple scattering, (c) inelastic scattering, and (d) non-zero temperature effects[21]. This chapter is mainly concerned with the results of LEED that can be understood using the kinematic theory, and will not elaborate on these aspects of the dynamical theories.

2.7 LEED Beam Profiles

When studying the order of adsorbed phases, there is additional
information in the intensity profile of the superlattice beams. It is
possible, under good experimental conditions, to extract parameters
which relate to the extent of ordered overlayer domains, the amount of
long-range order in the system, and the existence of residual
short-range order in a disordered phase. In the case of a simple
overlayer which undergoes an order-disorder transition, the LEED
intensity of a superlattice beam scattered from an adsorbed overlayer
consists of a Bragg diffraction term which describes the long-range
order in the system, and an additional term which describes the
short-range correlation near a critical temperature[23]. If an adsorbed
phase disorders via a continuous transition, then the critical
properties can be deduced from the temperature dependence of the peak
value and angular distribution of the superlattice beams. At $T<T_c$, the
superlattice intensity is dominated by the Bragg term, which decays as
$|T-T_c|^{2\beta}$, where β is the critical exponent associated with long-range
order. At $T>T_c$, there is short-range order scattering at the same
positions, for which the intensity varies as $|T-T_c|^{-\gamma}$, and the full width
at half maximum of the beam varies as $|T-T_c|^{\nu}$, where γ and ν are critical
exponents associated with susceptibility and short-range order,
respectively[24]. These are the quantities which provide the comparison
between statistical mechanical theories and experiments, as described,
for instance, in section 1.8.

In LEED, multiple scattering can make significant contributions to
diffracted intensities. The largest contribution to this arises from a
double scattering, once from the substrate and once from the overlayer.
Since this includes only one diffraction from the overlayer, the
temperature dependence should be the same as single scattering from the
overlayer as long as the substrate has not also altered. Events which
involve more than one scattering from the overlayer are very much
weaker. Therefore, multiple scattering is unlikely to greatly affect
the measurement of the temperature dependence of superlattice
intensities[23].

The determination of domain size from the beam profile is more
complicated and, strictly speaking, requires a model for the domain
structure in order to get quantitative results. In some systems, the
width of the superlattice beam is simply inversely proportional to the
average domain dimension on the surface[25]. In other systems a more
complicated relationship is necessary to extract domain-size
information[25,26]. In LEED experiments on domain-growth kinetics, an
ideal experiment would measure diffraction beam profiles as a function
of time. This is usually difficult to achieve, however, since it
requires the capture of large amounts of data over very short times.
Background intensity measurements should also be made since the
disordered phase contributes to the background intensity. Since
measuring all of these parameters can be cumbersome, some simplifying
assumptions can be made in some cases. One such assumption is that in
the case of islands which have only short-range positional correlation,
they can be treated as independently diffracting domains(i.e.

incoherent). It has been rigorously proven that the interference between islands has only a minor influence on the beam profiles[27]. In fact, it has been shown that even in the case of islands or domains with long-range correlation, the interference effects between islands are not measurable in the beam profiles with the usual detectors[27]. In this case, the peak intensity of a superlattice beam is proportional to NL^4, where N is the number of islands and L is the average diameter of the islands. For the case of constant coverage, where the number of islands is inversely proportional to their average area ($N \propto L^{-2}$), the peak intensity varies as L^2 and is directly proportional to the long-range order parameter.

3. EXAMPLES OF LEED STUDIES

3.1 Nitrogen Physisorbed on Graphite

When N_2 adsorbs on graphite below about 48K, it orders in a commensurate ($\sqrt{3} \times \sqrt{3}$)R30° structure, very similar to that shown in Figure 5 for commensurate Kr/graphite[28]. This commensurate structure is formed because the spacing of the the ($\sqrt{3} \times \sqrt{3}$)R30° structure is very close to the "size" of the N_2 molecules, and so the molecules are only about 6% expanded from a close-packed plane in solid 3D nitrogen. Above about 27K, the molecules are orientationally disordered, i.e. their molecular axes are pointing in random directions at any particular time, but below about 27K, the molecular axes order in a "herringbone" arrangement, shown in Figure 13[29]. The LEED pattern due to the orientationally ordered structure (T<27K) is shown in Figure 14b. The filled spots in the pattern arise from the ($\sqrt{3} \times \sqrt{3}$)R30° periodicity of the N_2 on graphite, and the open spots are superlattice spots which arise from the orientationally-ordered structure which has a larger period on the surface. There are "systematic absences" in the LEED pattern which can be explained by an extra symmetry in the herringbone structure called glide-line symmetry. These two perpendicular glide lines, shown in Figure 13, cause certain spots, shown as x's, to be absent in the LEED pattern along the same directions as the glide lines. Figure 14a shows the LEED pattern expected from the structure shown in Figure 13. But because the herringbone structure has three equivalent rotational directions on the graphite lattice, the observed pattern, Figure 14b, is actually a superposition of three of these patterns rotated with respect to each other, as a result of all three types of rotations being present on the surface.

Once the structure of this orientationally ordered phase was determined, the next question to ask was how does it disorder. As described in section 2.7, a measure of the amount of order in a superlattice structure is given by the intensity of the superlattice beam. In these experiments, as the temperature was increased through the order-disorder transition, the integrated intensity of the superlattice beams decreased, as shown in Figure 15, but the width remained constant to within the limits of the apparatus[30].

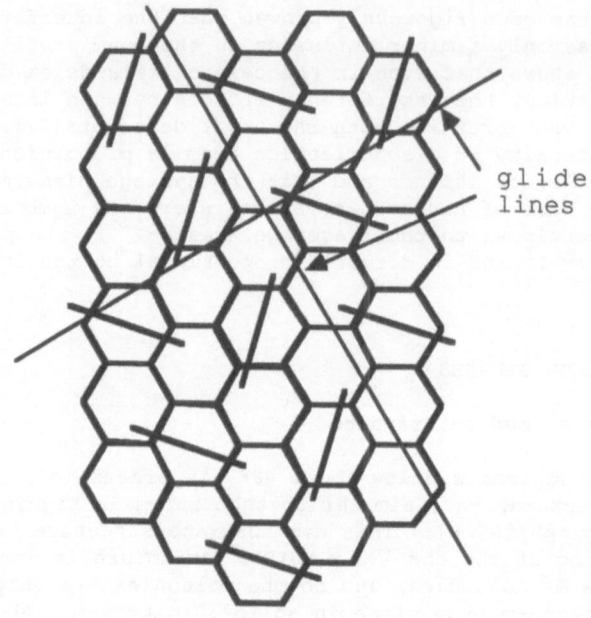

glide
lines

Figure 13. Herringbone Structure of N$_2$ on Graphite. The lines
represent the long axes of the N$_2$ molecules.

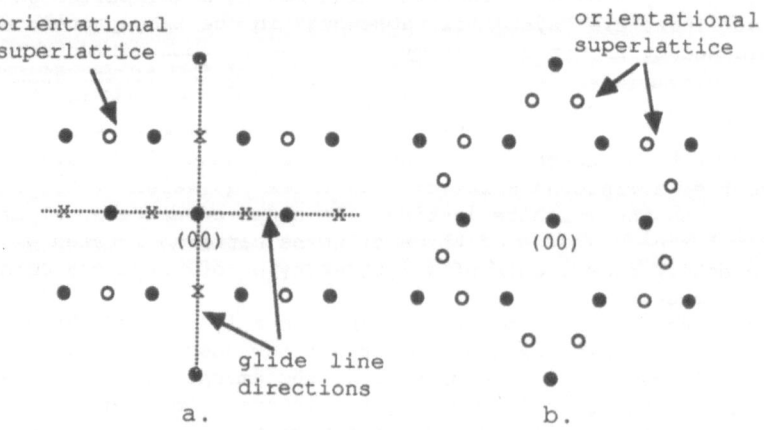

Figure 14. (a) Expected LEED pattern from one rotational domain of
orientationally ordered N$_2$ on graphite, and (b) the observed LEED
pattern, from three rotational domains.

The transition from the herringbone structure to the orientationally disordered structure is predicted to be in the universality class of the 2D Heisenberg model with face-oriented cubic anisotropy, and renormalization group results have determined this transition to be first-order[31], which would show up as a discontinuity in the graph of superlattice intensity vs. temperature. However, molecular dynamics simulations done on finite lattices show that there is a certain amount of rounding of the curves due to size effects[32], and so the experimental data is in fact consistent with a first-order transition.

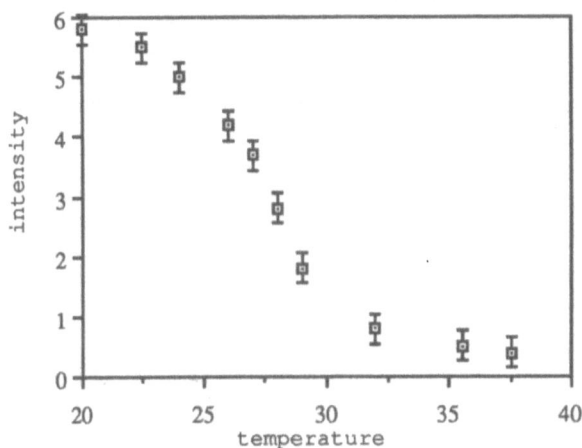

Figure 15.   Integrated superlattice spot intensity as a function of temperature.

The dynamics of order-disorder transitions such as this one have been of considerable theoretical interest for the last few years[33,34]. There have been few experimental studies so far with which to test these theories, and the dynamics of the domain growth of the orientationally ordered phase of N_2 on graphite has not been experimentally determined. Several computer simulations[33,34] disagree as to whether this system obeys the classical Allen-Cahn kinetics for domain growth[35], which predicts that the rate of domain growth varies as $t^{1/2}$. A Monte Carlo study performed at many quenching temperatures indicates that the domain-growth exponent does not remain at the same value as the quench temperature is increased, but passes through the Allen-Cahn value of 1/2 [33]. One parameter which affects these calculations is the "softness" of the walls of the orientationally-ordered domains, but this parameter is difficult to determine experimentally. On the other hand, spatial dimensionality seems to be irrelevant to these kinetics. An experimental study could resolve some of the uncertainties of these calculations. From the LEED structural studies it is quite clear that the kinetics of this transition are very fast and an experimental study might be difficult in terms of precise temperature control. However, in light of the LEED study to be discussed in the next section, the LEED technique itself is probably not a limitation.

3.2 Oxygen Chemisorbed on the W{112} Surface

These experiments were carried out by Gwo-Ching Wang and coworkers at Oak Ridge National Laboratory and Rensselaer Polytechnic Institute and they constitute probably the most detailed measurements of surface kinetics using LEED. The crystal structure of tungsten is body-centered cubic, and the W{112} surface has a rectangular symmetry. When an ordered submonolayer of oxygen is adsorbed on the clean surface, a p(2x1) LEED pattern is observed, which is shown in Figure 16[36]. From this LEED pattern and energy calculations, the atomic arrangement for the ordered oxygen overlayer was deduced and is shown in Figure 17[36].

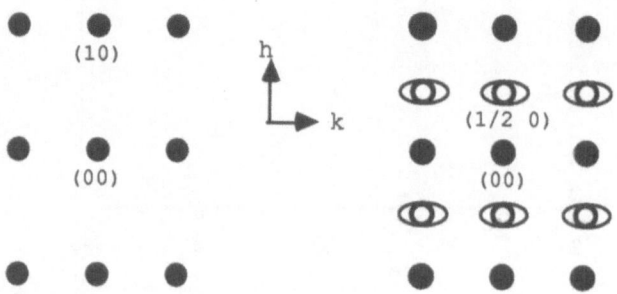

Figure 16. LEED patterns from clean W{112} and from the oxygen p(2x1) structure[36].

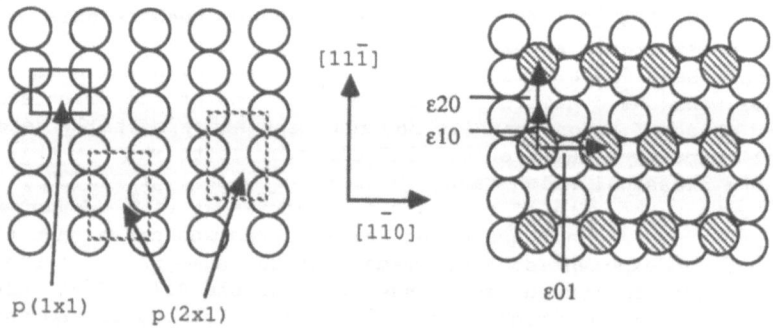

Figure 17. Atomic arrangement of clean W{112} and of the p(2x1) oxygen overlayer[36]. The dashed rectangles show two antiphase domains of the p(2x1) structure. The parameters ε are interaction energies for a pair of oxygen atoms at the specified sites and are given in the text.

At low oxygen coverages(<0.2 monolayers), the half-order beams are elongated in the k direction and are about 2-3 times wider in the k direction than in the h direction. This implies that there are islands in the oxygen structure which are longer in the h direction([11$\bar{1}$]) than in the k direction([1$\bar{1}$0]). This is probably because the adsorbed oxygen atoms can more easily migrate along the [11$\bar{1}$] direction. As the coverage increases, the spot size gradually decreases in both directions and reaches a minimum when the coverage is 0.5 monolayers. (The coverage in this case is the ratio of oxygen atoms present on the surface to the number of W atoms in the surface layer.) This corresponds to the maximum island size. At 0.5 monolayers, which is in fact full coverage for the p(2x1) structure, the surface has been determined from the LEED pattern spot profiles to be covered with antiphase domains having an average size of 110Å x 110Å[36].

The phase boundary of the p(2x1) structure of O/W{112}, shown in Figure 18, was determined by measuring the intensity of the superlattice beams as a function of temperature, and taking the inflection points of the isotherms as the transition temperature. The disordered phase is a 2D gas phase in which the O atoms are at any instant randomly distributed among the surface sites. The phase boundary is a second-order phase boundary associated with the nearest-neighbor repulsive interactions among the oxygen atoms.

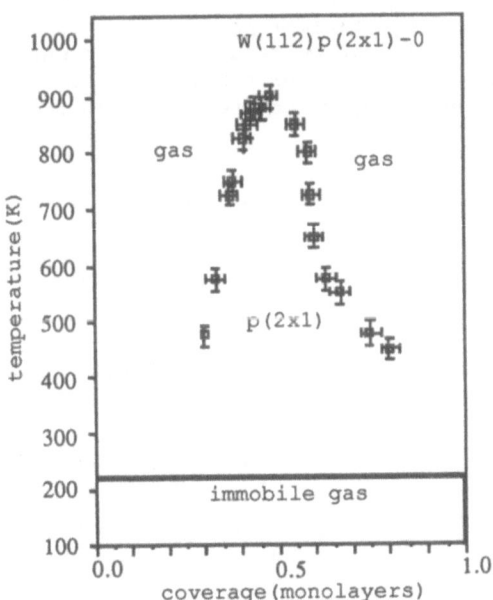

Figure 18. Phase Diagram for O/W{112}[36].

The fact that the p(2x1) structure has already been formed at coverages as low as 0.1 monolayers, however, implies that the next-nearest

462

neighbor interactions must be attractive along both the [11$\bar{1}$] and the [1$\bar{1}$0] directions. Interaction energies estimated from this LEED study (see Figure 17) are ε_{10}=0.18eV/atom, ε_{01}=-0.09eV/atom and ε_{20}=-0.16eV/atom[36]. At temperatures below 225K, the overlayer was completely immobile at all coverages, and if the oxygen was adsorbed at these temperatures, it would not order. This result allowed the following dynamics measurements to be made by using an "up-quenching" technique[37]. This technique is analogous to the normal quenching technique where a high-temperature disordered phase is quickly cooled to below the order-disorder transition temperature, except that in this case a low-temperature immobile disordered structure is quickly heated to a region where the surface mobility is higher but the equilibrium structure is ordered.(The equilibrium structure of the immobile phase is probably also ordered, but it cannot attain order due to its immobility.) In the study of the dynamics of the evolution of the anti-phase domains, 0.5 monolayers of oxygen was adsorbed on the W{112} surface at T=170K. No superlattice beams were visible because the structure was disordered. The crystal was then "up-quenched" to a temperature above 260K, after which superlattice beams would appear. The peak intensity of a superlattice beam was measured with a Faraday cup. An example of the (1/2,0) superlattice intensity as a function of time is shown in Figure 19[37].

According to the Allen-Cahn domain growth law, the linear size of the domains should grow after quenching as $t^{1/2}$. The domain area, which is proportional to the long-range order parameter, should grow linearly with time. From Figure 19, it can be seen that the experimental data do in fact increase linearly with time for the initial growth of the domains. This is the case for up-quench temperatures ranging from 260K to 343K.

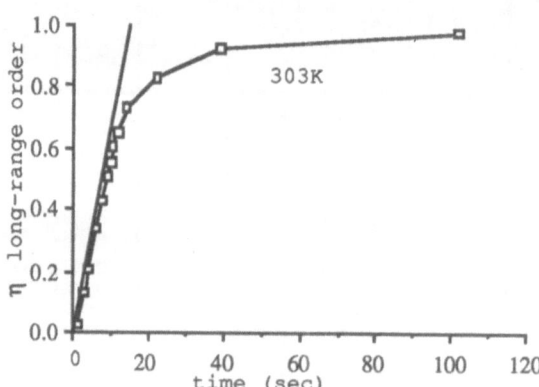

Fig 19. Long-range order parameter vs. time. Not all data points have been shown(see reference 37). The straight line represents the linear time dependence predicted by the Allen-Cahn domain-growth law.

The curves cease to be linear when the domains get close to their ultimate size. In addition, the growth rate of the order parameter is dependent on the temperature at which the system is up-quenched. The behavior is seen to describe an Arrhenius-type growth rate given by

$$d\eta/dt \propto \exp(-a/kT)$$

where η is the long-range order parameter and a is the activation energy for atomic migration. These data give a=0.14 $\pm$ 0.02 eV[37].

4. CONCLUSIONS

It is clear from the LEED studies of Wang et al.[37] that LEED is potentially a very useful technique for studying the dynamics of surface ordering. Up till now, studies of the ordering kinetics in 2D systems have been largely theoretical, the main issue being to determine the possible universal aspects of growth processes[33]. Due to the paucity of experimental studies which is largely a result of the difficulty of the experiments themselves, the interpretation of theoretical studies has been hindered[33]. It has been demonstrated now, though, that many of the experimental obstacles can be overcome. Indeed, an experimental program to study the details of domain-growth kinetics using LEED has been initiated by G. -C. Wang at Rensselaer Polytechnic Institute[38]. Other similar studies are being carried out at the University of Wisconsin-Madison[39,40]. In addition, experiments have been performed in Sam Fain's laboratory at University of Washington to test the feasibility of measuring the kinetics of the compositional ordering of the physisorbed mixture of CO and Ar on graphite[41].

5. ACKNOWLEDGMENTS

I would like to thank Sam Fain and Eberhard Riedel for their informative and stimulating graduate courses at University of Washington on surface physics and critical phenomena on surfaces, from which much of the material for this chapter has been adapted. I am grateful to Gwo-Ching Wang for sending me her papers on LEED dynamics studies and to Sam Fain, T.-M. Lu and Max Lagally for useful discussions about LEED intensities. I would also like to thank Kevin Purcell, Flemming Hansen, and Hak Taub for helpful comments concerning this manuscript and John Stevens for his assistance with the drawings.

6. REFERENCES

1. T. L. Einstein, CRC Crit. Rev. Solid State Sci., 1978, 7, 261.
2. J. G. Dash, Films on Solid Surfaces, (Academic Press:New York)1975.
3. O. E. Vilches, Ann. Rev. Phys. Chem., 1980, 31, 463; S. C. Fain, in Chemistry and Physics of Solid Surfaces IV, ed. R. Vanselow and R. Howe, (Springer-Verlag:Berlin)1982.

464

4. W. Steele, *Surf. Sci.*, <u>1973</u>, **36**, 317; S. C. Fain, M. F. Toney, and R. D. Diehl, in *Proc. of the Ninth International Vacuum Congress and Fifth International Conference on Solid Surfaces*, ed. J.L. de Segovia, (Imprenta Moderna:Madrid)1983.
5. M. D. Chinn and S. C. Fain, *J. Vac. Sci. Technol.*, <u>1977</u>, **14**, 314.
6. M. D. Chinn and S. C. Fain, *Phys. Rev. Lett.*, <u>1977</u>, **39**, 146.
7. S. C. Fain, M. D. Chinn, and R. D. Diehl, *Phys. Rev. B*, <u>1980</u>, **21**, 4170.
8. L. D. Roelofs and P. J. Estrup, *Surf. Sci.*, <u>1983</u>, **125**, 51.
9. C. J. Barnes, M. Q. Ding, M. Lindroos, R. D. Diehl, and D. A. King, *Surf. Sci.*, <u>1985</u>, **162**, 59.
10. P. Hofmann, S. R. Bare, and D. A. King, *Surf. Sci.*, <u>1982</u>, **117**, 245.
11. H. Conrad, G. Ertl, J. Koch, and E. E. Latta, *Surf. Sci.*, <u>1974</u>, **43**, 462.
12. L. D. Schmidt and R. Gomer, *J. Chem. Phys.*, <u>1966</u>, **45**,1605.
13. R. L. Park, A. R. Kortan, T. L. Einstein, and L. D. Roelofs, in *Ordering in Two Dimensions*, ed. S. K. Sinha (North-Holland, New York)1980, p.17.
14. L. D. Roelofs, A. R. Kortan, T. L. Einstein, and R. L Park, *Phys. Rev. Lett.*, <u>1981</u>, **46**, 1465.
15. M. Schick, *Phys. Rev. Lett.*, <u>1981</u>, **47**,1347.
16. C. Davisson and L. H. Germer, *Phys. Rev.*, <u>1927</u>, **30**,705.
17. M. G. Lagally and J. A. Martin, *Rev. Sci. Instrum.*, <u>1983</u>, **54**, 1273.
18. E. A. Wood, *J. Appl. Phys.*, <u>1964</u>, **35**, 1306.
19. C. Kittel, *Introduction to Solid State Physics*, 2nd edition (Wiley:New York)1956.
20. R. L. Gerlach and T. N. Rhodin, *Surf. Sci.*, <u>1969</u>, **17**, 32.
21. *Low-energy Electron Diffraction*, M. A. Van Hove, W. H. Weinberg, and C. -M. Chan, Springer Series in Surface Science **6**, (Springer-Verlag:Berlin) 1986.
22. D. P. Woodruff in *The Chemical Physics of Solid Surfaces and Heterogeneous Catalysis* 1, ed. D. A. King and D. P. Woodruff, (Elsevier:Amsterdam) 1981.
23. T. -M. Lu in *Ordering in Two Dimensions*, ed. S. K. Sinha (North-Holland, New York)1980, p.195.
24. G. -C. Wang, *Phys. Rev. B*, <u>1985</u>, **31**, 5918.
25. T. -M. Lu and M. G. Lagally, *Surf. Sci.*, <u>1980</u>, **99**, 695.
26. J. M. Pimbley, T. -M. Lu and G. -C. Wang, *J Vac. Sci. Technol.*, <u>1986</u>, **A4**, 1357.
27. T. -M. Lu, L. -H. Zhao, M. G. Lagally, G. -C. Wang and J. E. Houston, *Surf. Sci.*, <u>1982</u>, **122**, 519.
28. J. K. Kjems, L. Passell, H. Taub, and J. G. Dash, *Phys. Rev. Lett.*, <u>1974</u>, **32**, 724; R. D. Diehl, C. G. Shaw, S. C. Fain, and M. F. Toney, in *Ordering in Two Dimensions*, ed. S. K. Sinha (North-Holland, New York)1980, p.199.
29. R. D. Diehl, M. F. Toney, and S. C. Fain, *Phys. Rev. Lett.*, <u>1982</u>, **48**, 177.
30. R. D. Diehl and S. C. Fain, *Surf. Sci.*, <u>1983</u>, **125**, 116.
31. B. Nienhuis, E. K. Riedel and M. Schick, *Phys Rev. B*, <u>1983</u>, **24**, 5625.
32. J.Talbot, D.Tildesley, and W.A.Steele, *Molec.Phys.*, <u>1984</u>, **51**, 1331.

33. O. G. Mouritsen, *Phys. Rev B*, <u>1985</u>, **32**, 1632; <u>1983</u>, **28**, 3150.
34. K. Kaski, S. Kumar, J. D. Gunton, and P. A. Rikvold, *Phys. Rev B*, <u>1984</u>, **29**, 4420.
35. S. M. Allen and J. W. Cahn, *Acta Metallurgica*, <u>1979</u>, **27**, 1085.
36. G. -C. Wang and T. -M. Lu, *Phys Rev. B*, <u>1983</u>, **28**, 6795.
37. G. -C. Wang and T. -M. Lu, *Phys. Rev. Lett.*, <u>1983</u>, **50**, 2014.
38. G. -C. Wang, private communication.
39. M. Tringides, P. K. Wu, W. Moritz and M. G. Lagally, *Ber Bunsenges. Phys. Chem.*, <u>1986</u>, **90**, 277.
40. P. K. Wu, J. H. Perepezko, J. T. McKinney, and M. G. Lagally, *Phys. Rev. Lett.*, <u>1983</u>, **51**, 1577.
41. H. You and S. C. Fain, *Phys. Rev. B*, <u>1986</u>, **34**, 2840.

CHAPTER 27

NEUTRON SCATTERING STUDIES OF THE STRUCTURE, PHASE TRANSITIONS, AND
DYNAMICS OF COMMENSURATE HERRINGBONE MONOLAYERS PHYSISORBED ON
GRAPHITE

H. Taub
Department of Physics and Astronomy
University of Missouri-Columbia
Columbia, Missouri   65211
U.S.A.

ABSTRACT. Monolayer structures in which molecules lie in a
commensurate, two-sublattice herringbone (HB) arrangement on the
graphite basal plane surface are common to a number of nonspherical
molecules. Despite a variety of molecular sizes and shapes, these HB
monolayer phases exhibit similarities in their structure and phase
transitions which are striking--even when compared to rare gas
monolayers on graphite. This paper will review neutron scattering
studies of the structure and dynamics of the HB monolayer phases.
After illustrating the solution of the monolayer structures (including
the molecular orientations) by elastic neutron diffraction, we shall
examine several common features of these phases: 1) instability to
uniaxial incommensurate structures; 2) inhibition of multilayer
growth; 3) a two-stage melting process whose first stage depends on
the length of the molecule--an orientationally-disordered phase for
"short" molecules and possibly a liquid-crystal-like monolayer phase
for "long" molecules; and 4) a coupling between out-of-plane
translational and librational modes of the monolayer. We conclude
with some speculations on why the HB monolayer structure is so
pervasive on graphite.

1. INTRODUCTION

Monolayers bound to solid surfaces by weak van der Waals forces have
been the object of intense experimental and theoretical study for over
15 years [1]. The interest in these "physisorbed" monolayers stems
from their condensed phases which are, in many respects, two-
dimensional (2D) analogues of bulk molecular solids, fluids, and
gases. It is generally believed that phenomena such as crystal
growth, structural transformations, and melting will be easier to
understand in these quasi 2D systems than in more familiar bulk
matter. A theory of melting, for example, may be easier to formulate
conceptually in 2D [2]. In other cases, a model may simply be easier
to implement mathematically in less than three dimensions.
     Among substrates, graphite is especially desirable because it

G. J. Long and F. Grandjean (eds.), The Time Domain in Surface and Structural Dynamics, 467–497.

provides homogeneous surfaces having an exceptionally long coherence length (up to 10,000 Å) for diffraction studies [3]. It is also available in various high-quality powders which have sufficient surface area to permit complementary thermodynamic measurements. Another prototypical substrate for physisorption is the silver (111) surface [4]. It has the advantage of being smoother than the graphite basal plane; i.e., the adatom-substrate potential is less corrugated. However, the coherence lengths are generally smaller ($\lesssim 500$ Å) than for graphite.

It is noteworthy that, to date, only three different submonolayer structures are known to form at low temperature on the graphite basal plane surface. (Other structures are found at and above monolayer completion where compression of the first adsorbed layer can occur). The first group of films includes the rare gases [3] and nearly spherical molecules such as methane [5] whch close pack in simple hexagonal lattices. In the second group are nonspherical molecules such as oxygen [6] and butane [7] which are believed to form centered rectangular submonolayer phases. The third group is composed of nonspherical molecules which adsorb in a commensurate, two-sublattice structure on the graphite (0001) surface. The monolayer unit cell contains two molecules which lie on their side in a herringbone (HB) arrangement. Perhaps surprisingly, this is the largest of the three groups of structures--at least at the present time. Five molecules are known to form HB structures on graphite: N_2 [8,9], C_2H_6 (ethane) [10,11], $Fe(CO)_5$ [12], C_6H_{14} (hexane) [13,14], and N_2O_2 [15].

The quest to realize ever simpler condensed phases has led naturally to an emphasis on rare gas monolayers [3,4] for which the intermolecular interactions are most easily described. Nevertheless, due to subtle effects of corrugation in the adatom-substrate potential, there are essential differences in the phase diagrams of Kr and Xe on graphite [3]. In particular, the weakly incommensurate phase of Kr is believed to be a fluid at low temperature. Thus, even for the rare gases, it is impossible to construct a generic phase diagram for monolayers on a graphite substrate.

Considering the variety in their molecular size and shape, the HB monolayer phases on graphite are more similar in their structure and phase transitions than might be expected from the behavior of the rare gases. Our purpose here is to discuss these monolayers as a group thereby emphasizing these similarities. We shall begin in the next section by illustrating the solution of these HB monolayer structures (including the molecular orientations) by neutron diffraction techniques. Some discussion of the growth modes of the films will be included showing that the formation of multilayers of the HB structure is difficult. We will then go on in Sec. 3 to describe common features in the phase transitions of these systems: (i) the presence of uniaxial commensurate-incommensurate transitions; and (ii) a two-stage melting process. In Sec. 4, we discuss the lattice dynamics of an HB monolayer phase using ethane as an example. Finally, in Sec. 5 we speculate on why the HB monolayer phase is so pervasive among physisorbed monolayers on graphite as well as pose some questions requiring further study.

2. DETERMINATION OF HB MONOLAYER STRUCTURES BY ELASTIC NEUTRON DIFFRACTION

In this section, we illustrate the use of neutron diffraction techniques to solve HB monolayer structures by concentrating on two examples: nitrogen and ethane adsorbed on graphite. The monolayer structure of N_2 is simplest to describe and doing so will facilitate discussion of the $Fe(CO)_5$ monolayer. Likewise, considering ethane first will simplify discussion of the hexane monolayer structure. We shall see that neutron diffraction and low-energy electron diffraction (LEED) are complementary techniques which when combined can yield a nearly complete structure solution. We will also show how the experimentally determined monolayer structures can be tested by comparison with those calculated from models of the intermolecular and molecule-substrate interactions.

2.1 Nitrogen and $Fe(CO)_5$ on graphite

The simplest example of a commensurate HB phase on graphite is that of nitrogen [8,9] shown in Fig. 1. As discussed by Diehl in a previous chapter, [16] all the molecules lie in a plane with the N-N bond parallel to the surface. The a and b lattice constants are, respectively, $\sqrt{3}$ and 3 times that of graphite corresponding to a superlattice in which every third carbon hexagon on the basal plane surface is occupied by a nitrogen molecule. We shall denote this rectangular commensurate unit cell simply $\sqrt{3}x3$.

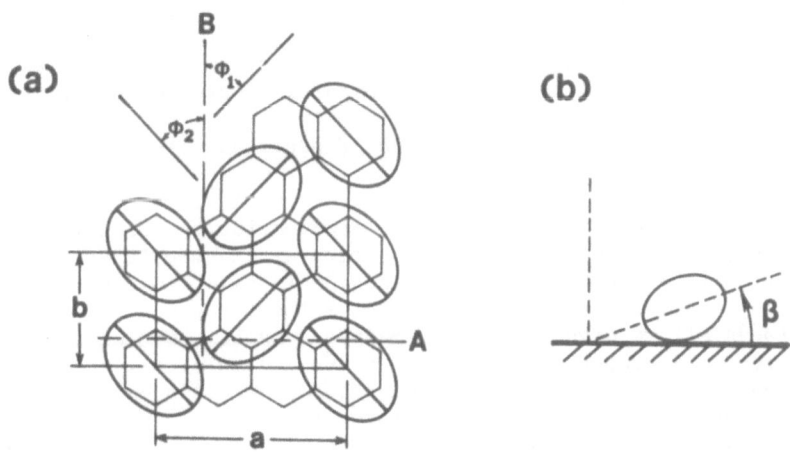

Figure 1. Definition of the structural parameters for the commensurate $(\sqrt{3}x3)$ HB phase of N_2 on graphite. (a) Projection of the unit cell (a=7.38 Å, b=4.26 Å) on the graphite (0001) surface; and (b) side view showing tilt of the molecular axis.

Theoretical analysis indicates that the orientational ordering of the N_2 molecules in an HB structure results from their appreciable quadrupole-quadrupole interaction. The HB structure is predicted in mean-field calculations [17] which consider a system of point quadrupoles on a hexagonal lattice subject to a crystal field of the substrate. It has also been found in computer simulations of N_2 monolayers which are based on detailed modeling of the intermolecular and molecule-substrate potentials [18,19]. These calculations indicate that each molecule is situated at the center of a graphite carbon hexagon.

Experimental investigations of the N_2 monolayer structure have been reviewed briefly in Ref. 9. The first neutron diffraction experiments [20] provided evidence of a commensurate phase having the same density as that of the structure in Fig. 1(a). However, since only a single Bragg peak was observed, no determination of the molecular orientation could be made. Subsequent neutron experiments [21] observed a second Bragg peak below 30 K consistent with a doubling of the unit cell in at least one direction. This strongly suggested the presence of an orientationally-ordered phase.

The first evidence of an HB structure of N_2 on graphite came from the LEED experiments of Diehl, et al. [8]. They were able to index the diffraction spots in the LEED pattern using the unit cell in Fig. 1(a) and to explain systematic absences of the (h0) and (0k) reflections with h and k odd by the existence of glide lines labeled A and B. The glide lines are consistent with an HB rather than a pinwheel monolayer structure [17] and imply that the azimuthal parameters $\phi_1 = \phi_2 = \phi$. However, the glide line symmetry alone does not yield a value of ϕ nor the tilt angle β defined in Fig. 1(b). Values of these parameters have been obtained in recent neutron diffraction experiments [9] which were performed out to larger Q values than previously [20,21]. Before discussing these experiments, we briefly review some salient features of neutron diffraction from adsorbed monolayers.

Due to the weak interaction of neutrons with matter, neutron diffraction experiments on monolayers require a high-surface-area substrate rather than a single crystal as in LEED. On the other hand, multiple scattering is less of a problem for the weakly interacting neutron probe so that the intensities of the monolayer Bragg reflections can be analyzed by conventional kinematic theory [7]. Also, scattering from point nuclei rather than the electron charge clouds allows higher-order Bragg peaks to be observed whose intensities depend sensitively on the molecular orientation. We now show how these advantages of neutron diffraction can be exploited in solving the HB monolayer structure of N_2.

Neutron diffraction patterns from a commensurate monolayer of N_2 on an exfoliated graphite substrate (Papyex) are shown in Figs. 2 and 3 [9]. Scattering from the graphite substrate has been subtracted. Data is missing from Q regions of intense graphite peaks plotted in Fig. 2(b). The reader is referred to Ref. 9 for details of the measurement technique and the method of calculating diffraction profiles for model structures.

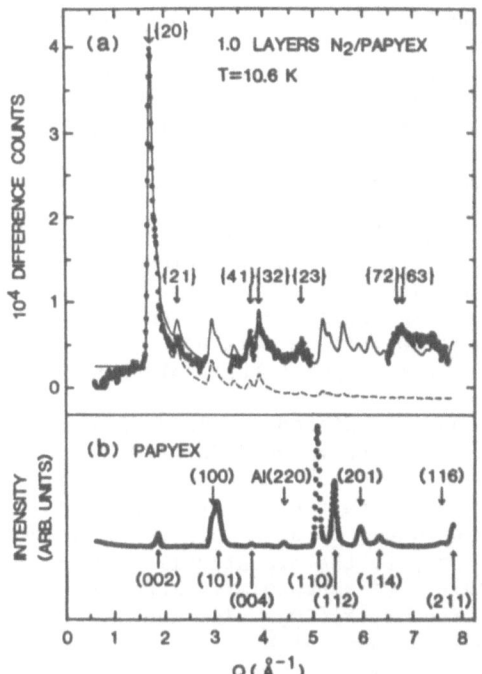

Figure 2. (a) Neutron diffraction pattern of the commensurate HB phase of N_2 on graphite after subtraction of the Papyex background [9]. The solid curve has been calculated for the cell parameters $\phi_1 = \phi_2 = 40°$ and $\beta = 0$ defined in Fig. 1. A mean-square displacement $\langle u^2 \rangle = 0.01$ $Å^2$ in the Debye-Waller factor and a coherence length L=105 Å were assumed. Only the most intense peaks {hk} have been labeled. For the dashed curve, all of the parameters remain the same except $\langle u^2 \rangle = 0.1$ $Å^2$ as reported in Ref. 22. (b) Diffraction pattern of the bare Papyex. In addition to the graphite Bragg peaks, the Al(220) peak from the sample cell can be seen.

Figure 3. Detail of the diffraction pattern of the commensurate HB phase of N_2 on graphite (Fig. 2) in the Q range most sensitive to the orientational ordering. Profiles have been calculated for several values of the in-plane azimuthal orientation ϕ assuming $\beta=0$ and $\langle u^2 \rangle = 0.01$ $Å^2$: $\phi = 55°$ (solid curve), $\phi = 45°$ (dotted curve), and $\phi = 35°$ (dashed curve). The difference in the {51} and {13} peak positions is unresolved. Again, only the most intense {hk} peaks have been labeled.

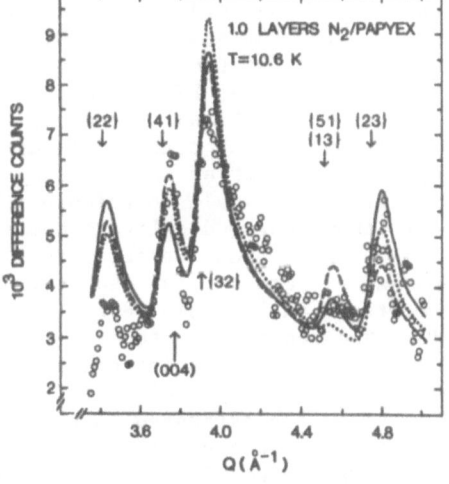

The monolayer peak positions in the neutron pattern can be indexed by the same $\sqrt{3}\times3$ unit cell found in the LEED experiments [8]. Most striking are the higher-order Bragg peaks shown on an expanded scale in Fig. 3. Due to attenuation by the molecular form factor, these peaks could not be observed in x ray diffraction experiments [22]. The diffraction pattern in Fig. 3 is compared with three profiles calculated for ϕ values of 35°, 45°, and 55°, respectively, and an in-plane molecular orientation ($\beta=0$). The calculated profiles assume a Warren lineshape for each reflection [7,9] and have been scaled to give the observed intensity for the {2,0} peak--the most intense peak in the pattern.

It is clear from Fig. 3 that the calculated profiles are quite sensitive to the azimuthal orientation ϕ. From a qualitative comparison of the profiles with the observed pattern [9], it is possible to define a range of ϕ values from 40° to 50° which best fits the data. For a fixed value of ϕ in this range, the tilt angle β could be increased to ~15° without a discernible degradation of the fit. Analysis of the neutron diffraction pattern does not give the position of the unit cell relative to the graphite basal plane nor the height of the molecules above the surface.

Molecular dynamics simulations have predicted ϕ values of 44.5° [18] and 42.8° [22] which lie well within the range inferred experimentally. Values of ϕ from 44.1° to 44.9° found in ground state calculations [19] also agree well with the neutron experiments.

Neutron diffraction has also been used to solve the structure of $Fe(CO)_5$ adsorbed on graphite [12]. Despite the dramatically different shape and size of the $Fe(CO)_5$ molecule, its submonolayer structure is similar to that of N_2. As shown in Fig. 4, $Fe(CO)_5$ also has a rectangular commensurate HB unit cell but one which is larger ($\sqrt{7} \times \sqrt{21}$) than that of N_2 in proportion to the size of the molecule. We also see in Fig. 4(b) that each molecule lies with one face of its bipyramid structure parallel to the surface. Note that registry with the graphite (0001) surface requires the lattice vectors to be rotated 19° from their direction in the N_2 unit cell.

The calculated $Fe(CO)_5$ diffraction profiles are sensitive to the azimuthal orientational parameter ψ defined in Fig. 4(a). This is because the neutron scattering amplitudes of the C and O nuclei are comparable to that of Fe, and they are located at the molecule's extremities. A small rotation of the molecule about an axis through the Fe atom (perpendicular to the surface) will then result in C and O displacements which are an appreciable fraction of a lattice constant. There is less sensitivity to the α rotation (Fig. 4(b)) since the C and O nuclei are closer to the molecule's principal axis. The sensitivity to a small tilt angle β is also less than for ψ rotations because the integrated peak intensities are essentially determined by the projection of the molecule onto the surface. An error of ±5° is estimated in the angles α and β which determine the orientation of the pyramid face relative to the surface.

There are small differences between the HB monolayer structures of N_2 and $Fe(CO)_5$. The experimental uncertainty in the azimuthal angles is small enough to exclude the case $\psi_1 = \psi_2$ so that the glide line symmetry present in the N_2 monolayer is broken. Also, the fit to the diffraction pattern can be improved if one of the molecules is displaced slightly from the cell center.

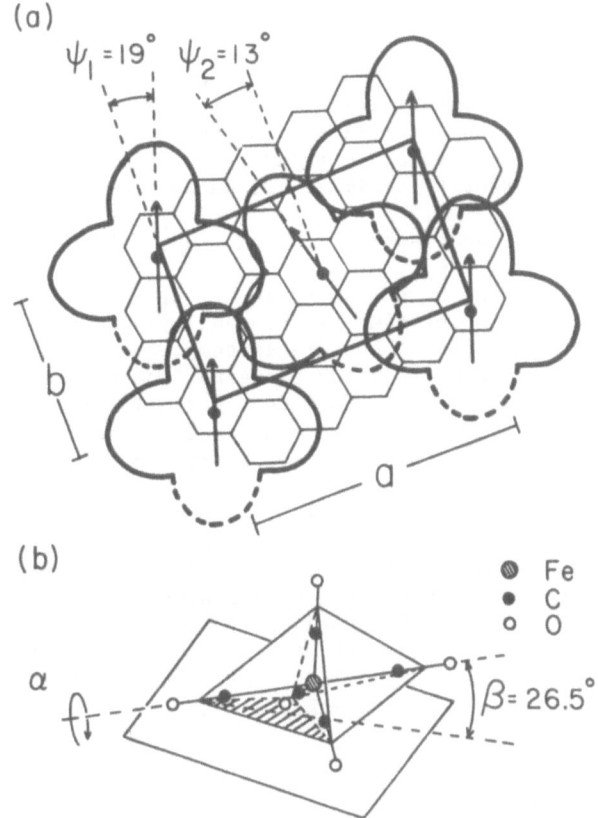

(a)

$\psi_1 = 19°$ $\psi_2 = 13°$

b

a

(b)

⊘ Fe
● C
○ O

α

$\beta = 26.5°$

Figure 4. Structure of the submonolayer commensurate HB phase of
Fe(CO)$_5$ on graphite [12]. (a) Projection on the graphite basal
plane. The ($\sqrt{7}$ x $\sqrt{21}$) R19° unit cell (a=6.51 Å, b=11.27 Å) is drawn
with its corners at the center of a graphite carbon hexagon.
Molecules are outlined by their approximate van der Waals radii
(dashed line at the end of molecule indicates CO group furthest from
the surface). Arrows are projection of molecule's principal axis
through the Fe atom (●). (b) Orientation of molecule relative to
surface (face-parallel configuration).

As in the case of N$_2$, analysis of the neutron diffraction pattern
does not yield the adsorption sites of the two Fe(CO)$_5$ molecules in
the unit cell. In Fig. 4(a), the corner molecule has been placed at
the site calculated from empirical atom-atom potentials to be most
favorable for a single molecule on a basal plane surface [12].
Potential energy calculations with an 18-molecule monolayer cluster
[12] confirm all of the basic structural features inferred
experimentally: (i) the face-parallel configuration of Fig. 4(b);
(ii) the absence of glide line symmetry; and (iii) the off cell-center
location of one of the molecules. A qualitative explanation for the

features (ii) and (iii) above may be that the intermolecular
interaction favors a slightly <u>smaller</u> unit cell than required for
registry in the $\sqrt{7}\times\sqrt{21}$ structure. The corrugation in the molecule-
substrate interaction may stabilize a rectangular cage of four
molecules within which a molecule relaxes off of a symmetric site at
the cell center in order to minimize its energy.

2.2 Ethane and hexane on graphite

In the discussion of the N_2 monolayer in the previous section, we have
seen how the symmetry properties of the monolayer inferred from LEED
experiments on single-crystal substrates complement the intensity
analysis of the neutron diffraction patterns from the same monolayer
adsorbed on a polycrystalline substrate. A similar interplay between
the two techniques has occurred in structural studies of ethane
(CH_3CH_3) on graphite. In this case, however, the intensity analysis
of the neutron diffraction patterns preceded the LEED experiments. In
order to eliminate the large incoherent scattering from hydrogen the
completely deuterated molecule is used in the neutron experiments.
The scattering amplitudes of the C and D nuclei are virtually equal.
Reasonable fits to the submonolayer ethane neutron diffraction pattern
in Fig. 5 were found for both an oblique unit cell containing one
molecule [10,11,24] and a commensurate HB structure [10,11,25]. The
LEED experiments [11] were decisive in favor of the HB structure,
although, as we have seen for N_2, they did not yield the molecular
orientations.

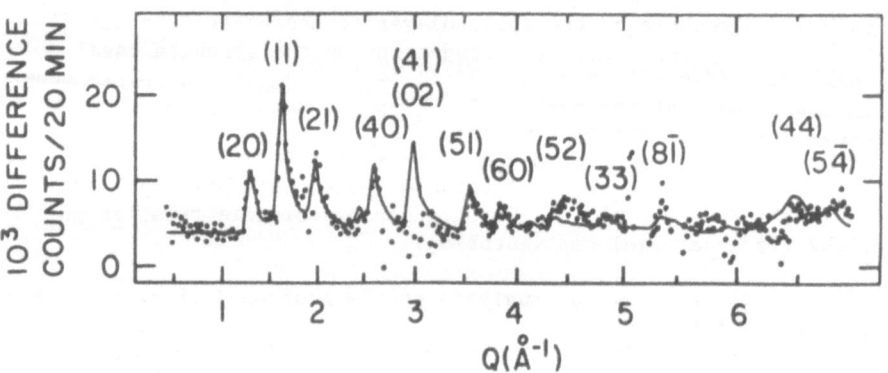

Figure 5. Neutron diffraction pattern from a 0.8-layer ethane (C_2D_6)
film adsorbed on Grafoil at 8.6 K [10]. The solid curve is the
diffraction profile calculated for the commensurate HB phase S1 shown
in Fig. 6.

The unit cell of the submonolayer or so-called S1 phase of ethane on graphite is shown in Fig. 6 [10,26]. It is a rectangular commensurate ($\sqrt{3}$x4) cell which has the same glide line symmetry as the N_2 submonolayer. It differs from N_2 only in that the **a** lattice vector has length 4a instead of 3a where **a** is the graphite lattice constant. The two ethane molecules are assumed rigid and in the staggered configuration. They are oriented with their C–C bond at an angle of 10° to the surface. In this way, three H atoms (numbered 3, 7, and 8 in Fig. 6) lie nearly in the same plane next to the surface. The diffraction profile calculated for this S1 structure is shown by the solid curve in Fig. 5.

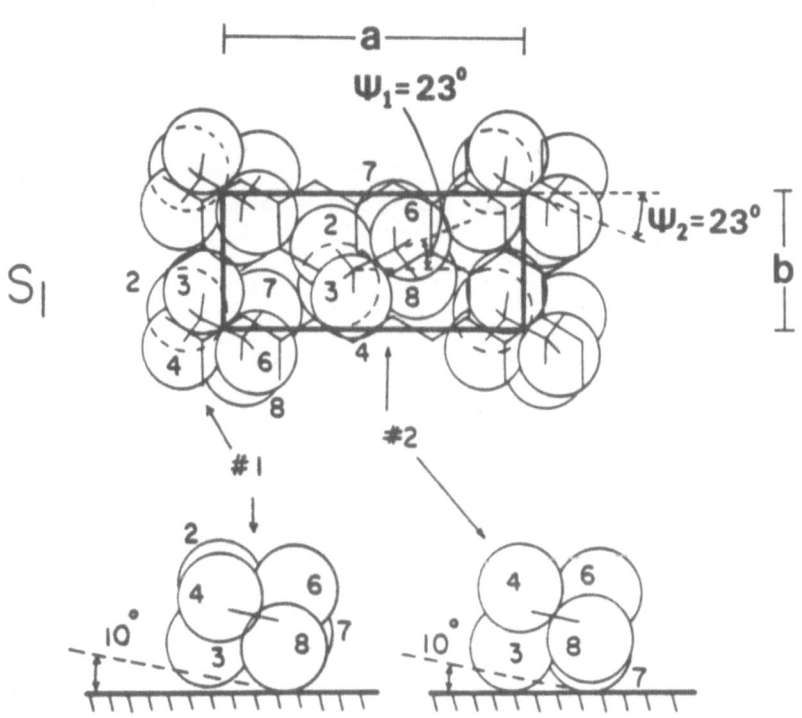

Figure 6. Structure of the commensurate HB phase S1 of ethane on graphite. Top: projection of the $\sqrt{3}$x4 unit cell (a=9.84 Å, b=4.26 Å) on the graphite basal plane. Bottom: side view showing orientation with respect to the surface.

As in the case of $Fe(CO)_5$, the experimentally determined ethane
S1 structure has been compared with that calculated by minimizing the
potential energy of an 18-molecule monolayer cluster [10,26,27].
Pairwise atomic C-C, C-H, and H-H potentials of the form $E = Ar^{-6} +$
$Bexp(-\alpha r)$ with Kitaigorodskii's [28] values for the parameters A, B,
and α were used to represent the intermolecular and molecule-substrate
interactions. The (0001) graphite surface was simulated by a single
honeycomb layer of carbon atoms since deeper layers lie beyond the
cutoff lengths of the potentials. During the energy minimization, the
ethane unit cell was constrained to be rectangular. The lattice
constants predicted in these calculations were about 7% smaller than
observed for the commensurate S1 phase, but the molecular orientations
were in excellent agreement with those determined from profile

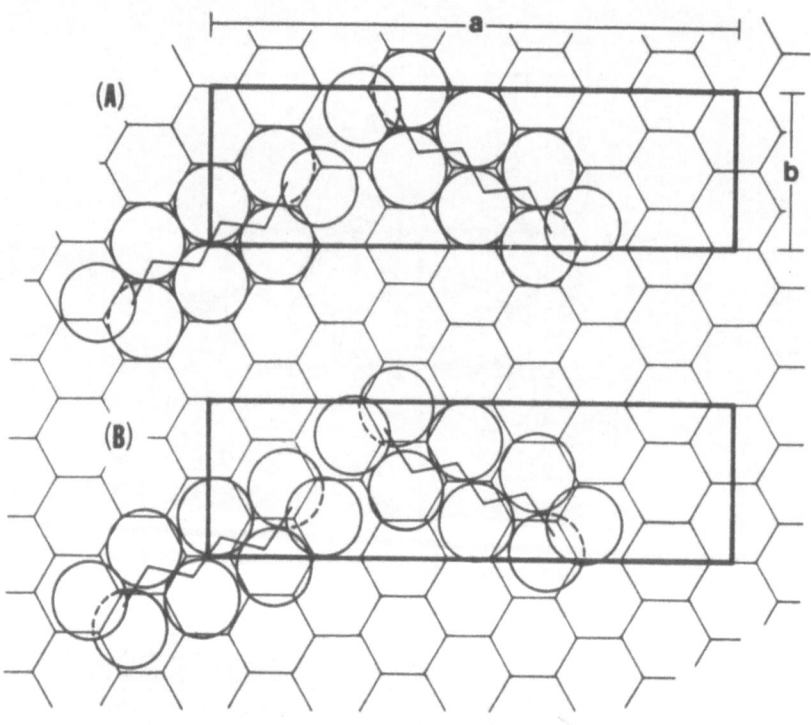

Figure 7. Projection of the commensurate HB monolayer structure of
hexane on the graphite basal plane. The unit cell is $2 \times 4\sqrt{3}$
($a=17.04$ Å, $b=4.92$ Å). (A) Molecular configuration in which H atoms
nearest the surface lie at the center of the graphite carbon
hexagons. Principal axis of the molecule makes an angle $\psi = 30°$ with
the a lattice vector. (B) Hexane monolayer structure inferred from
neutron diffraction experiments (Ref. 14). It is the same structure
as (A) but with $\psi = 23°$. The locations of the molecules on the
graphite surface are not determined experimentally.

analysis of the neutron diffraction pattern. While neither the experiment nor the calculations give the location of the monolayer unit cell on the graphite surface, we note that the structure drawn in Fig. 6 allows both molecules in the cell to occupy symmetry equivalent sites.

Hexane, a longer hydrocarbon chain in the paraffin series $(CH_3(CH_2)_4CH_3)$, also condenses in an HB structure on the graphite (0001) surface. The LEED pattern [13] indicated a rectangular commensurate $(2x4\sqrt{3})$ unit cell with the same glide line symmetry as submonolayer N_2 and ethane. As shown in Fig. 7, the lattice vectors are rotated $90°$ about a surface normal from their direction in the ethane unit cell (Fig. 6).

Again, neutron diffraction on a monolayer of the completely deuterated material has been used to determine the orientations of the two molecules in the unit cell [14]. The n-hexane molecule can be regarded as consisting of three planes of atoms--a carbon skeleton sandwiched between two layers of H (or D) atoms. In Fig. 7(A), we see that it is possible for the molecule to lie on its side with the carbon skeletal plane parallel to the surface such that each H (D) atom nearest the surface occupies the center of a graphite carbon hexagon. In this orientation, the principal axis of each molecule makes an angle $\phi = 30°$ with the longest lattice vector **a**. The sensitivity of the neutron diffraction pattern to a ϕ rotation is sufficiently great to show that ϕ is actually $23°$ (see Fig. 7(B)). Apparently, the intermolecular interaction overcomes the corrugation in the molecule-substrate potential, causing the molecule to twist out of an orientation in which each H (D) atom is in registry with the graphite surface. This feature is confirmed by potential energy calculations on a monolayer cluster [14] similar to those performed for ethane on graphite.

2.3 What happens to HB monolayers at higher coverages?

In the next section, we shall discuss in detail the uniaxial commensurate-incommensurate transitions which occur in N_2 and hexane as the coverage approaches monolayer completion. Before doing so, we would like to describe more generally the effect of increasing coverage on the HB monolayer structure and, in particular, to consider the structural changes which occur above monolayer completion.

The vast majority of physisorbed films (the rare gases Ar, Kr, and Xe are exceptions) incompletely wet a substrate [29]. That is, only a finite number of layers adsorb before the nucleation of bulk particles. $Fe(CO)_5$ [30] and hexane [14] represent an extreme limit of incomplete wetting in which only a single solid layer adsorbs on a graphite basal plane surface before bulk nucleation. In the case of hexane, the bulk coexists with the commensurate HB phase while for $Fe(CO)_5$ the monolayer first compresses isotropically into an incommensurate HB structure before bulk is nucleated. What makes the layering behavior of $Fe(CO)_5$ even more unusual is that immediately above the monolayer melting point only a single fluid layer wets the graphite surface. This feature together with an appreciable density decrease between the incommensurate solid monolayer and the monolayer fluid result in an interesting delayering effect: upon melting of the incommensurate phase, there is a migration of molecules from the monolayer fluid to the coexisting bulk particles [30].

The dependence of the ethane film structure on coverage differs somewhat from that of $Fe(CO)_5$ and hexane [10,11]. At low temperature, the HB monolayer structure (S1) first undergoes a transition to an intermediate phase S2 which persists in a narrow coverage range near monolayer completion. Between 1.0 and 1.5 layers (unity coverage is a complete layer of the S1 phase), the S2 phase coexists with a simple $\sqrt{3}x\sqrt{3}$ commensurate structure (S3) in which the molecules stand on their methyl tripod with the C-C bond perpendicular to the surface. The pure S3 phase is not reached until a coverage of 1.6 layers. It does not have an HB structure and appears to support a second solid layer [31]. However, the bilayer structure is unknown.

Nitrogen on graphite exhibits a succession of structural transitions at low temperature which bear some resemblance to those of ethane [8,9]. Just below monolayer completion the HB phase transforms into a uniaxial incommensurate (UI) phase. Between ~1.2 and ~1.4 layers (unity coverage is a complete layer of the commensurate HB structure), the UI phase coexists with a triangular incommensurate (TI) phase which is complete at 1.67 layers [9]. Since the density of the TI phase is ~10% greater than that of the commensurate HB structure, a partial second layer, presumably fluid or amorphous [9], must be present at 1.67 layers. Growth of the second layer continues at higher coverages until the bilayer crystallizes between 2.6 and 3.3 layers followed closely by bulk nucleation at ~3.7 layers [32]. Whether the bilayer is an incommensurate "two-out" HB structure with the molecules tilted out of the plane or a pinwheel structure is not yet clear [32].

A general conclusion from the above discussion is that an HB monolayer on graphite provides a very poor template for the growth of a second layer. We have seen that no second layer grows at all for $Fe(CO)_5$ and hexane. The bilayer crystallizes for ethane but only after the first layer has transformed out of the HB structure. In the case of N_2, a bilayer crystallizes just before the nucleation of bulk but appears to be quite unstable. For example, there is neutron diffraction evidence [32] suggesting that the melting point of the second layer is at 23 K where a large heat capacity peak is observed for the N_2 bilayer [33] This is a far lower temperature than the melting point of the commensurate monolayer at ~80 K [20]. In the more extreme case of $Fe(CO)_5$, even a fluid second layer is unstable (to bulk) just above the monolayer melting point. The cause may be a highly correlated fluid first layer in which the molecules retain their face-parallel configuration (see Fig. 4(b)) [30].

The common feature of $Fe(CO)_5$ and hexane which hinders second layer growth may be the difference between their HB monolayer structure and that of the densest packed plane of bulk. Because of their shape and their strong attraction to the substrate, these molecules lie flat on their side in the HB monolayer phase whereas in the bulk the molecules' principal axis is tilted with respect to the densest plane. Nitrogen could be an intermediate case where the fully compressed monolayer--either the two-out HB or the pinwheel structure--is similar enough to the (111) plane of bulk α-N_2 to allow crystallization of a second layer.

3. PHASE TRANSITIONS IN THE HB MONOLAYERS

3.1 Uniaxial Commensurate-incommensurate Transitions

Diehl and Fain [34] were the first to observe a uniaxial compression of the N_2 monolayer on graphite at low temperature and coverages just above completion of the commensurate HB phase. Their LEED measurements indicated a maximum compression in the uniaxial incommensurate (UI) phase of 5% in the direction of the a lattice vector in Fig. 1(a). Extinctions in the LEED pattern implied the same glide line symmetry as for the commensurate HB phase. Hence it was assumed that the UI phase preserved the HB ordering.

Neutron diffraction experiments [9] support the interpretation of an in-plane HB structure for the UI phase. The diffraction pattern from 1.13 layers of N_2 on a Papyex substrate (see Sec. 2.1) is fit well assuming a UI unit cell having a compression of only 0.7% along the a direction. Both the azimuthal angle ϕ (Fig. 1(a)) and the tilt angle β (Fig. 1(b)) specifying the molecular orientations lie in the same range inferred for the commensurate HB phase: $40° < \phi < 50°$ and $0 < \beta < 15°$. At a coverage of 1.27 layers, the compression along a increases to 3.3% but, again, with no observable change in the ϕ or β angles. Also, some evidence of coexistence of the UI and TI phases (see previous section) is seen.

There have been two kinds of computer simulations of the UI phase. In some molecular dynamics [23] and Monte Carlo [35] calculations, a one-phase uniaxially compressed system has been considered. In-plane ordering ($\beta \approx 0$) with $\phi = 41.0°$ is reported in Ref. 23 for a model with 5% compression. This value of ϕ is within the range inferred in the neutron experiment [9] at a coverage of 1.27 layers where a 3.3% compression is observed.

There have also been Monte Carlo calculations [35,36] with large enough systems to simulate modulated structures in which commensurate domains are separated by incommensurate domain walls. In the most recent work, [36] evidence of a UI phase which is modulated in the a direction of Fig. 1(a) is found for coverages between 1.0 and 1.06 layers. This monolayer structure consists of stripes of molecules which have the $\sqrt{3}\times3$ HB structure alternating with UI stripes of higher density. A striped UI phase is characterized by weak satellite diffraction peaks close to the principal Bragg peaks of the commensurate phase. Unfortunately, limited Q resolution and intensity in the LEED and neutron diffraction patterns have made it impossible to distinguish experimentally between uniform and striped UI models for N_2 on graphite.

A preliminary analysis of recent neutron diffraction experiments suggests that a striped UI phase does exist for hexane adsorbed on the graphite (0001) surface [14]. Earlier LEED experiments [13] had interpreted the solid monolayer phase at coverages below 0.9 layers as being an unmodulated UI structure corresponding to a uniform expansion of the $2 \times 4\sqrt{3}$ HB phase (see Sec. 2.2) along the **b** direction in Fig. 7. The commensurate-to-incommensurate transition (CIT) was thought to be continuous as a function of coverage at low temperature.

In contrast, the recent neutron experiments [14] indicate that, in a narrow coverage range between 0.9 and 1.0 layers, the UI and commensurate HB phases coexist. No further expansion of the monolayer is observed below 0.9 layers. This behavior is consistent with a first-order CIT. Furthermore, the neutron diffraction pattern of the UI phase is not fit well assuming a uniformly expanded unit cell. The fit is improved using a striped model having UI domains of lower density ("light walls") than the $2 \times 4\sqrt{3}$ HB phase. Work is now in progress to investigate different wall structures.

It is not known whether the other two commensurate HB monolayers which we have discussed, $Fe(CO)_5$ and ethane, are also unstable to a UI phase. Above completion of the $\sqrt{7} \times \sqrt{21}$ phase of $Fe(CO)_5$ (see Sec. 2.1), x-ray experiments indicate that the monolayer compresses nearly isotropically into an incommensurate structure [30]. Due to difficulties in sample preparation (the measurements are done on sealed cells), it has not been possible to study the CIT in detail. Neither the order of the transition nor whether it procedes via a uniaxially compressed phase have been determined.

We have considered the possibility that the S2 structure of ethane (see Sec. 2.3) is also a striped UI phase. A rather large superlattice $(2\sqrt{3} \times 10)$ was found to be consistent with both the LEED and neutron diffraction patterns of the S2 phase [11]. In addition, the series of structural transformations of monolayer ethane at low temperature, S1 → S2 → S3, described in Sec. 2.3 bears some resemblance to the commensurate → UI → TI series of N_2 on graphite [8,9]. Both the ethane S2 and nitrogen UI structures appear as intermediate phases in a narrow range of coverage above completion of the commensurate monolayer. The inelastic neutron scattering experiments to be discussed in Sec. 4 also provide some insight into the S2 structure. They reveal similarities in the vibrational spectra of the S1 and S2 phases, suggesting that S2 may have an HB structure in which the ethane C-C bond is nearly parallel to the surface. For example, it could be a striped phase in which commensurate HB regions are separated by narrow domain walls.

Admittedly, the evidence for the ethane S2 structure being a striped phase is not yet compelling; however, in the context of the other HB monolayers, the possibility seems sufficiently attractive to try to reanalyze the S2 neutron diffraction pattern [10] with a striped model. We caution, though, that another model of the S2 phase has been found to be consistent with both the LEED and neutron data [37,38].

3.2 Two-stage Melting Transitions

3.2.1. "Short" molecules. As already described by Diehl in a previous chapter [16], the commensurate HB phase of N_2 on graphite undergoes a molecular orientational-disordering (OD) transition upon heating to ~27 K. Above the OD transition, there is still long-range translational order in the monolayer but each molecule rotates about its center of mass. The translational order is lost in the monolayer at a much higher temperature. Neutron diffraction experiments indicate that the complete commensurate phase melts at ~80 K [20].

The large temperature difference between the OD and melting transitions of the N_2 monolayer may be related to the degree of corrugation in the adatom-substrate potential. Diehl has pointed out [16] that the lattice constants of the commensurate HB phase of N_2 on graphite are expanded by about 6% from those calculated for the monolayer on a smooth surface [19]. By itself, the substrate corrugation would tend to stabilize the monolayer against translational disorder and hence raise the melting point above its value on a smooth surface. With the attendant lattice expansion, molecular rotation will also be facilitated thereby lowering the OD transition temperature.

We saw in Sec. 2.1 that the submonolayer structure of $Fe(CO)_5$ was similar to that of the commensurate HB phase of the smaller N_2 molecule except that the sublattice unit mesh was scaled up from $\sqrt{3}a$ to $\sqrt{7}a$ where a is the graphite lattice constant (see Fig. 4). Upon heating, the commensurate HB structure of $Fe(CO)_5$ also undergoes a transformation to an OD phase prior to melting. The signature of the OD transition in the neutron diffraction pattern is a small change in the first and strongest Bragg peak and the complete disappearance of all higher order peaks [12]. The shift in the first peak position corresponds to a 1.5% lattice expansion in the OD phase. Also, an increase in the peak width implies a decrease of the monolayer coherence length L [20] from a value of 120 Å in the HB solid to 80 Å in the OD phase. The magnitude of L in the HB phase is limited by the graphite particle size in the Grafoil substrate. (Grafoil [20] is a recompressed exfoliated graphite similar to Papyex).

The principal difference with the OD transition in the N_2 monolayer is that for $Fe(CO)_5$ it occurs just a few degrees below the monolayer melting point at ~170 K. The proximity of the OD transition to melting probably results from steric hindrance to rotational motion of the bipyramid-shaped molecule. Although small, the 1.5% lattice expansion in the OD phase could greatly lower the barrier to rotation. This expansion may not be possible until close to the melting point of the HB phase. Presumably, steric effects in the OD phase cause the $Fe(CO)_5$ molecule to rotate about a surface normal in a nearly face-parallel configuration (Fig. 4(b)).

The ethane molecule is rod-shaped with a longer length-to-width ratio than N_2. As discussed in Sec. 2.2, this results in an HB phase at low temperature having a rectangular commensurate unit cell ($\sqrt{3}\times4$) which is longer than that of N_2 in the a direction by one

graphite lattice constant (see Fig. 6). Like N_2 and $Fe(CO)_5$, this S1
phase of ethane undergoes an OD transition at ~64 K as we now discuss.
  The neutron diffraction patterns in Fig. 8 are from a 0.8-layer
sample of C_2D_6 on the same Grafoil substrate used to obtain the
pattern in Fig. 5. Since the neutron wavelength is 4.07 Å, only the
first three Bragg peaks of the ethane S1 phase can be observed at a
temperature of 60.7 K. At 64.1 K, the diffraction patterns show
hysteresis. They change qualitatively depending on whether this
temperature is approached from below (Fig. 8(b)) or from above (Fig.
8(c)). The latter pattern contains two broad peaks: one at the same
position as the (11) peak of the S1 phase ($Q=1.61$ Å^{-1}) and another at
the position ($Q=1.5$ Å^{-1}) of the single peak in the 70.5 K pattern.
This behavior was interpreted as indicating coexistence between the S1
phase and a higher temperature phase, I1 [10]. Together, the
hysteresis and two-phase coexistence imply that the S1 to I1

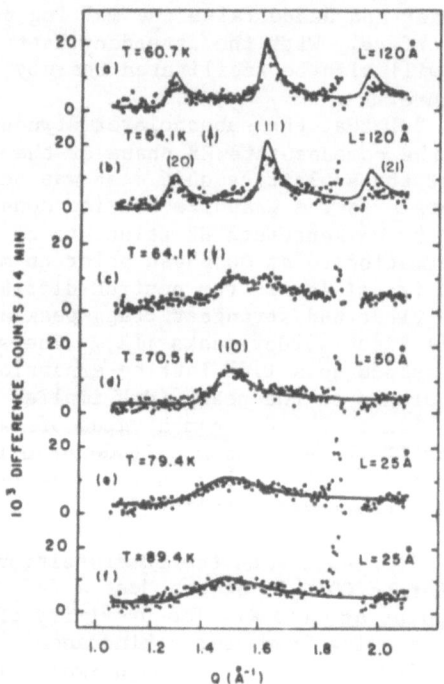

Figure 8. Temperature dependence of the neutron diffraction patterns
of a 0.8-layer ethane (C_2D_6) film adsorbed on Grafoil showing the
S1→I1 transition near 64 K [10]. The solid curves represent a fit of
a Warren lineshape (see text) to each diffraction peak with the
designated coherence length L. The arrows in (b) and (c) indicate
whether the temperature is increasing (↑) or decreasing (↓).

transition is first order as is the OD transition of submonolayer N_2 [38]. (The order of the OD transition in the commensurate HB phase of $Fe(CO)_5$ is unknown) [12].

The position of the broad peak in the I1 phase diffraction pattern (Fig. 8(d)) is inconsistent with a commensurate $\sqrt{3}x4$ structure which is orientationally-disordered. It is, however, consistent with a 2x2 monolayer structure as was later confirmed by LEED [11,40]. Quasi-elastic neutron scattering (QENS) experiments [41] have indicated that the 2x2 I1 phase is orientationally-disordered with the ethane molecules performing isotropic rotation about their center of mass. Moreover, they suggest that the onset of the OD transition is actually in the HB S1 phase where uniaxial rotation of the ethane molecule about its C-C bond begins somewhere between 10 K and 53.5 K. This feature may be inconsistent with neutron diffraction data which show the S1 pattern at 60.5 K to be virtually unchanged from that of 9 K [10]. However, this discrepancy might be due to the use of the protonated and deuterated molecules in the QENS and diffraction experiments, respectively. The smaller moment of inertia of the protonated molecule about the C-C bond could result in the onset of rotation at a lower temperature.

The conclusion reached from the QENS experiments [41] is that the ethane monolayer melts at the S1 to I1 transition. The I1 phase is then a highly correlated fluid in which the molecules are translationally diffusing between equivalent sites on the graphite surface. The onset of some translational motion at the S1 to I1 transition would be consistent with the decrease observed in the monolayer coherence length L from 120 Å in the S1 phase (Fig. 8(a)) to L=50 Å in the I1 phase at 70.5 K (Fig. 8(d)). It is also consistent with LEED studies [40]. As the I1 phase is heated, the single peak in the neutron diffraction patterns (Figs. 8(e) and 8(f)) continues to broaden, suggesting a progressive loss of translational order in agreement with the LEED experiments [40]. The QENS experiments [41] indicate that the molecules continue to lose both their positional and rotational order as the temperature is raised.

Generally speaking, there appears to be greater coupling between the rotational and translational degrees of freedom in the $Fe(CO)_5$ and ethane monolayers than for N_2. While the question of the onset of the ethane OD transition is not completely resolved, it is clear that the OD transition of both $Fe(CO)_5$ and ethane is delayed until close to the melting point of the monolayer. Evidently, the energy barrier to rotational motion is higher for these less spherical molecules than for N_2. In the case of ethane, the rotational hindrance may also derive from the lower symmetry of the HB unit cell which is rectangular rather than hexagonal. We note that both the S1 ($\sqrt{3}x4$) and I1 (2x2) ethane phases have an area of 21 Å^2 per molecule. However, the full rotational motion of the molecules is not achieved until reaching the hexagonal I1 phase.

As the length-to-width ratio of the adsorbed molecules increases, it should become increasingly difficult for orientational disorder to develop in the monolayer. Without promoting molecules to a second layer, there is simply not enough space available in the first layer to accommodate molecules rotating about a surface normal. Although the low-temperature solid is not believed to be an HB structure [7], the melting of a butane $(CH_3(CH_2)_2CH_3)$ monolayer on graphite seems to support this conjecture. In Fig. 9, we show the temperature dependence of the neutron diffraction patterns from 0.8 layers of completely deuterated butane (C_4D_{10}) adsorbed on Grafoil [10]. The neutron wavelength is 4.07 Å as in Fig. 8. Upon increasing the temperature, the intensity of the diffraction peaks begins to change slightly near 108 K. By 118.5 K, they have completely disappeared and have been replaced by a very broad peak attributed to the monolayer fluid. Diffraction scans taken at more closely spaced temperature intervals show the peak intensities to continue to decrease slowly above 108 K until they abruptly disappear at 116 K. Hysteresis is observed in the patterns within a few tenths of a degree about 116 K.

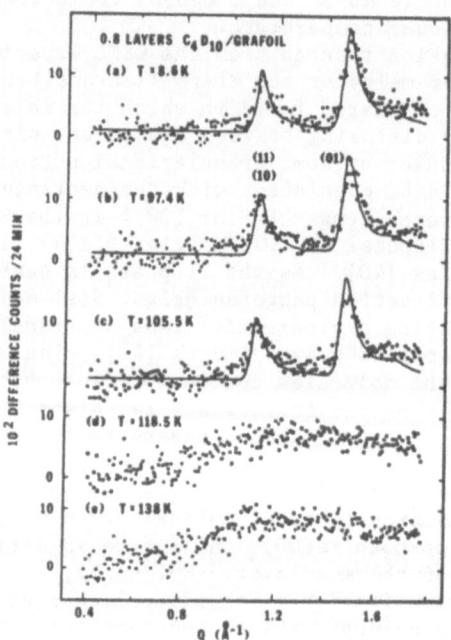

Figure 9. Temperature dependence of the neutron diffraction patterns showing the melting of a 0.8-layer butane (C_4D_{10}) film on Grafoil [10]. The solid curves are profiles calculated for a centered rectangular monolayer structure (Refs. 7 and 10).

We interpret the diffraction patterns in Fig. 9 as indicating first-order melting of the butane monolayer at 116 K. Quite clearly, there is neither an OD solid nor even a highly correlated fluid phase intervening between the orientationally-ordered monolayer at low temperature and the high-temperature fluid phase. In the next section, we shall examine the monolayer melting of a still longer molecule in the paraffin series, hexane. It will suggest that the butane monolayer represents a crossover to a qualitatively different melting behavior.

3.2.2. A "long" molecule. As a rule of thumb, submonolayer physisorbed films melt at a temperature 50-70% of the bulk triple point [1]. This is generally attributed to the fewer number of nearest neighbors in 2D which reduces steric hindrance to translational disorder. For some time, we have been interested in a related question of how the steric properties of molecules may affect the ease with which orientational and translational disorder is achieved in monolayers. It seems plausible that, for sufficiently strong binding to a surface, the monolayer melting point may be more sensitive to the molecular shape than that of bulk.

Consider a rigid rod-shaped molecule whose length is several times its width and which adsorbs with its long axis parallel to the surface. The potential energy barrier to rotation about an axis perpendicular to the surface may be so large that the monolayer melting point is very close to or even exceeds that of bulk. The reason is that in the bulk phase the molecules are not confined to a single plane. Orientational and translational disorder can be achieved more easily than in a monolayer by exchanging molecules between neighboring layers. Similarly, we would expect the melting point of a monolayer to decrease if the substrate binding were reduced to the point where the rod-shaped molecules could be thermally promoted into a second layer.

The qualitative arguments given in the preceding paragraph have motivated us to examine the melting behavior of a hexane monolayer on graphite. The structural studies described in Sec. 2.2 showed that hexane adsorbed on graphite in the trans conformation with a plane of H atoms nearest the surface (Fig. 7). In this configuration, the molecule is fairly rigid with a length roughly twice its width.

We have performed some preliminary neutron diffraction experiments [14] which suggest the melting of the hexane monolayer to differ from both that of ethane and butane (see previous section). Neutron diffraction patterns of a deuterated hexane (C_6D_{14}) monolayer on Paypex are shown as a function of temperature in Fig. 10. There is little qualitative change in the patterns up to 155 K where the (02) and (06) peaks of the commensurate HB structure begin to weaken. By 175 K, these peaks have completely disappeared and the pattern is dominated by a single broad peak at $Q=1.5 \text{ Å}^{-1}$ corresponding to a d-spacing in the monolayer of 4.3 Å. Measurements at higher temperatures have shown that this peak is still observable at room temperature [14].

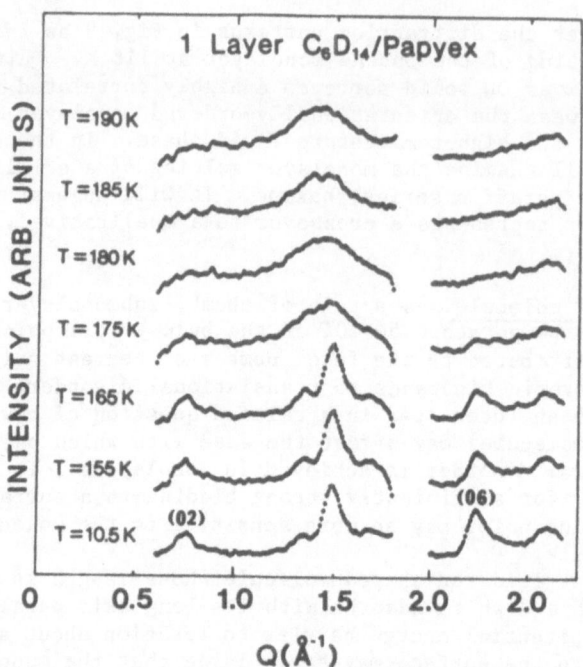

Figure 10. Temperature dependence of the neutron diffraction patterns of a 1.0-layer hexane (C_6D_{14}) film on Papyex. The transition observed near 170 K is from the commensurate HB structure (Fig. 7(B)) to possibly a liquid-crystal-like phase (see text).

The diffraction patterns in Fig. 10 raise some interesting questions concerning the melting of long rod-shaped molecules. The broad peak at 175 K is similar to that observed for the OD II phase of ethane. However, due to the length of the hexane molecule, it seems quite unlikely that the molecule could be rotating about its center of mass either isotropically or about an axis perpendicular to the surface. This conclusion is also supported by the absence of an OD transition in the monolayer of the shorter butane molecule. As an alternative interpretation, we have considered the possibility of a highly correlated liquid phase in which there is preferential alignment of the hexane molecules parallel to their long axis. The single broad diffraction peak above 175 K would then correspond to the distance between parallel lines of hexane molecules in this 2D liquid-crystal-like phase. The absence of any other Bragg peaks could be explained by the lack of translational order parallel to these lines as illustrated in Fig. 11.

Clearly, further experiments are required to confirm the presence of a 2D liquid-crystal phase of the hexane overlayer. For example, it would be desirable to perform a quantitative analysis of the single peak lineshape as a function of temperature. In conjunction with these experiments, molecular dynamics simulations are planned to study the effect of chain length on the melting of paraffin monolayers adsorbed on the graphite (0001) surface.

Assuming the melting of the commensurate HB phase of hexane at ~170 K as inferred from Fig. 10, it is interesting to compare its monolayer and bulk melting points. This is done for hexane as well as ethane and butane in Table I. Note that the melting point of the ethane monolayer is taken at the S1 → I1 transition. Also, the 2D melting point of butane is at a coverage of 0.8 layers where the monolayer is not believed to have a commensurate HB structure [7].

Table I. Comparison of monolayer and bulk melting points for some short-chain paraffin molecules.

Molecule	$T_t(2D)$	$T_t(3D)$	$T_t(2D)/T_t(3D)$
Ethane (CH_3CH_3)	64 K[a]	90 K	0.71
Butane ($CH_3(CH_2)_2CH_3$)	116 K[b]	135 K	0.86
Hexane ($CH_3(CH_2)_4CH_3$)	170 K[c]	178 K	0.96

[a]Ref. 10
[b]Ref. 7
[c]Ref. 14

As anticipated at the beginning of this section, we see from Table I that the ratio of the monolayer to bulk melting points increases with the length of the molecule, approaching unity for hexane. For even longer molecules, it may actually exceed unity provided that the substrate binding is strong enough and the molecules are sufficiently rigid. To increase the rigidity, it may be necessary to use unsaturated hydrocarbon chains rather than paraffins.

The melting behavior which has been observed for the commensurate HB monolayers on graphite can be summarized with the aid of the schematic diagram in Fig. 11. For the "short" molecules, N_2, $Fe(CO)_5$, and ethane, there is a two-stage melting process with a transition first to an intermediate OD phase before translational order is lost. We have discussed how ethane differs somewhat from N_2 and $Fe(CO)_5$ due to the change in translational symmetry and large drop in the monolayer coherence length at the OD transition. For the longer hexane molecule, we have proposed another type of two-stage melting in which the intermediate phase is liquid-crystal-like rather than orientationally disordered. However, more supporting evidence is required in order to establish the existence of the monolayer liquid-crystal phase. Butane, a molecule of intermediate length, has an abrupt, first-order melting of the monolayer and appears to represent a crossover between the two paths depicted in Fig. 11.

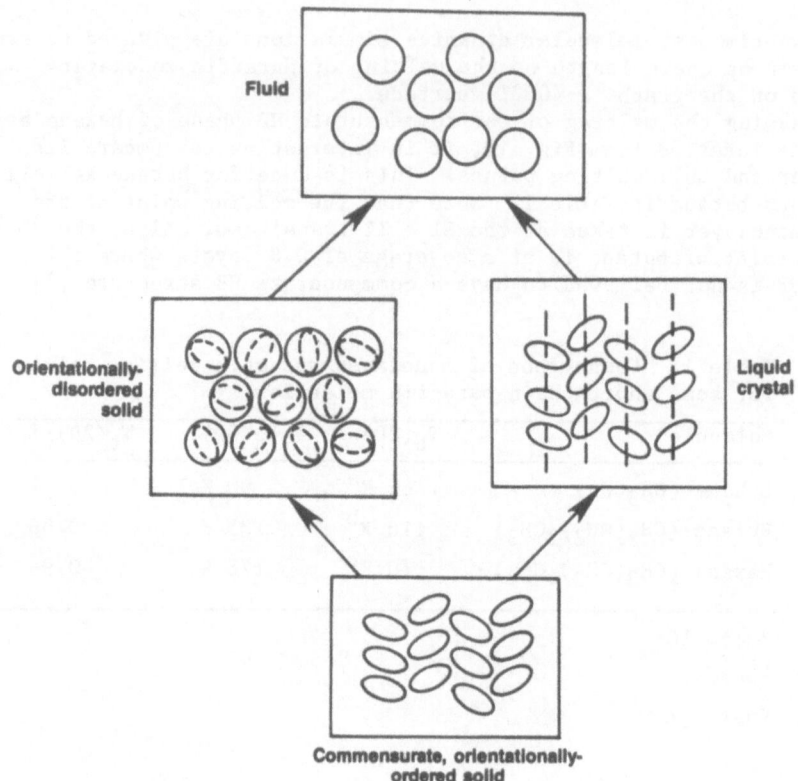

Figure 11. Illustration of two different melting paths for a commensurate HB monolayer.

We conclude this section on melting of HB monolayers by commenting that both two-stage processes described here involve some decoupling of the translational and rotational motion of the molecules. Depending on the length of the molecule, either orientational or translational disorder occurs first as the monolayer is heated. Thus the intermediate stage in the melting is not the hexatic phase of current theories of dislocation mediated melting in 2D [2,42,43]. A discussion of these theories is beyond the scope of this paper. We note, though, that their extension to anisotropic layers [44,45] may provide a useful framework for treating melting of the commensurate HB phases.

4. LATTICE DYNAMICS OF AN HB MONOLAYER PHASE: ETHANE S1

The orientational ordering of molecules in an HB structure creates elastic anisotropy in the monolayer. One manifestation of this essential anisotropy is the uniaxial commensurate-incommensurate transitions discussed in Sec. 3.1. There, the anisotropy in the adatom-substrate corrugation also plays a crucial role; and it is difficult, either by experiment or computer simulation, to separate the effect of these two contributing interactions.

Anisotropy in the intermolecular interactions can be probed by studying the collective vibrational excitations of the HB monolayers by inelastic neutron scattering (INS) [46,47]. The effect of adatom-substrate corrugation on the INS spectra is less easily seen [26]. Because of the high phonon density of states at the Brillouin zone boundaries, these modes dominate the INS spectra. Their frequencies are determined primarily by intermolecular forces and the average binding energy to the substrate. Molecule-substrate corrugation will be most important at the Brillouin zone center where it causes a small energy gap to develop in the acoustic branches (the corrugation provides a restoring force for a uniform translation of the monolayer parallel to the surface).

The weak neutron-film interaction requires INS experiments to be performed with high-surface-area substrates. The need is even more acute than for elastic neutron diffraction because the neutron-phonon cross sections are generally one to two orders of magnitude smaller than the elastic ones. Use of a polycrystalline substrate precludes direct measurement of the monolayer phonon dispersion relations as has been possible in some cases with inelastic electron [48] and helium atom [49] scattering. Nevertheless, INS offers several advantages over these other probes [26]: (1) Neutron-phonon cross sections can be calculated relatively easily; (2) selection rules are absent; (3) ultrahigh vacuum is not required; and (4) there is sensitivity to the motion of light atoms, particularly hydrogen. The last feature greatly facilitates observation of librational modes of adsorbed hydrocarbons in which H atoms at the molecules' extremities undergo large amplitude displacements.

To exploit the large incoherent cross section of hydrogen for thermal neutrons, it is desirable to select one of the hydrocarbons, ethane or hexane, for INS experiments on a commensurate HB phase. Due to the larger number of low-frequency internal modes of the longer hexane molecule [50], ethane was selected for this purpose [26,27]. Although the scattering is much weaker, recent experiments have demonstrated the feasibility of obtaining INS spectra from the HB phase of N_2 on graphite as well [51].

INS spectra for the S1 (HB), S2, and S3 monolayer phases of ethane on graphite (see Sec. 2.3) at a temperature of 10 K are shown in Fig. 12. The spectra are taken in a neutron-energy-loss mode, and the inelastic scattering from the Grafoil substrate has been subtracted. Further details of the experimental technique are given in Ref. 27.

The Grafoil substrate has preferential orientation of the graphite particle c axes perpendicular to its sheets. This permits spectra to be obtained in two different scattering geometries: the momentum transfer Q either parallel or perpendicular to the Grafoil sheets [46] (denoted $Q_\parallel$ and $Q_\perp$, respectively, in Fig. 12). In practice, little difference is observed in the two Q configurations, suggesting that most of the available surface area is contributed by smaller particles which are oriented isotropically [27].

We see in Fig. 12 that the spectra of the S1 phase at θ=0.80 (layers) is qualitatively similar to that of the S2 phase at θ=0.97. As we remarked in Sec. 3.1, this suggests a similarity in the S1 and S2 structures with S2 possibly being a striped phase containing commensurate HB regions. The spectra at θ=1.6 of the commensurate √3x√3 phase, S3, in which the molecules stand on end clearly differs from those of S1 and S2.

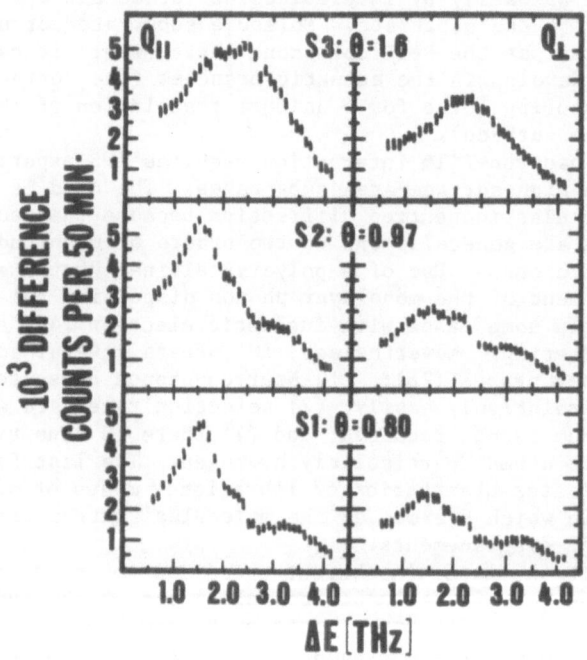

Figure. 12. Neutron energy-loss spectra from the S1, S2 and S3 monolayer phases of ethane (C_2H_6) adsorbed on Grafoil at 10 K [27]. θ is the coverage in layers. The inelastic scattering from the bare Grafoil has been subtracted. The spectra on the left have been taken with Q parallel to the Grafoil sheets and those on the right in the Q-perpendicular configuration (see text).

In order to identify the modes contributing to the S1 and S3 spectra in Fig. 12, calculations of the monolayer lattice dynamics and one-phonon thermal neutron cross sections were undertaken [27]. These calculations were performed on the same zero-temperature structures obtained by minimizing the potential energy of a monolayer cluster as described in Sec. 2.2. There we saw that the computed structure reproduced well the molecular orientations inferred by neutron diffraction but was not sufficiently accurate to predict the monolayer commensurability on the graphite surface.

The procedure for representing the intermolecular contribution to the monolayer potential energy and for calculating the dynamical matrix was similar to that developed for 3D molecular crystals [52]. Consistent with the cut-off lengths assumed for the atom-atom (Kitaigorodskii [28]) potentials, the intermolecuar interactions included nearest and next-nearest couplings for the rectangular S1 structure. Each molecule was coupled to the substrate by six force constants corresponding to its three rotational and three translational degrees of freedom. Unlike the structure calculations, the molecule was not assumed perfectly rigid in that torsional motion of the methyl (CH_3) groups was allowed about the ethane C–C bond. This gives a total of seven molecular degrees of freedom as listed in the table at the bottom of Fig. 13: the methyl torsion, the rigid rotations ($\theta_x, \theta_y, \theta_z$), and the translations (t_x, t_y, t_z). All of the

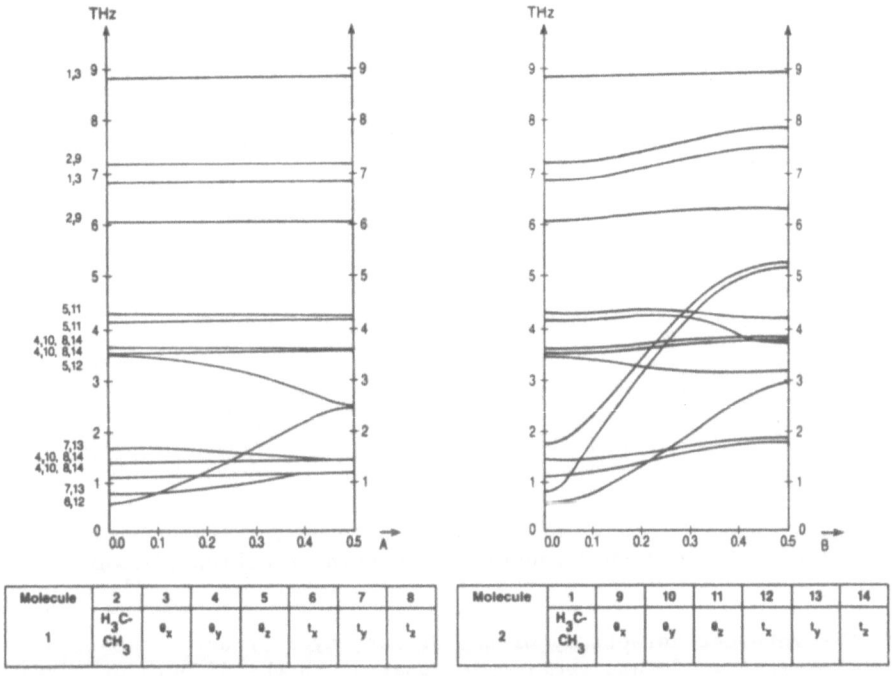

Molecule	2	3	4	5	6	7	8
1	H_3C- CH_3	θ_x	θ_y	θ_z	t_x	t_y	t_z

Molecule	1	9	10	11	12	13	14
2	H_3C- CH_3	θ_x	θ_y	θ_z	t_x	t_y	t_z

Figure 13. Phonon dispersion relations calculated for the commensurate HB phase S1 of ethane on graphite [27]. The numbers at left refer to the table at the bottom of the figure listing the types of motion for each molecule. θ_i is a rotation of the molecules about the ith axis, and t_i is a translation of the molecule in the direction of the ith axis. The x,y,z axes are fixed to the substrate with z normal to the surface and with x and y in the directions of **a** and **b** respectively, in Fig. 6.

force constants except that of the methyl torsion were calculated from the atom-atom potentials [27].

The dispersion curves calculated [27] for the Sl phase in the **A** and **B** directions (see Fig. 6) are shown in Fig. 13. With two molecules per unit cell and seven degrees of freedom per molecule, there are 14 branches in all. Due to the low symmetry of the Sl phase, the character of the modes is rather complex. The numbers labeling the branches at the left refer to the table at the bottom of the figure where the types of motion are listed for each molecule. The large number of branches is reflected in the rich structure of the phonon density of states in Fig. 14 [27]. The four uppermost branches are nearly flat and separated well enough to be seen as individual peaks.

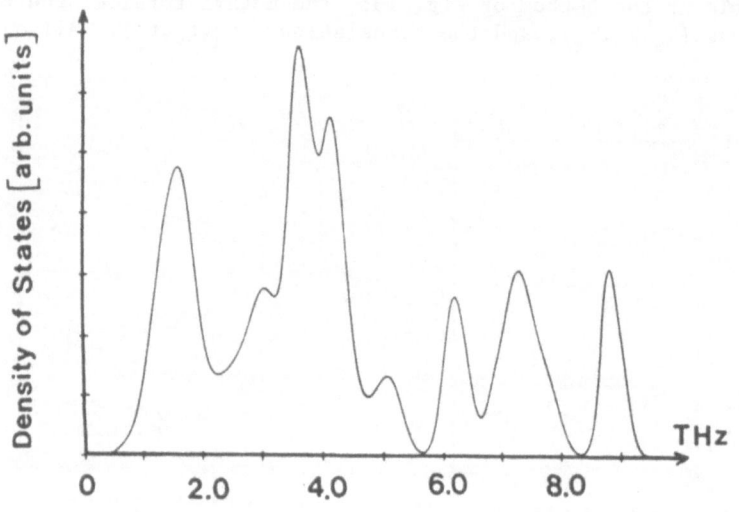

Figure 14. Phonon density of states calculated for the ethane Sl phase after folding with a gaussian resolution function of fwhm = 1.8 THz [27].

The inherent anisotropy of the HB monolayer structure mentioned at the beginning of this section is apparent in the dispersion curves of Fig. 13. In the **A** direction, we see that the longitudinal modes labeled 6,12 are more strongly dispersed than the transverse modes labeled 7,13. This effect is much more pronounced in the **B** direction--the direction of the nearest-neighbor bonds. (Note that in the **B** direction modes 7,13 are longitudinal while 6,12 are transverse). Yet, it is difficult on the basis of the dispersion relations alone, to predict a "soft" direction in the Sl structure along which a uniaxial distortion could occur (see Sec. 3.1). To address this question, the frequency and polarization of the modulation wave in the incommensurate phase must be known. One must also consider the anisotropy in the corrugation of the molecule–substrate potential. The atom–atom potentials used to model the corrugation are probably not accurate enough for this purpose [27].

The calculated incoherent neutron cross sections for one–phonon creation are compared with the INS spectra of the ethane S1 phase in Fig. 15. An isotropic substrate was assumed in the calculations, since, as remarked earlier, little difference is observed between the scattering in the Q parallel and perpendicular configurations. Coherent scattering can be neglected because of the dominance of the incoherent scattering from hydrogen. Other details of the calculations are given in Ref. 27. The INS spectrum of the S1 phase observed in the $Q_\parallel$ configuration (Fig. 12) has been reproduced in Fig. 15(a) to facilitate comparison with the theoretical spectra.

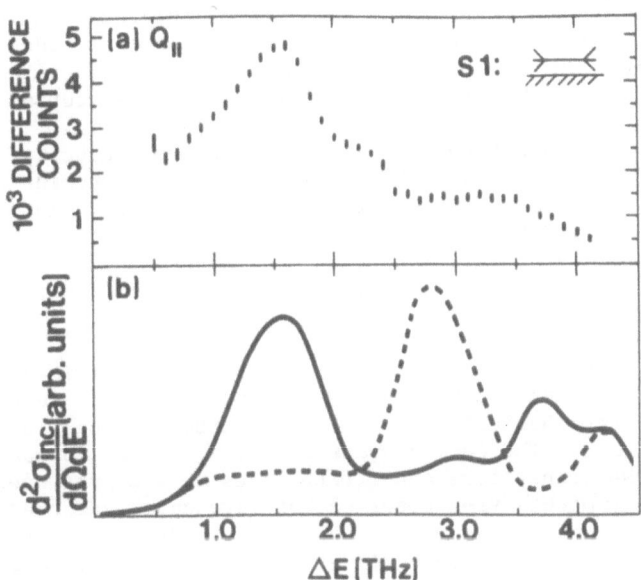

Figure 15. Comparison of observed and calculated inelastic neutron spectra for the S1 phase of ethane at 10 K [27]. (a) Neutron energy-loss spectrum observed at constant $Q = 4.0$ Å^{-1} parallel to the Grafoil sheets (replotted from Fig. 12). (b) Calculated one–phonon cross sections for the S1 phase. See text for the distinction between the dashed and solid curves.

We first discuss the calculated spectrum represented by the dashed curve in Fig. 15(b). After folding with the instrumental resolution function, it is dominated by broad band at ~2.8 THz contributed mainly by two modes: the translation of the molecules normal to the surface (t_z) and the libration about a symmetry axis perpendicular to the ethane C–C bond and parallel to the surface

(A_y). The agreement with the observed spectrum is poor with the calculated peak position a factor of two higher than that of the intense band observed at ~1.5 THz.

Rather than assume large errors in the force constants calculated for both the translational and librational modes, the effect of a coupling between these modes was investigated [27]. Such a coupling is plausible since the large distance of the CH_3 groups from the librational axis could result in a dependence of the librational restoring forces on the molecular height above the surface. The solid curve in Fig. 15(b) represents the best fit to the observed spectrum which could be obtained by adjusting this coupling constant. The coupling splits the band at 2.8 THz into two peaks, the more intense of which reproduces well the band observed at 1.5 THz except for the shoulder at ~2.3 THz. The weaker peak predicted at 3.7 THz is at a somewhat higher frequency than the band observed at 3.2 THz. The dispersion curves in Fig. 13 have been calculated assuming the same coupling between the translational and librational modes. Without it, the nearly flat branches labeled 4,10,8,14 collapse into a narrow band centered at ~2.8 THz (cf. peak position of dashed curve in Fig. 15(b)).

We would expect the coupling between the "bouncing" and "rocking" modes inferred for ethane to be a feature common to the dynamics of other HB monolayers of rigid rod-shaped molecules. In the case of N_2, the molecule may be too spherical to see the effect.

5. CONCLUSION

A combination of neutron diffraction and LEED experiments have revealed commensurate HB monolayer structures for a surprisingly large number of nonspherical molecules physisorbed on the graphite basal plane surface. The commensurate HB films display common features in their phase transitions and growth modes to a degree unexpected from the variety of behavior observed for rare gases adsorbed on graphite.

The pervasiveness of the HB phase suggests it to be insensitive to the details of the molecule structure. From the variety of molecular electronic properties represented, we conclude that mechanisms other than the quadrupole-quadrupole interaction proposed for N_2 can stabilize the HB structure.

We do not yet understand the apparently delicate interplay between the intermolecular interactions and the corrugation in the adatom-substrate potential to produce these commensurate HB phases. The corrugation amplitude for the HB phases on graphite is the most poorly known interaction parameter. Possibly, it is higher than predicted by atom-atom potentials as has been suggested for Kr on graphite [53,54]. However, this does not seem to be the case for N_2 on graphite [19].

Experimental determination of the corrugation in the molecule-substrate potential would be desirable. Efforts are now underway to measure the energy gap in the lowest lying translational mode of the N_2 commensurate HB phase on graphite [51]. If successful, this should provide an upper bound on the N_2-graphite corrugation amplitude. More qualitative measures of the corrugation amplitude are also needed. For example, it is unknown whether the HB structure persists on surfaces less corrugated than the graphite basal plane. It would be interesting to investigate the monolayer structure of the same molecules considered here, especially N_2, on the smoother Ag(111) surface [4]. More definitive evidence of incommensurate HB structures at submonolayer coverages on graphite [15] would be equally informative.

In the pursuit of the simplest monolayer phases, one seeks adsorbates for which the effect of substrate corrugation is minimal. From this standpoint, the HB phases may have advantages over the rare gases on graphite. If the intermolecular interaction is sufficiently anisotropic it could eventually dominate the effects of substrate corrugation. The hexane monolayer offers some supporting evidence of this. The intermolecular interaction is strong enough to rotate the molecules out of perfect registry with the graphite substrate (Fig. 7). Moreover, the highly correlated fluid phase of hexane may signal quasi one-dimensional melting in still more anisotropic monolayers.

6. ACKNOWLEDGMENTS

The author wishes to express his gratitude to F.Y. Hansen for his collaboration on interpreting our neutron scattering experiments on adsorbed films over the past decade. He is indebted to S.C. Fain, Jr. for many illuminating discussions of physisorbed films and for a critical reading of this manuscript. H. Shechter introduced the author to the $Fe(CO)_5$/graphite system and has provided assistance and encouragement for many years. He would also like to thank J.P. Biberian, J. Suzanne, and J.P. Coulomb for their collaboration on the ethane monolayer studies and R.D. Diehl for helpful comments while writing this manuscript. The neutron scattering experiments at the University of Missouri would not have been possible without the collaboration of G.J. Trott, R. Wang, H.R. Danner, J.C. Newton, S.-K. Wang, and J.R. Dennison. These experiments were supported by National Science Foundation Grants Nos. DMR-7905958, INT-8012228, and DMR-8304366, Israel-U.S. Binational Science Foundation Grant No. 2687, the Danish Natural Science Foundation, the Petroleum Research Fund administered by the American Chemical Society, and a Dow Chemical Grant of the Research Corporation.

7. REFERENCES

1. See, e.g., J.G. Dash, Films on Solid Surfaces, (Academic:New York) 1975.
2. J.M. Kosterlitz and D.J. Thouless, J. Phys. C, 1973, 6, 1181.
3. R.J. Birgeneau and P.M. Horn, Science, 1986, 232, 329.

4. J. Unguris, L.W. Bruch, E.R. Moog, and M.B. Webb, Surf. Sci., 1979, 87, 415.

5. G. Bomchil, A. Hüller, R. Rayment, S.J. Roser, M.V. Smalley, R.K. Thomas, and J.W. White, Philos. Trans. Roy. Soc. London, Ser. B, 1980, 290, 537.

6. M.F. Toney, R.D. Diehl, and S.C. Fain, Jr., Phys. Rev. B, 1983, 27, 6413. At submonolayer coverages, the unit cell is slightly oblique (centered-parallelogram).

7. G.J. Trott, H. Taub, F.Y. Hansen, and H.R. Danner, Chem. Phys. Lett., 1981, 78, 504.

8. R.D. Diehl and S.C. Fain, Jr., Surf. Sci., 1983, 125, 116.

9. R. Wang, S.-K. Wang, H. Taub, J.C. Newton, and H. Shechter, Phys. Rev., 1987, 35, 5841.

10. G.J. Trott, Ph.D. thesis, University of Missouri-Columbia, 1981 (unpublished).

11. J. Suzanne, J.L. Seguin, H. Taub, and J.P. Biberian, Surf. Sci., 1983, 125, 153.

12. R. Wang, H. Taub, H. Shechter, R. Brener, J. Suzanne, and F.Y. Hansen, Phys. Rev., 1983, 27, 5364.

13. J. Krim, J. Suzanne, H. Shechter, R. Wang, and H. Taub, Surf. Sci., 1985, 162, 446.

14. J.C. Newton, J.R. Dennison, S.-K. Wang, R. Wang, H. Taub, E. Conrad, and H. Shechter, Bull. Am. Phys. Soc., 1987, 32, 467, and to be published.

15. N_2O_2 is believed to form an HB structure which is incommensurate with the graphite basal plane (J.P. Coulomb, J. Suzanne, M. Bienfait, M. Matecki, A. Thomy, B. Croset, and C. Marti, J. Phys. (Paris), 1980, 41, 1155.

16. R.D. Diehl, 'Adsorption Studies Using Low-energy Electron Diffraction,' in this proceedings.

17. A.B. Harris and A.J. Berlinsky, Can. J. Phys., 1979, 57, 1852.

18. J. Talbot, D.J. Tildesley, and W.A. Steele, Mol. Phys., 1984, 51, 1331.

19. L.W. Bruch, J. Chem. Phys., 1983, 79, 3148.

20. J.K. Kjems, L. Passell, H. Taub, J.G. Dash, and A.D. Novaco, Phys. Rev. B, 1976, 13, 1446.

21. J. Eckert, W.D. Ellenson, J.B. Hastings, and L. Passell, Phys. Rev. Lett., 1979, 43, 1329.

22. K. Morshige, C. Mowforth, and R.K. Thomas, Surf. Sci., 1985, 151, 289.

23. J. Talbot, D.J. Tildesley, and W.A. Steele, Surf. Sci., 1986, 169, 71.

24. J.P. Coulomb, J.P. Biberian, J. Suzanne, A. Thomy, G.J. Trott, H. Taub, H.R. Danner, and F.Y. Hansen, Phys. Rev. Lett., 1979, 43, 1878.

25. J.P. Biberian, J.P. Coulomb, J. Suzanne, G.J. Trott, H. Taub, F.Y. Hansen, and H.R. Danner, in Proc. ICSS-4 and ECOSS-3, Cannes 1980 (Le Vide, Les Couches Minces 201, 1980, I, 126).

26. F.Y. Hansen, R. Wang, H. Taub, H. Shechter, D.G. Reichel, H.R. Danner, and G.P. Alldredge, Phys. Rev. Lett., 1984, 53, 572.

27. F.Y. Hansen and H. Taub, J. Chem. Phys., in press.

28. A.E. Kitaigorodskii, <u>Molecular Crystals</u>, (Academic:New York) 1973.
29. See, e.g., M. Bienfait, <u>Surf. Sci.</u>, **1985**, **162**, 411, and references cited therein.
30. J.R. Dennison, H. Taub, F.Y. Hansen, H. Shechter, and R. Brener, <u>Bull. Am. Phys. Soc.</u>, **1987**, **32**, 434 (1987), and to be published.
31. J.-M. Gay, J. Suzanne, and R. Wang, <u>J. Chem. Soc., Faraday Trans. 2</u>, **1986**, **82**, 1669, (Faraday Symposium 20).
32. S.-K. Wang, J.C. Newton, R. Wang, H. Taub, J.R. Dennison, and H. Shechter, <u>Bull. Am. Phys. Soc.</u>, **1987**, **32**, 433, and to be published.
33. Q.M. Zhang, H.K. Kim, and M.W.H. Chan, <u>Phys. Rev. B</u>, **1986**, **33**, 413.
34. R.D. Diehl and S.C. Fain, Jr., <u>Phys. Rev. B</u>, **1982**, **26**, 4785.
35. C. Peters and M.L. Klein, <u>Phys. Rev. B</u>, **1985**, **32**, 6077.
36. B. Kuchta and R.D. Etters, preprint.
37. J.W. Osen and S.C. Fain, Jr., <u>Phys. Rev. B Rapid Comm.</u>, in press.
38. J.W. Osen, <u>Ph.D. thesis, University of Washington</u>, **1987** (unpublished).
39. M.H.W. Chan, A.D. Migone, K.D. Miner, and Z.R. Li, <u>Phys. Rev. B</u>, **1984**, **30**, 2681.
40. J.M. Gay, J. Suzanne, and R. Wang, <u>J. Physique Lett.</u>, **1985**, **46**, L425.
41. J.P. Couloumb and M. Bienfait, <u>J. Physique</u>, **1986**, **47**, 89.
42. D.R. Nelson and B.I. Halperin, <u>Phys. Rev. B</u>, **1979**, **19**, 2457.
43. A.P. Young, <u>Phys. Rev. B</u>, **1979**, **19**, 1855.
44. S. Ostlund and B.I. Halperin, in <u>Ordering in Two Dimensions</u>, edited by S.K. Sinha, (North Holland:New York) 1980, p. 343.
45. Ibid., <u>Phys. Rev. B</u>, **1981**, **23**, 235.
46. H. Taub, K. Carneiro, J.K. Kjems, L. Passell, and J.P. McTague, <u>Phys. Rev. B</u>, **1977**, **16**, 4551.
47. H. Taub, in <u>Vibrational Spectroscopies for Adsorbed Species</u>, edited by A.T. Bell and M.L. Hair, <u>ACS Symposium Series, No. 137</u> (Am. Chem. Soc.:Washington, D.C.) 1980, p. 247.
48. J.M. Szeftel, S. Lehwald, H. Ibach, T.S. Rahman, J.E. Black, and D.L. Mills, <u>Phys. Rev. Lett.</u>, **1983**, **51**, 268.
49. K.D. Gibson and S.J. Sibener, <u>Phys. Rev. Lett.</u>, **1985**, **55**, 1514.
50. R. Wang, H.R. Danner, and H. Taub, in Ref. 44, p. 219.
51. J.R. Dennison, S.-K. Wang, H. Taub, J.Z. Larese, L. Passell, and J.M. Hastings (unpublished).
52. H.L. McMurry and F.Y. Hansen, <u>J. Chem. Phys.</u>, **1980**, **72**, 5540.
53. R.J. Gooding, B. Joos, and B. Bergersen, <u>Phys. Rev. B</u>, **1983**, **27**, 7669.
54. G. Vidali and M.W. Cole, <u>Phys. Rev. B</u>, **1984**, **29**, 6376.

CHAPTER 28

NON-BONDED INTERMOLECULAR INTERACTIONS AND THEIR MODIFICATION IN THE PRESENCE OF A SURFACE

F.Y. Hansen
Fysisk-Kemisk Institut 206 DTH
DK 2800 Lyngby
Denmark

ABSTRACT. The representation of non-bonded molecular interactions are discussed, and the physical nature of the London dispersion energy is considered in detail. It is expressed in terms of molecular- and electric field susceptibilities at imaginary frequencies. The analysis is based on elementary quantum mechanics and classical electrodynamics and assumes no knowledge of quantum field theory, which is an alternative approach to the problem. The method, suggested by A.D. McLachlan, gives a very good physical insight into the nature of the interaction. General expressions for the dispersion energy, which include retardation effects, are given for two and three molecules. The modification of the intermolecular dispersion energy, caused by scattering of radiation from a solid dielectric, is studied and McLachlan's result for isotropic molecules is extended to axial anisotropic molecules. The effect is relevant for molecules adsorbed on surfaces, and the results for ethane and butane molecules adsorbed on the basal planes of graphite are given. They show that the modification is rather small at typical intermolecular distances and therefore only of minor importance for the film structures and the excitations in the films.

I. INTRODUCTION.

The interpretation of experimental data on the structures or excitations in various condensed matter often depends on a theoretical analysis and computation of the observed data. This requires a knowledge of the interactions between the particles in the systems. A particular class of interactions are the relatively weak van der Waals energies between non-bonded atoms or molecules. Since this kind of interaction determines the properties of a large group of systems, much work has been done to determine van der Waals parameters of various atom combinations by, for example, molecular mechanics calculations [1].

In this paper we review an elementary and interesting de-

G. J. Long and F. Grandjean (eds.), The Time Domain in Surface and Structural Dynamics, 499–533.
© 1988 by Kluwer Academic Publishers.

rivation [2] of the van der Waals dispersion interaction, which gives a good physical insight into the nature of the interaction. It is based on elementary quantum mechanics and classical electrodynamics and requires no knowledge of quantum field theory, which is an alternative approach to the problem. The interaction is treated as a perturbation and linear response theory is used to derive a relation between the second order perturbation energy and the molecular polarizability at imaginary frequencies. This relation is of great interest, since it shows how the second order perturbation energy of the system may be determined from the frequency dependent polarizability, which can be measured experimentally. At the same time it shows how the polarizability may be calculated theoretically. The relation is used to determine the dispersion energy of two molecules and expressions are given for distances, where retardation effects are unimportant and for distances, where they are important. The interaction between three atoms is also considered, and we find the well known result that the energy may be written as a sum of pair interactions plus a "third body" energy term. Finally, we use the formalism to derive a perhaps less well known expression for the modification of the intermolecular interaction of two molecules, when they are adsorbed on a surface. This effect was first investigated by McLachlan [2] and is formally equivalent to the "third body" effect mentioned above, where the solid now plays the role of the third body. The intermolecular pair potential is modified, and the effect is either repulsive or attractive depending on the relative molecular positions. The modification may be as large as 2/3 of the intermolecular pair potential and can therefore be important for the structures of and excitations in 2 dimensional films of adsorbed molecules and atoms. We have extended McLachlan's result for isotropic molecules to anisotropic molecules like ethane and butane and give expressions for the modification of the intermolecular potential for different molecular orientations. Numerical calculations of the effect for the two hydrocarbons adsorbed on graphite are presented.

II. THE REPRESENTATION OF NON-BONDED INTERMOLECULAR INTERACTIONS.

The dominant interactions between non-polar and non-bonded atoms are the van der Waals interactions. Because of their special nature rather simple models for the interaction energy as a function of the distance between the atoms can be constructed. Various analytical expressions have been adopted to represent the atom-atom potential, $\phi(r)$, and the most common ones are included in the following expression [3]

$$(II.1) \quad \phi(r) = \frac{v_o(n+\lambda)}{(n+\lambda)-6} *$$

$$* \left[- (r_o/r)^6 + \frac{6}{(n+\lambda)} (r_o/r)^n \exp(\lambda(1-r/r_o)) \right]$$

r is the distance between two atoms and the four parameters v_o, r_o, n and λ have a direct physical meaning. It is easy to show that v_o is the depth of the potential well, r_o the equilibrium distance and n,λ regulate the steepness of the repulsive part of the potential. It is common to all analytic expressions of the van der Waals energy that the leading term in the attractive part of the energy has a r^{-6} dependence on r whereas the repulsive part has been given a variety of different representations as indicated in (II.1). It is a result of electron overlap of the two atoms, when they are brought close together, and quantum calculations seem to favor an exponential dependence on r. The r^{-6} dependence of the attractive part is a result of induced dipole-dipole interactions, when retardation effects are unimportant, which is often the case. Induced higher multipole interactions decay more rapidly.

When $\lambda \neq 0$ and $n \neq 0$ the r dependence of the repulsive part of the potential is described by a combination of a power law and an exponential function. If $\lambda=0$ the term varies like r^{-n} and with n=12, we get the well known Lennard-Jones potential

$$(II.2.a) \quad \phi_{ij}(r) = \varepsilon_{ij} [(r_{oij}/r)^{12} - (r_{oij}/r)^6]$$

With n=0 we obtain the Buckingham (exp-6) potential, where the repulsive part of the potential is described by an exponential

$$(II.2.b) \quad \phi_{ij}(r) = B_{ij} \exp(-C_{ij} r) - A_{ij} r^{-6}$$

The parameters ε_{ij}, r_{oij} and A_{ij}, B_{ij}, and C_{ij} are characteristic for the pair of atoms i,j.

The physical significance of the parameters in the Buckingham potential is obscured, and it may be useful to express them in terms of the more physical meaningful parameters of (II.1). From (II.1) and (II.2.b) we find

$$(II.2.c) \quad A = \frac{v_o \lambda r_o^6}{\lambda - 6} \quad ; \quad B = \frac{6v_o}{\lambda - 6} \exp(\lambda) \quad ; \quad C = \frac{\lambda}{r_o} \quad ;$$

Intermolecular interactions are often calculated by breaking up the molecule into force centers, usually with each atom in the molecule as a center, and adding the interactions between the centers in different molecules. On this basis many atom-atom pair parameters (II.2) have been determined by the so called molecular packing calculations or molecular mechanics calculations, where a variety of structural and thermodynamic data of molecular systems with the relevant atoms have been used to determine the parameters by least squares fit calculations [1], [3]. When many different systems have been used in a determination one expects that the derived parameters are transferable, at least to some degree, to other systems than those used in their determination. This of course is a highly desirable property and makes the parameters very useful in the analysis of a given system. A less favorable feature is that different systems are treated on an equal basis although there may be individual differences. Recently we came across an example [4]. In an analysis of adsorbed monolayers of ethane on graphite it was speculated that electrostatic forces might be of importance for the structures. The evidence of the charge distribution in ethane is given by the non-zero value of the quadrupole moment, hexadecapolemoment etc. (no dipole moment due to the molecular symmetry). We used a set of atomic parameters determined from molecular packing calculations with the potential [5]

$$(II.3) \qquad \phi_{ij}(r) = B_{ij}\exp(-C_{ij}r) - A_{ij}/r^6 + q_iq_j/r$$

where the electrostatic interactions are represented by point charges q_i at the atom sites. Our analysis showed that an atom site point charge representation of the charge distribution in the molecule is inadequate and gave a quadrupole moment of the wrong magnitude and sign and, even worse, gave a repulsive contribution to the intermolecular energy rather than an attractive contribution, when the proper electrostatic interaction was evaluated. The parameters in (II.3) are correlated and there is a certain capacity for compensation of the erroneous model for the electrostatic interactions. This demonstrates that different sets of parameters should not be combined and that deductions other than structural from potentials like (II.3) should be considered with great care. We shall in the following study the attractive part of the potentials.

The perturbation energy of two quantum systems.

Let two quantum systems A and B interact with an interaction Hamiltonian H'

$$(II.4) \qquad H' = - \sum_i x_1 y_i$$

where x_i and y_i are physical quantities belonging to A and B respectively. It is assumed that H' is a small perturbation and that the systems prior to the perturbation are in their ground states $|a_o\rangle$ and $|b_o\rangle$, where the expectation values of x_i and y_i are zero

$$(II.5) \qquad \langle a_o|x_i|a_o\rangle = (x_i)_{a_o a_o} = 0; \quad \langle b_o|y_i|b_o\rangle = (y_i)_{b_o b_o} = 0;$$

From ordinary time independent perturbation theory [6], the first order perturbation energy is given by

$$(II.6) \qquad \langle a_o b_o|H'|a_o b_o\rangle = -\sum_i \langle b_o|y_i|b_o\rangle\langle a_o|x_i|a_o\rangle = 0$$

according to (II.5). The first non-zero and leading contribution to the interaction energy is given by the second order perturbation energy w_2 [6]

$$(II.7) \qquad w_2 = \sum_{a,b} \frac{\langle a_o b_o| \sum_i x_i y_i |ab\rangle\langle ba| \sum_k x_k y_k |a_o b_o\rangle}{E_{a_o} + E_{b_o} - E_a - E_b}$$

$$= -\sum_{a,b} \frac{[\sum_i (x_i)_{a_o a} (y_i)_{b_o b}][\sum_k (x_k)_{aa_o} (y_k)_{bb_o}]}{\hbar\omega_{aa_o} + \hbar\omega_{bb_o}}$$

This expression requires a knowledge of the ground states and excited states of the systems, data which are not easily accessible. It is therefore of great interest and importance that w_2 can be expressed in terms of system susceptibilities, which may be determined experimentally. Linear response theory is used to derive this relation, which also shows how the susceptibilities may be calculated theoretically, see for example ref 7.

Linear Response Theory.

The aim of linear response theory is to calculate the response, as a function of time, of all observable quantities of a system, when it is exposed to a time-dependent or static perturbation. Let the Hamiltonian of the unperturbed system be H_o and the total Hamiltonian H(t)

$$(II.8) \qquad H(t) = H_o + H'(t)$$

where H'(t) is small and linear in the perturbation strength $F_i(t)$, which couples to system variable x_i

(II.9) $\qquad H'(t) = - \sum_i x_i \, F_i(t)$

Then, in the framework of linear response theory [8], the average response $\overline{x_i(t)}$ at time t is

(II.10) $\qquad \overline{x_i(t)} = \sum_k \int_0^\infty d\tau \; \phi_{ik}(\tau) \, F_k(t-\tau)$

where $\phi_{ik}(t)$ is the response function and given by the commutator relation [8]

(II.11) $\qquad \phi_{ik}(t) = \dfrac{i}{\hbar} \, \langle [\hat{x}_i(t+\tau), \hat{x}_k(\tau)] \rangle$

$\hat{x}_i$ is the operator corresponding to x_i and the bracket indicates a statistical average over a canonical ensemble of the unperturbed system. It is inferred in (II.11) that the response-function is stationary and therefore independent of τ, a property easily proven by developing the commutator. If the perturbation $F_i(t)$ is resolved into Fourier components

(II.12) $\qquad F_i(t) = \int_{-\infty}^\infty d\omega \, f_{i\omega} \, e^{-i\omega t}$

the response of the system may be written

(II.13) $\qquad \overline{x_i(t)} = \sum_k \int_{-\infty}^\infty d\omega \, \alpha_{ik}(\omega) \, f_{k\omega} \, e^{-i\omega t}$

where

(II.14) $\qquad \alpha_{ik}(\omega) = \int_0^\infty d\tau \, \phi_{ik}(\tau) \, e^{i\omega\tau}$

is the frequency dependent susceptibility of the system. It is important to note that it is given as a "one sided" fourier transform of the response function, a direct reflection of causality (cf. (II.10)). Let us suppose that the unperturbed system is in a pure state $|n\rangle$. Then the susceptibility is given by (II.11) and (II.14)

$$(II.15) \quad \alpha_{ik}(\omega) = \frac{i}{\hbar} \sum_m \int_0^\infty dt \ [(x_i)_{nm}(x_k)_{mn} \ e^{i(\omega_{mn}-\omega)t} -$$

$$(x_k)_{nm}(x_i)_{mn} \ e^{i(\omega_{nm}-\omega)t}]$$

where the closure property $\sum_m |m\rangle\langle m| = 1$ has been used. The integration over t may be carried out, when we introduce the integral representation of the "one sided" deltafunction $\delta_-(x)$

$$(II.16) \quad \delta_-(x) = -\frac{i}{\pi} P(1/x) + \delta(x) = \frac{1}{\pi} \int_0^\infty dk \ e^{-ikx}$$

and we find

$$(II.17) \quad \alpha_{ik}(\omega) = \frac{P}{\hbar} \sum_m [\frac{(x_i)_{nm} \ (x_k)_{mn}}{\omega_{mn} - \omega} + \frac{(x_k)_{nm} \ (x_i)_{mn}}{\omega_{mn} + \omega}] +$$

$$+ \frac{i\pi}{\hbar} \sum_m [(x_i)_{nm}(x_k)_{mn}\delta(\omega_{nm}+\omega) - (x_k)_{nm}(x_i)_{mn}\delta(\omega_{mn}+\omega)]$$

$$= \alpha'_{ik}(\omega) + i\alpha''_{ik}(\omega)$$

P refers to the Principal value. If we let the initial pure state $|n\rangle = |a_0\rangle$, the ground state of system A, we may write

$$(II.18) \quad \alpha_{ik}^{A''}(\omega) = \frac{\pi}{\hbar} \sum_a (x_i)_{a_0 a} (x_k)_{aa_0} \ \delta(\omega + \omega_{a_0 a})$$

since $\omega_{aoa} < 0$. For B we find

$$(II.19) \quad \alpha_{ik}^{B''}(\omega) = \frac{\pi}{\hbar} \sum_b (y_i)_{b_0 b} (y_k)_{bb_0} \ \delta(\omega + \omega_{b_0 b})$$

This is now introduced in (II.7), and we find

$$(II.20) \qquad w_2 = - \frac{\hbar}{\pi^2} \sum_i \sum_k \int_0^\infty d\omega_1 \int_0^\infty d\omega_2 \; \frac{\alpha_{ik}^{A''}(\omega_1) \; \alpha_{ik}^{B''}(\omega_2)}{\omega_1 + \omega_2}$$

which expresses the perturbation energy w_2 in terms of the imaginary parts of the susceptibilities.

w_2 may also be expressed in terms of the susceptibilities themselves, when the basic symmetry relations for the causal functions ϕ and α are used [8], [9]. We summarize these in the following. From (II.11) it is seen that

$$(II.21.a) \; \phi_{ik}(t) = - \phi_{ki}(-t)$$

and if the operator $\hat{x}_i$ is Hermitian

$$(II.21.b) \; \phi_{ik}(t) = - \frac{2}{\hbar} \sum_m Im \; \{(x_i)_{nm} \; (x_k)_{mn} \; e^{i\omega_{nm}t}\}$$

which shows that the response function of a physical system is real as expected.

The definition of the susceptibility in (II.14) for real frequencies ω may be extended to complex frequencies ξ by

$$\alpha_{ik}(\xi) = \int_0^{\pm\infty} d\tau \; \phi_{ik}(\tau) \; e^{i\xi\tau}$$

$+$ for ξ in the upper half of the complex plane
$-$ for ξ in the lower half of the complex plane

Hence

$$(II.21.c) \; \alpha_{ik}(\xi) = \alpha_{k1}(-\xi)$$

and for a Hermitian operator ($\phi_{1k}(t)$ is real)

$$(II.21.d) \; \alpha_{ik}^*(\xi) = \alpha_{ik}(-\xi^*) = \alpha_{k1}(\xi^*)$$

which shows that α_{ik} is a real function along the imaginary axis. The Cauchy integral formula [10] may be used to derive the well known Kramers-Kronig relations between the real and imaginary parts of α and of particular interest here is the relation

$$(II.21.e) \quad \alpha_{ik}(\xi) = \frac{1}{\pi} \int_{-\infty}^{\infty} \frac{\alpha_{ik}''(\omega)}{\omega - \xi} \, d\omega$$

which shows that α has a discontinuity on crossing the real axis

$$(II.21.f) \quad \alpha_{ik}(\omega+i\varepsilon) - \alpha_{ik}(\omega-i\varepsilon) \xrightarrow[\varepsilon \to 0]{} 2i\alpha_{ik}''(\omega)$$

This completes the review of the symmetry properties of the response function and the susceptibility. (II.21.f) is now used to replace $\alpha_{ik}^{B}(\omega_2)$ in (II.20) with α_{ik}^{B} and the integration of the resulting two terms over ω_2 is done as a contour integration in the 1. and the 4. quadrant. The contour is taken along the positive ω_2 axis and along respectively the positive and negative parts of the imaginary axis and closed at infinity, where the integrand vanishes. The result is an integral of the form

$$\int_0^\infty d\omega_1 \int_0^\infty d\omega \frac{\alpha_{ik}^{A}{}''(\omega_1) \, \alpha_{ik}^{B}(i\omega)}{\omega_1 + i\omega}$$

$$- \int_0^\infty d\omega_1 \int_0^\infty d\omega \frac{\alpha_{ik}^{A}{}''(\omega_1) \, \alpha_{ik}^{B}(i\omega)}{-\omega_1 + i\omega}$$

ω_1 is now substituted by $-\omega_1$ in the latter term, and with (II.21.d), it may be combined with the first term to give an integral identical to the one in (II.21.e). We find

$$(II.22.a) \quad w_2 = - \frac{\hbar}{2\pi} \sum_i \sum_k \int_0^\infty d\omega \, \alpha_{ik}^{A}(i\omega) \, \alpha_{ki}^{B}(i\omega)$$

and in tensor notation

$$(II.22.b) \quad w_2 = - \frac{\hbar}{2\pi} \int_0^\infty d\omega \, Tr[\underline{\alpha}^{A}(i\omega) \, \underline{\alpha}^{B}(i\omega)]$$

This is the central relation between the second order perturbation energy and the system susceptibilities. Note, they are taken along the imaginary axis, where they are real according to (II.21.b).

The dispersion energy of two molecules.

Let us apply the expression for w_2 to the interaction between a non-polar molecule and a classical electric field. In the dipole approximation for the interaction, the perturbation H'(t) is

(II.23) $H' = -p(t) \cdot E(r) = -\sum_i p_i E_i(r)$

where $p(t)$ is the molecular dipole and $E(r,t)$ the electric field at the molecular center r. If we in (II.4) identify p_i with x_i and E_i with y_i, $\alpha_{ik}^A(i\omega)$ in (II.22) is just the molecular polarizability and $\alpha_{ki}^B(i\omega)$ the electric field susceptibility. The physical significance of the molecular polarizability and the electric field susceptibility is sketched in fig.1. The molecular polarizability determines the dipole moment p induced by the electric field E, and the electric field susceptibility determines the electric field at a point r from a dipole p at r_0. All relations are given for imaginary frequencies.

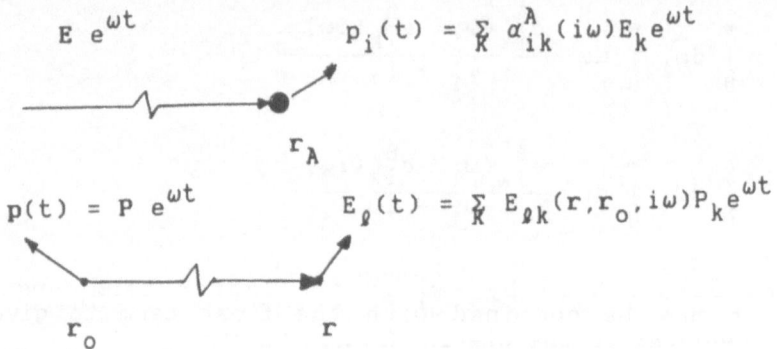

Figure 1. Definition of the polarizability $\underline{\alpha}$ of a molecule at r_A and the electric field susceptibility $\underline{E}$.

When a polarizable molecule A is introduced at r_A, the electric field susceptibility is changed. This is seen by determining the electric field at r from a dipole at r_0. We find from fig.2

1) $E_\ell(r,r_0) = \sum_k E_{\ell k}(r,r_0,i\omega) P_k(r_0)$

2) $P_i(r_A) = \sum_j \alpha^A_{ij}(i\omega)E_j(r_A,r_o)$

$= \sum_j \sum_k \alpha^A_{ij}(i\omega)E_{jk}(r_A,r_o,i\omega)P_k(r_o)$

3) $E_\ell(r,r_A) = \sum_i E_{\ell i}(r,r_A,i\omega) P_i(r_A)$

$= \sum_i \sum_j \sum_k E_{\ell i}(r,r_A,i\omega) \alpha^A_{ij}(i\omega) E_{jk}(r_A,r_o,i\omega) P_k(r_o)$

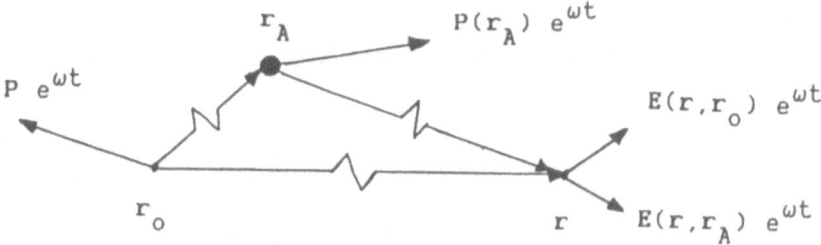

Figure 2. The electric field at r with atom A at r_A.

In the absence of A, the field susceptibility is seen to be

$$E_{\ell k}(r,r_o,i\omega)$$

and in the presence of A

$$E_{\ell k}(r,r_o,i\omega) + \sum_i \sum_j E_{\ell i}(r,r_A,i\omega) \alpha^A_{ij}(i\omega) E_{jk}(r_A,r_o,i\omega)$$

These results are now used to calculate the dispersion energy of two molecules. The energy of A in its own field (the self energy) is by (II.22)

(II.24) $w_A = - \dfrac{\hbar}{2\pi} \sum_k \sum_\ell \int_0^\infty d\omega\; \alpha^A_{k\ell}(i\omega) E_{\ell k}(r_A,r_A,i\omega)$

When another molecule B is introduced the field susceptibility at r_A is changed to

$$E_{\ell k}(r_A, r_A, i\omega) + \sum_i \sum_j E_{\ell i}(r_A, r_B, i\omega)\, \alpha^B_{ij}(i\omega)\, E_{jk}(r_B, r_A, i\omega)$$

and the self energy of A is

$$(II.25) \qquad w_A = -\frac{\hbar}{2\pi} \sum_\ell \sum_k \int_0^\infty d\omega\ \alpha^A_{k\ell}(i\omega)\, E_{\ell k}(r_A, r_A, i\omega) -$$

$$-\frac{\hbar}{2\pi} \sum_\ell \sum_k \sum_i \sum_j \alpha^A_{k\ell}(i\omega) E_{\ell i}(r_A, r_B, i\omega)\alpha^B_{ij}(i\omega)E_{jk}(r_B, r_A, i\omega)$$

The dispersion energy w_{AB} is now the difference between the two self energies of A in (II.25) and (II.24)

$$(II.26) \qquad w_{AB} = -\frac{\hbar}{2\pi}\int_0^\infty d\omega\ \mathrm{Tr}[\underline{\underline{\alpha}}^A(i\omega)\underline{\underline{E}}(r_A, r_B, i\omega)\ *$$

$$*\ \underline{\underline{\alpha}}^B(i\omega)\underline{\underline{E}}(r_B, r_A, i\omega)]$$

This is the general expression for the dispersion energy between two molecules with arbitrary polarizability tensors and retardation effects included. These considerations may easily be extended to any number of particles. For three molecules A,B and C, we find after extensive manipulations

$$(II.27.a)\quad w = w_{AB} + w_{AC} + w_{BC} + w_{ABC}$$

where the non-additive three body interaction w_{ABC} is given by

$$(II.27.b)\quad w_{ABC} = -\frac{\hbar}{2\pi}\int_0^\infty d\omega\ \mathrm{Tr}\big[\underline{\underline{\alpha}}^C(i\omega)\underline{\underline{E}}(r_C, r_B, i\omega)\ *$$

$$*\ \underline{\underline{\alpha}}^B(i\omega)E(r_B, r_A, i\omega)\underline{\underline{\alpha}}^A(i\omega)\underline{\underline{E}}(r_A, r_C, i\omega) +$$

$$+\ \underline{\underline{\alpha}}^C(i\omega)\underline{\underline{E}}(r_C, r_A, i\omega)\ *$$

$$*\ \underline{\underline{\alpha}}^A(i\omega)\underline{\underline{E}}(r_A, r_B, i\omega)\underline{\underline{\alpha}}^B(i\omega)\underline{\underline{E}}(r_B, r_C, i\omega)\big]$$

This is the well known result that the interaction between three atoms is not just given as a sum over pair interactions but a term depending on all molecules must be added. The various terms in (II.27.a) are sketched in fig.3.

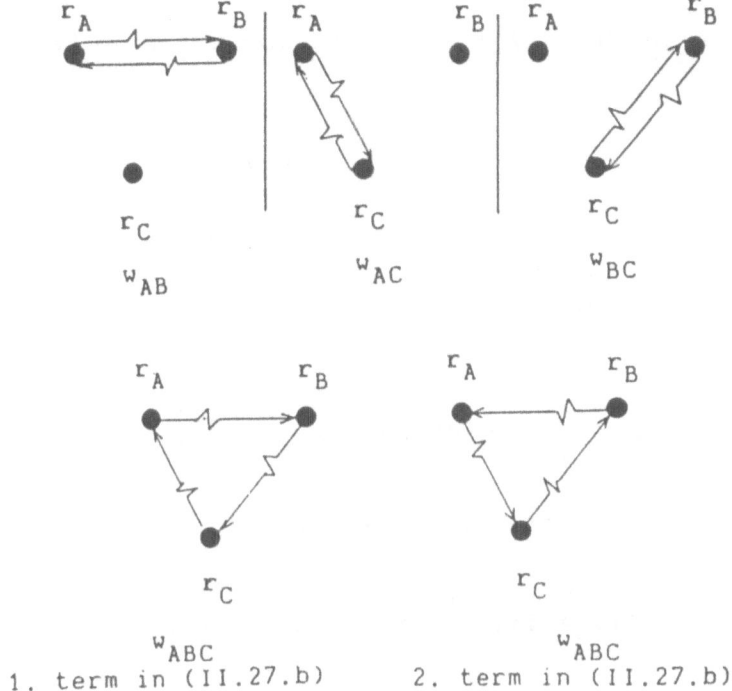

1. term in (II.27.b)   2. term in (II.27.b)

Figure 3. A sketch of the various terms in (II.27)

The electric field susceptibility.

In addition to molecular polarizabilities the electric field susceptibility tensor $\underline{\underline{F}}$ must be known. It is found by a calculation of the electric field at a point r from a dipole $p(t)$ at r_o
(see fig.1). When the dipole distribution

$$(II.28) \quad p(t) = P\, e^{\omega t}\, \delta(r - r_o)$$

is known, it is convenient to use Maxwell's equations in a form
[11], where the current density j and charge distribution ρ are
given in terms of the moment vector $p(t)$

$$j = \frac{\partial p(t)}{\partial t} \qquad \text{and} \qquad \rho = -\text{div } p(t)$$

and the vector potential **A** and scalar potential ϕ in terms of the Hertz vector **Π**

$$A = \frac{1}{c^2} \frac{\partial \Pi}{\partial t} \qquad \text{and} \qquad \phi = -\text{div } \Pi$$

In this formulation the equation of motion for **Π** is

$$(II.29) \qquad \nabla^2 \Pi - \frac{1}{c^2} \frac{\delta^2 \Pi}{\delta t^2} = -4\pi P \, e^{\omega t} \, \delta(r-r_0)$$

where the dipole distribution directly appears in the equation. The electric field is given by

$$(II.30) \qquad E = \text{curl curl } \Pi - 4\pi P \, \delta(r-r_0)$$

(II.29) is solved by the standard Greens function method [12], and we find

$$\Pi = \frac{e^{-kR}}{R} P \, e^{\omega t}$$

where $\qquad k = \omega/c \qquad$ and $\qquad R = |r - r_0|$

The solution is used in (II.30), and the electric field is

$$(II.31) \qquad E_1 = \sum_j [\nabla_i \nabla_j - \delta_{ij} \nabla^2] \frac{e^{-kR}}{R} P_j$$

and from fig.1

$$(II.32) \qquad E_{ij}(r,r_0,i\omega) = [\nabla_i \nabla_j - \delta_{ij} \nabla^2] \frac{e^{-kR}}{R}$$

$$= [\hat{x}_i \hat{x}_j (3+3kR+k^2R^2) - \delta_{ij}(1+kR+k^2R^2)] \frac{e^{-kR}}{R^3}$$

where $\hat{x}_i$ are the direction cosines of the R vector.

This result may be used in the equations for w_{AB} (II.26) and w_{ABC} (II.27.b) to give a general expression for the dispersion energy of two and three molecules. For isotropic molecules, where the polarization tensor is diagonal with identical elements, we find

$$(II.33) \quad w_{AB} = - \frac{\hbar}{2\pi} \frac{1}{R^6} \int_0^\infty d\omega \; \alpha^A(i\omega) \; \alpha^B(i\omega) \; *$$

$$* \; [\; 2(kR)^4 + 4(kR)^3 + 10(kR)^2 + 12(kR) + 6] \; e^{-2kR}$$

In the limit of small R, such that kR<<1 when $\alpha(i\omega) \neq 0$, the dispersion energy is given by

$$(II.34.a) \quad w_{AB} = - \frac{3\hbar}{\pi} \frac{1}{R^6} \int_0^\infty d\omega \; \alpha^A(i\omega) \; \alpha^B(i\omega)$$

which is the well known expression for the dispersion energy of two isotropic molecules with neglect of retardation effects (see (II.1) and (II.2)). At the other extreme, where R is large, such that kR>>0, we may replace the dynamical polarizabilities with the static values. By a simple integration of (II.33) we find

$$(II.34.b) \quad w_{AB} \xrightarrow{R \to \infty} - \frac{23 \; \hbar \; c}{4\pi} \frac{\alpha^A(0) \; \alpha^B(0)}{R^7}$$

When retardation effects are unimportant we find the well known R^{-6} dependence of the dispersion energy (II.34.a) and a R^{-7} dependence (II.34.b), when they are important, a result in agreement with a quantum field calculation [13] of the interaction. The expression for w_{ABC} in (II.27.b) may be evaluated under the same assumptions, which led to (II.34.a) for the pair interaction, and we find after lengthy manipulations the well known expression,

$$(II.35.a) \quad w_{ABC} = - \frac{\hbar}{\pi} \frac{(3 + 9\cos\theta_1 \cos\theta_2 \cos\theta_3)}{R_1^{\;3} \; R_2^{\;3} \; R_3^{\;3}} \; *$$

$$* \int_0^\infty d\omega \; \alpha^A(i\omega) \; \alpha^B(i\omega) \; \alpha^C(i\omega)$$

where the parameters are defined in fig.4.

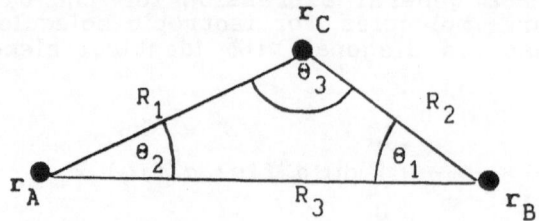

Figure 4. Definition of the variables in (II.35.a).

We note that the three body energy decays faster than the pair potential and is normally small. An expression for w_{ABC} with the same assumptions, which led to (II.34.b) for the pair interaction, is rather complicated and not given here. Of particular interest is the R-dependence of the expression, and it is found to be of the type

$$(II.35.b) \quad \frac{R_1^i \; R_2^j \; R_3^\ell}{R_1^3 \; R_2^3 \; R_3^3} \qquad \frac{1}{(R_1 + R_2 + R_3)^{i+j+\ell+1}}$$

with

$$i, j, \ell \;=\; 0, 1, 2.$$

III. THE DISPERSION ENERGY IN THE PRESENCE OF A SOLID SURFACE.

It was shown that the dispersion energy is related to the change in the electric field susceptibility caused by scattering of radiation. We shall now determine this change, when the radiation is scattered from a solid isotropic semi-infinite dielectric. The system is sketched in fig.5

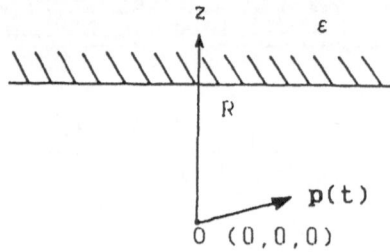

Figure 5. The geometrical arrangement of a dipole p(t) at the origin and a semi-infinite dielectric.

At the origin O is placed a point dipole p(t), and the solid with dielectric constant ε extends from z = R to infinity. The scattered radiation is found as the radiation from the dipole reflected by the dielectric. The amplitude of the reflected dipole radiation is conveniently expressed in terms of the Fresnel coefficients [12]. They are given for plane waves incident on the surface as shown in fig.6.

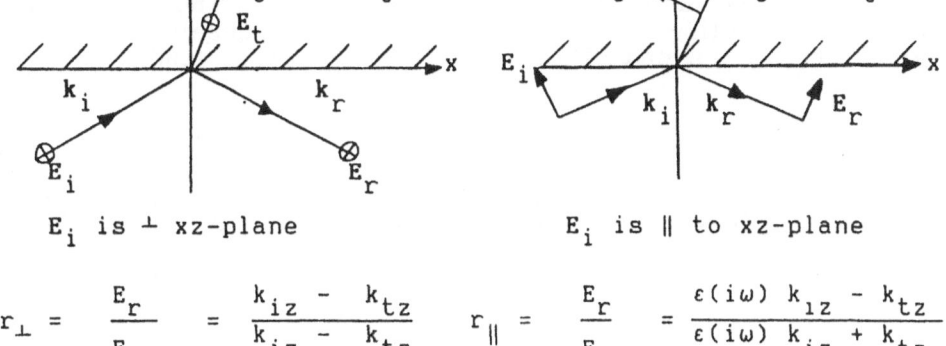

$$r_\perp = \frac{E_r}{E_i} = \frac{k_{iz} - k_{tz}}{k_{iz} - k_{tz}} \qquad r_\| = \frac{E_r}{E_i} = \frac{\varepsilon(i\omega) k_{iz} - k_{tz}}{\varepsilon(i\omega) k_{iz} + k_{tz}}$$

E_i is $\perp$ xz-plane E_i is $\|$ to xz-plane

Figure 6. Definition of the Fresnel coefficients $r_\perp$ and $r_\|$.

Accordingly, we resolve the point dipole at O into fourier-components

$$(III.1) \qquad p = P\, e^{i(q_x x + q_y y)}\, e^{\omega t}\, \delta(z)$$

which is equivalent to a sheet of dipoles in the z = 0 plane. $P\exp(\omega t)$ is the point dipole at O. The electric field at $r(x,y,z)$ from the (q_x, q_y) fourier component of the dipole distribution is (fig.1)

$$(III.2) \qquad E_i(r, q_x, q_y) = \sum_k \tilde{E}_{ik}(q_x, q_y, z, i\omega)\, P_k\, e^{i(q_x x + q_y y)}$$

The ~ indicates a fourier component of the susceptibility. A superposition of the fourier components leads to the field

$$(III.3) \quad E_i(r) = \frac{1}{4\pi^2} \sum_k \int_0^\infty \int_0^\infty dq_x dq_y \tilde{E}_{ik}(q_x,q_y,z,i\omega)P_k e^{i(q_x x+q_y y)}$$

which shows that the field susceptibility is given by

$$(III.4) \quad E_{ik}(r,i\omega) = \frac{1}{4\pi^2} \int_0^\infty \int_0^\infty dq_x dq_y \tilde{E}_{ik}(q_x,q_y,z,i\omega)e^{i(q_x x+q_y y)}$$

We start with a calculation of $\tilde{E}_{ik}$ and use the equation for the Hertz vector (II.29) with the dipole distribution given in (III.1). To simplify the derivations we limit ourselves to points along the z-axis, and with q along the x-axis we find

$$(III.5) \quad \tilde{\underline{\underline{E}}}(q,0,z,i\omega) = \begin{bmatrix} -k^2 & 0 & 0 \\ 0 & q^2 - k^2 & 0 \\ 0 & 0 & q^2 \end{bmatrix} \frac{2\pi\, e^{-kz}}{k}$$

where k is given by the dispersion relation:

$$(III.6) \quad k = \sqrt{q^2 + \omega^2/c^2}$$

With the wavevector of the incoming radiation k_i in the x-z plane, the Fresnel coefficient $r_\parallel$ is associated with the x,z components and $r_\perp$ with the y component of the electric field. The change in the electric field susceptibility caused by the surface is then easily found to be

$$(III.7) \quad \tilde{\underline{\underline{E}}}'(q,0,z,i\omega) = \begin{bmatrix} k^2 r_\parallel & 0 & 0 \\ 0 & -\omega^2/c^2\, r_\perp & 0 \\ 0 & 0 & (k^2 - \omega^2/c^2)r_\parallel \end{bmatrix} \frac{2\pi\, e^{(-2kR + kz)}}{k}$$

where $r_\parallel$ and $r_\perp$ are given in fig.6. To simplify the result further, let us consider the change in susceptibility at the origin

(0,0,0) and integrate over all fourier components:

(III.8) $\underset{=}{E}'(0,i\omega) = \dfrac{1}{4\pi^2} \int\limits_{-\infty}^{\infty}\int\limits_{-\infty}^{\infty} dq_x \, dq_y \, \underset{=}{\tilde{E}}'(q_x,q_y,0,i\omega) =$

$$\dfrac{1}{4\pi} \int\limits_0^{\infty} \begin{bmatrix} (k^2 r_{\parallel}+(q^2-k^2)r_{\perp}) & 0 & 0 \\ 0 & (k^2 r_{\parallel}+(q^2-k^2)r_{\perp}) & 0 \\ 0 & 0 & 2q^2 r_{\parallel} \end{bmatrix} \dfrac{2\pi\, e^{-2kR}}{k} q\, dq$$

It is not possible to find an analytic expression for the integral in the general case. It is noted, however, that the integrand is only significantly different from zero for k values in the range 0 to 1/2R due to the exponential function. Also, the most important frequencies ω are of the order of magnitude ω_0, the frequency where the dielectric function of the solid shows strong absorption (the $r_{\parallel}$ and $r_{\perp}$ coefficients). If it is assumed that

(III.9) $R/c \ll 1/\omega_0$ or equivalently $\omega \ll kc$, since $k \approx 1/R$

the integrand in (III.8) may be simplified. In table I we have cal-culated some values of R/c as a function of R. Since the charac-teristic time scales of dielectric usually are in the range of 1-10^{-3} psec, we note that the condition in (III.9) is satisfied at typical molecular distances R (1-10 Angstrom) from a surface.

TABLE I.

R/A	1	5	10	50	100
R/c psec	$3.3\ 10^{-7}$	$1.7\ 10^{-6}$	$3.3\ 10^{-6}$	$1.7\ 10^{-5}$	$3.3\ 10^{-5}$

The dispersion relation (III.6) now shows that the z composant of the wavevector in both the vacuum, $k = k_{iz}$, and the dielectric, k_{tz} is approximately equal to q so

(III.10) $r_{\parallel} \approx \dfrac{\varepsilon(i\omega) - 1}{\varepsilon(i\omega) + 1}$; $r_{\perp} \approx 0$;

and the integral in (III.8) may easily be evaluated analytically

$$(III.11) \quad \underline{\underline{E}}'(0,i\omega) = \begin{bmatrix} r_{\parallel} & 0 & 0 \\ 0 & r_{\parallel} & 0 \\ 0 & 0 & 2r_{\parallel} \end{bmatrix} \frac{1}{(2R)^3}$$

corresponding to a reflected field at the origin

$$(III.12) \quad \begin{bmatrix} E'_x \\ E'_y \\ E'_z \end{bmatrix} = \frac{r_{\parallel}}{(2R)^3} \begin{bmatrix} 1 & 0 & 0 \\ 0 & 1 & 0 \\ 0 & 0 & 2 \end{bmatrix} \begin{bmatrix} P_x \\ P_y \\ P_z \end{bmatrix}$$

This is just the field from the image dipole of the one at 0 with the surface of the solid as the mirror plane. A charge q at (0,0,0) will have an image charge q' at (0,0,2R), where [12]

$$q'_i = - \frac{\varepsilon - 1}{\varepsilon + 1} q$$

The components P'_i of the image dipole are:

$$P'_x = - \frac{\varepsilon - 1}{\varepsilon + 1} P_x = -r_{\parallel} P_x$$

$$P'_y = - \frac{\varepsilon - 1}{\varepsilon + 1} P_y = -r_{\parallel} P_y$$

$$P'_z = \frac{\varepsilon - 1}{\varepsilon + 1} P_z = +r_{\parallel} P_z$$

and the field generated by this dipole is according to (II.32) with $kR \to 0$ (neglect of retardation effects) just the field in (III.12). Therefore, for systems where retardation effects are of no importance, we may calculate the reflected field at r as if it is the field from the image of the given dipole. This field may be written

$$(III.13) \quad E'_i = \pm r_{\parallel} \sum_k E_{ik}(r,r',i\omega) P_k = \sum_k E'_{ik}(r,r',i\omega) P_k$$

where E_{ik} is the susceptibility of vacuum (II.32), $r_{\parallel}$ given in (III.10) and E'_{ik} the change in the electric field susceptibility given by

$$(III.14) \quad E'_{ik}(r,r',i\omega) = \pm \frac{\varepsilon(i\omega) - 1}{\varepsilon(i\omega) + 1} E_{ik}(r,r',i\omega)$$

$$+ : k = z \quad ; \quad - : k = x,y$$

This result was derived for the reflected field at the origin, and when retardation effects are unimportant it was shown that it may be calculated as the field of the image dipole. The results in (III.14) are therefore valid at __any field point__ r. If retardation effects are important it is necessary to go back and derive the results in (III.5 - III.8) at a general field point.

The dispersion energy of two molecules A and B in the presence of the solid is found by replacing the field susceptibilities in (II.26) with the new susceptibilities

$$\underline{\underline{E}}(r_A,r_B,i\omega) + \underline{\underline{E}}'(r_A,r'_B,i\omega)$$

and

$$\underline{\underline{E}}(r_B,r_A,i\omega) + \underline{\underline{E}}'(r_B,r'_A,i\omega)$$

where r' indicates the position of the image molecule. The dispersion energy is

$$
\begin{aligned}
W_{AB} = -\frac{\hbar}{2\pi}\Bigg[&\int_0^\infty Tr[\underline{\underline{\alpha}}^A(i\omega)\underline{\underline{E}}(r_A,r_B,i\omega)\underline{\underline{\alpha}}^B(i\omega)\underline{\underline{E}}(r_B,r_A,i\omega)]d\omega + \\
&+ \int_0^\infty Tr[\underline{\underline{\alpha}}^A(i\omega)\underline{\underline{E}}(r_A,r_B,i\omega)\underline{\underline{\alpha}}^B(i\omega)\underline{\underline{E}}'(r_B,r'_A,i\omega)]d\omega + \\
&+ \int_0^\infty Tr[\underline{\underline{\alpha}}^A(i\omega)\underline{\underline{E}}'(r_A,r'_B,i\omega)\underline{\underline{\alpha}}^B(i\omega)\underline{\underline{E}}(r_B,r_A,i\omega)]d\omega + \\
&+ \int_0^\infty Tr[\underline{\underline{\alpha}}^A(i\omega)\underline{\underline{E}}'(r_A,r'_B,i\omega)\underline{\underline{\alpha}}^B(i\omega)\underline{\underline{E}}'(r_B,r'_A,i\omega)]d\omega \Bigg]
\end{aligned}
$$

(III.15)

The first term on the right hand side is the ordinary dispersion energy of the two molecules (see (II.26)) and the additional terms represent contributions due to the surface when retardation effects are unimportant. The terms are sketched in fig.7.

520

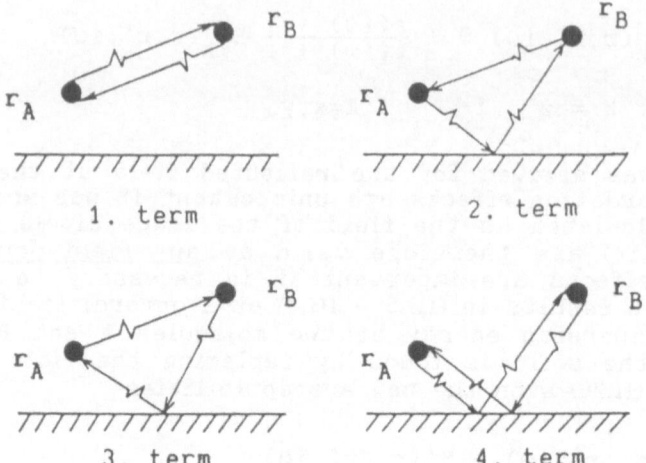

Figure 7. A sketch of the four terms in (III.15).

The result in (III.14) may also be used to calculate the dispersion energy, w_{AS}, of the adsorbed molecule, and we find

$$(III.16.a) \qquad w_{AS} = - \frac{\hbar}{2\pi} \int_0^\infty d\omega \ Tr[\underline{\underline{\alpha}}^A(i\omega)\underline{\underline{E}}'(r_A,r_A',i\omega)]$$

For an isotropic molecule we get

$$(III.16.b) \qquad w_{AS} = - \frac{\hbar}{4\pi R^3} \int_0^\infty d\omega \ \alpha(i\omega) \ \frac{\varepsilon(i\omega)-1}{\varepsilon(i\omega)+1}$$

which shows the same R dependence as when the energy is calculated by adding pair interactions [14].

The surface modification of the dispersion energy for anisotropic and isotropic molecules.

An expression for the surface modification of the dispersion energy of two isotropic molecules has been given by McLachlan [15]. We have extended these results to anisotropic molecules like ethane and butane molecules, where the polarizability tensor is characterized by two parameters $\alpha_\perp$ and $\alpha_\parallel$. $\alpha_\perp$ is the polarizability for directions perpendicular to a molecular axis along the C-atom backbone of the molecules and $\alpha_\parallel$ is the polarizability along that direction. Usually the static values of $\alpha_\perp$ and $\alpha_\parallel$ are known from experiments [16] or quantum calculations [17] and only the spherical average of the polarizability as a function of the frequency (see the next section). The spherical average, α_{av}, of the molecular polarizability may be written as

$$(III.17) \quad \alpha_{av} = \frac{1}{3} [\alpha_\perp + \alpha_\perp + \alpha_\|] = \frac{2 + \mu}{3} \alpha_\perp$$

with

$$(III.18) \quad \mu = \frac{\alpha_\|}{\alpha_\perp}$$

Since μ is unknown at non-zero frequencies, it is assumed to be independent of the frequency and given by the static value. We have considered three different orientations of the molecules to mimic the relative orientations in the ethane and butane films adsorbed on graphite [18]. For each orientation we have evaluated the latter three terms in (III.15) and obtained expressions for the surface modification of the dispersion energy, W_{AB}^S, of the molecules. The evaluation of the traces in the integrands involves rather lengthy manipulations, so we will just give the results here and omit details.

In fig.8a the molecular axes are parallel to each other and perpendicular to the surface.

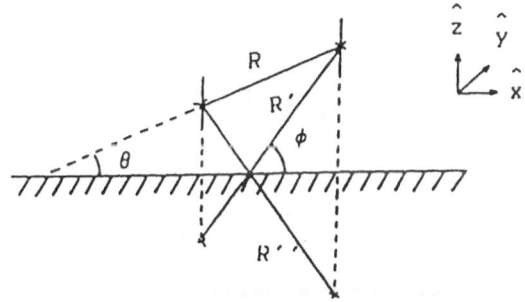

Figure 8a. Definition of the parameters in (III.20). R is the molecular separation and R' and R'' are the distances between one molecule and the image of the other.

The molecular polarizability tensors are given by:

$$\text{(III.19)} \quad \underline{\underline{\alpha}}^A = \begin{bmatrix} \alpha_\perp & 0 & 0 \\ 0 & \alpha_\perp & 0 \\ 0 & 0 & \alpha_\parallel \end{bmatrix} \quad ; \quad \underline{\underline{\alpha}}^B = \begin{bmatrix} \alpha_\perp & 0 & 0 \\ 0 & \alpha_\perp & 0 \\ 0 & 0 & \alpha_\parallel \end{bmatrix}$$

and the expression for the surface modification of the dispersion energy is found to be

$$\text{(III.20)} \quad W_{AB}^S = - \frac{\hbar}{\pi} \left[\frac{3}{2 + \mu} \right]^2 \int_0^\infty d\omega \, \alpha_{av}^2(i\omega) \, \frac{\varepsilon(i\omega) - 1}{\varepsilon(i\omega) + 1} \, *$$

$$\frac{[3\sin^2\theta + 3\sin^2\phi - 4 + (\mu^2 - 1)(9\sin^2\theta\sin^2\phi - 3\sin^2\theta - 3\sin^2\phi + 1)]}{R^3 R'^3} -$$

$$- \frac{\hbar}{2\pi} \left[\frac{3}{2 + \mu} \right]^2 \int_0^\infty d\omega \, \alpha_{av}^2(i\omega) \left[\frac{\varepsilon(i\omega) - 1}{\varepsilon(i\omega) + 1} \right]^2 \, *$$

$$* \, \frac{[6 + 18(\mu - 1)\cos^2\phi\sin^2\phi + (\mu^2 - 1)(9\sin^2\phi - 6\sin^2\phi + 1)]}{R'^3 R''^3}$$

In fig.8b the molecular axes are parallel to each other and parallel to the surface.

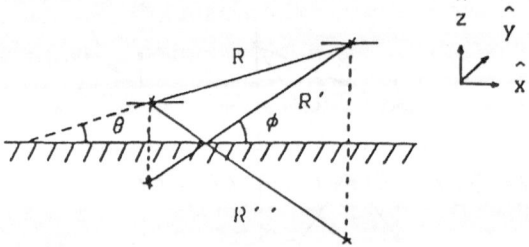

Figure 8b. Definition of the parameters in (III.22). R, R' and R'' are defined as in fig. 8a.

The molecular polarizability tensors are

$$\text{(III.21)} \quad \underline{\underline{\alpha}}^A = \begin{bmatrix} \alpha_\parallel & 0 & 0 \\ 0 & \alpha_\perp & 0 \\ 0 & 0 & \alpha_\perp \end{bmatrix} \quad ; \quad \underline{\underline{\alpha}}^B = \begin{bmatrix} \alpha_\parallel & 0 & 0 \\ 0 & \alpha_\perp & 0 \\ 0 & 0 & \alpha_\perp \end{bmatrix}$$

and the expression for the surface modification of the dispersion energy is

$$(III.22) \quad W_{AB}^S = - \frac{\hbar}{\pi} \left[\frac{3}{2 + \mu} \right]^2 \int_0^\infty d\omega \; \alpha_{av}^2(i\omega) \; \frac{\varepsilon(i\omega) - 1}{\varepsilon(i\omega) + 1} *$$

$$* \; \frac{[3\sin^2\theta + 3\sin^2\phi - 4 + (\mu^2 - 1)(-9\cos^2\theta\cos^2\phi + 3\cos^2\theta + 3\cos^2\phi - 1)]}{R^3 R'^3} -$$

$$- \frac{\hbar}{2\pi} \left[\frac{3}{2 + \mu} \right]^2 \int_0^\infty d\omega \; \alpha_{av}^2(i\omega) \left[\frac{\varepsilon(i\omega) - 1}{\varepsilon(i\omega) + 1} \right]^2 *$$

$$* \; \frac{[6 - 18(\mu - 1)\cos^2\phi\sin^2\phi + (\mu^2 - 1)(9\cos^4\phi - 6\cos^2\phi + 1)]}{R'^3 R''^3}$$

In fig.8c the molecular axes are perpendicular to each other and parallel to the surface.

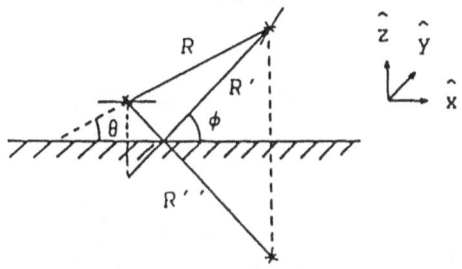

Figure 8c. Definition of the parameters in (III.24). R, R' and R'' are defined as in fig.8a

The molecular polarizability tensors are

$$(III.23) \quad \underline{\underline{\alpha}}^A = \begin{bmatrix} \alpha_{\parallel} & 0 & 0 \\ 0 & \alpha_\perp & 0 \\ 0 & 0 & \alpha_\perp \end{bmatrix} \quad ; \quad \underline{\underline{\alpha}}^B = \begin{bmatrix} \alpha_\perp & 0 & 0 \\ 0 & \alpha_{\parallel} & 0 \\ 0 & 0 & \alpha_\perp \end{bmatrix}$$

and the expression for the surface modification of the dispersion energy is

524

$$(III.24) \quad W_{AB}^S = - \frac{\hbar}{\pi} \left[\frac{3}{2+\mu} \right]^2 \int_0^\infty d\omega \; \alpha_{av}^2(i\omega) \; \frac{\varepsilon(i\omega) - 1}{\varepsilon(i\omega) + 1} \; *$$

$$* \; \frac{1}{R^3 R'^3} [3\sin^2\theta + 3\sin^2\phi - 4 + (\mu^2 - 1)(-9\cos^2\theta\cos^2\phi + 3\cos^2\theta +$$

$$+ 3\cos^2\phi - 9\cos\theta\sin\theta\cos\phi\sin\phi - 2] \; -$$

$$- \frac{\hbar}{2\pi} \left[\frac{3}{2+\mu} \right]^2 \int_0^\infty d\omega \; \alpha_{av}^2(i\omega) \left[\frac{\varepsilon(i\omega) - 1}{\varepsilon(i\omega) + 1} \right]^2 \; *$$

$$* \; \frac{[6 + (\mu - 1)(9\cos^2\phi - 6\cos^2\phi - 9\cos^2\phi\sin^2\phi + 2)]}{R'^3 R''^3}$$

The case studied by McLachlan [15] corresponds to $\alpha_\perp = \alpha_\parallel = \alpha$ or $\mu=1$, and it is readily verified that either of the expressions given above reduces to McLachlan's result for two identical molecules, which is

$$(III.25) \quad W_{AB}^S = - \frac{\hbar}{\pi} \; \frac{3\sin^2\theta + 3\sin^2\phi - 4}{R^3 R'^3} \int_0^\infty d\omega \; \alpha_{av}^2(i\omega) \; \frac{\varepsilon(i\omega) - 1}{\varepsilon(i\omega) + 1} \; -$$

$$- \frac{3\hbar}{\pi R'^3 R''^3} \int_0^\infty d\omega \; \alpha_{av}^2(i\omega) \left[\frac{\varepsilon(i\omega) - 1}{\varepsilon(i\omega) + 1} \right]^2$$

A qualitative discussion of the terms in W_{AB}^S is simplest in the isotropic case. The last term in (III.25) corresponds to the 4. term in (III.15) and always gives an attractive contribution to the dispersion energy. The first term is the sum of the identical 2. and 3. terms in (III.15), and the nature of this contribution depends on the molecular positions. If they are at the same vertical distance above the surface, the contribution is positive and dominates the attractive term, so the net effect is a repulsive contribution. If the molecules are on top of each other both terms in (III.25) gives an attractive contribution. The contributions may be considerable and in the limit of $\theta=0$ and $\phi=0$ they may be as much as 2/3 of the intermolecular dispersion energy, and they also decay slower with molecular distance than the three body contribution in (II.35.a).

Results for ethane and butane molecules adsorbed on graphite.

The polarizability of the molecules and the dielectric function of graphite at imaginary frequencies are required to calculate W_{AB}^S. Kramers-Kronig relations are used to determine the functions at imaginary frequencies from experimental measurements of the imaginary part of the functions (by absorption measurements) at real frequencies. It is found that (see II.21.e)

$$(III.26) \quad f(i\eta) = \frac{2}{\pi} \int_0^\infty d\omega \; \frac{\omega \; Im(f(\omega))}{\omega^2 + \eta^2}$$

where $f = \alpha$ or $(\varepsilon-1)$.

The dielectric tensor $\underline{\varepsilon}$ of graphite is anisotropic and characterized by two parameters $\varepsilon_\perp$ and $\varepsilon_{||}$, where $\varepsilon_\perp$ is the dielectric function for the direction perpendicular to and $\varepsilon_{||}$ parallel to the c-axis. The imaginary parts of $\varepsilon_\perp$ and $\varepsilon_{||}$ have been determined in the energy range 0-25 eV [19]. Although the derivations are based on an isotropic solid, the results can be applied to graphite if we use $\varepsilon = \sqrt{\varepsilon_\perp \varepsilon_{||}}$, the geometrical mean of $\varepsilon_\perp$ and $\varepsilon_{||}$ as the dielectric function. This is proved from the boundary conditions of the electromagnetic fields at the solid vacuum interface, and the function has been calculated by Bruch et al. [20]. The data are shown in fig.9.

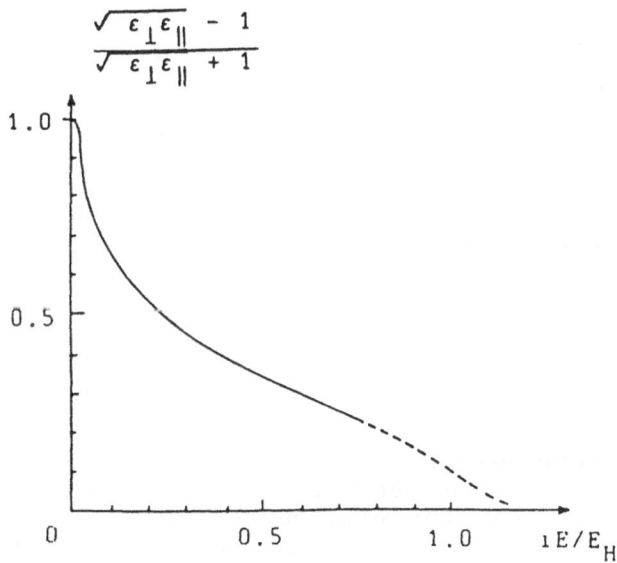

Figure 9. Data for graphite as function of the imaginary energy iE. E_H is one Hartree and $\varepsilon_\perp$, $\varepsilon_{||}$ are the dielectric constants of graphite.

The frequency dependent polarizability is calculated from gas phase data and only gives the spherical averaged function. W.J. Meath et al. [21] have analyzed various gas phase data, such as photo-absorption and high-energy electron scattering data, for a series of molecules and atoms, and derived molecular and atomic dipole oscillator strength distributions (DOSD). The molecular polarizability $\alpha(i\omega)$ is given in terms of the dipole oscillator strengths f_i by

$$(III.27) \quad \alpha(i\omega) = \frac{e^2}{m} \sum_i \frac{f_i}{\omega_i^2 + \omega^2}$$

where e is the electron charge, m the electron mass and ω_i a set of characteristic frequencies. When we replace the summation in (III.27) by an integration and introduce the DOSD (df/dω), we find

$$(III.28) \quad \alpha(i\omega) = \frac{e^2}{m} \int_{\omega_o}^{\infty} d\omega' \frac{(df/d\omega')}{\omega'^2 + \omega^2}$$

as the relation between α and the DOSD. The lower limit in the integration is set at the ultra violet absorption threshold to eliminate contributions from vibrational and rotational transitions. To facilitate the use of the DOSD data Meath et al. [22] have developed a discrete pseudospectral representation of the complete DOSD, so integrations over frequencies or energies are replaced by a discrete summation over a limited set of terms. If we change variables from the frequency ω to the dimensionless variable $E=\hbar\omega/E_h$, where E_h is one Hartree, we get

$$(III.29) \quad \alpha(iE) = a_o^3 \int_{E_o}^{\infty} dE' \frac{(df/dE')}{E'^2 + E^2} = a_o^3 \sum_{i=1}^{n} \frac{\tilde{f}_i}{\tilde{E}_i^2 + E^2}$$

where a_o is the Bohr radius. The $\tilde{f}_i$ and $\tilde{E}_i$ in the second relation are the discrete pseudospectral dipole oscillator strength parameters. For the hydrocarbons n=10, and we have used the data for ethane and butane in table I of ref. [22]. The polarizabilities at imaginary frequencies are easily found by replacing E with iE. The data are shown in fig.10.a for ethane and fig.10b for butane, and the results at zero frequency are in good agreement with the theoretical and experimental values [17], [18].

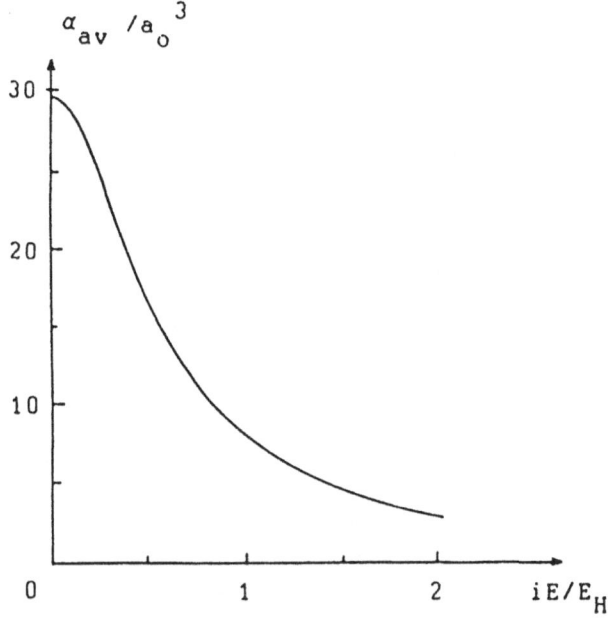

Figure 10.a. Spherical averaged polarizability of ethane as function of the imaginary energy iE. E_H is one Hartree and a_o the Bohr radius.

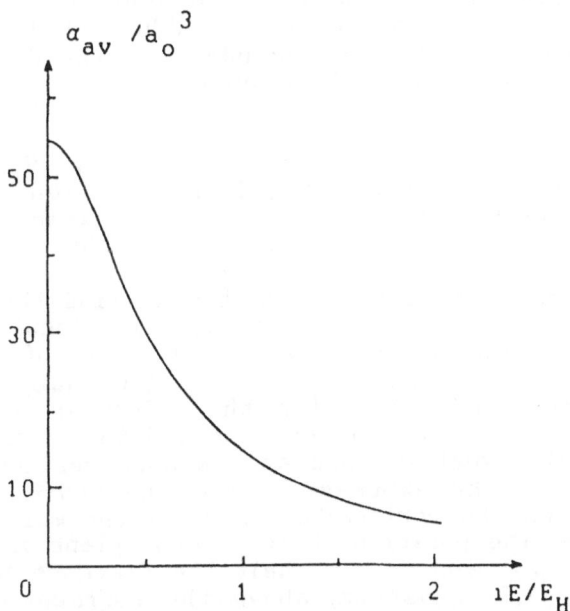

Figure 10.b. Spherical averaged polarizability of butane as function of the imaginary energy iE. E_H is one Hartree and a_o the Bohr radius.

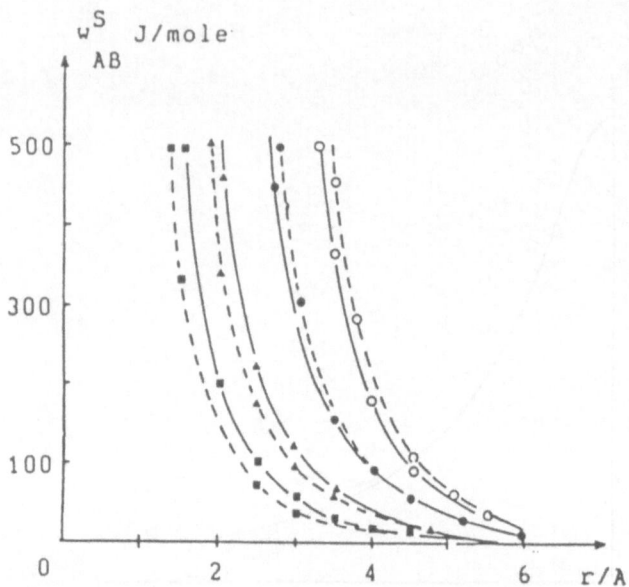

Figure 11.a. W_{AB}^S as a function of the intermolecular distance r
for two ethane molecules with the C-C bonds oriented like in
fig.8a. The molecules are at the same vertical distance above the
surface. Four different distances are considered, 1A: o ; 2A: • ;
3A: ▲ ; 4A: ■ ; The solid curves are the results when the
anisotropy $(\alpha_{\parallel}(0)/\alpha_{\perp}(0) = 1.14)$ is included and the dashed curves
are the results with neglect of the anisotropy.

The expressions for the surface modification, W_{AB}^S of the
intermolecular potential are valid, if retardation effects are
negligible, that is $R/c \ll 1/\omega_o$, where ω_o is a characteristic frequency for the solid. From fig.9 it is seen that $\varepsilon(i\omega)$ is significantly different from one in the energy range 0-1 Hartree,
so if we let ω_o correspond to one Hartre, we find that $c/\omega_o \approx 76$
Angstrom, which is much larger than typical molecule-substrate
distances of 3-4 Angstrom. Therefore, the validity of the expressions for W_{AB}^S is established. The ethane data has been used
to calculate the modification for the molecular orientations
sketched in fig.8.a and 8.c and the butane data for the orientation in fig.8.b. The molecules are at the same vertical distance
above the surface. The expressions for the modification show
that it depends on the vertical distance of the molecules above
the mirror plane. The position of the mirror plane is uncertain.
A good estimate [23] seems to be half the distance between the
basal planes, or 1.68 Angstrom, above the surface. Considering
the uncertaincy in the mirror plane position we have investigated the effect at different molecule-mirror plane distances. The
results for ethane are shown in fig.11 as a function of inter-

molecular distance r. We have used the quantum calculations of
Amos and Williams [17] to determine the static value of μ = 1.14.
The dashed curves are for the isotropic case (μ=1), and it is
noted that the effect of the anisotropy is different for the
two orientations but in both cases very modest. The vertical
distance of the molecules above the mirror plane has a much
more pronounced effect. The modification of the dispersion
energy becomes larger at a given intermolecular distance as the
vertical distance of the molecule above the mirror plane is
reduced. However, for typical intermolecular distances of 3-5
Angstrom it is concluded that the surface modification of the
dispersion energy is small no matter the position of the mirror
plane.

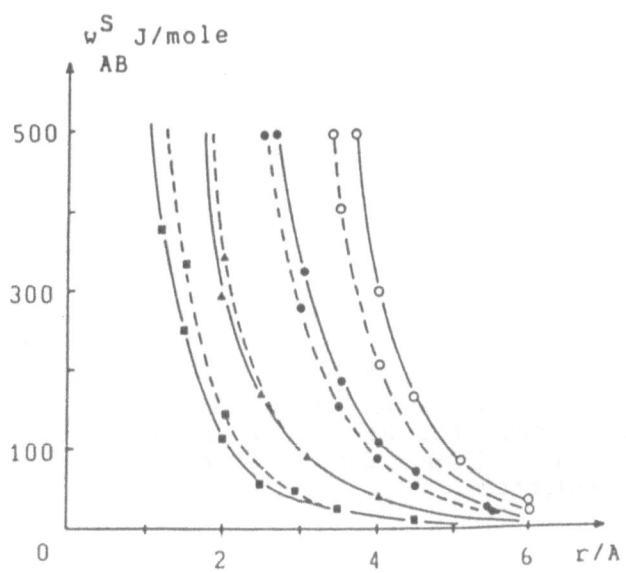

Figure 11.b. w_{AB}^S as a function of the intermolecular distance r
for two ethane molecules with the C-C bonds oriented like in
fig.8c. The molecules are at the same vertical distance above the
surface. Four different distances are considered, 1A: $\circ$; 2A: $\bullet$;
3A: $\blacktriangle$; 4A: $\blacksquare$; The solid curves are the results when the
anisotropy ($\alpha_{\|}(0)/\alpha_{\perp}(0)$ = 1.14) is included and the dashed curves
are the results with neglect of the anisotropy.

In fig.12 we have shown the results for butane molecules with
neglect of the anisotropy (μ=1), since the existing calculations
[18] of the polarizabilities are rather uncertain. They indicate

that the anisotropy is less than for ethane. The conclusions for butane are identical to those for ethane and the effect is small at typical intermolecular distances (6 - 7 Angstrom).

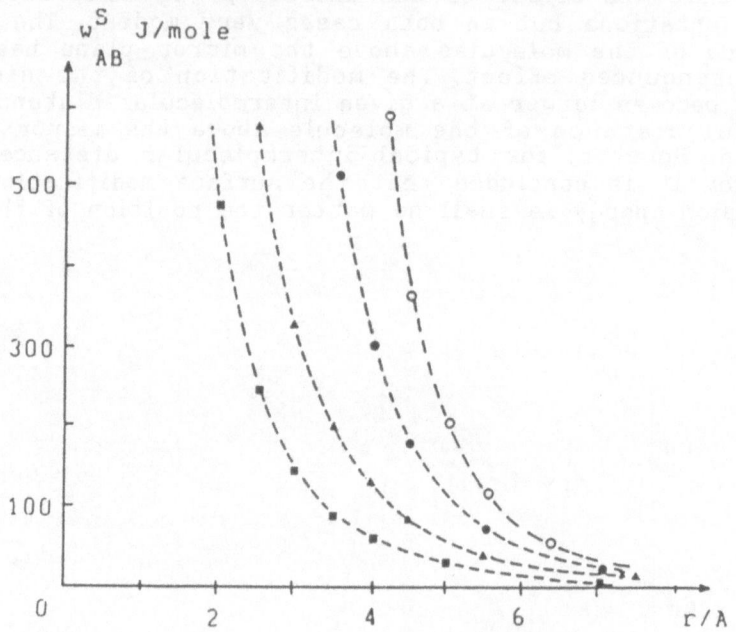

Figure 12. w_{AB}^S as a function of the intermolecular distance r for two butane molecules with the carbon backbone of the molecule oriented like in fig.8b. The molecules are at the same vertical distance above the surface. Four different distances are considered, 1Å: ○ ; 2Å: ● ; 3Å: ▲ ; 4Å: ■ ; The anisotropy of the molecule is neglected.

It appears that unless the anisotropy is more pronounced, the isotropic expression for the surface modification of the inter-molecular potential is a good approximation. Our calculations also show that the modification of the intermolecular potential for these molecules is very modest with little importance for the structures and dynamics of adsorbed monolayers of these molecules. Bruch [24] has calculated the effect for a series of isotropic noble gas atoms on different metal surfaces and shown that it is significant for a number of systems. J.M. Phillips [25] has included the effect in a work on the film structures of adsorbed methane molecules on graphite and shown that it is important for the calculated structures.

DISCUSSION

The transformation (II.22) of the standard expression for the second order perturbation energy (II.7) is of central importance for the present theory. The dispersion energy is expressed in terms of the molecular polarizability and the change in the electric field susceptibility caused by scattering of the radiation. The electric field susceptibility (II.32) is determined by a classical calculation of the electric field from a dipole. A quantum calculation of the susceptibility [6] gives exactly the same result. The reason is that the response function involves a commutator (II.11), which is the quantum analog of the classical Poisson bracket and, according to the correspondence principle [6], the commutator is proportional to the Poisson bracket of the classical quantities, which determines the classical response. When the change in the electric field susceptibility has been determined, it is easy to write down expressions for the dispersion energy, which include retardation effects.

The dispersion energy coefficients in the semiempirical expression of the van der Waals potentials (II.1) may be calculated independently by this theory. Since the coefficients are correlated and the expressions are approximate, one should not expect a very good agreement in general. The theory is, however, very useful when many body effects are studied, and it may be used to give good estimates of the importance of such effects for structure calculations and dynamical excitation calculations.

ACKNOWLEDGEMENT

I wish to thank professor L.W. Bruch for stimulating discussions and for providing the graphite dielectric constant data. I also want to thank professor H. Taub for the collaboration on the structure determination of monolayer films of hydrocarbon molecules on graphite, a work which motivated this investigation. The Danish Natural Science Foundation is acknowledged for economic support.

REFERENCES

1)   D.E. Williams, Trans. Am. Cryst. Ass. 1970, 6, 21.
     D.E. Williams, Acta Crystallogr. 1974, A30, 71.
     W.R. Busing, J. Am. Chem. Soc. 1982, 104, 4829.
     A.E. Kitaigorodskii, Molecular Crystals, Academic Press N.Y.
     1973.
     U. Burkert and N.L. Allinger, Molecular Mechanics, American
     Chemical Society Monograph 1982, 177.

2)   A.D. McLachlan, Proc. Roy. Soc. 1963, A 271, 387.
A.D. McLachlan Mol. Phys. 1963, 6, 423.
A.D. McLachlan, Proc. Roy. Soc. 1963, A 274, 80.

3)   K. Mirsky, Computing in Crystallography, ed. H. Schenk,
Delft University Press, Delft, 1978.

4)   F.Y. Hansen, G.P. Alldredge, L.W. Bruch and H. Taub,
J. Chem. Phys. 1985, 83, 349.

5)   B.E. Williams and T.L. Starr, Computers and Chemistry, 1979,
Vol 1, 173.

6)   L.I. Schiff, Quantum Mechanics, McGraw Hill Book Company,
1968.

7)   M. Jaszunski and R. McWeenig, Mol. Phys. 1982, 46, 863.

8)   R. Kubo, J. Phys. Soc. Japan 1957, 12, 570

9)   B.J. Berne and G.D. Harp, Advances in Chemical Physics,
1970, Vol XVII, 63.
L.D. Landau and E.M. Lifshitz, Statistical Physics, Pergamon
Press 1978, 343.

10)   M.A. Lawrentjew and B.W. Schabat, Metoden der Komplexen
Funktionstheorie, VEB Deutcher Verlag der Wissenschaften,
1967.

11)   H.G. Booker, Energy in Electromagnetism, The institution of
Electrical Engineers, London and N.Y., Peter Peregrinus Ltd.
1982.

12)   J.D. Jackson, Classical Electrodynamics, John Wiley and
Sons, N.Y. 1975.

13)   H.B. Casimir and D. Polder, Phys. Rev. 1948, 73, 360.

14)   W.A. Steele, The interaction of gases with solid surfaces,
Pergamon Press, 1974.

15)   A.D. McLachlan, Mol. Phys. 1964, 7, 381.

16)   M.P. Bogards, A.D. Buckingham, R.K. Pierens and A.H. White,
Faraday Trans. I, 1974, 74, 3008.
G.R. Alms, A.K. Burnham and W.H. Flygare, J. Chem. Phys. 1975,
63, 3321.
Landolt-Børnstein, Atom und Molekularphysik, Springer Ver-
lag, West Berlin, 1951, Vol I, part 3.

17)   R.D. Amos and J.H. Williams, Chem. Phys. Letters 1979, 66, 471.
D. Bhanmik, H.H. Jaffe and J.E. Mark, J. Mol. Struc., THEOCHEM,
1982, 87, 81.

18)   G.J. Trott, PhD. thesis 1981, Department of Physics and
Astronomi, University of Missouri-columbia Missouri, USA.
G.J. Trott, H. Taub, F.Y.Hansen and H.R.Danner, Chem. phys.
Letters 1981, 78, 504.

19)   E. Tossati and F. Bassani, Il Nuovo Cimento, 1979, 65 B, 161.
H. Venghaus, phys.stat.sol. (b), 1975, 71, 609.
J. Cazaux, Solid State Communications, 1970, 8, 545.

20)   L.W. Bruch and H Watanabe, Surface Science, 1977, 65, 619.

21)   G.D. Zeiss and W.J. Meath, Mol. Phys. 1977, 33, 1155.
G.F. Thomas and W.J. Meath, Mol. Phys. 1977, 34, 113.
G.D. Zeiss, W.J. Meath, J.C.F. MacDonald and D.J. Dawson,
Can. J. Phys. 1977, 55, 2080.
B.L. Jhanwar, W.J. Meath and J.C.F. MacDonald, Can. J. Phys.

1981, **59**, 185.

22) B.L. Jhanwar and W.J. Meath, <u>Mol. Phys.</u> **1980**, **41**, 1061.
23) N.D. Lang and W. Kohn, <u>Phys. Rev.</u> **1973**, **B7**, 3541.
     E. Zaremba and W. Kohn, <u>Phys. Rev.</u> **1976**, **B13**, 2270.
24) L.W. Bruch, <u>Surface Science</u> **1983**, **125**, 194.
25) J.M. Phillips, <u>Phys. Rev. B.</u>, **1984**, **29**, 5865.
     J.M. Phillips, <u>Phys. Rev. B.</u>, **1986**, **34**, 2823.

CHAPTER 29

TIME SCALE CONSIDERATIONS IN THE CHARACTERIZATION OF MELTING AND FREEZING IN MICROCLUSTERS

Heidi L. Davis, Thomas L. Beck, Paul A. Braier and
R. Stephen Berry
Department of Chemistry and the James Franck Institute
The University of Chicago
Chicago, Illinois 60637

ABSTRACT. In spite of their size, some microclusters exhibit distinct solid and liquid forms, according to diffraction and simulation studies. Both isothermal and isoergic simulations show that several free Ar_N clusters exhibit sharp but unequal freezing and melting temperatures, displaying a kind of phase change different from any conventional phase transition characteristic of bulk matter. Between the freezing and melting temperatures, two ''phases'' exist in dynamic equilibrium. This coexistence is consistent with a quantum statistical model of cluster melting and freezing. However no experiments have yet been devised to test this picture. Practical considerations aside, in devising appropriate experiments, one must take into account inherent time scale constraints on experiments capable of exhibiting the coexistence phenomenon without ambiguity. These constraints have been determined from computer simulations and are given explicitly for several 2- and 3-dimensional Ar_N systems. Extensions to other, experimentally more fruitful systems are also briefly discussed.

1. INTRODUCTION

Experimental and theoretical studies have suggested that very small clusters may exist in distinguishable solid- and liquid-like forms. Theoretical studies imply that, in at least some instances, the transition between these forms is not gradual. (See Figs. 1 and 2.) The possibility of sharp or nearly-sharp melting and freezing temperatures for microclusters raises several tantalizing and general questions about phase transitions. How little physics is required in the description of a system in order for us to predict that it will exhibit a detectable change of phase? What kind of equilibrium occurs between phases? How does the nature of the phase change depend on the size of the system?

Berry *et al.*[1] have proposed and applied to rare gas clusters a quantum statistical model, which identifies solid-like clusters with nearly rigid, conventional molecules and liquid-like clusters with very nonrigid molecules.

535

G. J. Long and F. Grandjean (eds.), The Time Domain in Surface and Structural Dynamics, 535–549.
© *1988 by Kluwer Academic Publishers.*

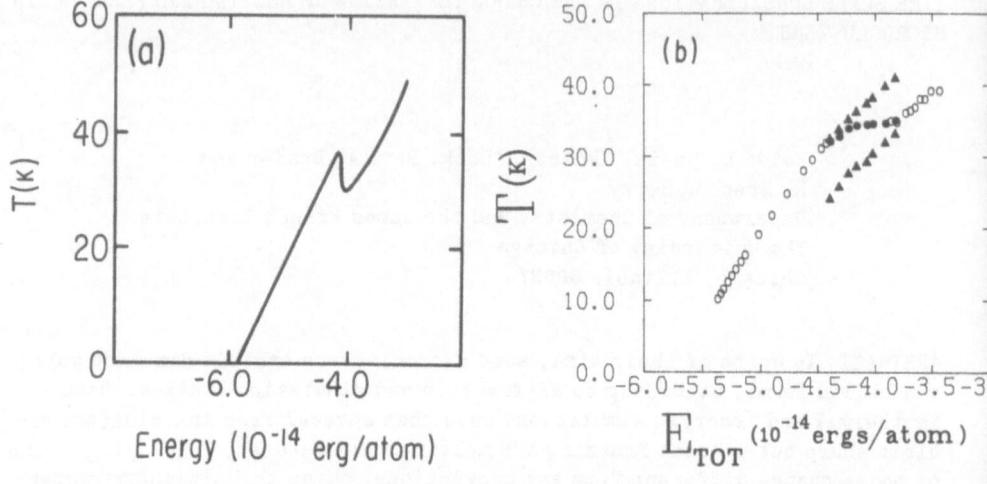

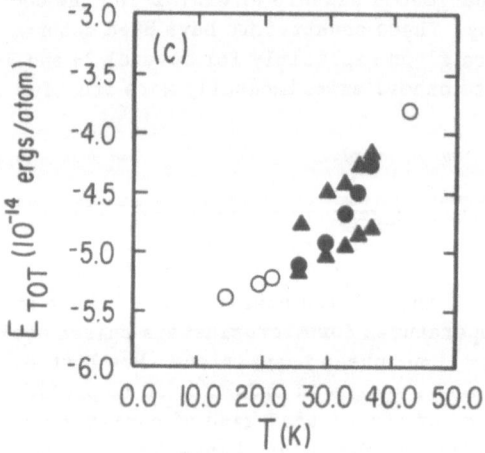

Fig. 1 a) Schematic representation of the Ar$_{13}$ temperature-total energy (caloric) curve obtained from iso-ergic MD simulations of Briant and Burton. Average temperatures were calculated for runs of length 10 ps. Longer MD simulations of Jellinek *et al.* produced curve b) for this system. The flattened portion corresponds to a coexistence region -- the triangles representing temperatures averaged over each form separately. Curve c) of total energy per atom vs temperature was obtained from MC simulations of isothermal Ar$_{13}$ by Davis *et al.* Closed circles represent average energies of both solid- and liquid-like components of the system at temperatures comprising the coexistence region. Triangles represent coexistence data calculated separately for single forms of the cluster.

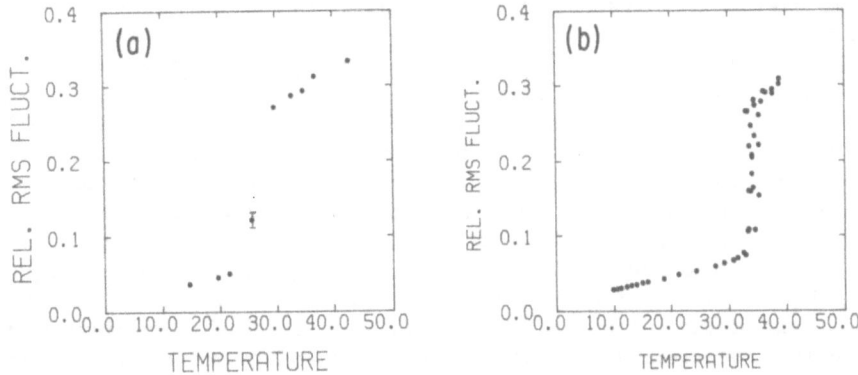

Fig. 2. Relative rms bond length fluctuations $\delta(T)$ for Ar_{13} from
MC studies (Ref. 19); b) $\delta(T)$ from MD simulations (Ref. 16) of
isoergic Ar_{13}.

The behavior of energy levels of a general model system between its idealized rigid and nonrigid limit suggests that the equilibrium between a solid
and a liquid cluster is unlike that of any bulk phase transition of any integral order, but is a phenomenon in its own right which becomes a first-order
solid-liquid transition in the limit of a large number of atoms comprising the
cluster. Under isobaric, isothermal conditions a cluster well-represented by
this model exhibits a sharp freezing temperature T_f below which no liquid is
thermodynamically stable, and a sharp melting temperature T_m above which no
solid is thermodynamically stable. The temperatures T_f and T_m are in general
different and depend on the number N of particles in the cluster. For a finite
range of intermediate temperatures, two forms, one near-rigid and solid-like,
the other presumably less rigid and liquid-like, may exist in thermal equilibrium.

To be described accurately by this picture, a cluster must satisfy several
conditions. 1) The system must be finite so that only a finite free energy
change is involved in the transition, even at temperatures at which the free
energies of the two forms are unequal. 2) The system must reside in a single
well at sufficiently low temperatures. The triatomic species Li_3 does not
meet this requirement because, in all likelihood, its zero-point energy exceeds the barriers separating its three equivalent potential minima.[2] A third
requirement for coexistence is that be a suitable separation of certain time
scales; this is a central subject of this discussion.

Computer simulations indicate that several Ar_N clusters do exhibit this
coexistence phenomenon.[3] Simulations also provide information about a cluster's dynamical behavior. Such information is not easily attainable now from

analytical studies, but simulations have made it increasingly clear that understanding the dynamics of very small clusters is crucial to understanding their melting and freezing behavior. The *simultaneous* coexistence accompanying first order transitions of bulk matter is restricted to the manifold on which the two phases have the same free energy. This is not the case for clusters. For them, the coexistence of two forms is a dynamic chemical equilibrium; the cluster can pass between solid and liquid whether or not they have the same free energy, in this respect, resembling an isolated system undergoing a unimolecular chemical reaction. (Unless otherwise specified, it is this dynamical phenomenon to which we refer when we speak of coexistence.)

Coexistence does not occur for all sizes of Ar_N clusters, and it may not be possible to observe readily the coexistence of two distinguishable phases even in some of the systems for which it does occur. Whether coexistence in a particular cluster can be detected in an experiment or in a simulation depends on the mean time the cluster spends in each coexisting form, on the time to establish equilibrated average properties of the individual phases and on the time characterizing the particular observation.

The available experimental data are consistent with those from computer simulations but are not conclusive. Solid-like, liquid-like and amorphous structures of several kinds of clusters have been inferred from diffraction experiments with molecular beams at various stagnation pressures.[4-6] Molecular beam diffraction experiments of Farges *et al.*[7] indicate that neutral Ar_N clusters exhibit short range icosahedral and pentagonal-bipyramidal order which is inconsistent with an fcc structure but is consistent with the computational results of Hoare and Pal,[8] and Farges *et al.*.[7] Photodissociation data from free Ar_NCHO^+ are consistent with the argon component having undergone a transition from a rigid to less rigid form as the stagnation stagnation nozzle pressure is reduced.[9] Infrared spectra of Ar_NSF_6 manifest frequency shifts and changes in width of the v_3 band as a function of stagnation pressure[10] which were interpreted initially as indications of melting and freezing of the argon,[11] but a new interpretation of these data has been given which would make them irrelevant to the question of coexistence of phases.[12] Several recent Monte Carlo (MC) computer simulations have given results, in addition to ours, that strongly support the idea of a finite temperature and energy range of coexistence for some clusters[13,14].

In this paper, we present exemplary systems for which experiments, provided they could be performed, might be able to detect and probe the coexistence phenomenon. More important, we demonstrate the usefulness of simulations in screening out experiments which are incapable of characterizing unambiguously the melting and freezing behavior of particular clusters. In the first section, we define the relevant time scales for a given system and provide criteria for the observability of the coexistence phenomenon. We focused our simulations mostly on free, Ar_N clusters because of the ease and reliability with which they can be modeled. Coexistence is as likely to occur with 2-dimensional systems as with 3-dimensional, and it may prove at least as convenient to conduct experiments with 2-dimensional systems. Hence we have now extended our work to include simulations of these systems. Some of the results are presented here.

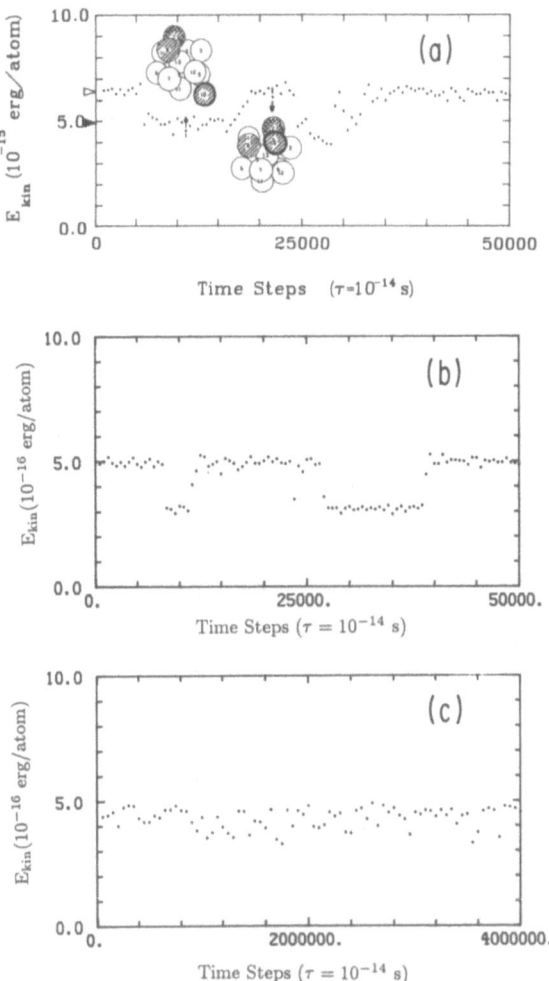

Fig. 3. Temperature averaged over a few breathing periods and plotted as a function of time for isoergic MD simulations of a) 3-dimensional Ar_{13} and b) 2-dimensional Ar_7. Total energy lies within the system's coexistence region; the rate of jumping between forms is dependent on total energy but is consistent throughout most of the coexistence region to within an order of magnitude. A breathing period is on the order of a ps in either system. Data in c) is the same as in b) except that temperature has been averaged over a time interval a hundred times longer. In this case our ''measurement" of temperature is too slow to resolve coexistence.

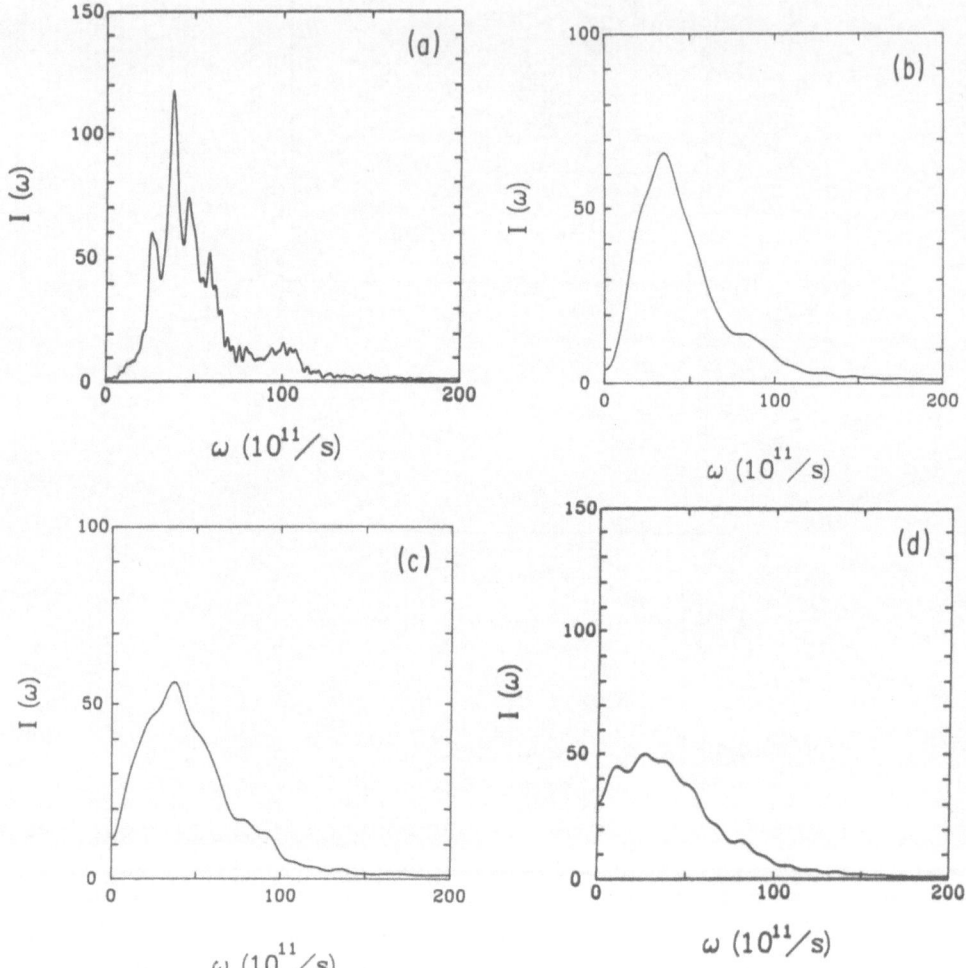

Fig. 4. Power spectra from MD simulations of isoergic Ar$_{13}$ in a) the single-phase, solid region of the caloric curve at E_{TOT}= -4.92x10^{-14} erg/atom. Spectra b) of the high-temperature form at a total energy (E_{TOT}= -4.13x10^{-14} erg/atom) in the co-existence range, and c) of the low-temperature form at the same energy. d) Spectra from the single-phase liquid portion of the caloric curve, E_{TOT}= -3.61x10^{-14} erg/atom.

2. 3-D COMPUTER SIMULATIONS OF Ar$_N$

Briant and Burton[15] were among the first to use computer simulations to study systematically the melting transition of microclusters. They obtained a caloric curve of average temperature $\langle T \rangle$ as a function of total energy E_{tot}, like that shown in Fig 1.a, for free Ar$_{13}$ using isoergic Molecular Dynamics (MD) calculations. The ''loop'' seemed indicative of a first order-like phase transition. Later, more extensive MD simulations have shown that the loop is an artifact of short runs and inadequate convergence.[16] The difficulty in obtaining a good average temperature $\langle T \rangle$ for each fixed E_{tot} in the ''loop region'' arises because two distinct forms of the cluster are present at these energies. When temperature[17] is averaged over a few of the cluster's breathing periods ($\tau_b \sim 1.5$ ps) and plotted against time (see Fig. 3.a), two distinct bands emerge, each representing a particular form of the N=13 cluster. The distribution of these short-time temperature averages is bimodal over a range of energies -4.4 x 10^{-14} ergs $< E_{tot} <$-3.8 x 10^{-14} ergs. Although both forms become less rigid as energy is increased within the coexistence region, the high-temperature form consistently exhibits very little or no density of soft modes in its power spectrum or rms bond length fluctuation data. The high-temperature form has the structure of a loose or ''hot'' icosahedron. This is not the case for the low-temperature, high-potential energy form, whose particles are typically arranged in a more random fashion. This form exhibits an appreciable diffusion constant, and its power spectrum $I(\omega)$, which exhibits little structure, has a significant intensity at $\omega = 0$. (See Fig. 4.)

A solid system typically oscillates in one potential energy well for times long relative to those required to establish its physical properties. Liquid behavior is associated in an essential way with diffusion and compliance. A liquid, then, cannot be characterized in terms of a particular well on the potential energy hypersurface but by its frequency of passage between wells. It has been conjectured[1a] and demonstrated [1b,18] that a requirement for the melting of a solid is that there be a breakdown of in the separation of the two time scales--the shorter one being the time scale for vibration of the system about a potential minimum; the longer one is the time scale for passage among wells of the potential surface. This criterion seems to apply for microclusters.[3] For coexistence to occur at a particular temperature two stable potential energy forms must be distinguishable as solid and liquid, in the above mentioned sense. For the coexistence phenomenon to be observed in either laboratory experiments or computer simulations requires a further time scale criterion be met. The cluster must stay in one ''phase'' long enough for the spectral densities, diffusion constants or other properties of that phase to be measured. This additional separability of time scales clearly occurs in isoergic simulations of Ar$_{13}$.

Although one cannot speak of time in a Monte Carlo calculation, it has been demonstrated for Ar$_{13}$ that the average number of configurations sampled before the system passes between solid- and liquid-like forms is much larger than the number of configurations required to establish a reasonable average

542

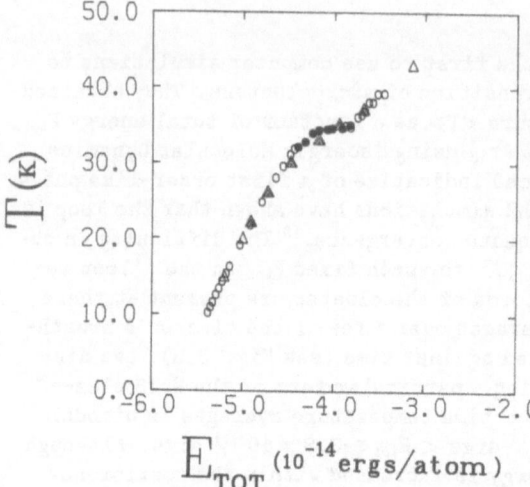

Fig. 5. Caloric curve of Ar₁₃
from isothermal simulations (△)
superimposed onto that obtained
from isoergic MD simulations
(o). All points represent averages over entire simulations;
for darkened points this implies
an average over two coexisting
forms.

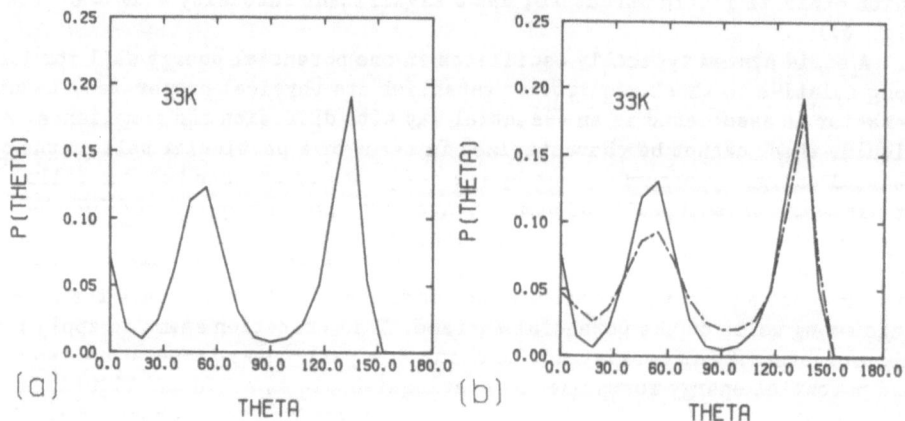

Fig. 6. Angular distribution functions for Ar₁₃ from MC calculations of Davis
et al. a) P(θ) fully averaged over both coexisting forms of the cluster at T=33K;
b) data obtained separately over each, the high- and low-potential energy form.
Dashed curve represents the distribution corresponding to the high-potential
energy (liquid-like) form.

of the potential energy U for each form.[19] There is a finite range of temperatures for which ''short-interval" averages of the potential energy $\overline{U}$, taken
over a few thousand configurations, separate into distinct bands when plotted against the total number of configurations n_c; a bimodal distribution of
$\overline{U}$ is obtained for 26K $\leq$T$\leq$37K. These data correspond well with isoergic MD
results for this system, as seen in Figs. 2 and 5. The bimodality of the distributions, the liquid-like character of the high-U form and the solid-like
character of the low-U form of the isothermal Ar_{13} cluster are confirmed from
constant temperature MD simulations. (See Ref. 19 for discussion.) Power
spectra for the solid- and liquid-like forms are similar to those obtained
from isoergic MD simulations. At T=33K the diffusion constant for the high-U
form is approximately 5 x 10^{-5} cm/sec^2.[20]

In earlier MC simulations of Ar_{13}, Etters and Kaelberer[21] had obtained
relative rms bond length fluctuation $\delta R/\Delta R$(T) data qualitatively similar
to those of Fig. 2. However they studied the system by taking short-interval
U averages over only 300 steps; this averaging produces a distribution of
$\overline{U}(n_c)$ too diffuse to resolve the two thermodynamically stable forms occurring
for temperatures in the coexistence region. (Compare the MD data in Figs.
3.b and 3.c.) Later MC calculations of Quirke and Sheng produced results for
Ar_{13} which, carried out without a consideration of coexisting forms, seemed
contradictory[22] to the $\delta R/\Delta R$(T) data of Etters and Kaelberer. The angular
distribution function P(θ), which gives the average probability of finding
three nearest neighbors about the center particle, is a sensitive, almost
qualitative probe of the structure of Ar_{13}. A rigid, icosahedral structure
produces sharp maxima in this function for $\theta = 60°$ and $\theta = 120°$ but will exhibit no probability of finding nearest neighbor angles of 90°. At 33K, which
is significantly above the temperature corresponding to the dramatic increase
in $\delta R/\Delta R$(T), a P(θ) like that shown in Fig. 6.a. was obtained by Quirke and
Sheng. From these data, and because the configurational heat capacity as a
function of temperature C(T) is smooth and broadly peaked ($\sigma \sim$10K) near melting, it was inferred that the melting of Ar_{13} occurs as the gradual softening
of a single stable form with temperature. The P(θ) in Fig. 6.a represents an
average over the entire distribution, and hence over two coexisting forms.
When the bimodal distribution of data for different forms are separated, as
in Fig. 6.b, P(θ) of the high-U form shows a probability at $\theta = 90°$ far above
zero; P(θ) of the low-U form does not. The width of C(T) is manifestation of
the coexistence phenomenon extending over a broad range of temperatures.[19]

Under the conditions of our theoretical model, a bulk system would have
an infinitely narrow coexistence region comprised of the single temperature
where the free energies of the solid and liquid phases are equal. For very
small clusters, for which this restriction does not apply, the temperature
width of the coexistence region ΔT_c is determined by the specifics of the particular cluster, which are strongly dependent on N. An isoergic MD study of
Ar_N for 7<N<33 has been carried out, in which the dynamics of the clusters
were probed using steepest decent quench techniques.[18,23,25] As seen in Fig. 7,
the behavior of ΔT_c(N) is quite size dependent;[24] for several values of N, the
Ar_N clusters do not exhibit the coexistence phenomenon.

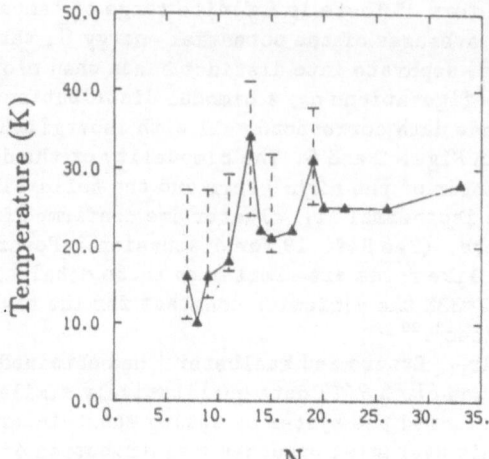

Fig. 7. The N-dependent melting temperatures and coexistence temperature ranges ΔT_c for Ar$_N$. Δ denote the long-time average temperature values at which $\delta(T)$ rise sharply (see Fig. 2.b). Dashed lines correspond to coexistence temperature ranges ($\Delta T_c=(T_f-T_m)$) for the clusters that display a bimodal form for the short-time averaged temperatures. Data from Ref. 23.

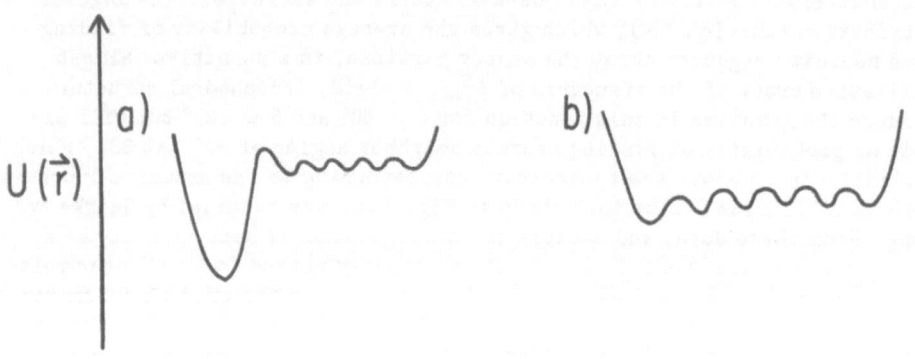

Fig. 8. Schematic representation of potential energy surface for a) a cluster like Ar$_{13}$, where the potential well characterizing the solid is significantly deeper than -- and separated by a relatively high energy barrier from -- those nearly equivalent wells accessed by the liquid. A system characterized by a potential energy region b), comprised of nearly-equivalent wells and shallow barriers, can not undergo coexistence; once the cluster has enough total energy to escape a well it will no longer be accurately described by that or any other single well for an amount of time long enough for a solid phase to be said to exist.

Of those clusters which do exhibit coexistence (N=7, 9, 11, 13, 15, 19), those whose potential energy minima correspond to either pentagonal bipyramidal (N=7), icosahedral (N=13) or double icosahedral (N=19) structures exhibit ΔT_c which are large relative to those for clusters nearest in size. (See Ref. 23.) For the N=13 and N=19 ''magic number'' clusters, the solid-like form, corresponding to a dynamically stable minimum, looks very different from the coexisting liquid-like cluster. The latter is described by a set of nearly equivalent--but not always equivalent--isomers which consist of one atom removed from the surface of the most stable structure. For Ar_7, for which the dynamic stability of the potential energy minimum is less than for Ar_{13} and Ar_{19}, ΔT_c is also very large; however the rise in $\delta R/\Delta R(T)$ occurs at a lower temperature than for it does for the other two ''magic number'' clusters.

For clusters of N=8, 14 or 20 particles the lowest-energy structure is that of the ''magic number'' Ar_{N-1} cluster with an additional atom on the surface. For N=14, there are 13 equivalent lowest-energy structures. Quench studies[18,25] suggest that the potential energy minima corresponding to these isomers are separated from each other and from other minima (corresponding to similar isomers) by low energy barriers; this implies that the region of the potential energy surface characterizing melting is like that sketched in Fig. 8.b, and might explain why coexistence is not observed in computer simulations of this system. At energies corresponding to <T>~26K, $\delta R/\Delta R(T)$ begins rising sharply and the cluster accesses its lowest-energy structures in a diffusive manner; at slightly higher energies, the cluster starts to sample potential minima corresponding to structures consisting of a 12-particle core and 2 surface atoms, and the diffusion constant increases significantly. For 8-, 14- and 20-particle Ar clusters, there is no range of E_{tot} for which a bimodal $\overline{U}$ distribution is observed.

It is conceivable that there exists another type of system which exhibits a range of energies at which the cluster is neither a pure liquid or pure solid nor consists of coexisting phases that interchange infrequently enough for either a solid or liquid phase to be dynamically stable. Such a system would appear as slush in any experiment, no matter what the characteristic time scale for measurement.

3. 2-D SIMULATIONS

For 2-dimensional systems, 7- and 31-particle clusters can be expected to have particularly stable, closed-shell structures. We have performed simulations on these systems, using both Lennard-Jones and Morse potentials and an algorithm similar to that described in Ref. 16. No attempt has been made to model the surface-cluster interaction or a 2- to 3- dimensional ordering transition. However, the 2-dimensional simulations of Ar_{13} display coexistence. Moreover the diagnostics of solid-like and liquid-like behavior yield results for the 2-d system that are similar to the 3-dimensional results.

The 7-particle cluster exhibits a clear coexistence phenomenon in plots of $\overline{T}$ vs time (see Fig. 3.b), the two forms differing in average temperature by

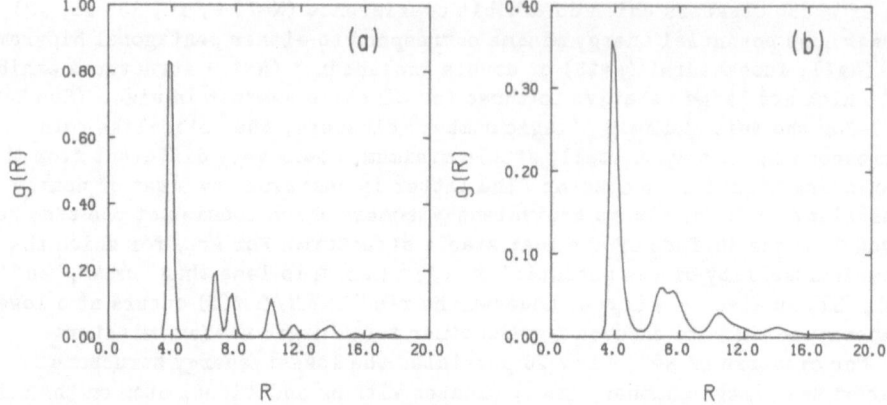

Fig. 9. Radial distribution functions calculated for a 2-dimensional, isothermal Ar$_{31}$ system a) before (<T>=10K) and b) after (<T>=34K) melting. Melting occurs at approximately 27K, as indicated by $\delta(R)/\Delta R$ and velocity autocorrelation data. A range of total energies for which coexistence occurs is not observed.

~10K. The velocity autocorrelation function and power spectra show that the high-temperature form is a solid, structured form, while the low-temperature form is less structured and liquid-like. The density of soft modes, I(0), is near zero for the former but is significantly greater than zero for the latter. The $\delta R/\Delta R$(T) data are similar to those obtained for 3-dimensional Ar$_7$, with the onset of large fluctuations occurring near 20K. The magnitude of fluctuations in the liquid-like cluster is also comparable to that obtained in three-dimensional calculations.

In preliminary studies of Ar$_{31}$, no coexistence has been observed, but melting is seen to occur (from $\delta R/\Delta R$(T) and spectral data) near 27K. The radial distribution function $g(R)$, shown in Fig. 9, indicates a well-structured cluster below freezing and one with significantly less structure for a temperature above 27K. This cluster appears to be less strongly bound than Ar$_7$, and above 30K atoms have a tendency to evaporate into a two-dimensional gas.

4. DISCUSSION

To summarize, there are several 3-dimensional and at least one 2-dimensional Ar$_N$ cluster which exhibit coexistence under idealized, free volume conditions. These model systems represent simplifications of those which can be studied experimentally; however, they illustrate several important and unique aspects of a phase change which we believe occurs for more general types of clusters. The coexistence phenomenon which may characterize the phase change of some microclusters is intimately connected with the cluster's dynamics; this imposes several restrictions on the type of experiments capable of detecting it. An experiment must be fast enough to allow measurements on the system undergoing coexistence before the system switches from one phase to the other. Too slow an experiment, measuring relaxation rates on a time scale comparable to that of conventional NMR experiments, will produce data (like $P(\theta)$ in Fig. 6.a and $\overline{T}(t)$ in Fig. 3.c) which are misleading, giving a long-term average over two forms. Alternatively, x-ray and other experiments with too high a time resolution, corresponding to a measurement time scale comparable to that of one of the system's vibrations within a single well, cannot allow us to differentiate between a liquid and a disordered solid cluster. The observation of a highly ordered structure in some measurements and of a disordered cluster in other instances at the same temperature or energy would be consistent not only with a dynamic coexistence of different phases, but also with a pure-liquid system accessing deep and less-deep potential energy wells in a diffusive manner. It is the detection of diffusive-type motion and of compliance for some clusters, and the observation of non-diffusive clusters under the same conditions by similar measurements taken at other, random times that would provide the strongest support of our model.

Our simulations indicate that experiments capable of measuring properties on a time scale shorter than 1000 ps and longer than 1 ps are necessary for the detection of coexistence in the Ar_N systems for which we have observed it. This time scale is not incompatible with the range of relaxation times which can be measured with He scattering experiments. Although a 7-particle argon ''island" is most likely too small a sample for a He scattering experiment, it is one system that our crude calculations indicate should display coexistence.

The time scale information from Ar_N simulations, taken as a rough guideline, suggests we should begin looking for coexistence in systems which can be probed by experiments capable of measuring processes on a 1-1000 ps interval. An IR or Raman active cluster, or even a fluorescent cluster, preferably one that can be size-selected, might be appropriate. From a theorist's viewpoint the cluster should be one that can be modeled reliably. This suggests an ionic salt, such as a cluster or a thallium or maybe a cuprous salt, possibly even an alkali halide if the cluster is fluorescent. A point will be reached, we hope, when simulations can be tailored to tentative experiments and it will be possible to find very small systems which are well-suited for an unambiguous characterization of the melting and freezing phenomenon. Recently the photofragmentation of Si_N microclusters has been reasonably modeled by a Stillinger-Weber potential.[26] A MD study, to determine whether coexistence accompanies the melting transition of these clusters, is now being considered by our group.

548

ACKNOWLEDGEMENTS

This research was supported by a Grant from the National Science Foundation. We would like to thank B. Gans and N. Quirke for helpful discussions, and Prof. S. Morup for assistance in preparing this manuscript.

REFERENCES

[1] a) R.S. Berry, J.Jellinek and G. Natanson, *Phys. Rev.*, 1981, A 30, 919; *Chem. Phys. Lett.*, 1984, 107, 227. See also, b) G.Natanson, F. Amar and R.S. Berry, *J. Chem. Phys.*, 1983, 78, 399.

[2] D. A. Garland and D. M. Lindsay, *J. Chem. Phys.*, 1983, 78, 2813; J. L. Martins, R. Car and J. Buttet, *J. Chem. Phys.*, 1983, 78, 5646.

[3] R.S. Berry, T.L. Beck, H.L. Davis and J. Jellinek, *Adv. in Chem. Phys.*, (in press), 70 B, I. Prigogine and S.A. Rice, eds. (Wiley, New York).

[4] R. K. Heenan, E. J. Valente, and L. S. Bartell, *J. Chem. Phys.*, 1983,78, 243.

[5] R. K. Heenan, L. S. Bartell, *J. Chem. Phys.*, 1983, 78, 1265.

[6] E. J. Valente and L. S. Bartell, *J. Chem. Phys.*, 1984, 80, 1451; *J. Chem. Phys*, 1984, 80, 1458.

[7] J. Farges, M. F. de Farauday, B. Raoult and G. Tourchet, *J. Chem. Phys.*, 1983, 78, 5067.

[8] M. R. Hoare and P. Pal, *Adv. Phys.*, 1971, 20, 161.

[9] A. J. Stace, *Chem. Phys. Lett.*, 1985, 113, 355.

[10] T. E. Gough, D. G. Knight and G. Scoles, *Chem. Phys. Lett.*, 1983, 97, 155.

[11] D. Eichenauer and R. J. LeRoy, *Phys. Rev. Lett.*, 1986, 57, 2920.

[12] R. J. LeRoy, (in press), *Proc. of the 20th Jerusalem Symposium*, B. Pullman and J. Jortner, eds.

[13] N. Quirke, 1987, private communication.

[14] Murty S. S. Chulla, D. P. Landau and K. Binder, *Phys. Rev.*, 1986, B 34, 1841.

[15] C.L. Briant and J.J. Burton, *J. Chem. Phys.*, 1975, 63, 2045.

[16] J. Jellinek, T. Beck and R.S. Berry, *J. Chem. Phys.*, 1986, 84, 2783.

[17] In the isoergic MD simulations, temperature has been approximated as $\frac{2}{kd_f} < E_K >$, where k is Boltman's constant, d_f is the number of degrees of freedom of the physical system and $< E_K >$ is the microcanonical average of the total kinetic energy.

[18] F. H. Stillinger and T. L. Weber, *Kinam*, 1981, 3A 159; *Phys. Rev.*, 1982, A 25, 978; *Phys. Rev.*, 1983, A 28, 2408.

[19] H. Davis, J. Jellinek and R. S. Berry, *J. Chem. Phys.*, 1987, 86, 6456.

[20] The diffusion constants appearing in Fig. 6 of Ref. 19 are incorrectly reported; the value given here has been obtained from dividing the previously reported value by a factor of 6. The self-diffusion coefficient for bulk, liquid Ar at 84K (near melting) is $\sim 2 \times 10^{-5}$ cm^2/sec. (S. A. Rice and P. Gray, *Monographs in Statistical Physics and Thermodynamics*,1964, 8, I. Prigonine, ed. (Wiley, NY), chapter 6.)

[21] R.D. Etters and J.B. Kaelberer, *J. Chem. Phys.*, 1977, 66, 5112; J.B. Kaelberer and R.D. Etters, *J. Chem. Phys.*, 1977, 66, 3233.

[22] N. Quirke and P. Sheng, *Chem. Phys. Lett.*, 1984, 110, 63.

[23]T. Beck, J. Jellinek and R. S. Berry, *J. Chem. Phys.*, 1987, 87, 545.

[24]Strictly speaking, the microcanonical coexistence region should be defined in terms of total energy. $\Delta T_c(N)$ has been obtained from the average temperatures which correspond to the energy boundaries of the coexistence range. As indicated in Figs. 2 and 5, for Ar_{13} where a comparison has been made, the microcanonical and canonical results are in agreement. (See Ref. 3.)

[25]T. Beck and R.S. Berry, in *The Physics and Chemistry of Small Clusters*, 1987, P. Jena, ed., (Plenum, New York); T. Beck and R. S. Berry, (in progress).

[26]B. P. Feuston, R. K. Kalia and P. Vashishta, *Phys. Rev.*, 1987, B 35, 6222.

572